IUTAM Symposium on Mesoscopic Dynamics of Fracture Process and Materials Strength

SOLID MECHANICS AND ITS APPLICATIONS
Volume 115

Aims and Scope of the Series

The fundamental questions arising in mechanics are: *Why?, How?,* and *How much?* The aim of this series is to provide lucid accounts written by authoritative researchers giving vision and insight in answering these questions on the subject of mechanics as it relates to solids.

The scope of the series covers the entire spectrum of solid mechanics. Thus it includes the foundation of mechanics; variational formulations; computational mechanics; statics, kinematics and dynamics of rigid and elastic bodies: vibrations of solids and structures; dynamical systems and chaos; the theories of elasticity, plasticity and viscoelasticity; composite materials; rods, beams, shells and membranes; structural control and stability; soils, rocks and geomechanics; fracture; tribology; experimental mechanics; biomechanics and machine design.

The median level of presentation is the first year graduate student. Some texts are monographs defining the current state of the field; others are accessible to final year undergraduates; but essentially the emphasis is on readability and clarity.

For a list of related mechanics titles, see final pages.

IUTAM Symposium on

Mesoscopic Dynamics of Fracture Process and Materials Strength

Proceedings of the IUTAM Symposium held in
Osaka, Japan, 6–11 July 2003.
Volume in celebration of Professor Kitagawa's retirement

Edited by

H. KITAGAWA

Department of Mechanical Engineering and Systems,
Osaka University,
Osaka, Japan

and

Y. SHIBUTANI

Department of Mechanical Engineering and Systems,
Osaka University,
Osaka, Japan

KLUWER ACADEMIC PUBLISHERS
DORDRECHT / BOSTON / LONDON

A C.I.P. Catalogue record for this book is available from the Library of Congress.

ISBN 1-4020-2037-6 (HB)
ISBN 1-4020-2111-9 (e-book)

Published by Kluwer Academic Publishers,
P.O. Box 17, 3300 AA Dordrecht, The Netherlands.

Sold and distributed in North, Central and South America
by Kluwer Academic Publishers,
101 Philip Drive, Norwell, MA 02061, U.S.A.

In all other countries, sold and distributed
by Kluwer Academic Publishers,
P.O. Box 322, 3300 AH Dordrecht, The Netherlands.

Printed on acid-free paper

Printed in the Netherlands.

CONTENTS

Preface xi

Scientific and Organizing Committee xv

Sponsors xvi

Participants xvii

Symposium Program xxv

Wei. Cai, Vasily V. Bulatov, Tim G. Pierce, Masato Hiratani, Moono Rhee, Maria Bartelt and Meijie Tang 1
Massively-Parallel Dislocation Dynamics Simulations

Yoshihisa Kaneko and Satoshi Hashimoto 13
Nondestructive Observation of Dislocation Structures Formed at Fatigued Copper and Stainless Steel Crystals

D. Weygand 23
A Discrete Dislocation Dynamics Investigation of Formation and Stability of Lomer Locks

Ronan Madec, Benoit Devincre and Ladislas Kubin 35
On the Use of Periodic Boundary Conditions in Dislocation Dynamics Simulations

Erik Bitzek, Daniel Weygand and Peter Gumbsch 45
Atomistic Study of Edge Dislocations in FCC Metals: Drag and Inertial Effects

Kisaragi Yashiro, Yasushi Tabata and YoshihiroTomita 59
Molecular Dynamics Study on the Characteristics of
Edge and Screw Dislocations in
Gamma/Gamma-Prime Microstructure in Ni-based Superalloy

R. Madec and L.P. Kubin 69
Dislocation Interactions and Symmetries in BCC Crystals

V. S. Deshpande, A. Needleman and E. Van der Giessen 79
Discrete Dislocation Predictions of Single Crystal Fatigue
Crack Growth

Yoshiteru Aoyagi and Kazuyuki Shizawa 87
A Crystal Plasticity Analysis for Accumulations of
Geometrically Necessary Dislocations and Dipoles around
Shear Band

Tetsuya Ohashi 97
A New Model of Scale Dependent Crystal Plasticity Analysis

Keiko Nakatani, Yoshihiko Sugiyama and Hiroshi Kitagawa 107
A Molecular Dynamics Study of Mechanical Properties and
Fractures of Amorphous Metal

Satoshi Izumi, Shotaro Hara, Tomohisa Kumagai and Shinsuke Sakai 117
Elastic Properties of the Surfaces and Interfaces of
Crystal and Amorphous Silicon

S. Brochard, J. Godet, L. Pizzagalli and P. Beauchamp 129
Atomistic Simulation of Dislocation Generation at
Surface Steps in Metals and Silicon

Pirouz Pirouz, Shanling Wang, Ming Zhang and Jean-Luc Demenet 139
On the Plasticity and Fracture of Semiconductors

Kenji Higashida and Masaki Tanaka 153
HVEM/AFM Studies on Crack Tip Plasticity in Si Crystals

C. Pokor, Y. Brechet, P.Dubuisson, J.P. Massoud and D. Rodney 163
Irradiation Hardening in Austenitic Steels:
Experiment and Simulation

D.J. Bacon, Yu. N. Osetsky, Z. Rong and K. Tapasa 173
Dislocation Glide in the Presence of either
Solute Atoms or Glissile Loops

M.A. Shehadeh, H. M. Zbib, T. Diaz de la Rubia and V. Bulatov 183
Multiscale Analyses of High Strain Rate Deformation:
Anisotropic Effects

Yu. N. Osetsky and D. J. Bacon 193
Atomic-Level Interaction of an Edge Dislocation with
Localized Obstacles in Fcc and Bcc Metals

Krystyn J. Van Vliet, Ju Li, Ting Zhu, Yoonjoon Choi Sidney Yip and Subra Suresh 203
Defect Nucleation: Predictions through
Nanoscale Experiments and Computations

T. Tsuru and Y. Shibutani 213
Formation of Prismatic Dislocation Loop of
Single Crystalline Aluminum under Nanoindentation

Jinpeng Chang, Ting Zhu, Ju Li, Xi Lin, Xiao-Feng Qian and Sidney Yip 223
Multiscale Modeling of Defect Nucleation and Reaction: Bulk to Nanostructures

S. Groh, B. Devincre, F. Feyel, L. Kubin, A. Roos and J.-L. Chaboche 235
Discrete-Continuum Modeling of Metal Matrix Composites Plasticity

Yoshihiro Tomita and Makoto Uchida 245
Computational Evaluation of Micro- to Macroscopic Deformation Behavior of Amorphous Polymer with Slightly Heterogeneous Distribution of Initial Shear Strength

Yoshikazu Higa, Hiroshi Kitagawa and Yoshihiro Tomita 255
Computational Modeling and Characterization of Materials with Periodic Microstructure using Asymptotic Homogenization Method

Igor L. Maksimov 269
Critical Behavior near the Crack/Dislocation Depinning Threshold

Sergey V. Dmitriev, Nobuhiro Yoshikawa and Aleksey A. Vasiliev 279
Topological Soliton Dynamics in Media with Microscopic Rotations

Jeffrey W. Kysar 289
Dependence of Ductile and Brittle Response on Initial Energy Dissipation Mechanism at Crack Tip

Kikuo Kishimoto and Masaki Omiya 301
A Cohesive Zone Model and Interfacial Crack Problems

S. Kubo and M. Misaki 311
Molecular Dynamic Simulation of Influence of
Grain Boundary on Near-Threshold Fatigue Crack Growth

Y. Nakasone, S. Kasumi and Y. Iwasaki 321
Plasticity-Induced Martensitic Transformation around
Semi-elliptical Surface Cracks in Fatigue of
an Austenitic Stainless Steel

Y.S. Kim, S.O. Choi, S.R. Lee and J. Kim 331
An Atomic Simulation of AFM-based Nano Lithography
Process for Nano Patterning

Ken-ichi Saitoh, Syuichi Nagase and Noboru Shinke 343
Molecular Dynamics Study on Morphology and Strength of
Copper Atomic-cluster-assembled Structure

Jakob Schiotz and Soren L. Frederiksen 355
Competing Deformation Mechanisms in Nanocrystalline Metals

A. Nakatani, T. Shimokawa, R. Matsumoto and H. Kitagawa 365
Atomistic Study on Ideal Strength of Nanocrystal and
Deformation Induced Nanostructures

Tadashi Hasebe 381
Continuum Description of Inhomogeneously
Deforming Polycrystalline Aggregate based on Field Theory

Yoshitaka Umeno and Takayuki Kitamura 391
Study on Strength of Microscopic Material by Simulations with
Atom and Electron Models

Shigenobu Ogata, Ju Li, Yoji Shibutani and Sidney Yip **401**
Ab initio Study of Ideal Shear Strength

Christophe Domain and Alexandre Legris **411**
Investigation of Glide Properties in Hexagonal Titanium and Zirconium: An Ab-initio Atomic Scale Study

Shingo Tanaka, Rui Yang and Masanori Kohyama **421**
Computational Study of the Mechanical Properties of Alumina - Copper Interfaces: Ab initio Calculations and Combination with Mesoscopic Simulations

Shuji Ogata **431**
Development of Hybrid Electronic - Density - Functional/ Molecular - Dynamics Simulation Schemes for Ceramics and Semiconductors

PREFACE

This volume contains the papers presented at the IUTAM Symposium of "Mesoscopic Dynamics of Fracture Process and Materials Strength", held in July 2003, at the Hotel Osaka Sun Palace, Osaka, Japan.

The Symposium was proposed in 2001, aiming at organizing concentrated discussions on current understanding of fracture process and inhomogeneous deformation governing the materials strength with emphasis on the mesoscopic dynamics associated with evolutional mechanical behaviour under micro/macro mutual interaction. The decision of the General Assembly of International Union of Theoretical and Applied Mechanics (IUTAM) to accept our proposal was well-timed and attracted attention. Driven by the development of new theoretical and computational techniques, various novel challenges to investigate the mesoscopic dynamics have been actively done recently, including large-scaled 3D atomistic simulations, discrete dislocation dynamics and other micro/mesoscopic computational analyses.

The Symposium attracted sixty-six participants from eight countries, and forty-two papers were presented. The presentations comprised a wide variety of fundamental subjects of physics, mechanical models, computational strategies as well as engineering applications. Among the subjects, discussed are (a) dislocation patterning, (b) crystal plasticity, (c) characteristic fracture of amorphous/nanocrystal, (d) nano-indentation, (e) ductile-brittle transition, (f) ab-initio calculation, (g) computational methodology for multi-scale analysis and others.

We would like to thank all the contributors for making this Proceedings which should be a milestone of research of computational fracture mechanics and express our sincere appreciation to all of the persons and associations concerned in the Symposium: the members of the Scientific Committee who have given us appropriate advices concerning selection of the invited speakers and conduct of the Symposium, the members of Local Committee who assisted in the construction of the symposium plans which we holed and the dedicated staffs of the secretariat.

Financial support to the Symposium was provided by the IUTAM by which we could invited a lot of active young scientists. In addition, generous supports are gratefully acknowledged for Graduate School of Engineering, Osaka University, Handai Frontier Research Center (FRC), Japan Society for the Promotion of Science (JSPS), Japan Society of Mechanical Engineers (JSME), the Japan Society of Materials Science (JSMS).

Finally, the cooperation of the Kluwer academic publishers for publication of the Proceeding is very much appreciated as well as the financial support.

October 2003
Editors Hiroshi Kitagawa, Yoji Shibutani

北川浩 澁谷陽二

We should like to express our deep appreciation for the great achievements of Prof. Kitagawa and his untiring contributions to the solid mechanics field over 35 years. All of the participants in this symposium would like to dedicate this volume to Prof. Kitagawa in celebration of his retirement.

October 2003
Yoji Shibutani

Prof. Hiroshi Kitagawa

SCIENTIFIC COMMITTEE

H. Kitagawa, Co-Chair (Osaka University, Japan)
Y. Shibutani, Co-Chair (Osaka University, Japan)
P. Gumbsch (Fraunhofer Institute for Mechanics of Materials, Germany)
L. P. Kubin (CNRS/ONERA, France)
A. Needleman (Brown University, USA)
S. Schmauder (University of Stuttgart, Germany)
S. Yip (MIT, USA)
L. B. Freund (Brown University, USA)

LOCAL ORGANIZING COMMITTEE

H. Kitagawa, Co-Chair (Osaka University, Japan)
Y. Shibutani, Co-Chair (Osaka University, Japan)
A.Nakatani, Secretary (Osaka university, Japan)
Y.Tomita (Kobe University, Japan)
S. Kubo (Osaka University, Japan)
T. Ohno (Nagoya University, Japan)
T. Kitamura (Kyoto University, Japan)
K. Kishimoto (Tokyo Institute of Technology, Japan)
S.Hashimoto (Osaka City University, Japan)
T.Ohashi (Kitami Institute of Technology, Japan)
K. Shizawa (Keio University, Japan)
K. Higashida (Kyushu University, Japan)
K. Saitoh (Kansai University, Japan)
S. Ogata(Osaka University, Japan)
Y. Higa (Osaka University, Japan)
K. Nakatani (Osaka Prefecture University, Japan)

SPONSORS

International Union of Theoretical and Applied Mechanics
Graduate School of Engineering, Osaka University
Handai Frontier Research Center (FRC)

Japan Society for the Promotion of Science (JSPS)
Japan Society of Mechanical Engineers (JSME)
Japan Society of Materials Science (JSMS)

LIST OF PARTICIPANTS

Yoshiteru Aoyagi Keio University, Graduate School of Science and Technology
3-14-1 Hiyoshi, Kohoku-ku Yokohama, Kanagawa 223-8522, Japan
Email: yoshiteru@shizawa.mech.keio.ac.jp

David J. Bacon The University of Liverpool, Department of Engineering
Brownlow Hill, Liverpool L69 3GH, U.K.
Email: djbacon@liverpool.ac.uk

Sandrine Brochard Lab. Métallurgie Physique, – Université de Poitiers – CNRS
SP2MI-BP30179 Futuroscope Chasseneuil, Cedex 86962, France
Email: sandrine.brochard@univ-poitiers.fr

Wei Cai Lawrence Livermore National Laboratory
L-353 7000 Est Ave Livermore, CA 94551, U.S.A.
Email: caiwei@llnl.gov

Benoit Devincre CNRS-ONERA, Laboratoire d'Etude des Microstructures (LEM)
29 Av de la Division Leclerc, 92322 Chatillon Cedex, France Email: devincre@onera.fr

Sergey V. Dmitriev Institute of Industrial Science, The University of Tokyo
4-6-1 Komaba, Meguro-ku, Tokyo 153-8505, Japan
Email: sergey@iis.u-tokyo.ac.jp

Christophe Domain EDF R&D Département MMC, Les Renardières
F-77250 Moret sur Loing, France
Email: christophe.domain@edf.fr

Peter Gumbsch Fraunhofer Institut für Werkstoffmechanik
Woehlerstr. 11, 79194 Freiburg, Germany
Email: Gumbsch@iwm.fhg.de

Tadashi Hasebe Doshisha University, Dept. of Mechanical and Systems Engineering
1-3 Tatara Miyako-dai, Kyotanabe 610-321, Japan
Email: thasebe@mail.doshisha.ac.jp

Yoshikazu Higa Osaka University, Department of Adaptive Machine Systems, Graduate School of Engineering
2-1 Yamadaoka Suita, Osaka 565-0871, Japan
Email: higa@ams.eng.osaka-u.ac.jp

Kenji Higashida Kyushu University, Dept. of Materials Science and Engineering
6-10-1 Hakozaki, Higashi-ku Fukuoka 812-8581, Japan
Email: higasida@zaiko.kyushu-u.ac.jp

Satoshi Izumi The University of Tokyo,, Department of Mechanical Engineering
7-3-1 Hongo Tokyo 113-8656, Japan
Email: izumi@fml.t.u-tokyo.ac.jp

Yoshihisa Kaneko Osaka City University, Dept. of Intelligent Materials Engineering
Faculty of Engineering
Sugimoto 3-3-138, Sumiyoshi-ku, Japan
Email: kaneko@imat.eng.osaka-cu.ac.jp

Young-Suk Kim Kyungpook National University

1370 Puk-Ku Taegu 702-701, Korea
Email: caekim@knu.ac.kr

Hajime Kimizuka The Japan Research Institute, Limited
1-4-17, Misaki-cho, Chiyoda-ku, Tokyo, 101-0061, Japan
Email: kimizuka.hajime@jri.co.jp

Kikuo Kishimoto Tokyo Institute of Technology, Dept. of Mechanical and Control Engineering
2-12-1, O-okayama, Meguro-ku, Tokyo 152-8552, Japan
Email: kkishimo@mep.titech.ac.jp

Hiroshi Kitagawa Osaka University, Department of Adaptive Machine Systems, Graduate School of Engineering
2-1 Yamadaoka Suita, Osaka 565-0871, Japan
Email: kitagawa@ams.eng.osaka-u.ac.jp

Takayuki Kitamura Kyoto University, Dept. of Engineering Physics and Mechanics, Graduate School of Engineering
Yoshida-hommachi, Sakyo-ku, Kyoto-shi, Kyoto 606-8501, Japan
Email: kitamura@kues.kyoto-u.ac.jp

Masanori Kohyama National Institute of Advanced Industrial Science and Technology, Special Division of Green Life Technology
1-8-31, Midorigaoka Ikeda, Osaka 563-8577, Japan
Email: m-kohyama@aist.go.jp

Ladislas P. Kubin CNRS-ONERA, Laboratoire d'Etude des Microstructures (LEM)
29, Av. de la Division Leclerc, BP 72, 92322 Chatillon Cedex, France
Email: kubin@onera.fr

Shiro Kubo Osaka University, Dept. of Mechanical Engineering and Systems, Graduate School of Engineering
2-1, Yamadaoka Suita, Osaka 565-0871, Japan
Email: kubo@mech.eng.osaka-u.ac.jp

Jeffrey Kysar Columbia University, Department of Mechanical Engineering
500 West 120th Street, MC 4703, New York, NY 10027, U.S.A.
Email: jk2079@columbia.edu

Ryosuke Matsumoto Kyushu Institute of Technology, Department of Mechanical Systems Engineering
680-4 Kawazu, Iizuka-City Fukuoka 820-8502, Japan
Email: matsumoto@mse.kyutech.ac.jp

Igor L. Maksimov Nizhny Novgorod University, Department of Theoretical Physics, Faculty of Physics, 23 Gagarin Ave. University
Nizhny Novgorod 603000, Russia
Email: ilmaks@phys.unn.ru

Ronan Madec CEA DAM DPTA SPMC
BP12 Bruyères-le-Châtel 91680, France
Email: madec@bruyeres.cea.fr

Yuji Nakasone Tokyo University of Science, Department of Mechanical Engineering, Faculty of Engineering
Kagurazaka 1-3, Shinjyuku-ku, Tokyo 162-8601 Japan
Email: nakasone@rs.kagu.tus.ac.jp

Akihiro Nakatani Osaka University, Department of Adaptive Machine Systems, Graduate School of Engineering
2-1 Yamadaoka Suita, Osaka 565-0871, Japan
Email: nakatani@ams.eng.osaka-u.ac.jp

Keiko Nakatani Osaka Prefecture University, Department of Aerospace Engineering, Collcgc of Engineering
1-1 Gakuen-cho, Sakai, Osaka 599-8531, Japan
Email: nakatani@aero.osakafu-u.ac.jp

Alan Needleman Brown University, Box D, Division of Engineering
Providence, RI 02912, U.S.A.
Email: needle@engin.brown.edu

Shigenobu Ogata Osaka University, Dept. of Mechanical Engineering and Systems, Graduate School of Engineering
2-1 Yamadaoka Suita, Osaka 565-0871, Japan
Email: ogata@mech.eng.osaka-u.ac.jp

Shuji Ogata Nagoya Institute of Technology
Gokiso, Showa-ku, Nagoya, Aichi 466-8555, Japan
Email: ogata@yamaguchi-u.ac.jp

Tetsuya Ohashi Kitami Institute of Technology
Koencho 165 Kitami, Hokkaido 090-8507, Japan
Email: ohashi@newton.mech.kitami-it.ac.jp

Yury N. Osetsky The University of Liverpool, Materials Science and Engineering, Department of Engineering
Liverpool L69 3GH, UK
Email: i.osetsky@liverpool.ac.uk
Oak Ridge National Laboratory, Metals and ceramics
Bldg. 4500S, MS-6138, P. O. BOX 2008, Oak Ridge, TN 37831-6158, U.S.A.
Email: osetskiyyn@email.cind.ornl.gov

Pirouz Pirouz Case Western Reserve University
10900 Euclid Avenue White, Room 510 Cleveland, Ohio 44106-7204, U.S.A.
Email: pxp7@cwru.edu

David Rodney INPG 101 rue de la Physique Domaine Universitaire
BP 46 Saint Martin d'Hères F38402 France
Email: david.rodney@gpm2.inpg.fr

Ken-ichi Saitoh Kansai University, Department of Mechanical Engineering
3-3-35 Yamate-cho, Suita-shi Osaka 564-8680, Japan
Email: saitou@ipcku.kansai-u.ac.jp

Tomotsugu Shimokawa Kanazawa University, Department of Mechanical Systems Engineering
2-40-20 Kodatsuno Kanazawa, Ishikawa 920-8667 Japan
Email: simokawa@t.kanazawa-u.ac.jp

Kazuyuki Shizawa Keio University, Department of Mechanical Engineering
3-14-1 Hiyoshi, Kohoku-ku Yokohama 223-8522, Japan
Email: shizawa@mech.keio.ac.jp

Jakob Schiotz Technical University of Denmark
DTU Building 307, Kongens Lyngby DK-2800
Denmark
Email: schiotz@fysik.dtu.dk

Siegfried Schmauder Universität Stuttgart, Staatliche Materialprüfungsanstalt (MPA)
Pfaffenwaldring 32, D-70569 Stuttgart, Germany
Email: Siegfried.Schmauder@mpa.uni-stuttgart.de

Yoji Shibutani Osaka University, Dept. of Mechanical Engineering and Systems, Graduate School of Engineering
2-1 Yamadaoka, Suita, Osaka 565-0871, Japan
Email: sibutani@mech.eng.osaka-u.ac.jp

Yoshihiro Tomita Kobe University, Graduate School of Science and Technology
Rokkodai 1-1, Nada-ku, Kobe 657-8501, Japan
Email: tomita@mech.kobe-u.ac.jp

Yoshitaka Umeno Kyoto University, Dept. of Engineering Physics and Mechanics, Graduate School of Engineering
Yoshida-hommachi, Sakyo-ku, Kyoto-shi, Kyoto 606-8501, Japan
Email:umeno@kues.kyoto-u.ac.jp

Krystyn J. Van Vliet Massachusetts Institute of Technology (M.I.T.), Dept of Materials Science and Engineering
Currently at the Department of Surgical Research, Children's Hospital, 300 Longwood Avenue Boston, MA 02115, U.S.A.
Email: krystyn@mit.edu

Daniel Weygand University of Karlsruhe IZBS
Kaiserstr. 12 Karlsruhe 76131, Germany
Email: Daniel.Weygand@izbs.uni-karlsruhe.de

Kisaragi Yashiro Kobe University, Department of Mechanical Engineering
1-1, Rokkodai, Nada, Kobe, 657-8501, Japan
Email: yashiro@mech.kobe-u.ac.jp

Sidney Yip Massachusetts Institute of Technology (M.I.T.), Department of Nuclear Engineering and Department of Materials Science and Engineering
77 Mass Ave Cambridge, MA 02139, U.S.A.
Email: syip@mit.edu

Hussein M. Zbib Washington State University, School of Mechanical & Materials Engineering
Pullman, WA 99164-2920, U.S.A.
Email: zbib@wsu.edu

Joey Crider U.S. Army NGIC
6385 Midway Road, Charlottesville, VA 22903, U.S.A.
Email: jfcrider@ceva.us

IUTAM Symposium on Mesoscopic Dynamics
of Fracture Process and Materials Strength
6 - 11, July 2003, Hotel Osaka Sun Palace

SYMPOSIUM PROGRAM

Sunday July 6

17:00 *Registration* starts at the NORTH ENTRANCE LOBBY of Hotel Osaka Sun Palace

18:00 *Reception* at the ORBIT HALL LOBBY of Hotel Osaka Sun Palace

Monday July 7

8:30 *Registration* at the Hotel Osaka Sun Palace

9:00 **Opening**

Session 1: DDD, Dislocation Patterning

[Chairperson: K. Shizawa]

9:30~10:00 Dislocation Patterning and Plasticity

W. Cai, V. V. Bulatov, T. Pierce, M. Hiratani, M. Rhee and M. Bartelt

10:00~10:30 Nondestructive Observation of Dislocation Structure Formed at Fatigued Copper and Stainless Steel Crystals

Y. Kaneko and S. Hashimoto

10:30~11:00 **Coffee Break**

11:00~11:30 Plasticity in Small Samples: A Discrete Dislocation Dynamics Description

D. Weygand

11:30~12:00 A Dislocation Dynamics Study of Some Constitutive Parameters for Plastic Flow

R. Madec

12:00~14:00 **Lunch**

Session 2: MD, DDD, Crystal Plasticity

[Chairperson: W. Cai]

14:00~14:30 Atomistic and DDD Studies of Inertial Effects on the Dynamics of Dislocations

E. Bitzek, D. Weygand and P. Gumbsch

14:30~15:00 Molecular Dynamics Study on the Characteristics of Edge and Screw Dislocations in Gamma/Gamma-Prime Microstructure in Ni-based Superalloy

K. Yashiro, Y. Tabata and Y.Tomita

15:00~15:30 **Coffee Break**

15:30~16:00 Multiscale Modeling of fcc Single Crystal Plasticity

L.P. Kubin

16:00~16:30 Discrete Dislocation Modeling of Fatigue Crack Growth in Single Crystals

V. S. Deshpande, A. Needleman and E. Van der Giessen

16:30~17:00 A Crystal Plasticity Analysis for Accumulations of Geometrically Necessary Dislocations and Dipoles around Shear Band

Y. Aoyagi and K. Shizawa

17:00~17:30 A New Model of Scale Dependent Crystal Plasticity Analysis

T. Ohashi

Tuesday July 8

8:30 *Registration* at the Hotel Osaka Sun Palace

Session 3: MD, Amorphous, Silicon, Fracture

[Chairperson: P. Gumbsch]

9:00~9:30 Molecular Dynamics Study on Mechanisms of Deformation and Fracture near a Crack Tip in Amorphous Metal

K. Nakatani, Y. Sugiyama and H. Kitagawa

9:30~10:00 Elastic Properties of the Surfaces and Interfaces of Crystal and Amorphous Silicon

S. Izumi, S. Hara, T. Kumagai and S. Sakai

10:00~10:30 **Coffee Break**

10:30~11:00 Atomistic Simulation of Dislocation Generation at Surface Defects in Metals and Silicon

S. Brochard, J. Godet, L. Pizzagalli, P. Beauchamp and J. Grilhe

11:00~11:30 On the Plasticity and Fracture of Semiconductors

P. Pirouz

11:30~12:00 HVEM/AFM Studies on Crack Tip Plasticity in Si Crystals

K. Higashida and M. Tanaka

12:00~14:00 **Lunch**

Session 4: MD, Dislocation, Obstacle, Indentation

[Chairperson: T. Ohashi]

14:00~14:30 Atomic and Mesoscopic Modeling of Irradiation Hardening in FCC Crystals

D. Rodney

14:30~15:00 Dynamics of an Edge Dislocation Glide in the Presence of Substitutional Solute Atoms and Glissile Interstitial Clusters

D.J. Bacon, Yu. N. Osetsky, Z. Rong and K. Tapassa

15:00~15:30 **Coffee Break**

15:30~16:00 Modeling the Dynamic Behavior of FCC Single Crystals under Shock Loading: Dislocation Dynamic Plasticity Analysis

H. M. Zbib, Mu'tasem, A. Shehadeh, Tomas. D. de la Rubia and V. Bulatov

16:00~16:30 Atomic-Level Interaction of an Edge Dislocation with Localized Obstacles in Fcc and Bcc Metals

Yu. N. Osetsky and D. J. Bacon

16:30~17:00 Understanding Defect Nucleation through Nanoscale Experiments and Computations

K. J. Van Vliet, J. Li, T. Zhu, S. Suresh and S. Yip

17:00~17:30 Dislocation Emission and Prismatic Dislocation Loop Formation of Single Crystalline Aluminum under Nanoindentation

T. Tsuru and Y. Shibutani

Wednesday July 9

8:30 *Registration* at the Hotel Osaka Sun Palace

Session 5: Multiscale Modeling

[Chairperson: T. Hasebe]

9:00~9:30 Multiscale Modeling of Materials Strength and Deformation

S.Yip

9:30~10:00 Parameter Link as an Approach to Hierarchical Modelling of Toughness Decrease of Steels

S. Schmauder, U. Weber, P. Binkele and P. Kizler

10:00~10:30 **Coffee Break**

10:30~11:00 Modelling Plasticity at Mesoscale with Dislocation Dynamics and Finite Elements Coupling

B. Devincre

11:00~11:30 Computational Evaluation of Micro- to Macroscopic Deformation Behavior of Amorphous Polymer with Slightly Heterogeneous Distribution of Molecular Chains

Y. Tomita and M. Uchida

11:30~12:00 Computational Modeling and Estimation of Materials with Periodic Microstructure Using Asymptotic Homogenization Method

Y. Higa, H. Kitagawa and Y. Tomita

12:00~14:00 **Lunch**

Session 6: Crack, DB Transition, Transformation

[Chairperson: S. Schmauder]

14:00~14:30 Critical Behavior near the Crack/Dislocation Depinning Threshold: Critical Indices and Landau-type Expansion

I. L. Maksimov

14:30~15:00 Domain wall and dislocation dynamics in media with microscopic rotations

S. V. Dmitriev, N. Yoshikawa and A. A. Vasiliev

15:00~15:30 **Coffee Break**

15:30~16:00 Dependence of Ductile and Brittle Response on Initial Energy Dissipation Mechanism at Crack Tip

J. W. Kysar

16:00~16:30 A Cohesive Zone Model and Interfacial Crack Problems

K. Kishimoto

16:30~17:00 Molecular Dynamic Simulation of Influence of Crystallographic Orientation and Grain Boundary on Near-Threshold Fatigue Crack Growth in Iron

S. Kubo and M. Misaki

17:00~17:30 Plasticity-Induced Martensitic Transformation around Semi-elliptical Surface Cracks in Fatigue of an Austenitic Stainless Steel

Y. Nakasone, S. Kasumi and Y. Iwasaki

Thursday July 10

8:30 *Registration* at the Hotel Osaka Sun Palace

Session 7: Nano-crystal, Poly-crystal

[Chairperson: Y. Nakasone]

9:00~9:30 An atomistic simulation of AFM-based nano lithography process for nano patterning

Y.-S. Kim, C.-Il Kim and S.-S. Lee

9:30~10:00 Molecular Dynamics Study on Morphology and Strength of Copper Atomic-cluster-assembled Structure

K. Saitoh, S. Nagase, H. Kitagawa and N. Shinke

10:00~10:30 **Coffee Break**

10:30~11:00 The Strength of Nanocrystalline Metals: An Optimal Grain Size

J. Schiotz

11:00~11:30 An Atomistic Study of Ideal Strength of Polycrystalline Metals

A. Nakatani, T. Shimokawa, R. Matsumoto and H. Kitagawa

11:30~12:00 Continuum Description of Inhomogeneously Deforming Polycrystalline Aggregates based on Field Theory

T. Hasebe

12:00~14:00 **Lunch**

Session 8: Ab-initio Calculation, Mutiscale Modeling

[Chairperson: Y. Shibutani]

14:00~14:30 Study on Strength of Microscopic Material by Simulations with Atom and Electron Models

Y. Umeno and T. Kitamura

14:30~15:00 Ab initio Study of Ideal Shear Strength

S. Ogata, J. Li, Y. Shibutani and S. Yip

15:00~15:30 **Coffee Break**

15:30~16:00 Ab initio Atomic-scale Simulation Investigation of the Plasticity in Zirconium and Titanium - Influence of Hydrogen

C. Domain and A. Legris

16:00~16:30 Computational Study of the Mechanical Properties of Alumina-Copper Interfaces: Ab initio Calculations and Combination with Mesoscopic Simulations

M. Kohyama, S. Tanaka and R. Yang

16:30~17:00 A Hybrid Electronic-Density-Functional/Molecular - Dynamics Simulation Scheme for Multiscale Simulation of Materials on Parallel Computers

S. Ogata

19:00 **Banquet at Banpaku Memorial Guest House (Geihinkan)**

Friday July 11

One-Day Tour to Himeji, Spectacular World Heritage

Massively-Parallel Dislocation Dynamics Simulations

Wei Cai, Vasily V. Bulatov, Tim G. Pierce, Masato Hiratani, Moono Rhee, Maria Bartelt and Meijie Tang
Lawrence Livermore National Laboratory, University of California, Livermore, CA 94551

Abstract. Prediction of the plastic strength of single crystals based on the collective dynamics of dislocations has been a challenge for computational materials science for a number of years. The difficulty lies in the inability of the existing dislocation dynamics (DD) codes to handle a sufficiently large number of dislocation lines, in order to be statistically representative and to reproduce experimentally observed microstructures. A new massively-parallel DD code is developed that is capable of modeling million-dislocation systems by employing thousands of processors. We discuss the general aspects of this code that make such large scale simulations possible, as well as a few initial simulation results.

Keywords: parallel computation, dislocation dynamics, plasticity

1. Introduction

It has been known for a long time that crystal plasticity is produced by the motion of many dislocation lines [1]. Consequently, *a priori* predictions of the strength of a single crystal against plastic deformation must be possible, at least in principle, by modeling the dynamics of dislocation lines under the influence of external stress and mutual interactions. Such has been a dream of the computational materials scientists for several decades. Yet, it remains a grand challenge even to date. The major difficulty lies in the fact that, to have a representative model of crystal plasticity, the dynamics of a *large enough* number of dislocations needs to be followed for a *long enough* time interval. The length and time scales required have remained beyond the reach of the existing simulation codes.

To understand why, let us consider a typical dislocation microstructure spontaneously developed in copper during plastic deformation [2]. The structure exhibits patterns over the length scale of microns. To model this behavior, a simulation box of about $L = 10\mu$m would be necessary. Given that the experimental estimates of dislocation density in such conditions are around $\rho = 10^{12}\text{m}^{-2}$, the total length of dislocation lines in the simulation box is about $\Lambda = \rho L^3 = 10^{-3}$m. In a dislocation dynamics (DD) simulation, dislocations are discretized into segments. If the average segment length is d = 1nm, then the total number of segments in this simulation would be $N = \Lambda/d = 10^6$, i.e., simultaneous

H. Kitagawa and Y. Shibutani (eds.), IUTAM Symposium on Mesoscopic Dynamics of Fracture Process and Materials Strength, 1–11.

treatment of a million segments is required. This is a rough estimate but serves to identify the order of magnitude of the complexity of simulation required to address the crystal plasticity problem. Here in this paper we refer to a simulation with million dislocation segments as our target problem.

Because the interactions between dislocation segments are complex and long-ranged, dislocation dynamics codes, when running on a single processor, can only handle up to 10^4 segments. Beyond that size the simulation becomes very slow and no longer useful. Notice that this is two orders of magnitude away from the target size stated above. Yet another computational limit exists: when a reasonable initial dislocation density is used, the total plastic strain one can accumulate using the sequential DD codes is on the order of 0.1%, another two orders of magnitudes below the levels of strain where dislocation patterning and strain hardening behaviors are observed (typically at around 10% plastic strain).

To extend our simulation capability in both length and time scales by two orders of magnitude and to meet the requirements for faithful modeling of crystal plasticity, massively parallel computing appears to be a natural solution. For example, imagine a simulation where 10,000 processors are used simultaneously, each handling on average only 100 dislocation segments. Because the load on each processors is relatively light, a million dislocation segments can be simulated at a reasonable speed in order to accumulate large enough plastic strain. However, devloping a DD code that is scalable up to $10^3 - 10^4$ processors is a highly nontrivial task. In this paper, we describe a few general features of our new massively-parallel DD code and present a few initial results from runs on up to 200 processors.

2. Simulation Methodology

The development of the `DD3d` code began at the Lawrence Livermore National Lab (LLNL) in 2001. To date (two years later) the first version is completed while further developments are still on-going, mainly focused on further enhancing the simulation efficiency and the accuracy of the physical models. The main objective for the `DD3d` code is to be able to take advantage of massively-parallel computers effectively. To achieve this goal, there are two basic design principles to which we have adhered during the entire development of `DD3d`. First, whenever possible, we choose algorithms that are *conceptually and logically simple.* Second, we intend to make this code as *generic* as possible. Keeping these design principles in mind should be helpful in understanding the aspects of code development described below.

The reason that we are only interested in conceptually simple algorithms is obvious. A complex algorithm with many *ad hoc* rules is not only aesthetically less appealing, but necessitates complex bookkeeping that can be disastrous if one tries to implement (and debug) it in a massively parallel setting. On the other hand, dislocations are known to be peculiar objects: they are topological line defects with a singular elastic field. Not surprisingly, the algorithms for simulating dislocations are necessarily more complex than those for simulating point objects, such as atoms in molecular dynamics (MD). Therefore, our choice of algorithms in `DD3d` is usually a compromise between conceptual simplicity and computational efficiency.

If the code is *generic* then it can be easily applied to simulate various materials after it is developed and fully tested for one test case. In our development work on `DD3d` we find that almost all elements of the algorithm deal with various generic issues that are independent of the specific physical system. These include, for example, meshing the dislocations into segments, computing driving forces, and communication between the processors. The system specific parts, on the other hand, can be grouped into one place — the mobility module (details later) that specifies how individual dislocations move in response to the driving force it sees. The separation of the system-specific parts from generic parts is rather similar to that in the commercially available finite element (FEM) codes. This way the code will be able to model a new material once the user defines his/her own material module of interest.

2.1. Data Structure

In `DD3d` dislocations are represented as a set of "nodes" connected with each other by straight line segments, as shown in Fig. 1. The position of nodes, together with their connectivity, is our fundamental degrees of freedom. If a node is connected with n other nodes, we call it a n-node, or a node with n arms, or n neighbors. In Fig. 1, node 1, 2 and 3 are 2-nodes, or "discretization" nodes, while node 0 is a "physical"-node, indicating the position where three dislocations meet.

The Burgers vectors are defined on every arm emanating from the node, with the line direction always pointing away from the node. For example, $\vec{b}_{01}$ is the Burgers vector of the arm going from node 0 to node 1, and $\vec{b}_{10}$ is the Burgers vector of the same arm going in the reverse direction. Therefore the sum rule $\vec{b}_{01} + \vec{b}_{10} = 0$ follows. Furthermore

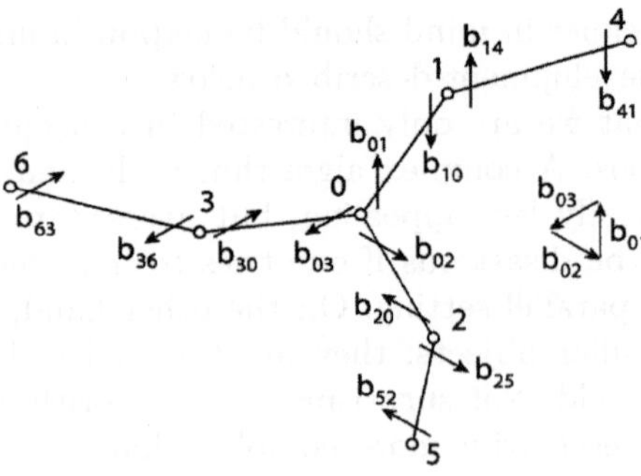

Figure 1. Dislocation network represented as a set of "nodes" (empty circles) interconnected by straight segments (see text).

the total Burgers vector of all arms going out of any given node is also zero, e.g., $\vec{b}_{01} + \vec{b}_{02} + \vec{b}_{03} = 0$.

Under this convention, an arbitrary dislocation network can be uniquely specified by a set of nodes $\{\mathcal{N}_i\}$, each described by its location $\vec{r}_i$, its connectivity and Burgers vectors of its arms, i.e.,

$$\mathcal{N}_i = \left[\vec{r}_i;\; I_{ij},\; \vec{b}_{ij},\; (j = 1, \cdots, n_i)\right], \tag{1}$$

where I_{ij} are the indices of the neighboring nodes of node i, and n_i is its total number of neighbors. The node set $\{\mathcal{N}_i\}$ is the data the code deals with.

2.2. Generic Algorithm

In general, a DD3d computational cycle goes as the following.

1. Compute driving force $\vec{f}_i$ on each node.

2. Compute velocity $\vec{v}_i$ of each node based on $\vec{f}_i$ and local dislocation character.

3. Determine suitable time step Δt.

4. Evolve all dislocation nodes to time $t + \Delta t$, handling topological changes occurring during $[t, t + \Delta t]$.

5. $t := t + \Delta t$. Go to 1.

Except step 2, all the steps above are generic aspects of DD simulations that are not dependent on the material of interest. They will be discussed in this section. The mobility module (step 2) will be discussed in the following section.

2.2.1. *Nodal Force Calculation*

The driving force on any given node i can be rigorously defined as (minus) the derivative of the total elastic energy of the dislocation network $E(\{\mathcal{N}_i\})$, with respect to a virtual displacement of the nodal position $\vec{r}_i$, i.e.,

$$\vec{f}_i = -\frac{\partial E(\{\mathcal{N}_i\})}{\partial \vec{r}_i} \tag{2}$$

In the elasticity theory of dislocations [1], the total elastic energy can be written as the sum of self energies between each segment pairs, such as

$$\begin{aligned} E(\{\mathcal{N}_i\}) &= W_S(01) + W_S(14) + W_S(02) + \cdots \\ &\quad + W_I(01,14) + W_I(01,02) + W_I(02,25) + \cdots \\ &= \sum_{\langle i,j \rangle} W_S(ij) + \frac{1}{2} \sum_{\langle i,j \rangle ; \langle k,l \rangle} W_I(ij,kl) \ , \end{aligned} \tag{3}$$

where $W_S(ij)$ is the self energy of segment (i,j), and $W_I(ij,kl)$ is the interaction energy between segments (i,j) and (k,l). Contrary to several earlier claims that the driving force could be infinite due to the existence of sharp corners at the nodes, it can be shown [3] that the driving force in Eq. (2) is well defined, well-behaved and numerically converges to the known analytical solutions for smooth dislocation curves as the discretization becomes finer. It is also shown in [4] that the contributions to nodal driving force $\vec{f}_i$ due to segment interactions, such as $W_I(ij,kl)$ can be evaluated by numerically integrating the stress field of segment (k,l) on segment (i,j), with proper weights. The contributions from self energies ($W_S(ij)$) on the other hand, are obtained by analytical differentiation. Most of the computational time in `DD3d` is spent on nodal force calculations, most of which is the evaluation of stress field of one segment on another segment (assuming isotropic linear elasticity). Because periodic boundary condition [5] is used, for every segment the stress field due to an infinite array of its images is also included. The image stress contribution is pre-computed and stored in a table for interpolation during the simulation [6].

2.2.2. *Moving the Nodes*

For simplicity, we integrate the first order equation of motion describing the over-damped motion of dislocations. This implies that there exists a mobility function ($\mathcal{M}$), which determines the instantaneous velocity

($\vec{v}_i$) of a node, given its instantaneous driving force and local geometry:

$$\dot{\vec{r}}_i \equiv \vec{v}_i = \mathcal{M}(\vec{f}_i) \tag{4}$$

The mobility function $\mathcal{M}$ will be discussed in more detail in the next section. For now let us simply assume that $\mathcal{M}$ is available and can be used to compute all the nodal velocities $\vec{v}_i$.

At this point, one can imagine the following simple algorithm for a DD simulation. With a pre-selected time step Δt, we can update the position of all nodes by (the forward Euler method),

$$\vec{r}_i := \vec{r}_i + \vec{v}_i \cdot \Delta t \tag{5}$$

However, in practice the velocities of the nodes could vary significantly during the simulation. For accuracy and numerical stability, it is better to use a variable Δt for each integration step. One approach is to put an upper bound ($r_{\max}$) on the distance any node is allowed to travel during one simulation step. Let $v_{\max} = \max_i |\vec{v}_i|$ be the maximum velocity of all nodes. Then the maximum allowed time step becomes $\Delta t = r_{\max}/v_{\max}$.

One more complication still remains. Remember that our nodes are not simple point objects, instead they are interconnected by dislocation segments. If we simply update nodal positions according to Eq. (5), certain dislocation segments may pass through each other without notice, which would be an unphysical artifact. The segment-segment collisions are accounted for in `DD3d` in the following way. For every pair of segments, e.g., $(1,2)$ and $(3,4)$, given the positions of all four participating nodes at time $t = 0$ ($\vec{r}_1$, $\vec{r}_2$, $\vec{r}_3$, $\vec{r}_4$), and assuming their respective velocities ($\vec{v}_1$, $\vec{v}_2$, $\vec{v}_3$, $\vec{v}_4$) remain constant, we developed an algorithm to predict whether or not these two segments will collide during period $[0, \Delta t]$, and if they do, when and where will the collision occur. Let us call this algorithm,

$$[\texttt{col}, t_p, \vec{r}_p] = \texttt{predictcollision}(\Delta t; \vec{r}_1, \vec{r}_2, \vec{r}_3, \vec{r}_4; \vec{v}_1, \vec{v}_2, \vec{v}_3, \vec{v}_4) \tag{6}$$

If there is a collision, `col` returns 1, and $t_p \in [0, \Delta t]$ and $\vec{r}_p$ are the predicted time and location of the collision, respectively. If there is no collision between the two segments, `col` returns zero.

At every time step, after we compute the velocities of all nodes, we use the `predictcollision` algorithm to check for possible collisions between all segment pairs. If there are no collisions at all during $[0, \Delta t]$, then we can safely update the positions of all nodes, and proceed to the next iteration. Otherwise, we have to perform a few sub-iterations to reach the desired time step Δt. Let δt be the time of the first collision. We will move all the nodes to time δt, ($\vec{r}_i := \vec{r}_i + \vec{v}_i \cdot \delta t$), and perform the necessary topological changes (details below) at that time. After that, we increment the time once again to the next collision time. This procedure is then repeated until the desired time step Δt is reached.

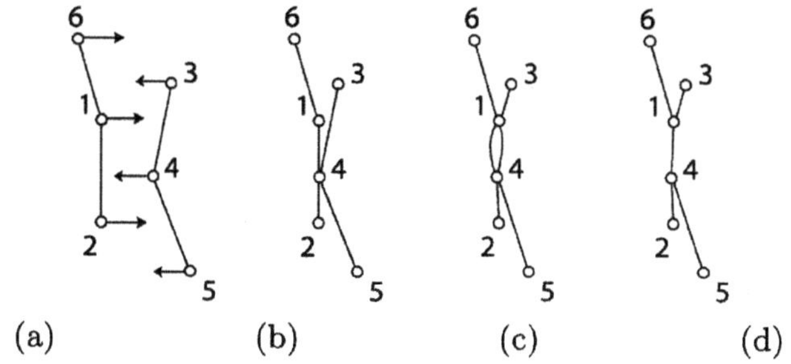

Figure 2. A simple approach to handling the topological changes during dislocation collision by creating a new node at the collision point. (a) Initial state at time 0. (b) At collision time t_p, we replace node 4 with a new node that connects to all four nodes $1-3$ and node 5. (c) At a later time segment 1-6 and 3-4 collide. Again we create a new node with 4 arms to replace node 1. At this time nodes 1 and 4 become doubly connected. (d) This is resolved by replacing the two arms connecting nodes 1 and 4 with a single arm with Burgers vector equal to the sum of that of the two original arms.

To take into account the topological changes when two dislocation lines meet each other, we adopt the following simple approach. A new node is created at the collision point that connects with all four nodes participating in the collision. Therefore, the new node has 4 arms, as shown in Fig. 2(b). It is interesting to note that by following this very simple algorithm, several different dislocation reaction scenario are reproduced naturally. To see this, let us follow this algorithm for a few more steps.

As shown in Fig. 2(c), the second collision occurs at a later time, between segments 1-6 and 3-4. Following the above procedure, we introduce a new node with 4 arms to replace node 1. However, this would result in a double connection between nodes 1 and 4, which is obviously redundant. If the sum of the Burgers vectors of these two arms is non-zero, we replace them with one arm with the Burgers vector equal to the sum. This makes a new dislocation segment (a junction), connecting two "physical" nodes – now each with three arms. If, on the other hand, the sum of the two Burgers vector is zero, nodes 1 and 4 are disconnected. Hence, dislocation annihilation occurs.

2.2.3. *Parallelism*

The all important feature of DD3d is its capability to utilize a large number of processors efficiently in parallel. To date, an efficient usage of 1500 has been demonstrated. To make this possible, all processors are treated equally during the simulation, i.e. there is no distinction such as "master" versus "slaves" between the processors. The total simulation box is divided into rectangular "domains", each assigned to one processor, as shown in Fig. 3. This way, the communications are mostly *local*, that is, each processor can obtain most of the information it needs by communicating with its nearest neighbors.

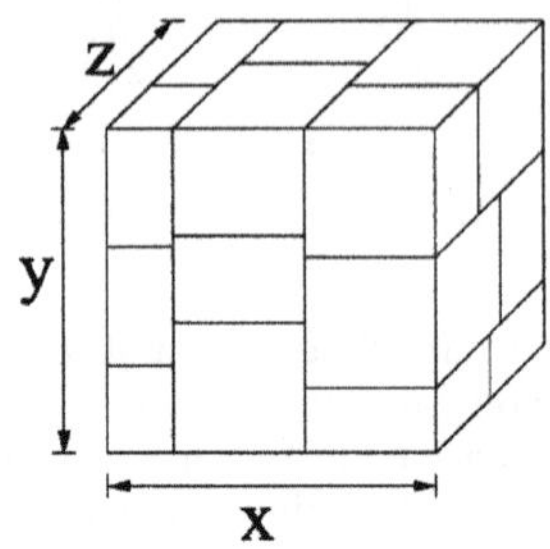

Figure 3. Decomposition of total simulation space into $3 \times 3 \times 2$ domains along x, y, z axes.

Because dislocation microstructures can be highly inhomogeneous, dividing the total simulation box into equally sized domains may lead to severe load imbalance, since some processors may contain a lot more nodes than others. To reach a better load balance, we use the following data decomposition procedure. The total simulation box is first divided into N_x domains along the x direction such that each domain contains equal number of nodes. Each domain is then further divided along y direction by N_y times, and the resulting domains again divided along z direction by N_z times. In the end, we obtain $N_x \times N_y \times N_z$ domains, all containing the same number of nodes, as shown in Fig. 3. However, because the dislocation structure evolves during the simulation, we need to re-partition the problem among processors from time to time, in order to maintain a good load balance. It is found that the optimal number of nodes per domain is in the range from 200 to 1000. In this case, the computational load on each processor is relatively light, while most of the computing time is still spent on computation instead of communication. If and when the total number of dislocation segments increases significantly (e.g. due to dislocation multiplication), we stop and restart the simulation with more processors, to maintain a reasonable simulation speed.

2.3. Mobility Module

The mobility module is the only material specific part of the DD3d code. It specifies how fast a node should move in response to its driving force. The effects of crystallography and temperature on dislocation mobility are both accounted for here. It is expect that the users will develop their own (possibly very sophisticated) mobility modules to simulate materials of their interest. Here as an example, we describe a simple mobility module that mimics the generic behavior of dislocation in body-centered-cubic (BCC) metals at high temperatures. We call it "pencil-glide" mobility module and use it in the simulation described in the next section.

The "pencil-glide" mobility module is specified by three parameters: the edge mobility M_e, the screw mobility M_s, and a critical angle θ_c. For simplicity, we will only discuss "discretization" nodes here, i.e., nodes with only two neighbors. Let $\vec{r}_1$ and $\vec{r}_2$ be the position of the two neighbors of node i, and let $L = |\vec{r}_2 - \vec{r}_1|/2$. Then $\vec{f}_i/L$ is the average Peach-Koehler force around node i. Unit vector $\vec{\xi} = (\vec{r}_2 - \vec{r}_1)/|\vec{r}_2 - \vec{r}_1|$ approximates the dislocation line direction. The dislocation character angle θ is defined through $\cos\theta = |\vec{\xi} \cdot \vec{b}|$. If $\theta < \theta_c$, the dislocation is locally "screw", otherwise it is "non-screw". The velocity of "screw" segments is simply $\vec{v}_i = M_s \vec{f}_i/L$. Because it follows the direction of the driving force and is not confined to any plane, this mobility function describes the well-known "pencil-glide" behavior observed in BCC metals at high temperatures. The velocity of "non-screw" segments, on the other hand, is confined within the glide plane, with normal vector $\vec{n} = \vec{b} \times \vec{\xi}/|\vec{b} \times \vec{\xi}|$. Let $\vec{\bar{v}} = [M_e \sin^2\theta + M_s \cos^2\theta]\vec{f}_i/L$, the velocity for "non-screw" dislocation is simply $\vec{v}_i = \vec{\bar{v}} - (\vec{\bar{v}} \cdot \vec{n})\vec{n}$.

3. Results

Here we describe the results of initial benchmark simulations using DD3d. The mobility law parameters chosen here are intended to mimic the behavior of BCC metal Mo. For example, screw dislocations have a lower mobility than edge dislocations. Specifically, $M_e = 10b \cdot (\mathrm{s} \cdot \mathrm{Pa})^{-1}$, $M_s = 1b \cdot (\mathrm{s} \cdot \mathrm{Pa})^{-1}$, $\theta_c = \arccos(0.95)$, where $b = 0.27$nm is the magnitude of the smallest Burgers vector in BCC Mo.

In these simulations, we used a cubic simulation box with edges along [100], [010] and [001] directions and 10μm in length. The initial configuration consists of 8 long screw dislocations (with Burgers vectors along $\frac{1}{2}\langle 111 \rangle$ directions) randomly positioned in the simulation box. The initial dislocation density is around $1.2 \times 10^{11}\mathrm{m}^{-2}$. A uniaxial

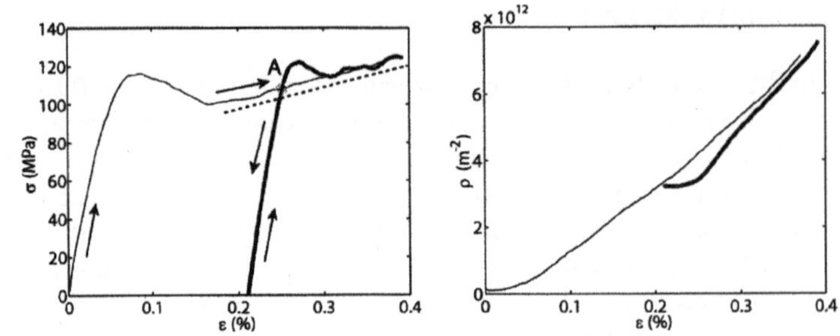

(a) (b)

Figure 4. (a) Stress-strain curves produced by DD3d simulations of two uniaxial loading tests along [100] with strain rate $\dot{\epsilon} = 1\text{s}^{-1}$ (see text for more details). The slope of the dashed line indicates the rate of strain hardening. (b) Dislocation density in these two simulations as a function of strain.

tension is then applied along [100] direction under a constant strain rate of $\dot{\epsilon} = 1\text{s}^{-1}$. The thin lines in Fig. 4 are the results from a 12-processor Linux Beowulf cluster after about 6 weeks of the wall-clock time. A total strain reaches about 0.4% at the end of this simulation, while the dislocation density has increased by more than 70 times. The stress-strain curve also exhibits three different behaviors. Initially the response is almost elastic and stress increases linearly. When the upper yield point around 120MPa is reached, the stress drops upon further loading, which indicates a strain softening behavior. Then, after a lower yield point around 100MPa is reached, the stress starts to increase again, this time exhibiting a strain hardening behavior. The slope of the stress-strain curve in this region, i.e., the strain hardening rate $d\sigma/d\epsilon$, as indicated by the dashed line, is around 10GPa.

The thick lines in Fig. 4 correspond to a separate simulation of the same specimen, but with different loading history. After the original specimen was deformed to point A [in Fig. 4(a)], it was unloaded and relaxed under zero stress. We then reloaded it using the same strain rate and temperature, and run the new simulation on a 200-processor Linux cluster for 3 days. It is interesting to note that the new simulation does not follow the original trajectory. Instead, it develops its own upper and lower yield points. However, after the lower yield point, the new simulation enters a strain hardening regime with about the same strain hardening rate as before. This indicates the robustness of the strain hardening behavior observed here.

4. Concluding Remarks

In this paper, we give a brief overview of the new massively parallel dislocation dynamics simulation code DD3d. Our description here is intended to be brief, so that the reader can get a general appreciation about the overall structure of the code, without being distracted by many technical details. To save the space, many general and important aspects are left out, such as remeshing, fast-multipole stress calculations, realistic mobility laws, patterning analysis of simulated dislocation micro-structures, etc. Other issues, such as more efficient parallel collision handling and time-stepping algorithms on 10^4 or more processors, are not completely resolved yet: DD3d is constantly evolving to better address these challenging problems. We expect that DD3d will soon become powerful enough to provide a statistically representative model for dislocation patterning and crystal plasticity. We hope that by exercising this explicit, large scale model, one can obtain new insights that will help the development of more reliable physics-based continuum theories of crystal plasticity.

This work was performed under the auspices of the U. S. Department of Energy by the University of California, Lawrence Livermore National Laboratory under Contract No. W-7405-Eng-48. Benchmark runs of DD3d on 200 to 1500 processors were performed on the MCR cluster of LLNL.

References

[1] J. P. Hirth and J. Lothe, *Theory of Dislocations*, (Wiley, New York, 1982).
[2] H. Mughrabi, T. Ungar, W. Kienle and M. Wilkens, Phil. Mag. A **53**, 793 (1986).
[3] Wei Cai, *Atomistic and Mesoscale Modeling of Dislocation Mobility*, Ph.D. Thesis, M.I.T., May (2001).
[4] Wei Cai, et al. unpublished.
[5] V. V. Bulatov, M. Rhee and W. Cai, Mater. Soc. Proc. **653**, Z1.3.1 (2001).
[6] Wei Cai, Vasily V. Bulatov, Jinpeng Chang, Ju Li and Sidney Yip, Philos. Mag. A, **83**, 539 (2003).

Nondestructive Observation of Dislocation Structures Formed at Fatigued Copper and Stainless Steel Crystals

Yoshihisa KANEKO and Satoshi HASHIMOTO

Department of Intelligent Materials Engineering, Faculty of Engineering, Osaka City University, Osaka 558-8585, Japan.

Abstract: Dislocation structures of fatigued metallic materials were investigated using electron channelling contrast imaging (ECCI) technique using a scanning electron microscope. By imaging electron channelling contrast resulting from dislocations, we could observe the dislocations lying close to crystal surface. At a fatigued copper single crystal of single slip orientation, the ECCI technique successfully imaged ladder-like dislocation structure peculiar to persistent slip band (PSB). In fatigue tests on copper bicrystals, effect of grain boundary on the PSB distribution was confirmed by the ECCI observation. This ECCI technique was attempted also to reveal the change in the dislocation structure of a stainless steel during fatigue.

Key words: fatigue, dislocation structure, electron channelling contrast imaging, copper, stainless steel, single crystal, bicrystal

1. INTRODUCTION

Meso-scale structures of dislocations are associated closely with deformation properties of fatigued metallic materials. Push-pull fatigue deformation gives rise to to-and-fro dislocation motions. Such motions in annealed materials would cause increase in dislocation density due to occurrence of multiplication mechanisms such as activation of the Frank-Read source. When the dislocation density exceeds a certain value, mutual annihilations of dislocations become frequent, and the dislocation density is saturated. The fatigue deformation affects not only the change in the dislocation density but also dislocation distribution which shift to be inhomogeneous. Resultant dislocation structures have been established by many fundamental works [1]. After the multiplication and annihilation processes are repeated sufficiently, the dislocations are self-organized into bundle structures called "vein" in crystals that have relatively high stacking-fault energy. The vein structure consists of high and low dislocation density regions of irregular shapes whose diameter is of a few micrometers. It has been recognized that evolution of this vein structure occurs during the cyclic hardening stage which is accompanied by an increase in dislocation density. Further fatigue cycling results in the formation of "persistent slip bands" (PSBs) in the vein structure if plastic strain amplitude is in a certain range. The morphology of the PSB dislocation structure is characterized by a ladder-like structure as shown

H. Kitagawa and Y. Shibutani (eds.), IUTAM Symposium on Mesoscopic Dynamics of Fracture Process and Materials Strength, 13–22.

schematically in Fig.1. The ladder-like structure is composed from dislocation walls which are considered as collections of edge dislocation multipoles. The spacing between the dislocation walls is about 1.3μm. It has been found that production of the PSBs gives rise to a change in the hysteresis loop shape —— the hysteresis loop tends to shifts from pointed shape to parallelogram —— in several materials although the stress amplitude is almost unchanged [2-5]. In this way, the mechanical properties of fatigued materials are affected by the internal dislocation structures that are organized at a few micrometer scales. Hence, for a better understanding of fatigue phenomena, it seems important to clarify the evolution of the dislocation structures during fatigue. In addition, because slip activities are concentrated at the PSBs which can accommodate larger plastic strain than the matrix vein structure, pronounced extrusions and intrusions are generated at intersections of the PSBs with the crystal surface. It has been recognized that intragranular fatigue cracks are nucleated preferentially at the PSBs causing rough surface [6-8]. Because the PSB formation precedes the fatigue crack nucleation, the detection of the PSB formation can correspond to the prediction of the fatigue crack nucleation. Hence, it can be said that observation of the dislocation structure is also desirable for the investigation of fatigue cracking.

The above-mentioned dislocation structures have been studied using transmission electron microscopy (TEM). However, conventional TEM observation has several disadvantages in studying the dislocation structure of fatigued materials. Since the TEM observation involves shaping of the material to a thin foil, the dislocation structure evolution throughout fatigue is impossible to be observed in one specimen. In addition, the TEM observation has a problem concerning viewable area of the thin foils.

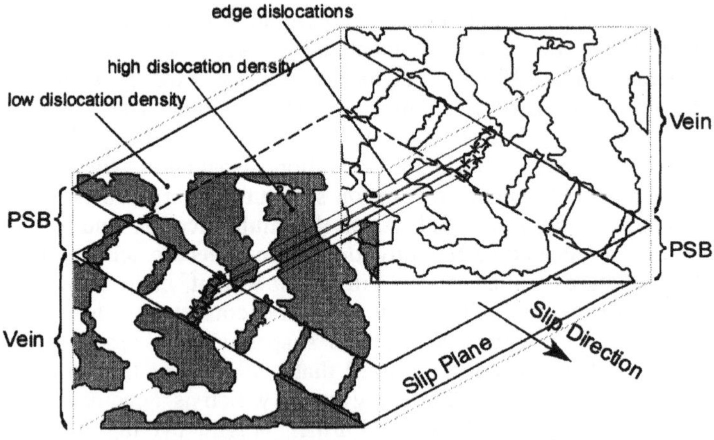

Figure 1: Schematic model of the dislocation structure of a fatigued copper crystal. The dislocation structure consists of the veins and the PSB having a ladder-like dislocation wall structure along slip planes.

Since the viewable area of the TEM is considerably restricted in comparison with that of an optical microscope or a scanning electron microscope (SEM), one can miss formation of special structures such as PSBs if their volume fractions are very low. Accordingly, a new method to observe dislocation structure is desired instead of the TEM.

Recently, a new technique called "electron channelling contrast imaging (ECCI)" has been developed to image dislocations (see review [9]). The ECCI has a characteristic feature that the dislocations lying near the crystal surface can be detected nondestructively using a SEM. In 1979, Morin et al [10] developed an SEM system with the field-emission gun and succeeded in imaging dislocations in a bulk silicon sample. Thereafter, the dislocations imaging by the ECCI have been reported in bulk samples of several materials [11-15]. The dislocation structures in several fatigued materials have also started to be observed by the ECCI [16-23]. The authors strongly expected that the ECCI technique is adequate to investigate the dislocation structure evolution during fatigue because the thinning process is avoided for the surface observation. Moreover, the ECCI technique using SEM enables us to the observe specimen surface at various magnifications compared to TEM observations. This must be convenient for the detection of unique dislocation structures which can be formed at special sites such as grain boundary vicinity.

In the present study, in order to confirm the validity of the ECCI technique, the authors first attempted to observe dislocation structures of fatigued copper single crystals which have been investigated systematically using TEM. The effect of grain boundaries on the dislocation structures was also investigated in fatigued copper bicrystals by taking advantage of the ECCI technique which has large viewing area. In addition, fatigued stainless steels were examined to know whether the ECCI technique is applicable to engineering materials or not.

2. EXPERIMENTAL PROCEDURE

2.1 Electron Channelling Contrast Imaging

By detecting intensity of the backscattered electrons in SEM, we could obtain several informations including electron channelling contrast. This electron channelling comes from a characteristic that the intensity of the backscattered electrons is sensitive to incident beam orientation relative to lattice planes as shown in Fig.2. It should be emphasized that the backscattered electron intensity changes rapidly in the vicinity of angles where the Bragg condition is satisfied. Booker et al [24] first suggested a possibility that dislocations are detectable in SEM by imaging the backscattered electron intensity. In the vicinity of the dislocation, lattice planes are slightly bent and their lattice spacings are also varied locally. This kind of lattice distortion gives rise to small changes in the Bragg condition around the dislocation. In order to observe the dislocations using the ECCI, crystal samples need to be tilted suitably such that the Bragg condition is satisfied between the incident beam and a certain set of lattice planes. Because the backscattered electron intensity changes rapidly near the Bragg angle, the channelling contrast would arise around dislocations which induce local change in the Bragg condition. Accordingly, we can image the dislocations lying close to the crystal surface by scanning the incident beam tilted to satisfy the Bragg angle and by detecting the backscattered electrons that escape from the surface.

In the present study, the ECCI observation was conducted in a JEOL JSM-6500F field-emission SEM at an acceleration voltage of 15kV and a probe current of about 3.5nA. In order to adjust the crystal orientation to achieve the Bragg condition, we utilized the electron-channelling pattern (ECP). Before the ECCI observation, specimens were electrolytically polished to obtain flat surface.

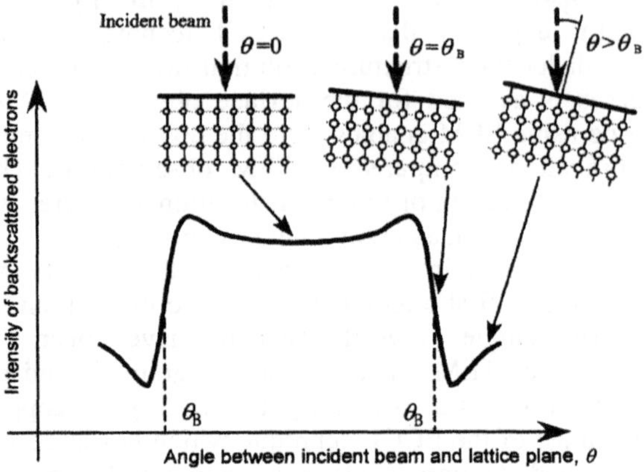

Figure 2: Schematic diagram of backscattered electron intensity as a function of beam angle relative to the crystal plane. The intensity changes rapidly near the Bragg angles.

2.2 Fatigue Tests

Copper single crystals having defined orientations were grown by the Brigdman technique from a copper material of 99.99% purity. The single crystals were shaped to strip specimens of 6mm gage length and 4x4mm^2 cross-section. We prepared four different stress axes along [123], [111], [100] and [211] directions. The stress axis of [123] corresponds to a single-slip orientation, and the others are multiple slip orientations.

Two copper bicrystal specimens —— compatible and incompatible bicrystals —— with a grain boundary parallel to the stress axis were also fabricated to investigate the effect of grain boundary on the dislocation structures. Since the stress axes of the bicrystals were oriented to <123> direction as shown in Fig.3, all constituent grains would deform in the same manner if we neglect grain boundary effects. The compatible bicrystal had primary slip systems whose intersections with grain boundary plane are parallel. Primary slip vectors of the constituent grains were oriented symmetrically towards the grain boundary. This kind of the bicrystal having such slip system geometry is regarded as plastically compatible. The intersections with grain boundary plane of the other bicrystal are not parallel each other.

Low-cycle fatigue tests on the copper single- and bicrystals were carried out in air at room temperature. During the fatigue deformation, the plastic shear strain amplitude of the specimen was controlled to be 2×10^{-3}.

Specimens of an austenitic stainless steel (Fe-19wt.%Ni-11wt.%Cr alloy) were also prepared. The stainless steel was shaped to strip specimens with rectangle gage shape of $3\times3\times6\,\text{mm}^3$. The specimens were annealed at 1473K for 1 hour. The average grain size was about 120μm. High-cycle fatigue tests were carried out at stress amplitudes of 110MPa and 150MPa at 5Hz. Yield stress of the stainless steel obtained in a monotonic tensile test was about 90MPa.

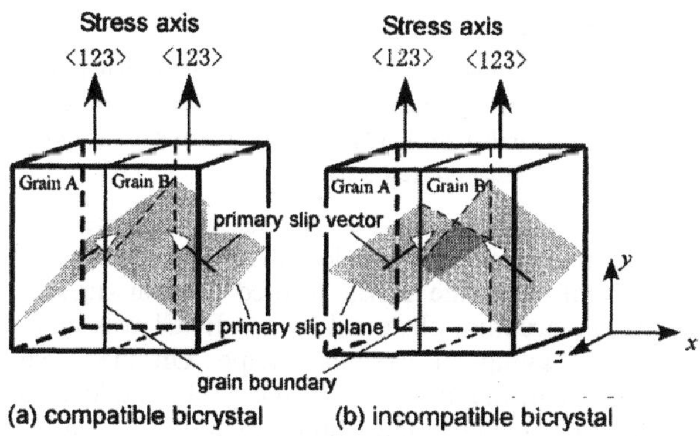

Figure 3: Primary slip systems of constituent grains of the copper bicrystals, where plastic strain is (a) compatible and (b) incompatible at the grain boundary. The stress axes of all constituent grains are along <123> direction.

3. RESULTS AND DISCUSSION

3.1 The Copper Single Crystals

The ECC images of the specimens having [123], [211], [111] and [100] stress axes are presented in Fig.4. Band-like structures composed from regularly-spaced walls are visible in the ECC images of the fatigued [123] and [211] single crystals. The walls in the [123] single crystal were arranged perpendicular to the band at an average spacing of about 1.4μm. This morphological feature is identical to the PSB of single-slip oriented crystals which have been observed using TEM [25,26] by taking account of contrast reversal in the ECCI. (Bright contrast corresponds to a region

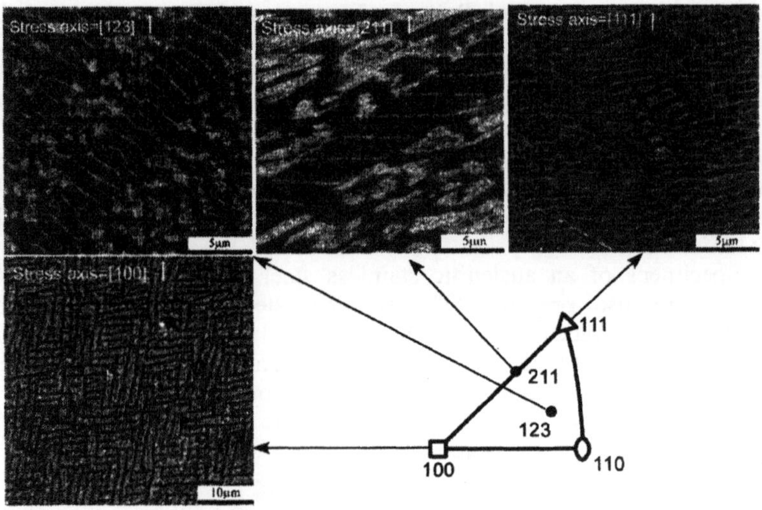

Figure 4: **Electron channelling contrast images showing orientation dependence of dislocation structures of the fatigued copper single crystals. Surface planes of the single crystals were parallel to $(1\bar{1}1)$, $(0\bar{1}1)$ $(1\bar{1}0)$ and $(0\bar{1}1)$ for the [123], [112], [111] and [100] stress axes, respectively.**

of high dislocation density in the ECCI, as opposed to the TEM observation.) Hence, it can be concluded that the band-like structure in the [123] single crystal were the PSBs formed at near-surface layer. The surrounding matrix structure observed by the ECCI was also identical to the vein structure having an irregular shape. On the other hand, the dislocation structure inside the band in the [211] single crystal differed from that of the PSBs of the single-slip orientation: the walls in the [211] specimen were perpendicular to the stress axis. This kind of unique wall arrangement has not been reported as far as we know, even in TEM observation.

The [111] and [100] specimens exhibited considerably different dislocation structures from the [123] crystal. In the [111] specimen, sinusoidal dislocation walls arranged perpendicular to the tensile axis are clearly visible. This structure is analogous to the TEM observation in the same orientation [27]. The dislocation structure of the [001] crystal observed by the ECCI is composed from two kinds of dislocation walls which intersect each other almost at right angle. This special structure has also been reported in TEM observations [28] and is called "labyrinth structure".

3.2 The Copper Bicrystals

At regions sufficiently distant from the grain boundaries, the dislocation structures observed by the ECCI consisted of PSB and vein structures which are common to the [123] single crystal. This is because all the constituent grains of both bicrystals were oriented to the <123> direction. In the vicinity of the grain boundaries, the dislocation structures depended on the character of grain boundary. Figure 5 shows ECC images of the grain boundary vicinities of the compatible and incompatible bicrystals. At the compatible bicrystal, the PSBs along $(\bar{1}11)$ and $(1\bar{1}1)$ primary slip planes of the respective grains reached the grain boundary. In this respect, the grain boundary of this bicrystal had no influence on self-organization of dislocations. On the other hand, complicated distribution of the PSBs appeared around the grain boundary of the incompatible bicrystal. Formation of PSBs along primary slip planes was limited to the regions that were more than 50μm distant from the grain boundary at both constituent grains. In the Grains A and B, the PSBs were formed along $(1\bar{1}1)$ and $(11\bar{1})$ planes at the region adjacent to the grain boundary instead of the primary systems.

In order to discuss such difference in the PSB formation around the grain boundaries, the authors paid attention to plastic strain compatibility between constituent grains. It is known that plastic strain due to slip deformation at a certain grain must be compensated at the grain boundary by the occurrence of slip deformation of adjacent grain such that the continuity at the grain boundary is maintained [29]. If the plastic deformation at the grain boundary is not accommodated solely by the

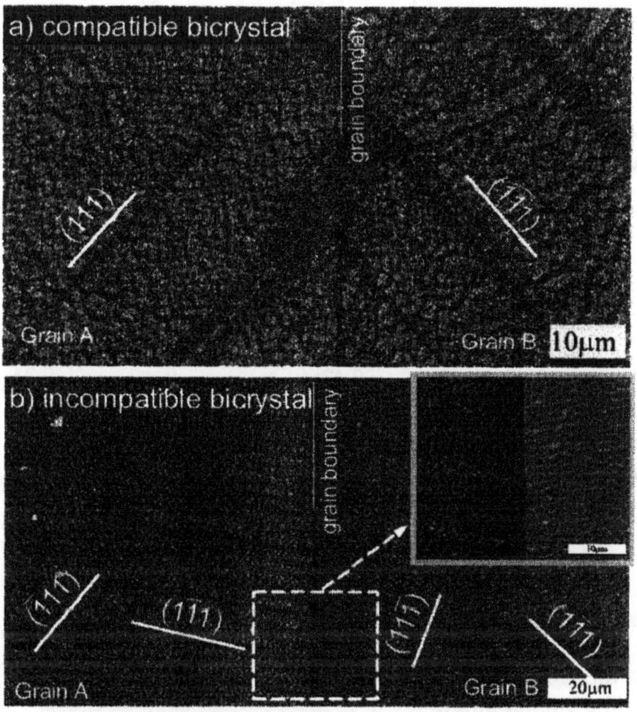

Figure 5: Electron channelling contrast images near grain boundaries of (a) compatible and (b) incompatible bicrystals. Because individual grains have different channelling conditions, each photograph was made by combining two images.

primary slip deformation of an adjacent grain, additional slip systems should be locally activated around the grain boundary. The plastic strain compatibility has usually been discussed by plastic strain components of slip shear, which are resolved on the grain boundary plane; i.e., *y-z* plane in Fig.3. At the grain boundary plane, the following relations between Grains A and B must be satisfied:

$$\varepsilon_y{}^A = \varepsilon_y{}^B, \quad \varepsilon_z{}^A = \varepsilon_z{}^B, \quad \gamma_{yz}{}^A = \gamma_{yz}{}^B$$

Table 1 presents the plastic strain components of the primary slip systems, which are resolved on the *y-z* grain boundary plane. In Grain A of the compatible bicrystal, if slip deformation of shear strain γ_A along primary plane occurs in the vicinity of grain boundary, the strain components resolved along the grain boundary plane become $\varepsilon_y{}^A=0.47\gamma_A$ and $\gamma_{yz}{}^A=0.25\gamma_A$. According to the calculation listed in Table 1, both the ε_y and γ_{yz} due to the primary slip in Grain A can be completely accommodated by the slip shear γ_B in Grain B. On the other hand, if the primary slip systems of both grains are activated so as to equalize the axial strain component ($\varepsilon_y{}^A$ and $\varepsilon_y{}^B$) at the incompatible bicrystal, the $\gamma_{yz}{}^A$ and $\gamma_{yz}{}^B$ components due to both primary slips become opposite to each other. For this reason, additional slip systems are considered to be activated at the grain boundary vicinity to maintain the grain boundary continuity.

Table 1. Plastic strain components of the primary slip systems resolved on the grain boundary plane, where γ_A and γ_B mean plastic shear strains due to slip deformation along the primary slip system of each grain.

	grain	Strain components resolved along GB plane		
		$\varepsilon_y{}^A$ or $\varepsilon_y{}^B$	$\varepsilon_z{}^A$ or $\varepsilon_z{}^B$	$\gamma_{yz}{}^A$ or $\gamma_{yz}{}^B$
compatible	Grain A	$0.47\gamma_A$	0	$0.25\gamma_A$
	Grain B	$0.47\gamma_B$	0	$0.25\gamma_B$
incompatible	Grain A	$0.47\gamma_A$	0	$-0.25\gamma_A$
	Grain B	$0.47\gamma_B$	0	$0.25\gamma_B$

3.3 Stainless Steel

The ECC image of the stainless steel fatigued at a stress amplitude of 140MPa until 10^4 cycles is presented in Fig.6. The ladder-like dislocation structures of the PSBs are visible in the ECC images, similar to the copper single crystals. Shapes of the dislocation walls are somewhat irregular in

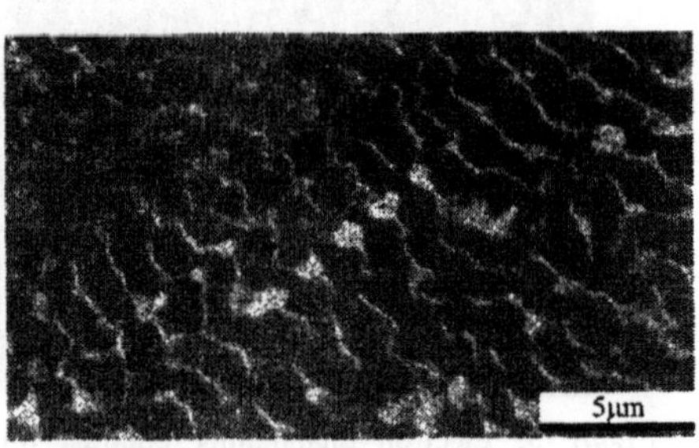

Figure 6: ECC image of the stainless steel fatigued at stress amplitude of 140MPa for 10^4 cycles

comparison with that of the copper.

It is apparent that the most important advantage of the ECCI is the nondestructive observation of dislocation structure without the need for a thinning process. This observation enables us to depict the dislocation structure by interrupting mechanical tests. This kind of the evaluation helps one to understand the relationship between mechanical properties and dislocation processes, but has been impossible as long as the TEM is used. For this reason, the authors tried to reveal the change in dislocation structure of the same grain, by repeating fatigue interruption and the following ECCI observation.

Figure 7 shows the changes in the dislocation structure during the fatigue tests at 110MPa and 140MPa stress amplitudes. At the fatigue test of 110MPa, no significant structures were generated even at 10^5 cycles. On the other hand, a ladder-like PSB structure was already generated at 10^4 cycles of the 140MPa fatigue test. In the ECC image of the 3×10^{-4} cycles, appearance of a cell structure having slight misorientation is detectable. It is likely that formation of the cell structure is caused by the activation of additional systems due to the grain boundary incompatibility.

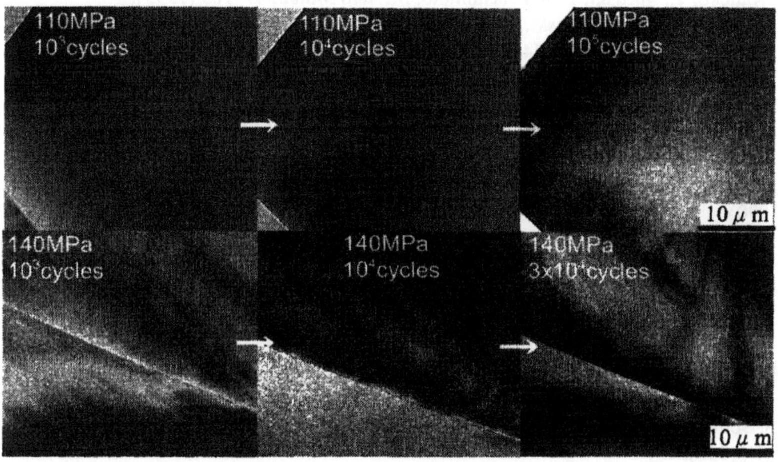

Figure 7: **Changes in the dislocation structure of the same region of a stainless steel samples fatigued at 110MPa and 140MPa stress amplitudes.**

4. SUMMARY

The ECC images of fatigued copper single crystals were identical to the results of conventional TEM observation. Hence, we think the ECCI offers an alternative method to observe the dislocation structures. In addition, since the ECCI observation has a large viewing area, the distributions of the PSBs in the grain boundary vicinity could be revealed in the copper bicrystals.

It was found that the ECCI technique was also applicable to the stainless steel. Because no thinning processes are required for the ECCI technique, we could observe the changes in the dislocation structures of the same grain during fatigue experiments. For the stainless steel sample in fatigue, the formation of PSBs and subsequent cell structures were successfully detected using the ECCI technique.

Acknowledgment

We would like to thank K.Fukui (Osaka City Univ.), A.Ono (JEOL Ltd.) and T.Yanagihara (JEOL Ltd.) for helping a part of the ECCI observation. This research is supported by JSPS Grants-in-Aid for Scientific Research (#15360372).

REFERENCES

1. Suresh S.: *Fatigue of Materials, Second Edition*, Cambridge University Press, Cambridge, p.39 (1998).
2. Abel, A.: *Mater.Sci.Eng.*, **36**, 117 (1978).
3. Mughrabi, H.: *Mater.Sci.Eng.*, **33**, 207 (1978).
4. Kaneko, Y., Morita, Y. and Hashimoto, S.: *Scripta Mater.*, **37**, 963 (1997).
5. Kaneko, Y., Mimaki, T. and Hashimoto, S.: *Acta Mater.*, **47**, 165 (1999).
6. Basinski Z.S. and Basinski, S.J.: *Scripta Metall.*, **18**, 851 (1984).
7. Polak J.: *Mater.Sci.Eng.*, **74**, 85 (1985).
8. Hunsch A. and Neumann P.: *Acta Metall.*, **34**, 207 (1986).
9. Wilkinson A.J. and Hirsch P.B.: *Micron*, **28**, 279 (1997).
10. Morin P., Pitaval M., Besnard D. and Fontaine G.: *Phil.Mag.*, **40**, 511 (1979).
11. Joy D.C.: *Mater.Res.Soc.Symp.Proc.*, **183**, 199 (1990).
12. Czernuszka J.T., Long N.J., Boyes E.D. and Hirsch P.B.: *Phil.Mag.Lett.*, **62**, 227 (1990).
13. Czernuszka J.T., Long, N.J. Boyes, E.D. and Hirsch P.B.: *Mater.Res.Soc.Symp.Proc.*, **209**, 289 (1991).
14. Ng. B.-C., Simkin M.A. and Crimp M.A.: Mater.Sci.Eng.A, **A239-240**, 150 (1997).
15. Simkin B.A. and Crimp M.A.: *Ultramicroscopy*, **77**, 65 (1999).
16. Zauter R., Petry F., Bayerlein M., Sommer C., Christ H.-J. and Mughrabi H.: *Phil.Mag.A*, **66**, 425 (1992).
17. Zhai T., Martin J.W., Briggs G.A.D. and Wilkinson A.J.: *Acta Mater.*, **44**, 3477 (1996).
18. Schwab A., Bretschneider J., Buque C., Blochwitz C. and Holste C.: *Phil.Mag.Lett.*, **74**, 449 (1996).
19. Zhai T., Martin J.W., Briggs G.A.D. and Wilkinson A.J.: *Acta Mater.*, **44**, 3477 (1996).
20. Melisova D., Weiss B. and Stickler R.: *Scripta Mater.*, **36**, 1061 (1997).
21. Ahmed J., Wilkinson A.J. and Robbert S.G.: *Phil.Mag.Lett.*, **76**, 237 (1997).
22. Ahmed J., Wilkinson A.J. and Roberts S.G.: *Phil.Mag.A*, **81**, 1473 (2001).
23. Kaneko Y. and Hashimoto S.: *J.Japan Inst.Metals*, **66**, 1297 (2002). (in Japanese)
24. Booker G.R., Shaw A.M.B., Whelan M.J. and Hirsch P.B.: *Phil.Mag.* **16**, 1185 (1967).
25. Ackermann F., Kubin L.P., Lepinoux J. and Mughrabi H.: *Acta Metall.*, **32**, 715 (1984).
26. Ma Bao-Tong and Laird C.: *Mater.Sci.Eng.A*, **102**, 247 (1988).
27. Lepisto T., Kuokkala V.-T. and Kettunen P.: *Scripta Metall.*, **18**, 245 (1984).
28. Jin N.Y. and Winter A.T.: *Acta Metall.*, **32**, 1173 (1984).
29. Livingston, J.D. and Chalmers, B.: *Acta Metall.*, **5**, 322 (1957).

A DISCRETE DISLOCATION DYNAMICS INVESTIGATION OF FORMATION AND STABILITY OF LOMER LOCKS

D. Weygand
Institut für Zuverlässigkeit von Bauteilen und Systemen (IZBS), University of Karlsruhe, Kaiserstr. 12, 76131 Karlsruhe, Germany
Daniel.Weygand@izbs.uni-karlsruhe.de

Abstract The simulation of dislocation motion at small length scale by means of the discrete dislocation dynamics (DDD) method shows that the detailed and thorough study of dislocation interaction, leading to the formation of junction, is of importance. The present study revisits the Lomer (Lomer-Cottrell) lock in order to comment on the ability to generate rules for junction stability for DDD simulations, to improve and accelerate the model. The present study finds some important deviation from other dislocation dynamics simulations, including other DDD and simple line tension model calculations with respect to the range of stability of the Lomer lock. Furthermore, the actual loading path or effect of the dislocation mass might also play a role on the stability of a lock.

Keywords: Lomer lock, Lomer-Cottrell lock, discrete dislocation dynamics simulation, stability, inertial effect

1. Introduction

In classical continuum mechanics, the sample size is much larger than the size of structures that dislocation may form, which allows for averaging over the dislocation micro-structure. The mechanical response of small scale devices on mechanical loading is strongly influenced by individual dislocation-dislocation interactions, leading to the formation of locks and by the presence of interfaces or surfaces.

The present investigation is revisiting the stability of the Lomer and the Lomer Cottrell lock, which has found quite some interest in the recent literature using different simulation methods ranging from atomistic (quasicontinuum method), discrete dislocation dynamics and line tension descriptions [1–4], and is serving as a benchmark for dislocation calculations. The yield surface, indicating the stability range of this particular lock, is recalculated, and the findings are discussed in light of the literature results. Furthermore the loading

H. Kitagawa and Y. Shibutani (eds.), IUTAM Symposium on Mesoscopic Dynamics of Fracture Process and Materials Strength, 23–33.

path for the calculation of the stability range of the Lomer lock is changed and the effect on the critical stress needed to break up a Lomer lock is discussed. It is demonstrated, that the usually neglected dynamic effect due to the mass of the dislocation has some effect on the critical stress needed to break a junction even at room temperature.

In section 2 of the paper the discrete dislocation dynamics method is briefly recalled. A short description of the adaption of the original code is given, allowing for partial dislocations and junction formation. The basics on the introduction of the dislocation mass in this formulation is given in [5]. Some supplementary consideration needed for junction formation are mentioned here.

In section 3 the configuration and loading conditions are described. In section 4 the results are presented and discussed.

2. Simulation Method

The discrete dislocation dynamics method as described in [6] is used for the study of the Lomer lock. In this model, the dislocation is discretised using straight lines, connected at nodal points. The degrees of freedom of the dislocation are the positions $\{\vec{R}_A\}$ and the velocities $\{\vec{V}_A\}$ of the N nodal points $\{A = 1, \ldots, N\}$. The equation of movement of the nodal points $\{A\}$ are derived using the principle of virtual work.

The DDD simulation method has been extended for this study to be able to handle partial dislocations and the corresponding stacking fault between dissociated dislocation. The stacking fault is accounted for by a stacking fault force term acting along the junction segment, which results in an contribution to the nodal force. From the topological point of view the formation of junctions using partial dislocation required some adaption in the junction formation and dissolution scheme, which explicitly introduces the junction segment in the simulation coupling thus two dislocations [6]. For partial dislocations, this scheme is replaced by a scheme, uniquely based on the elastic interaction between the partial dislocations involved in the junction formation process, where independent nodes are sliding along the junction direction. The scheme has been validated on the behaviour of locks formed by perfect dislocation.

The dislocation mass is introduced in the framework of the virtual work description for the derivation of the equations of movement. The derivation, valid for non intersecting dislocations, can be found in [5] in this volume. The handling of junctions in the used DDD framework, requires that the junction nodes have prescribed moving directions (sliding nodes), and therefore less degrees of freedom. The calculation of the nodal velocities including the inertial effect has to take this into account too. The velocity of the dislocation in the DDD simulation cannot exceed half of the transversal speed of sound. The dislocation-dislocation interaction is approximated by the stress fields of

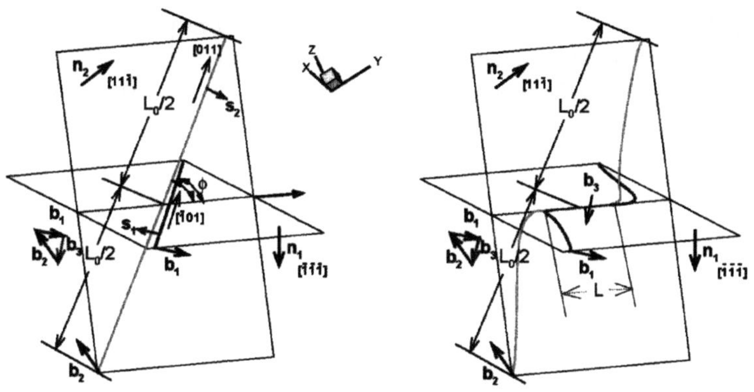

(a) (b)

Figure 1. Initial (a) and relaxed (b) configurations. The initial length of the dislocation between the two pinned endpoints is L_0. The vector $\mathbf{s}_i = \mathbf{t}_i \times \mathbf{n}_i$ is shown for both dislocations ($i = 1, 2$). The resolved shear stress τ_i for dislocation i is in direction of $\mathbf{s}_i$.

static dislocations. Furthermore, a constant time step is chosen to improve the numerical accuracy for the calculation of the acceleration of the dislocation.

3. Lomer-Lock and Lomer-Cottrell lock

A particular dislocation junction is studied using both perfect dislocations (Lomer lock) and dissociated dislocations (Lomer-Cottrell lock). The initial configuration for the perfect dislocation is shown in Fig. 1 and the parameter for the dislocations are summarised in table 1. The distance $L_0 = 30$nm and $\phi = 60°$ is chosen as in the references [3]. The scaling of the yield surface is obtained by increasing L_0 by more than one order of magnitude. For the

Table 1. The initial dislocation geometry: $\vec{n}$ is the glide plane normal; $\vec{b}$ is the Burgers vector in units of the lattice spacing a; $\vec{t}$ is the dislocation line direction. The line directions given here correspond to $\phi = 60°$.

dislocation i	$\vec{n}_i$	$\vec{b}_i[a]$	$\vec{t}_i$
1	$\frac{1}{\sqrt{3}}(\bar{1}\bar{1}\bar{1})$	$\frac{1}{2}(01\bar{1})$	$\frac{1}{\sqrt{2}}(\bar{1}01)$
2	$\frac{1}{\sqrt{3}}(11\bar{1})$	$\frac{1}{2}(101)$	$\frac{1}{\sqrt{2}}(011)$

initial study of perfect and partial dislocations the parameters of Al are used

($\mu = 27\text{GPa}$, $\nu = 0.347$, $\gamma_{sf} = 104\text{mJ/m}^2$), where μ resp. ν are the shear module resp. Poisson ratio and γ_{sf} is the stacking fault energy.

An external stress field is applied, leading to the resolved shear stress τ_1 resp. τ_2 on the dislocation 1 resp. 2

$$\begin{aligned} \tau_1 &= \frac{\sigma_0}{\sqrt{6}} \cos\theta \\ \tau_2 &= \frac{\sigma_0}{\sqrt{6}} \sin\theta \end{aligned} \tag{1}$$

where the direction of the resolved shear stress τ_i is given by $\vec{s}_i$ (i: dislocation number as defined in Fig. 1(a)).

The first type of loading consists of a quasi-static calculation, where the stress σ_0 is increased incrementally for a fixed loading direction θ, allowing for complete relaxation between the increments. The length of the junction is monitored and the breaking of the lock occurs for a junction length of about $3b$, where b is the length of the Burgers vector.

The second type of loading consist in keeping the stress σ_0 constant during the whole simulation for fixed loading direction θ. The third loading conditions is equivalent to the previous case, but the mass of the dislocation is taken into account during the calculation.

4. Result and Discussion

The Lomer and the Lomer-Cottrell lock are studied using the quasi-static loading procedure. The corresponding yield surfaces are determined. The starting configuration consists of a relaxed lock under zero applied stress, as shown in Fig. 1(b) for non-dissociated dislocations.

Lomer lock

Figures 2 (a) and (b) show the Lomer lock close to the breaking of the lock. In the configuration (a) with loading given by Eq. (1) for $\theta = 45°$ the lock is un-zipping from both sides and the centre of the lock does not move. The curvature of the dislocation line is determined by the externally applied load. A different opening mechanism is observed for the configuration of Fig. 2(b) for $\theta = 315°$. The lock has moved from its centre position to the left side, where a zipping mechanism is working at the leading node of the junction and an un-zipping mechanism occurs at the trailing node. The lock breaks, as the leading node is stopped by the dislocation arms leaving the lock at the leading node. Note that the curvature of the dislocation arm on glide plane 1 leaving the lock at the trailing node changes its sign, when approaching the lock of the dislocation. Far away from the lock the curvature is given by the externally applied shear stress τ_1. Close to the lock, the elastic interaction between the

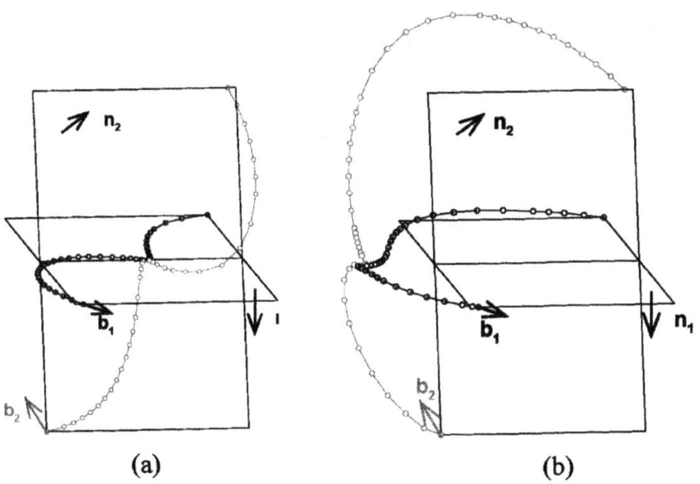

Figure 2. The Lomer Lock is shown for two loading conditions at critical loading. (a) configuration (I) $\tau_1 = \tau_2$ ($\theta = 45°$ in Eq. (1)). The centre of gravity of the Lomer lock has not moved. (b) configuration (II) $\tau_1 = -\tau_2$ ($\theta = 315°$). The Lomer lock has considerably moved to the left side.

dislocation arms is strong enough to overcome the externally applied shear stress, leading to an inversion of curvature.

Figure 3 shows the yield surface of the Lomer lock. An interesting finding is that the configuration with $\theta = 45°$ (geometry shown in figure 2(a)) is less stable than the $\theta = 315°$ configuration. This is in contradiction to the DDD and line tensions results in [3, 2], where the opposite behaviour has been found. The reason for this discrepancy has not yet been found [7]. The results in [3] have been rationalised using the line tension model, which supports the higher stability of the $\theta = 45°$ compared to the $\theta = 315°$ loading condition. When comparing the dislocation geometries obtained by a line tension model [4] and the present DDD model for the Lomer lock, the most striking difference is in the shape of the dislocation arms. In the DDD model, a strong variation or even inversion of the curvature is found, whereas the line tension model dislocation geometry consists of sectors of an ellipses (anisotropic line tension model) which are connected to a straight junction line. There is no reason that these two dislocation configurations should lead to similar breaking stresses for the Lomer lock.

In Fig. 3 the critical stress needed to activate a single Frank-Read source of arm length L_0 on each glide plane is shown. Note that the strength of the Lomer lock under $\theta = 45°$ loading is identical within the numerical error

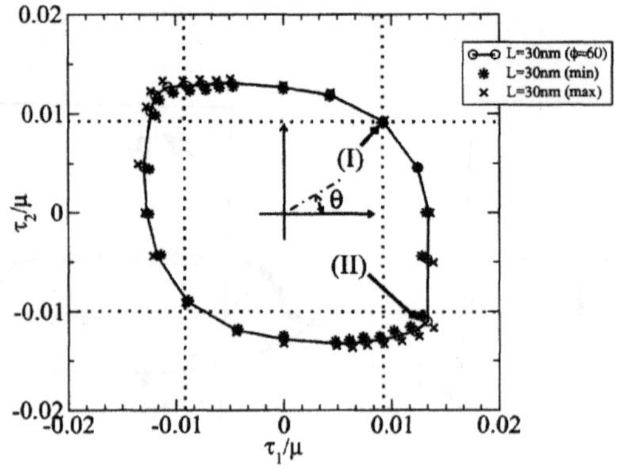

Figure 3. The yield surface of the Lomer lock is of elliptical shape. The point marked with (I:$\theta = 45°$) and (II:$\theta = 315°$) show two resolved shear stresses for the configurations shown in the previous figure. The dotted lines give the critical stress of a Frank-Read source with arm length L_0 in glide plane 1 or 2.

to the intersection of both dotted lines. Compared to a Frank-Read source, the Lomer lock (with $\phi = 60°$, see fig. 1(a)) does not increase the global hardening under this particular loading and boundary condition (fixed endpoints for the dislocations). For the $\theta = 315°$ loading (Eq. (1)), the critical load exceeds the one of the individual Frank-Read sources by about 30%. This finding is again in contradiction to the results of [3], where a critical stress is found for the $(I) : \theta = 45°$ loading, which exceeds the critical stress of the FR source by a factor of about 2. This has been explained in [3], assuming that the free arm length is reduced by 2 due to the lock. The present simulation suggests that the junction nodes cannot be considered as "fixed" endpoints, but rather as sliding nodes.

Lomer-Cottrell lock

For sake of completeness a detailed study of this lock using partial dislocations is performed. The influence of the partial splitting on the yield-surface of the Lomer-Cottrell lock, when compared to the Lomer lock, is shown to be rather small. Both the partial dislocation configurations, as shown in Fig. 4, and the yield surface (Fig. 5) are very close to the results obtained for perfect dislocations. For high stacking fault materials DDD simulations based on perfect dislocations are justified.

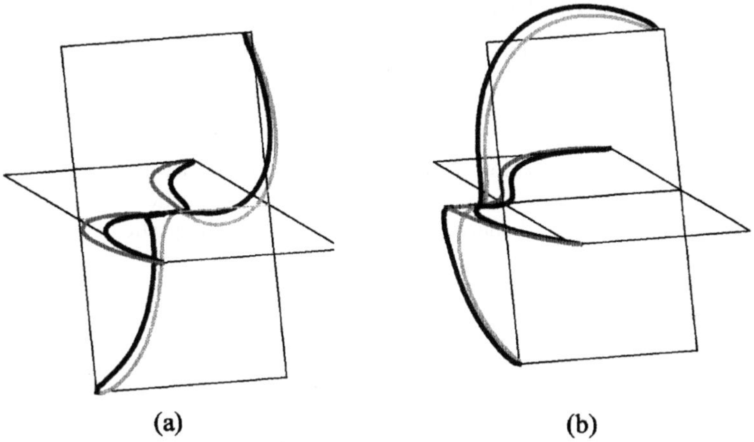

Figure 4. The Lomer-Cottrell lock is shown for two loading conditions close to the critical loading. (a) $\tau_1 = \tau_2$ ($\theta = 45^\circ$ in Eq. (1)); (b) $\tau_1 = -\tau_2$ ($\theta = 315^\circ$).

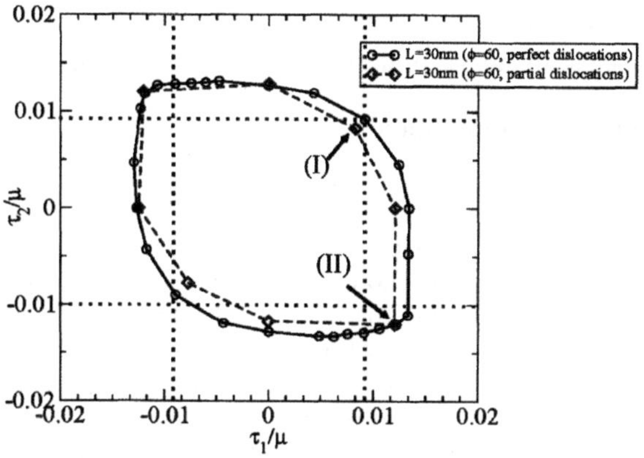

Figure 5. The yield surface of the Lomer-Cottrell (dashed line) shows the same shape and almost same critical resolved shear stresses for all loading conditions (θ) as the one for perfect dislocations (solid line). The point marked with (I:$\theta = 45^\circ$) and (II:$\theta = 315^\circ$) show two resolved shear stresses for the configurations of the previous figure.

Scaling of the yield surface

A simple scaling relation for the yield surface with the arm length L_0 has been found

$$\tau \propto \frac{b}{L_0} \ln \frac{L_0}{\rho} \tag{2}$$

where ρ is the regularisation parameter for the self-interaction of the dislocation [6] (explored range: $L_0 = 30\text{nm}, \ldots, 600\text{nm}$). The critical stress, needed to activate a Frank-Read source, has the same scaling properties [8, 6]. In [3] a $\tau \propto 1/L_0$ scaling has been proposed.

Influence of loading path

In DDD simulation with many dislocation interactions, the loading conditions for individual dislocations and the resulting locks are quite different compared to the one used for calculating the yield surfaces of the Lomer lock shown in Figs. 3 and 5. More realistic loading conditions are probably better described by a constant externally applied shear stress and the dynamics process of lock formation has to be considered too. It has been found that on one side, for loading conditions, where the dislocations run into each other ($\theta = 45°$, large relative velocity of dislocations), the yield point remains within the range of numerical error of the static calculation. On the other side, for loading conditions where both dislocations are pushed in the same direction (relative velocity is approximately zero), the question whether or not the lock forms depends critically on the initial configuration. Once a lock has formed, the results are close to the one of the corresponding static calculation.

Inertia effect on Lomer lock

The importance of the mass per unit length of a dislocation on the stability of the Lomer lock is investigated for two particular loading conditions ($\theta = 45°$ leading to a high relative velocity between the dislocations and $\theta = 315°$ leading to a low relative dislocation velocity) for Ni, where the dislocation mass per unit length and the drag coefficient used in the DDD framework have been taken from MD simulations [5] (mass per unit length=1.1 10^{-16} kg/m, at $T = 300$ K: drag coefficient $B = 15\ 10^{-6}$ Pas; elastic properties: $\mu = 74.9$ GPa, $\nu = 0.36$). The dislocation velocity in the DDD model is limited to half of the transversal speed of sound. Due to the mass of Ni atoms, it is expected that the effect of inertia even at room temperature is more pronounced than for instance in Al under the same conditions.

The starting configuration consists of two circular dislocation sectors, where the fixed endpoints are located at the same position as shown in Fig. 1. This configuration increases the distance between the dislocations before lock formation occurs. The configuration is chosen such, that the dislocation can gain some velocity and therefore kinetic energy and the interaction between the dislocation can change the shape of the dislocation line for the $\theta = 45°$ loading. The dislocations move with a high relative velocity. For the second loading ($\theta = 315°$), where both dislocations are pushed in the same direction, the initial distance between the dislocation has been reduced such that lock formation at the beginning of the simulation occurs. This setup allows to evaluate the in-

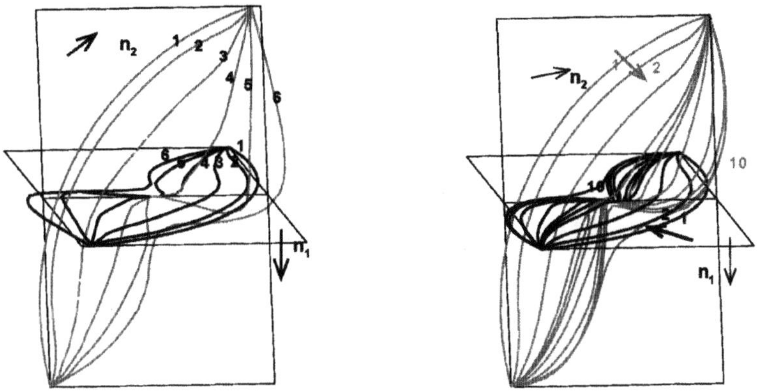

(a) (b)

Figure 6. A series of snapshots: (a) $L_0 = 30$nm the applied shear stress is half of the static values. The configurations (1) to (6) show the evolution from the initial to the depinned stage. In the intermediate stage (5) an important overshooting due to the dislocation mass is observed. (b) $L = 300$nm the applied shear stress are set to the static values. There is no overshooting visible, only some wave propagation on the dislocation due to inertia. The thick arrows indicate the direction of time evolution.

ertial effect in the yield surface, without aiming at a complete parameter study of the problem. The figures 6(a) and (b) show each a series of dislocation configuration ($L_0 = 30$nm in (a) and $L_0 = 300$nm in (b); loading for both configurations in direction $\theta = 45°$). The series (a) shows the overshooting of the dislocation leading to the breaking of the lock, where the applied resolved shear stresses on both glide planes are half of the static critical resolved shear stresses. The effect of the increasing the length $L_0 = 300$nm on the pinning is shown in Fig. 6(b), where resolved shear stresses have to reach the static values, in order to break the lock. The inertial effect is only found for small L_0, which can be explained by equating the available kinetic energy and the energy supplied by the applied shear stress, both proportional to the length L_0, with the energy dissipated during the bowing out of the dislocation, which is approximately proportional to the swept area ($\approx\propto L_0^2$) of the dislocation. As the dissipation scales with a higher exponent of L_0 compared to the available energy the effect of inertia is less important for large L_0.

Table 2. The critical stresses τ_0 needed to break the lock are summarised for two loading conditions: (i) $\theta = 45°$ $\tau_1 = \tau_2 = \tau_0$; with increasing length L_0 the inertial effect vanishes; (ii) $\theta = 315°$ $\tau_1 = -\tau_2 = \tau_0$; the lateral dislocation movement associated with this loading mode, damps out the inertial effect.

L_0	static τ_0 $[\mu]$ $(\theta = 45°)$	inertial τ_0 $[\mu]$	static τ_0 $[\mu]$ $\theta = 315°$	inertial τ_0 $[\mu](\theta = 315°)$
30nm	0.0065	0.0028	0.008	0.008
150nm	0.0016	0.0013	0.0018	0.0018
300nm	0.001	0.001	0.0011	0.0011

Table 2 shows critical stresses for the studied configurations. With decreasing length L_0 the inertial effect is gaining of importance for the $\theta = 45°$ loading direction, whereas the loading with $\theta = 315°$ is almost unaffected. In the second loading, both dislocations are moving into the same direction. The area swept by both dislocations is much larger than in the first case and the relative velocity of the dislocations is negligible. An oscillatory motion is superposed in the overall lateral movement of the dislocation. As the critical configuration is reached, the kinetic energy has been dissipated to a large extend. Therefore the critical shear stresses to break the lock are identical to the ones obtained by static calculations.

5. Conclusion

The Lomer lock has been studied using a discrete dislocation model using either the assumption of an overdamped dislocation motion or including the mass of dislocations. The results for the overdamped case are presented and compared to results from literature, where the following differences have been found: (i) the Lomer lock under loading, where both dislocations move in the same direction, is found to be stronger than the case, where the centre of the lock is not moving; (ii) the strength of the studied Lomer lock does not exceed by more than 30-40% the critical resolved shears stress of the corresponding FR source of arm length L_0; (iii) the scaling $1/L_0 \ln(L_0/\rho)$ instead of $1/L_0$of the yield surface has been found, similar to the results of the FR source.

Furthermore the influence of loading path and inertia on the yield surface is shown. The loading path is found to play a minor role, as long as it allows for lock formation. The inertia is found to have a considerable effect on the stress needed to break the lock for small L_0 and $\theta = 45°$ loading conditions even at room temperature. Interpolating the results, one can state that the smaller half axis of the yield surface decreases whereas the larger half axis is unaffected, leading to a more needlelike yield surface. For future work the influence of the velocity on the dislocation stress fields should be included.

The present results suggest that the development of rules to replace expensive calculations, due to the occurrence of formation and destruction of locks is limited in the case of small structures and distances. According to this model, the stability is strongly determined by the full three dimensional interaction of the dislocation, forming the lock. A simple estimate, using line-tension models behaves quantitatively and qualitatively different. A possible way to solve this problem, consists of subdividing the calculation in DDD models in short range and long range interactions, where lock formation processes can be calculated in more detail, without compromising too much the overall performance of the simulation tool.

Acknowledgments

This research was partly carried out under the project number MS97006B in the framework of the Strategic Research Programme of the Netherlands Institute for Metals research (www.nimr.nl).

References

[1] D. Rodney and R. Phillips. Structure and strength of dislocation junctions: An atomistic level analysis. *Phys. Rev. Lett.*, 82:1704–1707, 1999.

[2] C. Shin, D. Rodney et al. Formation and strength of dislocation junctions in fcc metals: A study by dislocation dynamics and atomistic simulations. *J. de. Physique*, 2001.

[3] V. Shenoy, R. Kukta and R. Phillips. Mesoscopic analysis of structure and strength of dislocation junctions in fcc metals. *Phys. Rev. Lett.*, 84:1491–1494, 2000.

[4] M. F. L. Dupuy. A study of dislocation junctions in fcc metals by an orientation dependent line tension model. *Acta Mater.*, 50:4873–4886, 2002.

[5] E. Bitzek, D. Weygand and P. Gumbsch. Atomistic study of drag and inertial effects of edge dislocations in fcc metals. *this volume*, 2003.

[6] D. Weygand, L. Friedman et al. Aspects of boundary-value problem solutions with three-dimensional dislocation dynamics. *Modelling Simul. Mater. Sci. Eng.*, 10:437–468, 2002.

[7] R. Phillips. personal communication, 2002.

[8] A. Foreman. The bowing of a dislocation segment. *Phil. Mag.*, 15:1011–1035, 1967.

ON THE USE OF PERIODIC BOUNDARY CONDITIONS IN DISLOCATION DYNAMICS SIMULATIONS

Ronan Madec, Benoit Devincre[*] and Ladislas Kubin[*]

DPTA, Commissariat à l'Energie Atomique, BP12, 91680 Bruyères-le-Châtel, France
[*]*LEM, CNRS-ONERA, 29 Av. de la Division Leclerc, BP 72, 92322 Châtillon Cedex, France*

Abstract: The use of periodic boundary conditions in dislocation dynamics simulations results in spurious self-annihilation events. Solutions are presented for avoiding this artefact and obtaining realistic dislocation mean free-paths.

Key words: dislocation dynamics simulation, periodic boundary conditions.

1. INTRODUCTION

The present work is dealing with periodic boundary conditions (PBCs) in dislocation dynamics (DD) simulations. For the so-called mass simulations, PBCs are needed in order to ensure the continuity of the dislocations lines and their associated fields through the boundaries of the simulation cell and to balance incoming and outgoing dislocation fluxes. Artificial dislocation losses due to free surfaces and undesirable size effects due to finite dimensions are then avoided.

Bulatov, Rhee and Cai[1] have discussed the application of PBCs to DD simulations. In short, dislocations glide in a periodic array formed by a simulation cell and its replicas. Every time a portion of dislocation line

H. Kitagawa and Y. Shibutani (eds.), IUTAM Symposium on Mesoscopic Dynamics of Fracture Process and Materials Strength, 35–44.

crosses a boundary between two cells, which it does simultaneously in all cells, it emerges in all cells at the equivalent position on the opposite boundary. A balance of fluxes is then established through all the internal interfaces of the simulated volume. However, the application of PBCs to linear objects is known to lead to spurious self-interactions[2]. In the case of dislocations, portions of loops may self-annihilate with replicas having emerged after a certain number of boundary crossings. Self-annihilation may reduce the mean free-path of dislocations; this can have drastic consequences that have not been considered so far in the context of DD simulations. A too short effective mean free-path affects the density of mobile dislocations and their storage rate and, hence, both the arrangement of the microstructure and the strain hardening properties. The object of the present study is to discuss several methods that have been developed in order to control this artifact. These methods apply to DD simulations in which, like in real crystals, the slip plane spacings obey crystallographic relations.

In part 2, a few definitions are given. For particular slip geometries, self-annihilation occurs after a single boundary crossing and two methods are proposed for dealing with such situations. In part 3, it is shown that self-annihilation distances can be controlled by adequately tailoring the dimensions of orthorhombic simulation cells. Two model situations of practical interest are considered, those of isotropic and strongly anisotropic dislocation loops. Concluding remarks are presented in part 4.

2. DIRECT SELF-ANNIHILATION

2.1. Self-Annihilation: Definitions

Figure 1 shows a simple two-dimensional example of PBCs applied to a square cell of side L. Every dislocation line crossing the boundary of the simulation cell, enters the cell and all replicas at positions defined by translation vectors parallel to the cell sides, $\boldsymbol{L}_x$ or $\boldsymbol{L}_y$ (in what follows, bold face denotes a vector). These vectors are called modulo translations, as their linear combinations determine the respective positions of all the replicas. This method has several advantages[1]. In particular, it fulfills the requirement of continuity of the dislocation lines at the cell boundaries. Then, the origin of the cells can be shifted in an arbitrary manner with respect to the dislocation microstructure.

As the array of cells is periodic, its section by a slip plane is also periodic, albeit with a different period. A portion of dislocation loop, then, necessarily meets with a replica of the same loop after a certain free path. For instance in the square cell of Fig. 1-*a* and for a trajectory going along the second diagonal, self-interaction occurs after two boundary crossings only. The simplest way to control these undesirable spatial correlations consists in introducing orthorhombic distortions of the simulation cells.

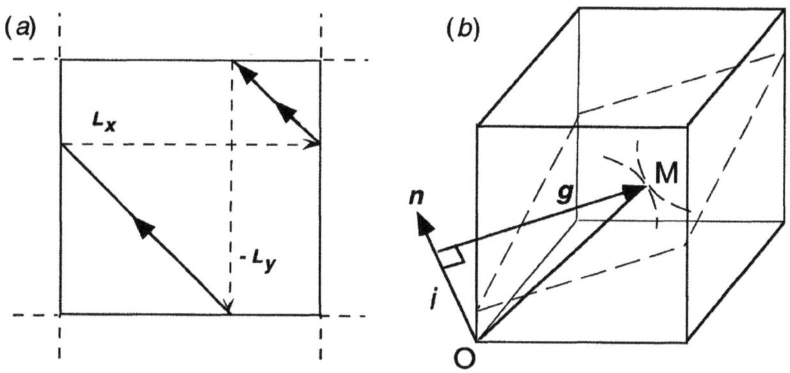

Figure 1. a - In a square cell, a trajectory along the direction parallel to the second diagonal meets with a replica after two modulo translations. *b* – Self-annihilation in three-dimensions and in a slip plane of normal ***n***.

We now consider the replicas of a three-dimensional trajectory in the primary cell (cf. Fig. 1-*b*). In what follows, the primary simulation cell has {100} faces and its dimensions are defined by the vectors ($\boldsymbol{L}_x$, $\boldsymbol{L}_y$, $\boldsymbol{L}_z$). The slip plane normal is denoted $\boldsymbol{n}(h, k, l)$, and the position of a point along an expanding dislocation loop is defined by the vector ***OM***. The origin of the expansion, O, can be selected in an arbitrary manner, as mentioned above, and is taken at the origin of the cell. The vector ***OM*** can be decomposed into a sum of two components, one along the glide plane (***g*** in Fig. 1-*b*) and the other along the slip plane normal. The last component allows distinguishing between slip planes of different heights in the simulation cell. For this purpose, we define a slip plane index, *i*, such that (Fig. 1-*b*):

$$i = \boldsymbol{OM}.\boldsymbol{n} \tag{1}$$

Upon self-annihilation, the two interacting segments have same slip plane index, as they are in the same slip plane. In the case depicted in Fig. 1-*b*, self-annihilation occurs in a plane with non-zero index, which is not that of the loop expanding from the origin. It can also occur in the plane $i = 0$. Finally, we define the total translation, ***T***, as the sum of the whole set of modulo translations upon self-annihilation:

$$\boldsymbol{T} = u\boldsymbol{L}_x + v\boldsymbol{L}_y + w\boldsymbol{L}_z, \tag{2}$$

In this expression, *u*, *v*, and *w* are integers counting the number of cell boundary crossings before self-annihilation. The total translation $\boldsymbol{T}$ has a glide component and a component along the normal to the slip plane. Thus, a necessary condition for self-annihilation of a segment in a slip plane with index i is written $\boldsymbol{T.n} = i$, or, in developed form:

$$huL_x + kvL_x + lwL_x = i \tag{3}$$

Finally, all the slip plane indexes encountered after the successive boundary crossings are a linear combination of the quantities hL_x, kL_y and lL_z. The greatest common divisor of these three quantities defines, along the normal $\boldsymbol{n}$, a one-dimensional lattice that includes the intersections with all the possible slip planes available to the considered dislocation loop. This property is, actually, the only one that is needed to apply the general solutions developed below to three-dimensional DD simulations. It also defines the smallest dipole height that can be met in a given slip system.

Before coming back to the general calculation, we examine particular slip geometries leading to very short annihilation distances.

2.2 Direct Self-Annihilation

In general, the modulo translations are not in a slip plane. This can, however, happen for particular slip geometries. As schematically depicted in Fig. 2-a, when a dislocation line propagates along a direction parallel to a modulo translation, self-annihilation occurs after one boundary crossing, irrespective of the cell shape, cubic or orthorhombic. This type of situation is not uncommon and is met typically with {100} and {110} slip planes in some crystallographic structures (bcc structures in the last case), which contain respectively two and one modulo translations.

A solution to this problem consists in shifting a set of parallel boundaries by a translation δ_{ij} (Fig. 2-a). The first index refers here to the shift direction and the second one to the boundary normal. The set of shift vectors must preserve the continuity of the simulated material, i.e., they should not introduce voids or matter overlaps. As a consequence, normal shifts δ_{ii} are forbidden, as well as simultaneous symmetrical shears like δ_{ij} and δ_{ji} and the three possible non-zero shears cannot be in three different faces. Thus, in three dimensions and for multiple slip conditions, another solution may be needed for one of the faces, depending upon the slip geometry.

The alternative solution consists in tilting the crystallographic axes with respect to the primary cell (Fig. 2-b), in order to tilt the slip planes away from the direction of the modulo translation. In cubic or orthorhombic structures, a convenient solution consists in rotating the crystallographic axes around simple crystallographic directions, for instance <001>. This method can be employed alone, as one rotation can be applied along each <100> direction or in combination with shifts. The optimization of the cell dimensions presented below can be carried out in the presence of shifts and rotations (cf. section 3.5).

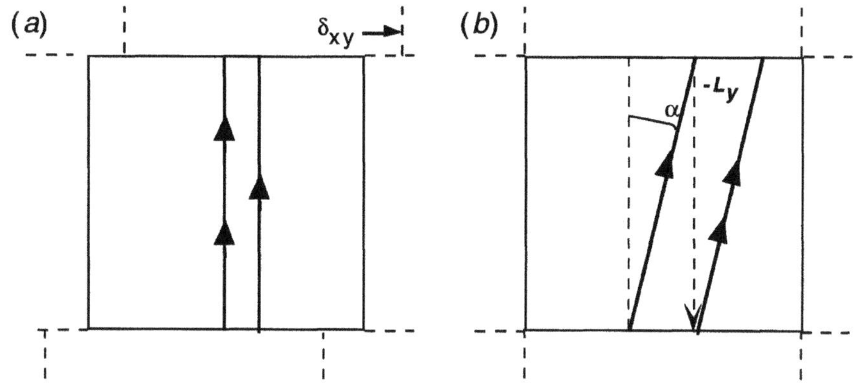

Figure 2. Two-dimensional sketches of direct self-annihilations after a single boundary crossing. *a* – A vertical slip direction is transformed into itself after one boundary crossing (single arrow). The problem is solved by shifting the neighboring cells with respect to each other (here by δ_{xy}) so that the direction and its replica (double arrow) do not coincide. *b* – The same problem is solved by introducing a rotation α of the crystallographic axes with respect to the primary cell.

3. CONTROLLING SELF-ANNIHILATIONS

3.1. Isotropic and Anisotropic Dislocation Loops

In this part, relations are established between the cell dimensions, slip geometry and self-annihilation distances, which allows controlling the spurious self-annihilations induced by PBCs. Two model situations are considered in what follows, which correspond to common configurations found in the practice. In fcc crystals, dislocation loops expand with rather isotropic shapes and mean-free paths of the order of a several tens of microns[4]. At low temperatures in bcc crystals, and for prismatic slip in hcp crystals, screw dislocations experience a strong lattice friction and non-screw segments are comparatively highly mobile. Non-screw segments can have long mean-free paths, perhaps up to one hundred of microns or more in pure crystals; their propagation results in the formation of loops with very anisotropic shapes[5].

3.2. Geometry in the Extended Slip Plane

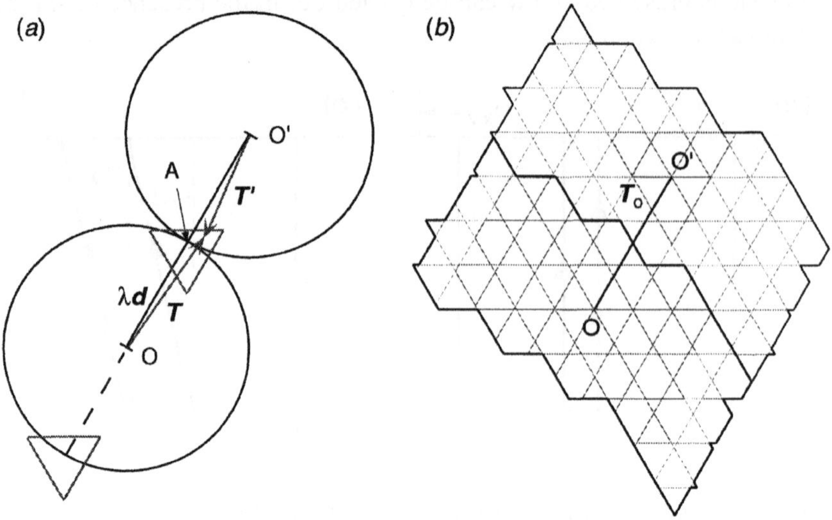

Figure 3. a - A circular dislocation loop expands from the origin O in a slip plane of normal ***n***. Along a direction ***d***, and for a certain radius λ, the first self-annihilation occurs in A with a replica of the opposite portion of loop (see text for other details). *b* - Tiling of a (111) slip plane in the fcc structure by polygonal intersections with a periodic array of orthorhombic simulation cells. The latter have here dimensions proportional to the set of incommensurate numbers (40, 30, 31). Two supercells are shown, of which the boundaries were computed from the condition that each supercell only contains polygons with different indexes.

Figure 3-*a*, schematically depicts the self-annihilation of an isotropic dislocation loop expanding by glide from the origin O of the primary cell (initial index $i_o = 0$). As indicated in section 2.1 the origin can be chosen in an arbitrary manner without loss of generality. The first self-annihilation event on the expanding loop occurs along a certain direction $\boldsymbol{d}(d_x, d_y, d_z)$ that will be calculated below. It corresponds to a minimum critical glide distance as a function of orientation, λ, which is called the self-annihilation radius.

The intersections of a slip plane with the periodic array of simulation cells define a tiling of contiguous polygons with various indexes and shapes. This tiling is necessarily periodic since there is a finite number of possible indexes i , which corresponds to an equivalent number of distinct polygonal intersections. One can thus define in the slip plane a periodic array of two-dimensional supercells, of which the periodicity is defined by the periodicity of the indexes of the polygonal intersections. The supercells have complex shapes, as shown in the example given in Figure 3-b. This example also illustrates the problems met with non-optimal cell dimensions: the self-annihilation radius is rather small in this case (1.43 times a cell diagonal) and the supercell is rather anisotropic, whereas one would prefer to have rather isotropic effective mean-free paths.

In Fig. 3-*a*, self-annihilation occurs when a portion of the expanding loop meets in the same polygonal intersection with the replica of a portion propagating in the opposite direction. The two segments meet with a common tangent and a common normal parallel to the propagation direction $\boldsymbol{d}$. Thus, the first self-annihilation event on the dislocation loop occurs along the direction of the shortest distance between the origins of the supercells, $\boldsymbol{T}_o = \boldsymbol{OO'}$ (cf. Fig. 3-*b*). The self-annihilation diameter is then $2\lambda = \boldsymbol{T}_o$, and the self-annihilation radius, λ, is half this value (Fig. 3-*a*). As already mentioned, the two total translations $\boldsymbol{T}$ and $\boldsymbol{T'}$ (gray lines in Fig. 3-*a*) are not in the slip plane. They do not necessarily correspond to symmetrical sequences of modulo translations. For instance, a portion of loop expanding in the primary cell starts with a glide sequence, whereas its replicas start by crossing cell boundaries, which implies one additional modulo translation.

The geometrical considerations developed in the previous sections are now used to calculate a relation between the self-annihilation radius and the cell dimensions. The practical applications are outlined in section 3.5.

3.3 Self-Annihilation of Isotropic Loops

The direction of first self-annihilation is that of the smallest vector connecting the centers of two neighboring supercells, $\boldsymbol{T}_o$. Considering the total translations $\boldsymbol{T}$ and $\boldsymbol{T'}$, we see from Fig. 3-*a* that $\boldsymbol{T}_o = \boldsymbol{T} - \boldsymbol{T'}$ and, further, that $\boldsymbol{T}_o = 2\lambda\boldsymbol{d}$, so that the condition for first self-annihilation is written:

$$2\lambda\boldsymbol{d} = \boldsymbol{T}_o. \tag{4}$$

In projection along the slip plane normal, we have $\boldsymbol{T}_o . \boldsymbol{n} = 0$, or:

$$huL_x + kvL_x + lwL_x = 0 \tag{5}$$

In this expression, the unknowns are the numbers of boundary crossings u, v and w. Linear equations, of which the solutions searched for are integer numbers, like Eq. (5), are known under the name of first-degree Diophantine equations (after Diophantes of Alexandria, cf. Bashmakova 1997); their solutions can be obtained with the help of various algorithms. The first self-annihilation corresponds to the set of integer numbers (u, v, w) leading to the smallest translation vector between supercell origins, $\boldsymbol{T}_o$. The self-annihilation diameter is then obtained from Eq. (4):

$$2\lambda = \sqrt{(uL_x)^2 + (vL_y)^2 + (wL_z)^2} \tag{6}$$

3.4. Self-Annihilation of Anisotropic Loops

The fast moving portions of the anisotropic loops are approximated by segments of infinitesimal length propagating along the direction $\pm\boldsymbol{d}$. In contrast with the previous case, the direction $\boldsymbol{d}$ is known and may differ from that of $\boldsymbol{T}_0$, since it is fixed by a physical argument. Anisotropic loops or segments can, thus, cross several supercells before self-annihilating. As a consequence, their trajectories have one interesting property, which is not detailed here by lack of space. Every time the boundary of a supercell is crossed by a small segment, the latter emerges at another boundary of the supercell through the operation of the PBCs. The successive positions of these replicas are shifted by a constant amount, due to the fact that the propagation direction is not, in general, parallel to direction connecting the origins of neighboring supercells. It follows that the portions of the trajectories of anisotropic loops folded in a supercell are equidistant. Then, the condition for treating short segments as infinitesimal ones is that their length be smaller than the spacing between their trajectories in a supercell.

We consider an anisotropic dislocation source, for instance in a bcc crystal[5]. An edge segment is emitted along the screw direction and one of its replicas returns to the position of the source after some time interval. Different topological self-reactions can occur, depending upon whether the screw segments are sessile or not and whether, in the last case, another segment can be further emitted in the opposite direction. To simplify the geometry, we consider an expanding anisotropic loop and define, as above, its self-annihilation radius (λ) and diameter (2λ). Each segment starts from the origin ($i = 0$) and self-annihilates at another supercell origin with a replica of the opposite part of the loop. Equation (3) of section 2.1, taken with $i = 0$ applies to this situation. However, Eq. (3), which is also a first-order Diophantine equation, does not explicitly include the propagation direction $\boldsymbol{d}$ and yields solutions for all directions in a given slip plane. A more transparent solution can be derived as follows.

The position of the self-annihilation event in the extended plane is brought back into the primary cell by subtracting the total translation from the self-annihilation radius. Then (cf. section 2.1 and Fig. 1-*b*):

$$\boldsymbol{OM} = \lambda\boldsymbol{d} - \boldsymbol{T} \tag{7}$$

Upon self-annihilation at the origin, we have $\boldsymbol{OM} = 0$ and Eq. (7) reduces to $\lambda\boldsymbol{d} - \boldsymbol{T} = 0$, or, in developed form:

$$\lambda = \frac{uL_x}{d_x} = \frac{vL_y}{d_y} = \frac{wL_z}{d_z}, \tag{8}$$

Combining Eqs. (8) with the condition $\boldsymbol{d}.\boldsymbol{n} = 0$, which expresses that the direction $\boldsymbol{d}$ is in the slip plane of normal $\boldsymbol{n}$, one recovers the more general Eq. (5). The set of integer numbers (u, v, w) is obtained from the three equations at the right-hand side of Eq. (8):

$$v = u\frac{L_x d_y}{L_y d_x}; \quad w = u\frac{L_x d_z}{L_z d_x}. \tag{9}$$

These two equation have an infinite number of solutions. The first self-annihilation is obtained through the additional condition that it corresponds to the total translation vector $\boldsymbol{T}$ of smallest magnitude, i.e., to the smallest distance between supercell origins along the direction $\boldsymbol{d}$. The self-annihilation radius is then determined from the left-hand side of Eq. (8).

3.5 Implementation

In practice, several conditions are initially prescribed. The average dimension of the primary cell should be large with respect to the typical length scale of interest, for instance an average distance between dislocations or the wavelength of a dislocation pattern. This leads to average dimensions of typically 10 μm for standard mass simulations. The value of the self-annihilation radius is then prescribed according to the physical situation considered. In general, one prefers to work with orthorhombic distortions that are not too large and the cell dimensions should then be proportional to a set of three incommensurate numbers centered around an average value.

In the anisotropic case, several possible solutions can be obtained directly from Eqs. (8) and (9). In the isotropic case, the coupled equations (5) and (6) are solved numerically, which also leads to several possible solutions. In this last case, solutions yielding isotropic supercells are preferred in order to have isotropic annihilation radii.

Several situations related to self-annihilation can also be treated via first-order Diophantine equations. This includes, in particular, close dipolar interactions, from either a single or a few dislocation sources, which can be troublesome during simulations of simple elementary configurations. The condition for self-annihilation is then replaced by a condition on the indexes of two segments forming a dipole inside a same cell.

Finally, these solutions can easily be extended to the particular slip geometries examined in section 2. Rotations of the crystallographic lattice are cared of by carrying out calculations in the rotated lattice. Cell boundary shifts are accounted for by replacing in Eq. (2) the modulo translation $\boldsymbol{L}_i$ by $\boldsymbol{L}_i^* = (\boldsymbol{L}_i + \delta_{ji})$, incorporating the shift on the boundary (i).

4. CONCLUDING REMARKS

The self-annihilation of dislocation lines caused by the use of PBCs is an artifact that can be controlled. The methods outlined in the present work allow setting the values of the self-annihilation distances to values that are never smaller than the dislocation mean-free paths. This last length scale has a strong influence on the development of dislocation microstructures and its effective realization allows drawing quantitative estimates from DD simulations[3,7].

The two situations considered here, those of isotropic and anisotropic loops cover most of the cases of practical interest. To avoid severe artifacts occurring with particular slip geometries, two methods are proposed, which can be implemented in any DD simulation, discretized or not. The self-annihilation distances of dislocations are prescribed through a choice of the dimensions of an orthorhombic simulation cell, based on a quantitative analysis. To be implemented, the latter only involves the requirement that the slip planes available to the moving dislocations be located on a set of equidistant heights in the simulation cell.

ACKNOWLEDGEMENTS

Part of this work was conducted by one of the authors (RM) at the Centre Européen de Calcul Atomique et Moléculaire (CECAM) during a post-doctoral stay.

REFERENCES

1. Bulatov VV, Rhee M, Cai W. Periodic boundary conditions for dislocation dynamics simulations in three dimensions. *Multiscale Modeling of Materials 2000*, LP Kubin, R. Selinger, JL Bassani J.L. and K Cho eds.; Warrendale, PA: Materials Research Society. Symposium Proceedings Vol. 653, p. z1-3, 2000.
2. Schwarz K.W. Three-dimensional vortex dynamics in superfluid ^{4}He: homogeneous superfluid turbulence. Phys. Rev. B 1998; 38:2398-2417.
3. Madec R., Devincre B., Kubin L.P. From dislocation junctions to forest hardening. Phys. Rev. Lett. 2002; 89:255508.
4. Ambrosi P., Shwink Ch. Slip line length of copper single crystals oriented along [100] and [110]. Scripta Metall. 1978; 12:303-308.
5. Tang M., Kubin, L.P., Canova, G.R. Dislocation mobility and the mechanical response of bcc single crystals: a mesoscopic approach. Acta Mater. 1998; 46:3221-3235.
6. Bashmakova, I.G. *Diophantes and Diophantine Equations.* Washington DC: Math. Assoc. Amer., 1997.
7. Madec R., Devincre, B., Kubin, L.P. Simulation of dislocation patterns in multislip. Scripta Mater. 2002; 47:689-695.

ATOMISTIC STUDY OF EDGE DISLOCATIONS IN FCC METALS: DRAG AND INERTIAL EFFECTS

Erik Bitzek[1], Daniel Weygand[1,2] and Peter Gumbsch[1,2]

[1] *Universität Karlsruhe, Institut für Zuverlässigkeit von Bauteilen und Systemen, Kaiserstr. 12, 76131 Karlsruhe, Germany*

[2] *Fraunhofer Institut für Werkstoffmechanik, Wöhlerstr. 11, 79108 Freiburg, Germany*

Abstract Atomistic simulations of an accelerating edge dislocation were carried out to study drag and inertial effects. Using an embedded atom potential for nickel the Peierls stress, the effective mass and the drag coefficient of an edge dislocation was determined for different temperatures and stresses in a simple slab geometry. A dislocation intersecting a void is used as a model to demonstrate the importance of inertial effects for dynamically overcoming short range obstacles. Significant effects are found even at room temperature. Including inertial effects in discrete dislocation dynamics simulations allows to reproduce the atomistic results.

Keywords: Inertia; Obstacle; Atomistic Simulation; Dislocation Dynamics

1. Introduction

Understanding the motion of dislocations and their interaction with each other or with obstacles is essential for the description of plastic deformation of crystalline materials. The dynamics of the dislocations is of direct interest in high strain-rate deformation, dislocation interaction with short range obstacles, low temperature deformation or high frequency agitation.

In most treatments, the inertia of the dislocations is not explicitly taken into account. It can, however, assist overcoming of obstacles [1–6], thus leading to a lower effective critical stress for obstacle passing. This notion of inertially overshooting dislocations was first proposed to explain the enhanced plasticity in superconductors [1, 2], but also used in the description of plastic deformation of solid solutions [5] and precipitation hardened materials [4]. Compression tests on copper-single crystals strengthened by cobalt precipitates, for example, showed a positive dependence of the CRSS on temperature below 200 K due to dynamic dislocation effects [4]. These results were in semi-quantitative agreement with the computer simulations of Schwarz and Labusch [3]. Various

H. Kitagawa and Y. Shibutani (eds.), IUTAM Symposium on Mesoscopic Dynamics of Fracture Process and Materials Strength, 45–57.

other efforts where undertaken to include inertial effects and thermal activation in analytical models of dislocation motion through a random array of localized obstacles [6–9]. However, the direct observation of inertial effects in dislocation - obstacle interactions is only possible with atomistic simulations.

Molecular dynamics (MD) simulations automatically include dynamic effects in the dislocation motion. It is thus possible with MD simulations to determine the parameters governing the dynamics of dislocation motion. These parameters can then be used in discrete dislocation dynamics (DDD) simulations [10] which are often used to study plasticity at the mesoscale, but do not incorporate the dislocation mass so far.

After introducing the theoretical background we present a method to determine the effective mass and the drag coefficient of dislocations from atomistic simulations and show direct evidence for inertial effects in the interaction of dislocations with obstacles. These effects will be rationalized in a framework based on energetic considerations. DDD simulations including inertia are presented as a tool to asses the importance of dynamic effects for different situations that are not accessible to MD-simulations.

1.1 Dynamics of dislocations

The Peach-Koehler force $F_a = \tau b$ that an applied shear stress τ exerts on a dislocation with Burgers vector b leads to an acceleration of the dislocation if the resolved shear stress is larger than internal forces F_i arising from interactions with defects or from the intrinsic lattice resistance. The rate of acceleration is determined by the inertia (or effective mass) M of the dislocation. A deceleration comes from damping or drag effects which dissipate energy. They are characterized by the viscous drag coefficient B. For non-straight dislocations the dislocation self-interaction, which can for simplicity be regarded as a line tension Γ, acts in the way to minimize the line energy. One can therefore formulate the following equation of motion for a dislocation segment along the x-direction and moving with speed $v = \frac{\partial y}{\partial t}$ along the y-direction:

$$\frac{\partial}{\partial t}\left(M\frac{\partial y}{\partial t}\right) + B\frac{\partial y}{\partial t} - \Gamma\frac{\partial^2 y}{\partial x^2} = F_a + F_i \,. \tag{1}$$

The rest mass per unit length of a screw dislocation is given by [11]

$$m_0^s = \frac{E_0^s}{c_t^2} \approx \frac{\mu b^2}{4\pi c_t^2}\ln\frac{R}{r_0}\,, \tag{2}$$

where $c_t = \sqrt{\mu/\rho}$ is the transverse wave speed, μ is the shear modulus, ρ the mass density, b the Burgers vector, R the upper integration limit (crystal radius or half the typical distance between dislocations), and r_0 the core cut-off radius. The rest mass of an edge dislocation is given by [11]

$$m_0^\perp = \left(1 + \frac{c_t^4}{c_l^4}\right) m_0^s . \tag{3}$$

The rest masses of edge and screw dislocations are very similar, since the longitudinal wave speed $c_l = \sqrt{\frac{\lambda+2\mu}{\rho}}$ is usually about twice the transverse wave velocity c_t. Using the expressions for the energy of a screw dislocation (eq.2) and $c_t^2 = \mu/\rho$, one can estimate the rest mass of a dislocation segment of length b to be $m_0 \approx \frac{1}{2}\rho b^2$, roughly one half atomic mass per Burgers vector.

The velocity dependence of the effective mass of a screw dislocation is $m^s(v) = m_0^s \cdot (1 - v^2/c_t^2)^{-1/2}$, the effective mass $m^\perp(v)$ of fast moving edge dislocation is rather difficult to derive and was given by Sakamoto [12]. As displayed in fig. 1, the effective mass and the energy of moving dislocations are diverging at the transverse shear wave velocity c_t.

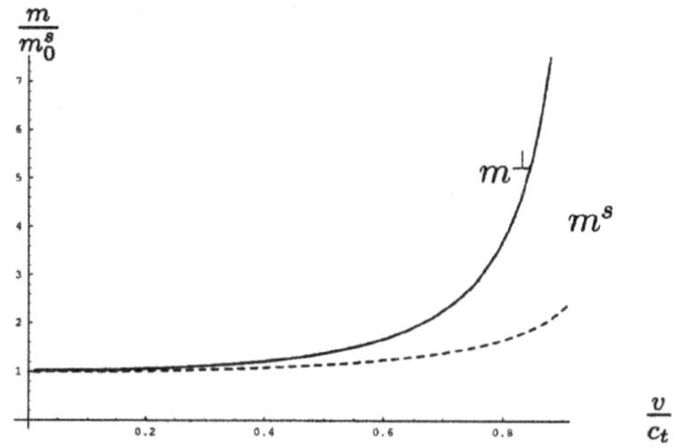

Figure 1. Velocity dependence of the mass of screw and edge dislocations for Ni (after [12]).

The drag coefficient B characterizes the drag forces on a moving dislocation which arise from its interaction with various elementary excitations like phonons and electrons [13]. In the intermediate temperature range, which will be of interest later, the dominating drag mechanism is the scattering of phonons. Leibfried [14] gave an estimation of the drag coefficient per unit length:

$$B_{\text{ph}} = \frac{3kTz}{20c_t b^2} , \tag{4}$$

where T is the temperature, k the Boltzmann factor and z the number of atoms per unit cell. It was later shown that a distinction should be made between different scattering mechanisms like the *phonon wind*, the *slow-phonon viscosity*, and the *flutter effect*. For a thorough treatment of these and other drag mechanisms the reader is referred to [13].

The reaction of a dislocation upon an imposed shear stress is thus fully determined by B and m_0. For small velocities $v < c_t/2$ the dislocation mass stays nearly constant, and thus the uniform steady state velocity can be calculated as $v_\infty = \frac{\tau b}{B}$. After the relaxation time $t_r = m_0/B$ the velocity of a dislocation accelerating from rest differs from the steady state velocity v_∞ by less then $1/e$. For typical values for B at room temperature, this time is in the range of several ps.

1.2 Dislocation - obstacle interaction and inertial effects

The resistance which a localized obstacle provides to dislocation motion can be represented by a point force F acting on the dislocation line. At an applied shear stress τ the dislocation bows out between neighboring obstacles. Balance of forces between line tension Γ at the obstacle and the force acting on the dislocation F in the isotropic line tension approximation yields

$$F = \tau b = 2\Gamma \cos \Phi \ . \tag{5}$$

The dislocation overcomes the obstacle if the maximal resisting (pinning) force F_p is reached. The obstacle strength is then characterized by the critical angle Φ_c at which the breakaway occurs (see fig. 6). At lower applied stresses, the equilibrium configuration is characterized by Φ_{eq}.

Moving dislocations can overcome obstacles by inertial overshooting [1, 2]. The overshooting results in a smaller cusp angle $\Phi_{\text{d}} < \Phi_{\text{eq}}$ and thus in an effective increase of the maximum dislocation force on the obstacle. Neglecting thermal activation, the obstacle is passed if the dislocation force is larger than the obstacle force, or $\Phi_{\text{d}} \leq \Phi_c$. Such a dynamic effect is expected when the dislocation carries significant kinetic energy and does not loose it to damping while bowing out between obstacles. In contrast, in the overdamped case any kinetic energy is dissipated during the bow out.

The following expression was proposed by Nadgornyi [13, 6] for the critical drag coefficient above which the dislocation motion becomes overdamped:

$$B_c = \frac{2\pi\Gamma}{c_t L} \ . \tag{6}$$

Inserting typical values for fcc metals like Cu and Al at room temperature the above expression gives critical damping coefficients which are in the range or slightly above the measured damping coefficients [13]. Therefore, the assumption of overdamped, viscous dislocation motion, usually made in DDD simulations may only be valid in certain specific cases and must be critically looked at in others.

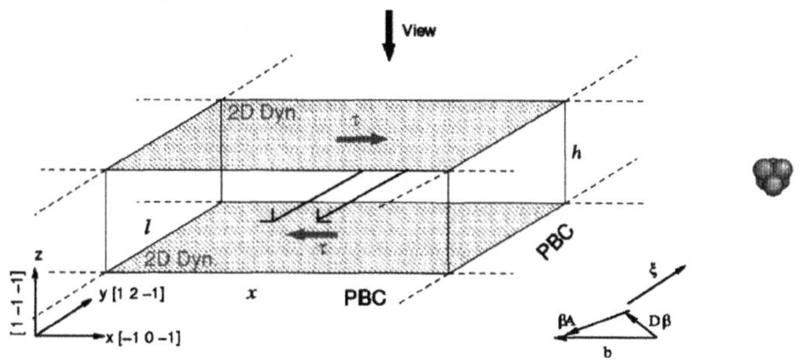

Figure 2. Sketch of the simulation box, the crystallographic orientation, and of the arrangement of the vacancies creating the obstacle. Periodic boundary conditions (PBC) are applied in the line- and glide-direction, whereas 2D-dynamic boundary conditions are applied in z-direction. Upon relaxation the edge dislocation splits in two partial dislocations $D\beta + \beta A$.

2. Simulation Methods

To obtain the parameters governing the dynamics of dislocations, Peierls stress τ_P, rest mass $m_0^{\perp}$ and drag coefficient B, an infinitely long edge dislocation was simulated in a molecular dynamics (MD) simulation with 2D periodic boundary conditions. The dynamics of dislocation - obstacle interaction was studied using a small void cluster often found in irradiated materials [15].

Nickel, described by the embedded atom method (EAM) potential of Angelo et al [16], was chosen as the model material for this study. A sketch of the simulation box and the boundary conditions is shown in fig. 2. Standard box sizes were x = 42 nm, l = 8 nm and h = 10 nm. The box dimensions were varied to assess the importance of finite size effects. Shear loading was applied incrementally by combining force boundary conditions on the outermost $(1\bar{1}\bar{1})$-layers with corresponding deformation of the sample.

The displacement field of a perfect edge dislocation was introduced in accordance with the boundary conditions. Upon relaxation the dislocation splits into two partial dislocations. A cluster of 6 vacancies (see fig. 2) was sometimes used as a localized attractive obstacle. The dislocation was always positioned at a distance of x_r=15 nm away from the obstacle. The obstacle spacing was also l=15 nm in this case.

Energy minimization was performed with a global version of the damped Newtonian algorithm by Beeler [17]: if the scalar product of the global force and momentum vectors is negative, all momenta are reset to zero. All systems were relaxed until the average force component of an atom was smaller then at least 10^{-17} N. Dynamic simulations were performed on equilibrated samples with a Nosé-Hoover thermostat [18]. The thermostat coupling was optimized to produce canonical temperature fluctuations about the desired thermostat temperature.

An inertia based equation of motion for dislocations is also implemented into the discrete dislocation dynamics model of Weygand et al. [19], which is based on a discretization of the dislocation line using straight segments connected at nodal points. The degrees of freedom of the dislocation are the positions $\{\vec{R}_A\}$ and the velocities $\{\vec{V}_A\}$ of the nodal points $\{A = 1, \ldots, N\}$. The equation of motion of the nodal points are derived using the principle of virtual work.

The equation of motion of a dislocation segment at point $\vec{r}(l)$ is given by

$$m_0 \ddot{r}_i + B \dot{r}_i = f_i^s(l) \tag{7}$$

where m_0 is the rest mass of the dislocation, f_i^s is the Peach-Koehler force in the glide plane of the dislocation and l is the curvilinear coordinate along the dislocation line. The velocity dependence of the dislocation mass is ignored. The equation of motion for the nodal points is obtained from the principle of virtual work [19]:

$$\oint_L f_i^s \delta v_i dl = \oint_L B v_i \delta v_i \, dl + \oint_L m_0 \dot{v}_i \, \delta v_i dl. \tag{8}$$

As in [19], a linear interpolation scheme for the velocities $\vec{v} = \vec{v}(l,t) = \sum_{A=1}^{N} N_A(l) \vec{V}_A(t)$ is employed. The acceleration is expressed by $\dot{\vec{v}}(l,t) = \frac{1}{\Delta t}(\vec{v}(l,t) - \vec{v}(l,t-\Delta t))$, where Δt is the time increment used in the integration scheme for the movement of the dislocation. After integration of equation (8) over the dislocation segments, the following linear system for the nodal velocities $\{V_{Ai}\}$ has to be solved:

$$\sum_{B=1}^{N} (K_{AB} + M_{AB}) V_{Bi} = F_{Ai} + \sum_{B=1}^{N} M_{AB} V_{Bi}(t - \Delta t) \tag{9}$$

$$F_{Ai} = \oint_L f_i^s(l,t) N_A(l) \, dl \tag{10}$$

$$K_{AB} = B \oint_L N_A(l) N_B(l) \, dl \tag{11}$$

$$M_{AB} = \frac{1}{\Delta t} \oint_L m_0 N_A(l) N_B(l) \, dl \tag{12}$$

The interaction between dislocations is calculated using linear elasticity theory, neglecting relativistic effects. The maximum velocity of a dislocation segment (not a nodal point) must therefore be restricted to about $c_t/2$.

The obstacles in the atomistic simulation consist of vacancy clusters with a diameter of about 2 atoms. In the DDD simulation, these vacancy clusters are translated into finite size obstacles of similar diameter. After detection of the pinning event, two pinned nodal points, representing the two intersection points of the dislocation with the vacancy cluster, are introduced on the dislocation line. The depinning event is controlled by a critical curvature/angle of the dislocation at the pinning center. Note that the minimum and maximum length of the dislocation segments has to be controlled such that the behavior is independent of the regularization scheme for the local self-interaction.

3. Results

The critical shear stress τ_c for initiating dislocation motion at 0 K was determined to lie between 0.12 and 0.14 MPa. This critical stress corresponds to $\tau_c^{\perp} \approx 10^{-6} c'_{44}$. Dislocation motion starting from rest under an applied shear stress was studied in the temperature range from 30 - 700 K and for different sizes of the simulation box. Fig. 3 shows the velocity $v(t_o)$ reached by the dislocation after a time $t_o \geq 100$ ps under various loads and temperatures together with the Rayleigh wave velocity c_R and the lower velocity of transverse waves c_t in the corresponding crystallographic orientation. Although the dislocation velocities were almost constant at the end of the simulation, the steady state velocities could not always be reached, especially in the case of relativistic dislocations which reach the limiting velocity only asymptotically. The damping coefficient B can be estimated via $B = \frac{\tau b}{v(t_o)}$ for different temperatures from the measured velocities (see fig. 4).

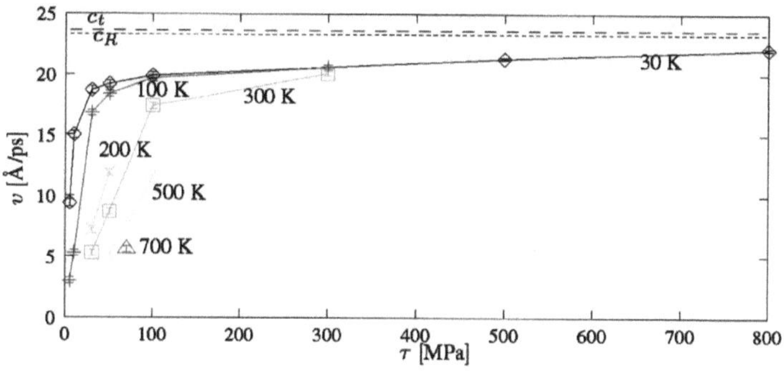

Figure 3. Velocities $v(\tau, t \geq 100$ ps) of the dislocations under applied shear stress τ at different temperatures as measured in MD simulations

The system including a void formed by six vacancies is studied in static and dynamic simulations at different applied shear stresses and temperatures to determine the lower and upper bound of the critical shear stress τ_c required for the dislocation to pass the obstacle. Figure 5 shows the pinned dislocation

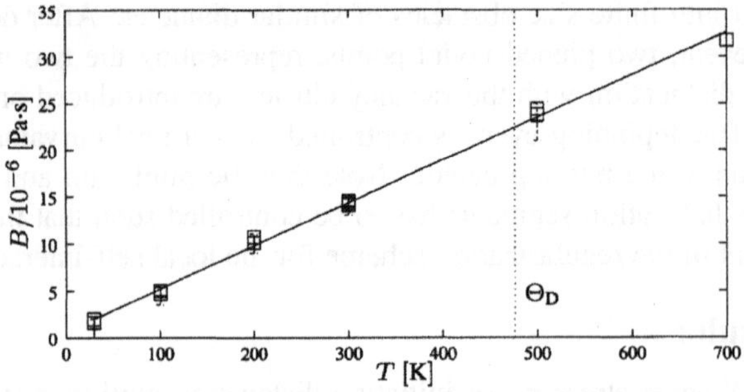

Figure 4. The damping coefficient $B(T)$ as determined from the "steady state velocities" $v(t \geq 100\ \text{ps}) \leq c_t/2$ of fig. 3 (squares) and from the fits of the equation of motion to the dislocation position (crosses) at different temperatures together with the Debye temperature Θ_D. The line is for a linear fit corresponding to Leibfrieds estimation of the damping coefficient (eq. 4).

for the static case at a stress level just below the critical stress τ_c and snapshots from the dynamic simulation at 100K. The results are summarized in table 1. Even at temperatures of 300 - 500 K there is a very significant lowering of the critical stress and therefore a significant effect of the inertia of the dislocation.

Table 1. Lower and upper bound for the stress required for the edge dislocation to pass an array of obstacles separated by $L = 15$ nm at different temperatures T. The obstacles in the MD simulations consist of 6 clustered vacancies (see fig. 2). For the DDD simulations a pinning center with similar diameter and critical angle was chosen (see text). Initially the distance between dislocation and obstacle was in both cases $x_r = 15$ nm.

Simulation	τ_c^{MD} [MPa]	τ_c^{DDD} [MPa]
static	75 - 80	75 $\pm$2
100 K	15 - 16	13 $\pm$1
300 K	30 - 33	30$\pm$2
500 K	40 - 45	45 $\pm$2

DDD Simulations are used to analyze the system simulated by the MD method. To mimic the atomistic results, the critical depinning angle used in the DDD simulation is set to $\Phi_c = 73.5°$. The size of the pinning centers is set to be $2a_0$, where a_0 is the lattice parameter. Comparing the motion of the overdamped and underdamped dislocation for the same resolved shear stress τ_c, the overshooting due to the mass of the dislocation is clearly visible in the figure 6. The thin black line corresponds to the critical configuration at $\tau_{c,\text{od}}$ for the overdamped dislocation. Note that the critical configuration for overdamped and underdamped calculations are almost identical. A critical re-

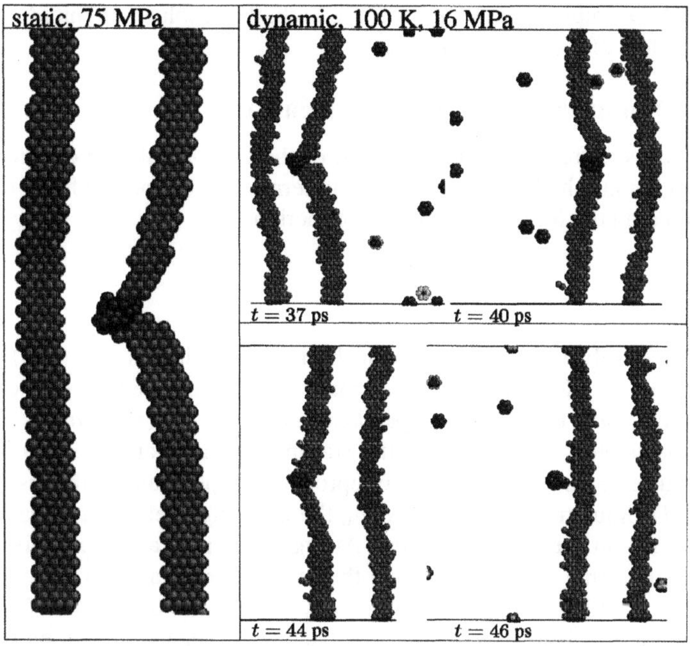

Figure 5. Left: Relaxed configuration of a edge dislocation pinned at 6 vacancy void at an applied shear stress of 75 MPa. Right: snapshots of the same configuration from the dynamic simulation at 100K and 16 MPa.

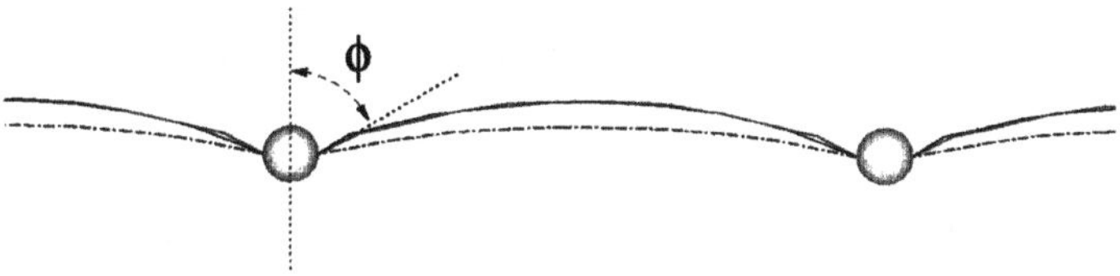

Figure 6. Results of the DDD simulations: three dislocation configurations are shown: the thick line corresponds to the configuration before unpinning of the underdamped dislocation at τ_d; the dashed line represents the final configuration of an overdamped dislocation at the same stress τ_d; the thin black line represents the critical configuration at $\tau_c > \tau_d$ for the overdamped dislocation.

solved shear stress of $\tau_c \approx 75$ MPa for the overdamped situation is obtained. The same configuration, including the inertia and $B = 15 \cdot 10^{-6}$ Pa s taken from the MD simulation results for 300 K (see fig. 4) leads to dynamic passing stress of $\tau_d \approx 30$ MPa. The results shown in table 1 are in excellent agreement with the MD simulations, with respect to both, the absolute magnitude and the relative effect of the dislocation mass.

4. Dynamics of Straight Dislocations

The obstacle-free accelerated motion of the two partial dislocations in the MD simulations shall be regarded here in terms of the motion of one dislocations based on the following equation of motion

$$m_0^{\perp} \cdot \frac{\partial}{\partial t} f(v)v - F + Bv = 0 \,. \tag{13}$$

Eq. 13 was solved numerically with the initial conditions $x(t = 0) = x_0$, $v(t = 0) = v_0$. Here the dislocation mass was decomposed into the constant rest mass $m_0^{\perp}$ and the velocity dependent term $f(v)$: $m^{\perp}(v) = m_0^{\perp} f(v)$.

The parameters m_0, B, F and the starting positions and velocities x_0, v_0 were adjusted by a least square fit to reproduce the positions of the dislocation in the MD simulations. With know m_0 the damping coefficient B can be determined for different temperatures. At high velocities additional dissipative mechanisms come into play [13], so that the drag resistance is no longer described by eq. 13. Therefore B is fitted only at velocities below $c_t/2$. The force per unit length F is in principle known, but can also be used as fitting parameter. The results of the fitting procedure are shown in table 2 for two exemplary cases at low temperature and low stress. From these and other fits the rest mass

Table 2. Parameter describing the dislocation motion as determined from the best fits of the numerical solution of the equation of motion (13) to the positions from the MD simulations of the dislocation at 30 K . Δx is the mean deviation of each data point from the fit.

System	Δx [Å]	m [$\frac{\text{pN}\cdot\text{ps}^2}{\text{Å}^2}$]	F [$\frac{\text{pN}}{\text{Å}}$]	B [10^{-6}Pa·s]
5 MPa	0.425	1.094	0.1214	1.086
7 MPa	0.281	1.096	0.1686	0.950

of an edge dislocation is determined to $m_0^{\perp} = 1.1 \pm 0.02$ pN · ps^2/Å^2 ≈ 0.28 atoms/b. Using eqs. (2) and (3) with $R = h$, $r_0 = b$ and $\mu = 74.6$ GPa gives $m_0 = 0.58$ atoms/b. The theoretical estimate is roughly twice the value obtained from the simulations, which is most probably due to the splitting in two partial dislocations.

The effective force F per unit length determined by the fits (see table 2) is slightly smaller than the applied force $F_a = \tau b$. The amount of the reduction can be interpreted as a dynamic friction stress τ_f. The value $\tau_f = 0.15 \pm 0.03$ MPa of this friction stress is similar to the Peierls stress determined by the static MD calculations.

The values of the damping coefficient as determined from the above mentioned fitting procedure are included in fig. 4. Like the values determined directly by the steady state velocity, they fall on a line with slope $s = (0.0457 \pm 0.0008) \cdot 10^{-6}$ Pa s K^{-1}. This value is in excellent agreement with the prefactor of Leibfried's estimation (eq. 4): $B(T) = 0.0462 \cdot 10^{-6}$ Pa s K^{-1}T as calculated with the corresponding values for Ni. This agreement with the rather crude estimation of Leibfried can be explained by the fact that within the temperature range studied here phonons with wavelengths of the order of a_0 dominate the phonon spectrum, which is the range of the scattering cross section assumed by Leibfried.

5. Dislocation - Obstacle Interaction

The stress τ_d needed for the dislocation to pass the void is significantly lower in the dynamic simulations than the critical stress τ_c in the static simulations (see table 1). The dynamical passing stress τ_d increases with increasing temperature, thus thermal activation can not account for this effect. In this section it will be shown how these inertial effects can be rationalized in a framework based on energetic considerations.

Passing the obstacle requires to surmount an energy barrier of height ΔE. In the line tension model this energy barrier corresponds to the increase of line energy at the critical configuration. The critical configuration for breakaway in the static case is characterized by the displacement profile $u(x, \tau_c)$ at the critical stress τ_c. The critical angle is related to the displacement profile via $\cot \Phi_c = u'(-L/2, \tau_c)$. In the static case, corresponding to the balance of forces (eq. 5) for Φ_c the energy required to pass the barrier is provided by the work W_c done on the dislocation by the applied stress τ_c.

If a fast moving dislocation has enough kinetic energy $E_{\text{kin}} = \frac{1}{2}m(v)v^2L$ to outweigh the energy E_{diss} which is dissipated during the process of the dislocation bow-out, the dislocation can overshoot its equilibrium position and eventually pass the energy barrier ΔE dynamically. Dynamical obstacle passing in a purely mechanical model is thus possible at stresses $\tau_d < \tau_c$ when the work W_d done by this stress is larger than the dynamically lowered critical work:

$$W_d > W_c - (E_{\text{kin}} - E_{\text{diss}}) \ . \tag{14}$$

The contributions of the terms in eq. (14) are path dependent and not easily accessible. However, the DDD simulations show that the critical depinning configuration is the same for the dynamic case and the static case and is characterized by the critical angle Φ_c. To reach this configuration the dislocation has to slide over the area A_c. The work done in the static case can thus be estimated to be $W_c^{\text{est.}} = \tau_c b A_c$. In the dynamic case, the work done by the dynamic passing stress τ_d to reach the critical configuration is then $W_d^{\text{est.}} = \tau_d b A_c$. The dissipated energy during bow-out can be estimated to $E_{\text{diss}}^{\text{est.}} = \frac{1}{2} v_i B A_c$, which, just like the kinetic energy, depends on the velocity $v_i(\tau_d, B)$ at which the dislocation impinges on the obstacle. This velocity can directly be determined by solving the equation of motion (eq. (13)) with $\tau_d, m_0^{\perp}, B$. The critical area A_c, however has to be determined from atomistic or DDD simulations.

Table 3. Estimated contributions to the energy necessary for the dislocation to pass the obstacle. In the static case the energy required to pass the energy barrier is provided by the work W_c done on the dislocation by the acting stress τ. In the dynamic cases there is an additional contribution by the kinetic energy of the dislocation E_{kin} which is decreased by the dissipated energy E_{diss} during the bowout.

Simulation	$\tau_{c,d}$ [MPa]	$W^{\text{est.}}$ [eV]	$E_{\text{kin}}^{\text{est.}}$ [eV]	$-E_{\text{diss}}^{\text{est.}}$ [eV]	$\sum_{\text{energies}}$ [eV]
static	75 - 80	2.19 - 2.34	0	0	2.19 - 2.34
100 K	15 - 16	0.42 - 0.45	1.99 - 2.19	0.15 - 0.15	2.26 - 2.48
300 K	30 - 33	0.84 - 0.92	1.41 - 1.72	0.39 - 0.42	1.87 - 2.23
500 K	40 - 45	1.12 - 1.26	0.98 - 1.27	0.52 - 0.59	1.58 - 1.94

The so estimated values of $W_{c,d}^{\text{est.}}$, $E_{\text{kin}}^{\text{est.}}$ and $E_{\text{diss}}^{\text{est.}}$ are shown in table 3. The sum of the contributions is very similar to the work done in the static case and thus provide a very simple way to estimate the magnitude of inertial effects for dislocation - obstacle interaction.

6. Conclusions

It has been shown that the parameters which determine the dynamics of dislocations (the dislocation rest mass m_0 and the damping coefficient B) can be extracted from atomistic simulations of accelerating straight dislocations. MD simulations of the interaction of edge dislocations with an array of vacancy clusters showed pronounced inertial effects even at room temperature. This can be explained by a simple model based on energy arguments. With the input from atomistic simulations, DDD simulations which include inertial effects can reproduce all the dynamic effects. They thus provide a tool to asses the importance of inertial effects in a wide range of situations including different temperatures, stress conditions, different obstacles arranged in various configurations.

References

[1] M. Suenaga and J. M. Galligan. Dislocation motion in the normal and the superconducting states. *Scr. Metall.*, 5:829–836, 1971.

[2] A. V. Granato. Dislocation inertial effects in the plasticity of superconductors. *Phys. Rev. B*, 4:2196–2201, 1971.

[3] R. B.Schwarz and R. Labusch. Dynamic simulation of solution hardening. *J. Appl. Phys.*, 49:5174–5187, 1978.

[4] K.-D. Fusenig and E. Nembach. Dynamic dislocation effects in precipitation hardened materials. *Acta Metall. Mater.*, 41:3181–3189, 1993.

[5] E. Vigueras-Santiago, A. A. Krokhin et al. Creep at low temperatures: unzipping of dislocations , inertia and criticality processes. *Physica A*, 258:11–16, 1998.

[6] M. Hiratani and E. M. Nadgorny. Combined modelling of dislocation motion with thermally activated and drag-dependent stages. *Acta Metall.*, 49:4337–4346, 2001.

[7] R. D. Isaac and A. V. Granato. Rate theory of dislocation motion: Thermal activation and inertial effects. *Phys. Rev. B*, 37:9278–9284, 1988.

[8] A. I. Landau. The effect of dislocation inertia on the thermally activated low-temperature plasticity of materials: I. theory. *Phys. Status Solidi A*, 61:555–563, 1980.

[9] V. L. Indenbom and V. Chernov. Dynamic waves along dislocations overcoming local obstacles. *Sov. Phys. Solid State*, 21:759–764, 1979.

[10] L. P. Kubin, G. Canova et al. Dislocation microstructures and plastic flow: A 3D simulation. *Solid-State Phys.*, 23&24:455–472, 1992.

[11] J. Weertman. High velocity dislocations. In P. G. Shewmon and V. F. Zackay, eds., *Response of Metals to High Velocity Deformation*, Metallurgical Society Conferences, pp. 205–247. Interscience, New York, 1961.

[12] M. Sakamoto. High-velocity dislocations: effective mass, effective line tension and multiplication. *Philos. Mag. A*, 63:1241–1248, 1991.

[13] E. M. Nadgornyi. Dislocation dynamics and mechanical properties of crystals. vol. 31 of *Progress in Materials Science*. 1988.

[14] G. Leibfried. Über den Einflußthermisch angeregter Schallwellen auf die plastische Deformation. *Z. Phys.*, 127:344–356, 1950.

[15] T. D. de la Rubia, H. M. Zbib et al. Multiscale modelling of plastic flow localization in irradiated materials. *Nature*, 406:871–874, 2000.

[16] J. E. Angelo, N. R. Moody and M. I. Baskes. Trapping of hydrogen to lattice defects in nickel. *Modelling Simul. Mater. Sci. Eng.*, 3:289, 1995.

[17] J. R. Beeler. *Radiation Effects Computer Experiments*, p. 27. North-Holland, Amsterdam, 1983.

[18] W. G. Hoover. Canonical dynamics: Equilibrium phase-space distributions. *Phys. Rev. A*, 31:1695–1697, 1985.

[19] D. Weygand, L. Friedman et al. Aspects of boundary-value problem solutions with three-dimensional dislocation dynamics. *Modelling Simul. Mater. Sci. Eng.*, 10:437–468, 2002.

MOLECULAR DYNAMICS STUDY ON THE CHARACTERISTICS OF EDGE AND SCREW DISLOCATIONS IN GAMMA/GAMMA-PRIME MICROSTRUCTURE OF NI-BASED SUPER-ALLOY

Kisaragi Yashiro
Department of Mechanical Engineering, Faculty of Engineering, Kobe University
yashiro@mech.kobe-u.ac.jp

Yasushi Tabata
Student of Graduate School of Science and Technology, Kobe University
tabata@solid.mech.kobe-u.ac.jp

Yoshihiro Tomita
Graduate School of Science and Technology, Kobe University
tomita@mech.kobe-u.ac.jp

Abstract Fundamental characteristics of edge and screw dislocations in the γ/γ' microstructure of Ni-based superalloys are investigated by using molecular dynamics simulations. Edge/screw dislocations are nucleated and glide in a slab cell of the Ni matrix involving an apex of a cuboidal Ni_3Al precipitate, that mimics a part of the idealized γ/γ' microstructure. The edge dislocation decreases its velocity at the γ' precipitate, showing dislocation pinning there, then penetrates it under the force from following dislocations. The screw dislocation runs through the precipitate without slowdown by shrinking the width between its Shockley partials. Detailed investigation of the stress distribution suggests that the constriction is due to interactions between the stress field around the precipitate and the partials: the stress causes a repulsive Peach-Koehler force on the leading partial and an attractive force on the trailing one since their edge components have opposite Burgers vectors.

Keywords: Molecular Dynamics, Edge and Screw Dislocation, γ/γ' Microstructure, Ni-Based Superalloy, Embedded Atom Method

H. Kitagawa and Y. Shibutani (eds.), IUTAM Symposium on Mesoscopic Dynamics of Fracture Process and Materials Strength, 59–68.

1. INTRODUCTION

Ni-based superalloys possess the characteristic microstructure where Ni_3Al-based phases (γ' phases) are precipitated with the size less than $1\mu m$ in the Ni-based matrix (γ phase). The small lattice mismatch between the γ and γ', fcc and $L1_2$ ordered alloy, makes the γ/γ' interfaces coherent; however, dislocations are affected by the γ/γ' interface and show complicated behavior depending on the morphology of the γ/γ' microstructure. The dislocation behavior in the array of γ' precipitates has attracted intense interest with respect to the understanding of the deformation mechanism of superalloys.

Numerous studies have been devoted to dislocations in the γ/γ' microstructure from the viewpoint of their role in the *rafting* process, or the directional coarsening of the precipitates under creep deformation. Experimental observations by means such as transmission electron microscopy have revealed that dislocation lines are bent along the γ' precipitates [1–3]. Finite-element analyses based on the geometrically necessary dislocation theory have suggested that dislocations are required geometrically around the cuboidal precipitates [4–7] . It is difficult, however, to explore further details of the interaction between the dislocation and the γ/γ' interface by experimental observations or FEM analyses, since the scale of the phenomena is less than one micrometer and of a nanoscale. Dislocation behavior in the length scale would be strongly affected by the atom configuration at the interfaces, so that atomistic study is desired to clarify the mechanism. We have conducted a series of molecular dynamics simulations of the dislocation behavior at the γ/γ' interfaces, with more than one million atoms [8–10]. In the present study, edge and screw dislocations approaching the cuboidal precipitate are simulated with a slab cell of Ni containing a Ni_3Al precipitate. The different behaviors of the edge and screw dislocations at the interface are investigated and the mechanisms are discussed.

2. SIMULATION PROCEDURE

The interatomic potential adopted is the embedded-atom method [12, 13], in which parameters are fitted to the properties of the Ni, Al, and Ni-Al binary systems by Voter and Chen [14,15]. The total energy, E_{tot}, is evaluated by

$$E_{\mathrm{tot}} = \frac{1}{2}\sum_{\alpha}\sum_{\beta(\neq\alpha)} \phi_{t_\alpha t_\beta}(r^{\alpha\beta}) + \sum_{\alpha} F_{t_\alpha}(\overline{\rho}_\alpha) \tag{1}$$

with

$$\overline{\rho}_\alpha = \sum_{\beta(\neq\alpha)} \rho_{t_\beta}(r^{\alpha\beta}), \tag{2}$$

where $r^{\alpha\beta}$ is the scalar distance between atom α and atom β, ϕ is the pairwise interaction between atoms, $F(\overline{\rho})$ is the "embedding function" of the "density" at atom α, $\overline{\rho}_\alpha$, that is given by the superposition of another pairwise interaction, $\rho(r)$, from neighboring atoms. t_α and t_β indicate the types of atoms α and β, respectively. The details of the functions and parameters are described in the original papers [14, 15] and our previous reports [8, 9].

The simulations are implemented with a slab cell that mimics a local part of an apex of cuboidal Ni_3Al precipitate in a Ni matrix, as schematically illustrated in Fig. 1. The cell has the size of 43.4 nm × 40.1 nm × 18.4 nm and the total number of atoms is 2,880,000. The precipitate is made by substituting Al for Ni in the matrix, thus there is no lattice mismatch between γ and γ' and the interfaces are coherent in the initial arrangement. The edges of the cubic are rounded off to have the radius of 3.5 nm. This initial configuration is relaxed by molecular dynamics calculation of 5000 fs in order to redistribute the internal strain around the precipitate. Here, the motion of atoms is restricted in the yz or xy planes at the x or z boundaries, respectively, while the periodic boundary condition is adopted along the y axis. The temperature is kept at 300 K during the simulation by the velocity scaling. After the initial relaxation, the cell is subjected to two different deformations, mode II type for edge dislocation and mode III type for screw dislocation, respectively.

<u>Simulation of edge dislocation</u>

Mode II type deformation is applied by controlling the displacement in the region of $x < 4.3$ nm and $z > 9.2$ nm, which is represented by the

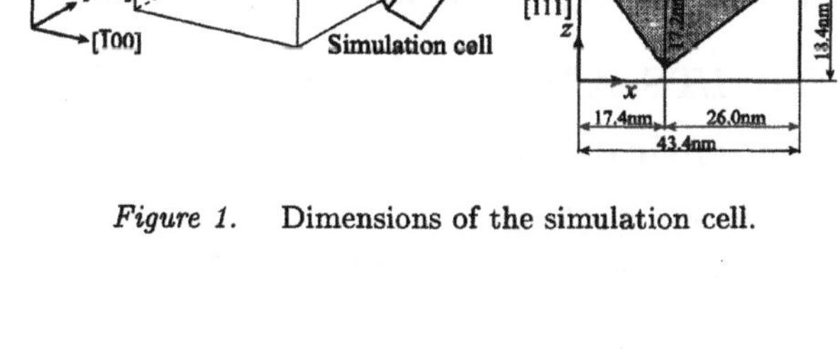

Figure 1. Dimensions of the simulation cell.

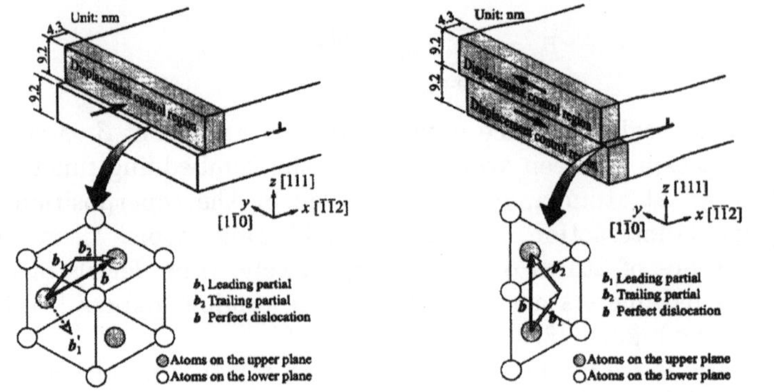

(a) Simulation of edge dislocation (b) Simulation of screw dislocation

Figure 2. Displacement control and crystallographic orientation.

light shaded region of the upper schematic in Fig. 2 (a). Atoms in this region are moved in the direction of $\mathbf{b}_1 = \frac{1}{6}[1\bar{2}1]$, the Burgers vector of a leading partial, then in the direction of $\mathbf{b}_2 = \frac{1}{6}[\bar{1}\bar{1}2]$ of a trailing partial, resulting in the total migration of $\mathbf{b} = \frac{1}{2}[0\bar{1}1]$ of a perfect dislocation, as shown in the lower schematic in Fig. 2 (a). The dislocation of Burgers vector $\mathbf{b}$ is not a pure edge but has a screw component. This condition is adopted in order to avoid the nucleation of two different leading partials, $\mathbf{b}_1$ and $\mathbf{b}_1' = \frac{1}{6}[\bar{2}11]$, simultaneously, with the displacement only in the x direction. A small displacement of $\mathbf{b}_1/2500$ or $\mathbf{b}_2/2500$ is applied at every step in order to nucleate a perfect dislocation with a relaxed core structure every 5000 fs. During the calculation, the atoms in the displacement control region are fixed in the x and y directions. The cell size is also controlled to keep the normal stress at zero.

<u>Simulation of screw dislocation</u>

The region of x<4.3 nm is subjected to the positive y-displacement in the upper half and the negative one in the lower half, as shown in the upper schematic of Fig. 2 (b). A perfect dislocation of the Burgers vector of $\mathbf{b} = \frac{1}{2}[1\bar{1}0]$ is nucleated every 5000 fs by applying a small displacement of $\mathbf{b}/5000$ at every step. The motion of atoms along the y direction is constrained in the displacement control region during the simulation, and the normal stress is kept at zero by changing the cell size.

3. RESULTS AND DISCUSSION

3.1 Dislocation Motion

Stacking faults between leading and trailing partials can be visualized by means of the common neighbor analysis (CNA) [16]. Figure 3 shows the motion of stacking faults, or extended dislocations, on the slip plane evaluated by the CNA in the simulation of edge dislocation. Ni atoms in the stacking faults are drawn with dark shade while Al atoms of the γ' are indicated with light shade in the figure. An extended dislocation composed of a pair of leading and trailing partials is nucleated at $t =$ 5000 fs in Fig. 3 (a). This dislocation, referred to as "Dislocation 1", has the width of about 2 nm or $6\,b$. Here, b is the magnitude of the Burgers vector or the average lattice length in the simulation cell. Then the second dislocation appears at t = 7000 fs of Fig. 3 (b) under the increase of the controlled displacement and pushes Dislocation 1 out. Dislocation 1 reaches the γ' precipitate at about $t = 8000$ fs and is bent along the interface during $t = 8000 \sim 9000$ fs (Figs. 3 (c) and (d)). It begins to penetrate into the precipitate from the upper corner of the triangular cross section, or an edge of the γ' cubic, during $t = 9500 \sim$ 10000 fs (Figs. 3 (e) and (f)). The distance between the Shockley partials becomes wider in the precipitate than in the matrix (Figs. 3 (g)$\sim$(i)).

The motion of the extended screw dislocations as evaluated by the CNA is illustrated in Fig. 4. The width of the first dislocation introduced at t = 5000 fs is about the same as that of the edge dislocation. The screw dislocation, however, decreases its width significantly as it approaches the precipitate under the force from the following dislocation, and becomes less than 1 nm at t = 7000 fs as shown in Figs. 4 (a)$\sim$(c). The dislocation reaches the precipitate in Fig. 4 (d) of t = 8000 fs and passes through the interface with no remarkable slowdown. The con-

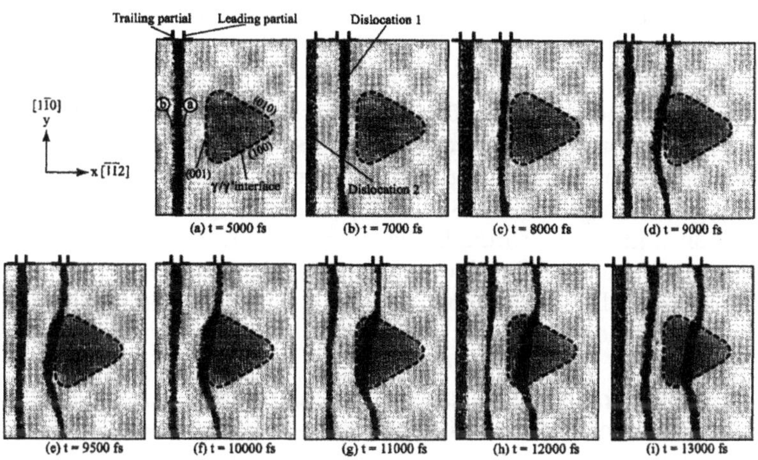

Figure 3. Motion of edge dislocations on the slip plane.

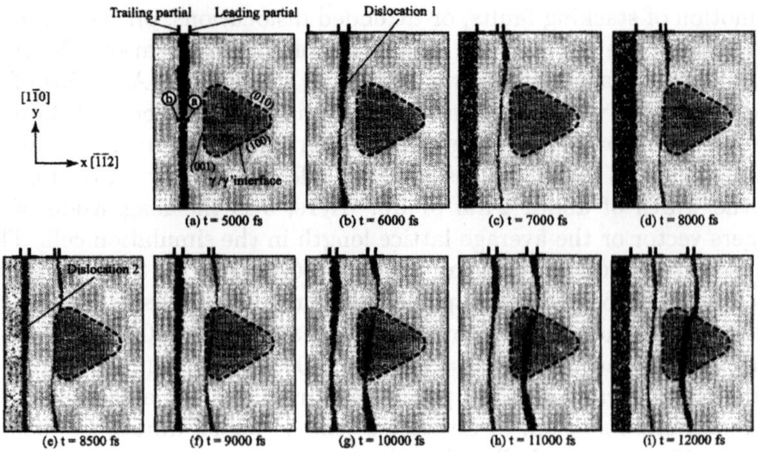

Figure 4. Motion of screw dislocations on the slip plane.

stricted dislocation expands again in the precipitate same as in the case of the edge dislocation.

3.2 Details near the γ/γ' Interface

Figure 5 shows the change in the position of each core of the leading and trailing partials, measured at Points Ⓐ and Ⓑ indicated in Figs. 3 and 4, respectively, to clarify the difference between the edge and screw dislocations at the γ/γ' interface. The position of the (001) face of the γ' cubic is also indicated with the solid line at $x = 11.4\,\mathrm{nm}$. The leading partial of the edge dislocation decreases its velocity at about $t = 8000\,\mathrm{fs}$ as shown in Fig. 5 (a). Here, the trailing partial shows the same deceleration while maintaining its distance to the leading partial. When the leading partial begins to penetrate into the precipitate and glide again during $t = 9000 \sim 9500\,\mathrm{fs}$, the trailing partial also resumes its motion and passes through the interface without slowdown. Thus it is revealed that the Shockley pair of an edge dislocation behaves as a dislocation belt with a constant width, and is stopped just at the interface, and then cuts into the precipitate.

In the case of the screw dislocation shown in Fig. 5 (b), the leading partial moves more slowly than the trailing one at $t < 7000\,\mathrm{fs}$, so that the latter catches up with the former and thus causes the extended dislocation to constrict. The trailing partial maintains a constant speed during the constriction. There is no distinct obstruction at the interface against either the leading or trailing partials when they cut into the

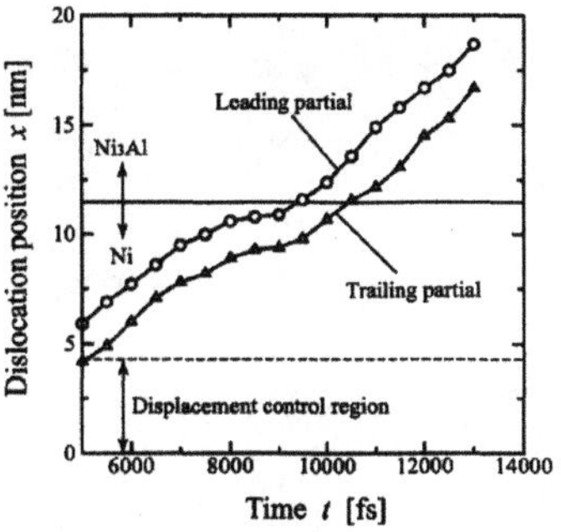

(a) edge dislocation

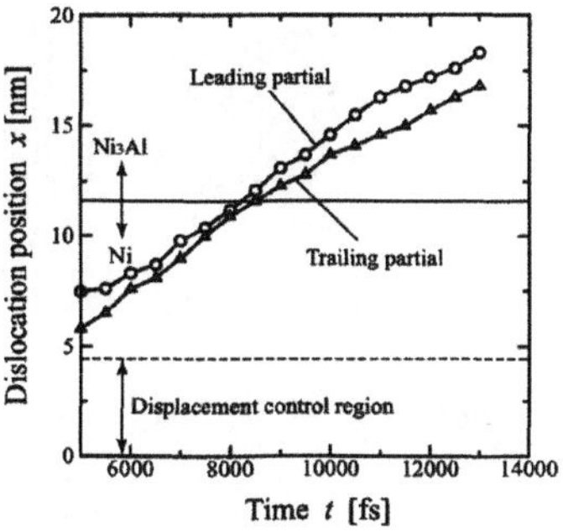

(b) screw dislocaton

Figure 5. **Change in the position of the leading and trailing partials.**

precipitate at $t = 8000 \sim 8500\,\mathrm{fs}$. Then the Shockley partials extend again in the precipitate as in the case of the edge dislocation, due to the slight decrease of the velocity of the trailing partial. Thus the remote stress field around the precipitate might affect the motion of the leading partial and cause the width of the Shockley pair to narrow.

3.3 Stress Field near the Interface

In order to clarify the different mechanism near the interface, the local stress, τ_{zx} for edge dislocations and τ_{yz} for screw dislocations, respectively, is evaluated at every position of atoms. These stress components generate the Peach-Koehler force in the x direction. Here, the local stress is defined as the the first order differential of the internal energy with regard to the virtual strain perturbation; its formulation in the EAM was presented in our previous papers [9, 11]. Figure 6 shows the distribution of the local stress of τ_{zx} on the slip plane indicated in Fig. 3, by representing all atoms with gray scale color depending on their magnitude. In the figure at $t = 5000\,\mathrm{fs}$, intense positive and negative stresses can be seen at the front and back of the core of the leading and trailing partials, respectively, as expected on the basis of the dislocation theory. The stress in the γ' precipitate is nearly uniform, so that it is also consistent with the ellipsoidal inclusion [17]. It is noteworthy that there is negative stress in the Ni matrix at the front of the (001) face of the γ' cubic, while there is positive stress at the back of it, or at the edge between the (010) and (100) faces. The negative τ_{zx} near the interface is expected to cause repulsive force on both the leading and trailing partial; however, the dislocation lines continue to glide toward the precipitates without significant resistance during $t = 5000 \sim 8000\,\mathrm{fs}$.

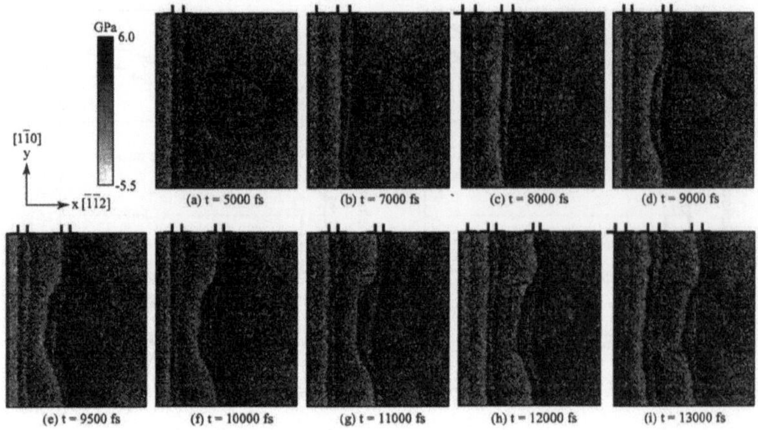

Figure 6. **Distribution of shear stress, τ_{zx}, on the slip plane (edge dislocation).**

This is because the negative τ_{zx} of the interface is far smaller than the positive τ_{zx} caused by the following dislocation, and because the repulsive force between the partials is large in the case of edge dislocation. The dislocation is blocked and bent when the core of the leading partial reaches the interface at $t = 8000\,\mathrm{fs}$. It is therefore concluded that the edge dislocation is not affected by the internal stress around the precipitate but stopped at the interface due to the difference of the atomic structure.

The distribution of the local shear stress, τ_{yz}, in the simulation of the screw dislocation is shown in Fig. 7. The maximal and minimal magnitudes in the vicinity of the dislocation cores are nearly equal to those of τ_{zx} in the edge dislocation. Positive and negative stresses can be recognized at the (010) and (100) faces of the precipitate, respectively. However, the contrast between γ and γ' is less clear than that observed for the edge dislocation and there is no distinctive stress at the (001) face. The distribution shows negligible change even in the process of the dislocation constriction during $t = 5000 \sim 7000\,\mathrm{fs}$. Thus it is deduced that the other stress component should give rise to the constriction.

The leading partial of a screw dislocation has the edge component opposite to that of the trailing partial, so that the stress, τ_{zx}, would generate the Peach-Koehler force on the leading and trailing partials in the opposite directions. Figure 8 illustrates the distribution of the local stress, τ_{zx}, evaluated at $t = 5000\,\mathrm{fs}$ in the simulation of screw dislocation. We can find the same negative stress at the front of the

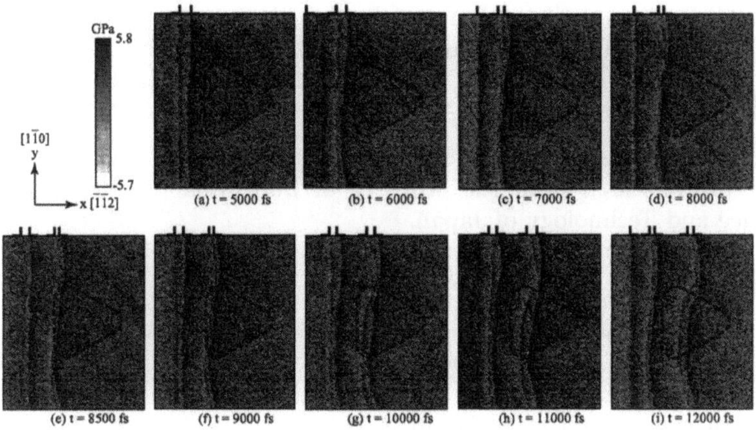

Figure 7. **Distribution of shear stress, τ_{yz}, on the slip plane (screw dislocation).**

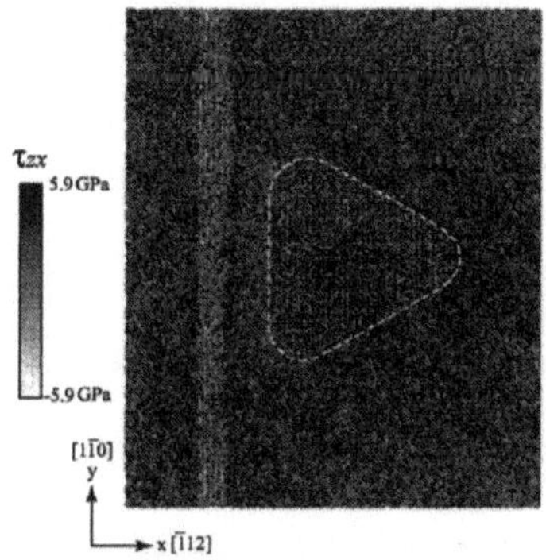

Figure 8. **Distribution of atomic shear stress, τ_{zx}, on the slip plane at t = 5000 fs (screw dislocation).**

(001) face of the precipitate as in the case of edge dislocation. Both the stress and the following dislocation generate pull-back force on the leading partial and push-out force on the trailing one, resulting in the dislocation constriction.

The constricted screw dislocation does not lead to a line step on the front face of the precipitate but an atomic step on the side face in the penetration process. The screw dislocation also has a lower strain energy than that of the edge dislocation by the factor of $1/(1-\nu)$ [18]. Thus the screw dislocation easily penetrates through the interface even though it forms the same antiphase boundary (APB) as the edge dislocation.

Acknowledgements

This work was supported financially in part by a Grant-in-Aid for Scientific Research from the Ministry of Education, Culture, Sports, Science and Technology of Japan.

References

[1] T. M. Pollock and A. S. Argon, *Acta Metall. Mater.*, **40**-1, (1992), 1–30.

[2] Benyoucef, M., Clément, N. and Coujou, A., *Phil. Mag.*, A, **72** (1995), 1043–1056.

[3] Coujou, A., Benyoucef, M., Legros, M. and Clément, N., *Solid State Phenomena Vols.*, **59**-60 (1998), 185–200.

[4] Busso, E. P., Meissonnier, F. T., O'Dowd, N. P. and Nouailhas, D., *Journal de Physique IV*, **8** (1998), Pr8–55–Pr8–61.

[5] Ohashi, T., *J. Phys. IV, France* **9**, Pr9, (1999), 279–286.

[6] Busso, E. P., Meissonnier, F. T. and O'Dowd, N. P., *Journal of the Mechanics and Physics of Solids*, **48** (2000), 2333–2361.

[7] Higa, Y. and Tomita, Y., *Advances In Computational Engineering & Sciences*, (Atluri, S. N. and Brust, F. W. Eds.), Vol.II, pp.1116-1121, Tech Science Press (2000).

[8] K. Yashiro, M. Naito and Y. Tomita, *Trans. JSME*, series A, **68**-668, (2002), pp.573-579.

[9] K. Yashiro, M. Naito and Y. Tomita, *Int. J. Mech. Sci.*, **44-9**, (2002), pp.1845-1860.

[10] K. Yashiro, M. Naito and Y. Tomita, Trans. JSME, series A, **69**-677, (2003), pp.210-217.

[11] K. Yashiro and Y. Tomita, *Journal de Physique* IV-11, (2001), pp.Pr5-3 - Pr5-10.

[12] Daw, M. S. and Baskes, M. I., *Phys. Rev. Lett.*, **50**-17 (1983), 1285–1288.

[13] Daw, M. S. and Baskes, M. I., *Phys. Rev.*, B, **29**-12 (1984), 6443–6453.

[14] Voter, A. F. and Chen, S. P., *Mat. Res. Soc. Symp. Proc.*, **82** (1987), 175–180.

[15] Voter, A. F. *Chapter 4 The Embedded Atom Method*, 77–90. John Wiley & Sons, (1994).

[16] Honeycutt, J. D. and Andersen, H. C., *J. Phys. Chem.*, **91**, (1987), 4950–4963.

[17] Eshelby, J. D., *Proc. R. Soc. Lond.*, **A241**, (1957), 376–396.

[18] Hirth, J. P. and Lothe, J., Theory of Dislocations, 2nd Edition, (1992), Krieger Publishing.

DISLOCATION INTERACTIONS AND SYMMETRIES IN BCC CRYSTALS

R. Madec[1], L.P. Kubin[2]

[1] *DPTA, Commissariat à l'Energie Atomique, BP 12, 91680 Bruyères-le-Chatel, France*

[2] *LEM, CNRS-ONERA, 29 av. de la division Leclerc, BP 72, 92322 Châtillon Cedex, France*

Abstract The symmetries of dislocation interactions in bcc crystals are examined by dislocation dynamics simulations with emphasis on the collinear interaction. The focus is on repulsive barriers that oppose the formation of configurations of minimum energy. An interaction matrix including the effect of these barriers is proposed.

Keywords: Dislocation junction, collinear interaction, bcc metals, interaction matrix.

1. INTRODUCTION

With the view of further establishing models for strain hardening, the aim of the present study is to determine the number of distinct types of mutual interactions of different strengths occurring between two slip systems in bcc metals at high temperatures. The interacting slip systems considered are the twelve 1/2<111>{110} slip systems and the twelve 1/2<111>{112} slip systems. For studies on strain hardening, it is convenient to construct an interaction matrix between these systems, in which each coefficient has a specific value representing the strength of the interaction. The number of distinct coefficients to be investigated is less than 144, as it is reduced by various symmetries. It can be deduced from the consideration of the geometrical variables involved in the interaction between two dislocation segments.

In the high temperature regime, above the so-called athermal temperature, the Peierls barriers are negligible and the dislocations achieve high mobilities, like in fcc crystals. The flow stress of bcc crystals in multiple slip conditions is then controlled by reactions between non-coplanar dislocations, as described by the "forest model"[1, 2].

Assuming, as is usually done, that the formation of these reactions is mainly a question of self-energy minimization [1, 3], the symmetries of the dislocation interactions depend only on the angles between the junction direction and

H. Kitagawa and Y. Shibutani (eds.), IUTAM Symposium on Mesoscopic Dynamics of Fracture Process and Materials Strength, 69–78.

the three Burgers vectors involved, namely those of the parent dislocations and that of the junction. The latter is always of the <100> type [4].

There is a particular type of reaction called the collinear interaction, which deserves a particular attention. This reaction is related to the partial annihilation of two non-coplanar dislocations of same Burgers vector, i.e., such that each slip plane is the cross-slip plane of the other. These annihilations always reduce the total self-energy but their occurrence for initially repulsive dislocations is conditioned by the overcoming of an energy barrier. As a consequence, it will be shown that an additional geometrical variable is involved, which affects not only the smmetries of the collinear annihilation, but also those of one type of junction-forming interaction.

In what follows, use is made of simplified elastic models to predict the number of different types of interactions to be accounted for. These results are further examined with the help of dislocation dynamics (DD) simulations for both junctions and collinear interactions. The consequences regarding strain hardening properties are then prospectively discussed.

2. SYMMETRIES RELATED TO SELF-ENERGY

A very complete study on junctions in bcc metals based on elastically anisotropic self-energies was performed by Püschl [4], who also introduced reaction mappings similar to those presented below. As was checked more recently by Wickham et al. from DD simulations on some interactions between 1/2<111> {110} slip systems [5], junction formation can be reasonably well predicted by a balance of self-energy. A simplified expression for the self-energy, E, is written:

$$E = \frac{\mu b^2}{4\pi(1-\nu)}(1-\nu\cos^2\alpha), \tag{1}$$

where α is the angle between the line direction and the Burgers vector. Using simple geometrical considerations (see [6] for details), one can show that the balance in self-energies only depends on the angles β_1, β_2 and β_j between the direction of intersection of the slip planes and the Burgers vectors of the two parent dislocations and of the junction. Considering only self-energies, there are thirteen distinct configurations for the twenty four 1/2<111>{110} and 1/2<111>{112} slip systems. Ten of them correspond to junctions and the three others are related to self-hardening, the coplanar interaction (coplanar slip systems with different Burgers vectors) and the collinear annihilation. The results are reported in Table 1; they include another symmetry effect, which is discussed in the next section. Table 1-a shows the interaction matrix, which eventually contains seventeen distinct coefficients. The crystallography of the interacting slip systems is detailed in Table 1-b. Table 1-c details the different

Table 1. Interaction matrix, slip systems and symmetries in bcc crystals.

(a)

a^i	1	2	3	4	5	6	7	8	9	10	11	12	13	14	15	16	17	18	19	20	21	22	23	24
1	1	2	3	3	4	5	5	6	4	5	6	5	7	8	9	9	10	11	12	13	10	11	13	12
2	2	1	3	3	5	4	6	5	5	4	5	6	8	7	9	9	11	10	13	12	11	10	12	13
3	3	3	1	2	5	6	4	5	6	5	4	5	9	9	7	8	12	13	10	11	13	12	10	11
4	3	3	2	1	6	5	5	4	5	6	5	4	9	9	8	7	13	12	11	10	12	13	11	10
5	4	5	5	6	1	3	2	3	4	6	5	5	10	12	11	13	7	9	8	9	10	13	11	12
6	5	4	6	5	3	1	3	2	6	4	5	5	12	10	13	11	9	7	9	8	13	10	12	11
7	5	6	4	5	2	3	1	3	5	5	4	6	11	13	10	12	8	9	7	9	11	12	10	13
8	6	5	5	4	3	2	3	1	5	5	6	4	13	11	12	10	9	8	9	7	12	11	13	10
9	4	5	6	5	4	6	5	5	1	3	3	2	10	12	13	11	10	13	12	11	7	9	9	8
10	5	4	5	6	6	4	5	5	3	1	2	3	12	10	11	13	13	10	11	12	9	7	8	9
11	6	5	4	5	5	5	4	6	3	2	1	3	13	11	10	12	12	11	10	13	9	8	7	9
12	5	6	5	4	5	5	6	4	2	3	3	1	11	13	12	10	11	12	13	10	8	9	9	7
13	7	8	9	9	10	12	11	13	10	12	13	11	1	14	15	15	4	16	16	17	4	16	17	16
14	8	7	9	9	12	10	13	11	12	10	11	13	14	1	15	15	16	4	17	16	16	4	16	17
15	9	9	7	8	11	13	10	12	13	11	10	12	15	15	1	14	16	17	4	16	17	16	4	16
16	9	9	8	7	13	11	12	10	11	13	12	10	15	15	14	1	17	16	16	4	16	17	16	4
17	10	11	12	13	7	9	8	9	10	13	12	11	4	16	16	17	1	15	14	15	4	17	16	16
18	11	10	13	12	9	7	9	8	13	10	11	12	16	4	17	16	15	1	15	14	17	4	16	16
19	12	13	10	11	8	9	7	9	12	11	10	13	16	17	4	16	14	15	1	15	16	16	4	17
20	13	12	11	10	9	8	9	7	11	12	13	10	17	16	16	4	15	14	15	1	16	16	17	4
21	10	11	13	12	10	13	11	12	7	9	9	8	4	16	17	16	4	17	16	16	1	15	15	14
22	11	10	12	13	13	10	12	11	9	7	8	9	16	4	16	17	17	4	16	16	15	1	14	15
23	13	12	10	11	11	12	10	13	9	8	7	9	17	16	4	16	16	16	4	17	15	14	1	15
24	12	13	11	10	12	11	13	10	8	9	9	7	16	17	16	4	16	16	17	4	14	15	15	1

(b)

System number	1	2	3	4	5	6	7	8	9	10	11	12
Schmid and Boas	B5	C5	D6	A6	B4	C3	D4	A3	B2	C1	D1	A2
Burgers vectors b	111	$11\bar{1}$	$1\bar{1}1$	$1\bar{1}\bar{1}$	111	$11\bar{1}$	$1\bar{1}1$	$1\bar{1}\bar{1}$	111	$11\bar{1}$	$1\bar{1}1$	$1\bar{1}\bar{1}$
Slip plane n	$1\bar{1}0$	$1\bar{1}0$	110	110	$10\bar{1}$	101	$10\bar{1}$	101	$01\bar{1}$	011	011	$01\bar{1}$
System number	13	14	15	16	17	18	19	20	21	22	23	24
Schmid and Boas	B5'	C5"	D6"	A6'	B4'	C3"	D4"	A3'	B2"	C1'	D1"	A2'
Burgers vectors b	111	$11\bar{1}$	$1\bar{1}1$	$1\bar{1}\bar{1}$	111	$11\bar{1}$	$1\bar{1}1$	$1\bar{1}\bar{1}$	111	$11\bar{1}$	$1\bar{1}1$	$1\bar{1}\bar{1}$
Slip plane n	$11\bar{2}$	112	$1\bar{1}\bar{2}$	$1\bar{1}2$	$1\bar{2}1$	$1\bar{2}\bar{1}$	121	$12\bar{1}$	$2\bar{1}\bar{1}$	$2\bar{1}1$	$21\bar{1}$	211

(c)

a_i	1	2	3	4	5	6	7	8	9
β_j	-	-	90.00	-	54.74	54.74	-	54.74	45.00
β_1	-	-	54.74	0.00	70.53	70.53	0.00	70.53	35.26
β_2	-	-	54.74	0.00	0.00	70.53	0.00	0.00	90.00
β_{12}	0.00	0.00	90.00	60.00	60.00	60.00	90.00	90.00	54.74
inter	self-hard	dip	junc	annihil	junc	junc	annihil	junc	junc
self-ener	1	2	3	4	5	6	4	5	7
mapping	-	-	a	Ω_1	b_1	c	Ω_2	b_2	d
a_i	10	11	12	13	14	15	16	17	
β_j	-	54.74	72.45	72.45	90.00	90.00	59.53	25.24	
β_1	0.00	70.53	79.98	29.50	90.00	39.23	28.56	58.52	
β_2	0.00	0.00	58.52	58.52	90.00	39.23	72.98	58.52	
β_{12}	30.00	30.00	73.22	73.22	70.53	48.19	80.41	33.56	
inter	annihil	junc	junc	junc	junc	junc	junc	junc	
self-ener	4	5	8	9	10	11	12	13	
mapping	Ω_3	b_3	e	f	g	h	i	j	

Interaction matrix, slip systems and symmetries in bcc crystals. (a):- The matrix coefficients represent the seventeen possible types of interactions between $\{110\}$ and $\{112\}$ slip planes (numbered from 1 to 12 and 13 to 24, respectively), of which thirteen are deduced from the consideration of the self-energy - (b): The indexes of all the slip systems and their denomination in the Schmid and Boas notation [7] are listed. - (c): For each coefficient of the interaction matrix, this Table gives the angles β_i (see text), the angle between interacting slip planes, β_{ij} and the nature of the interaction. The thirteen different configurations deduced from an energetic criterion and the fifteen types of reaction mappings shown in Fig. 1 are also listed.

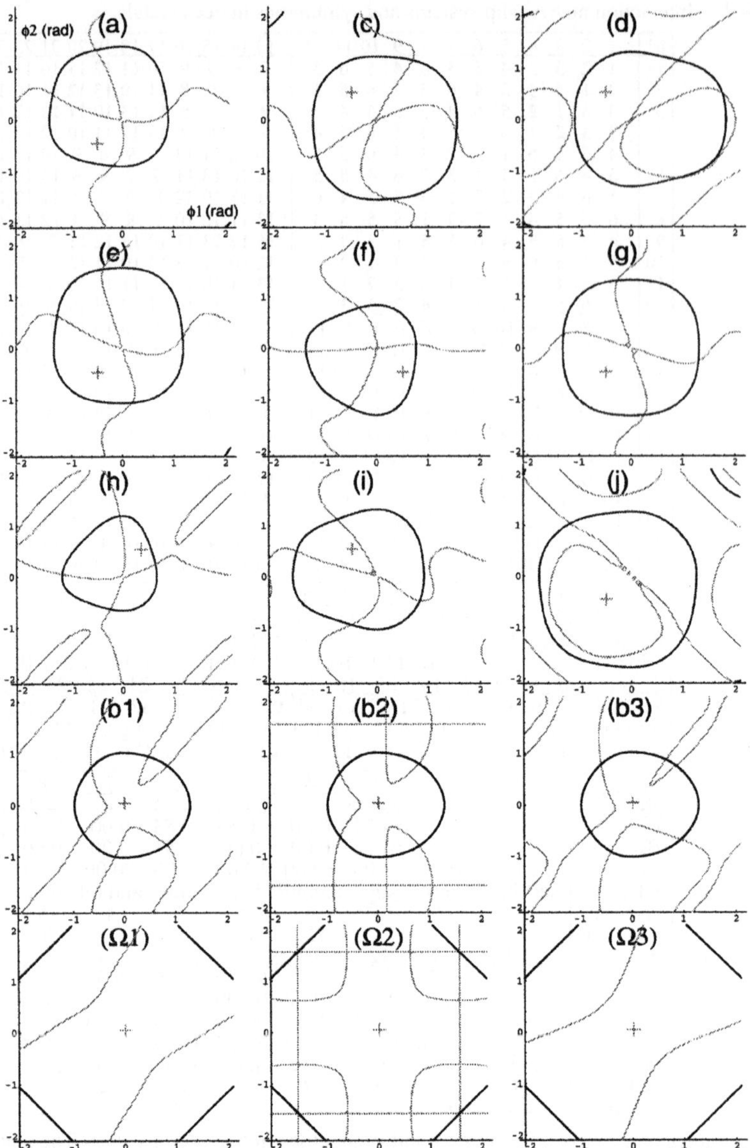

Figure 1. Reaction mappings in bcc crystals showing the calculated output for the fifteen possible reactions between dislocations in bcc crystals. ϕ_1 and ϕ_2 are the initial angles between the interacting segments and the direction of intersection of their slip planes. Only the central portion of the graphs is reproduced here. The label on each figure refers to Table 1, where the interacting slip systems are defined. Black curves: reaction lobes determined from an energy balance criterion. Gray curves: lines of zero initial interaction force between the segments, as determined from Kroupa's formula (see text). The sign (+) indicates a domain where the interaction force is attractive.

types of interactions and the corresponding values of the angular parameters.

For each reaction (i.e., omitting non-contact interactions like the self- and coplanar interactions), calculated angular reaction mappings are given in Fig. 1. These mappings show, in addition to other features discussed below, the domain of occurrence of the reactions for straight parent segments of same length and crossing each other at their midpoints, as calculated from an energy balance criterion. Neutral lines are plotted, which yield the border between the domain of attractive, junction-forming, configurations and repulsive crossings. The results are plotted as a function of the angles ϕ_1 and ϕ_2 between the parent dislocations and the direction of intersection of the slip planes. The angular conventions for these numerical investigations are detailed in [8]. With these conventions, two initially parallel segments with $\phi_1 = \phi_2 = 0$ are always attractive. The domain of junction formation takes the form of a lobe encircling the origin, plus its periodic duplicates (thick lines in Fig. 1, where only the central portion is shown). Since changing simultaneously the circulation along the dislocation lines does not modify the nature of the interaction, the set of lobes is periodic with period π along the two diagonals. One may notice, for instance by comparing Figs. 1-h and 1-j, that lobes may strongly differ in shape; this is due to the orientation-dependence of the line tension of the parent segments and of the <100> junction (cf. Eq. 1).

For the collinear interactions, the energy balance always favor annihilation, except along neutral lines along the two diagonals.

3. INITIAL INTERACTION FORCE AND BARRIER EFFECT

The intuitive notion that initially attractive dislocations make junctions and repulsive ones do not is sometimes in error. For instance, the so-called crossed-states, which were found by Wickham et al. [5] by DD simulations, are attractive configurations for which junction formation is energetically unfavorable [2, 6]. Thus, in addition to the energy balance criterion, it is necessary to consider an additional criterion based on the sign of the initial interaction force between the two interacting segments. A simplified analytical formulation for the interaction force along the shortest approach distance of two straight and infinite segments was given by Kroupa [9]:

$$\mathbf{F_{12}} \approx \frac{\mu}{|\xi_1 \times \xi_2|} \frac{\mathbf{R_{12}}}{R_{12}} \{ \frac{1}{2}(\mathbf{b_1}.\xi_1)(\mathbf{b_2}.\xi_2) - (\mathbf{b_1} \times \mathbf{b_2}).(\xi_1 \times \xi_2) + \frac{1}{1-\nu}[(\mathbf{b_1} \times \xi_1).\frac{\mathbf{R_{12}}}{R_{12}}][(\mathbf{b_2} \times \xi_2).\frac{\mathbf{R_{12}}}{R_{12}}] \}, \quad (2)$$

where the ξ_i are unit vectors along the dislocation lines of Burgers vectors $\mathbf{b_i}$ and μ and ν are, respectively, the shear modulus and the Poisson's coefficient.

$\mathbf{R}_{12}$ is the vector along the shortest approach distance between the two dislocations. The condition of null interaction force, which is drawn as gray curves in Fig. 1), delimitates a border between initially repulsive domains on the one hand and attractive states leading or not (crossed-states) to junction formation.

The self-energy balance and the initial interaction force are derived using different approximations and provide estimates for different quantities; the latter only deals with the initial state, whereas the former depends on both the initial and final states. For example, inside a lobe, one can find both attractive and repulsive initial configurations (cf. Fig. 1) and one cannot decide whether or not a junction can be formed. What is likely to happen is discussed in the next sections but, at this step, useful information can be drawn from this apparent indeterminacy. Since the force is the derivative of energy with respect to position, an initially repulsive interaction inside a junction lobe indicates the presence of a barrier to be overcome before reaching the final state of minimum energy. The physical origin of this barrier resides in the fact that the line tension opposes the bending of the dislocation lines under the effect of their interactions. With the collinear interaction, annihilation is predicted all through the reaction mapping because the self-energy balance is always favorable (except along the neutral lines). In contrast, as will be shown in Fig. 4 below, annihilation is not found to occur systematically. Thus, in this case, the effect of the repulsive barrier is the dominant one. In both cases, the calculations based on rigid dislocations cannot predict whether or not the elastic interactions are strong enough to overcome the line tension effects (cf. Fig. 2 for the reaction of initially repulsive dislocations).

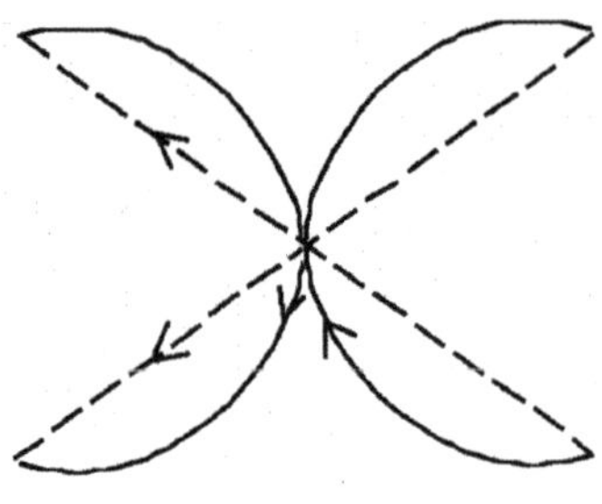

Figure 2. Barrier overcomming for initially repulsive dislocations. Two repulsive dislocations are in presence. The strongest initial repulsion is at the center, where the lines are the closest. The lines tend to avoid each other. As they are pinned at their ends, the local rotations at the center can be sufficiently large to bring them locally into an attractive angular region and reaction becomes possible. Thus, the barrier to be overcome originates in the line tension, which opposes the bending of the lines.

According to Eq. (2), the barrier effect involves, like for the self-energy criterion, the angles β_1 and β_2 between the Burgers vectors of the two parent dislocations and the direction of intersection of their slip planes, but also the angle between the two slip planes β_{12}. As a result, two configurations determined from the self-energy criterion, the collinear interaction and one type of junction, become triply degenerate (cf. Table 1). Then, the final number of predicted configurations amounts to seventeen: twelve junctions, three collinear interactions, plus the self-hardening and dipolar interactions.

4. DD SIMULATIONS OF JUNCTIONS AND COLLINEAR ANNIHILATIONS

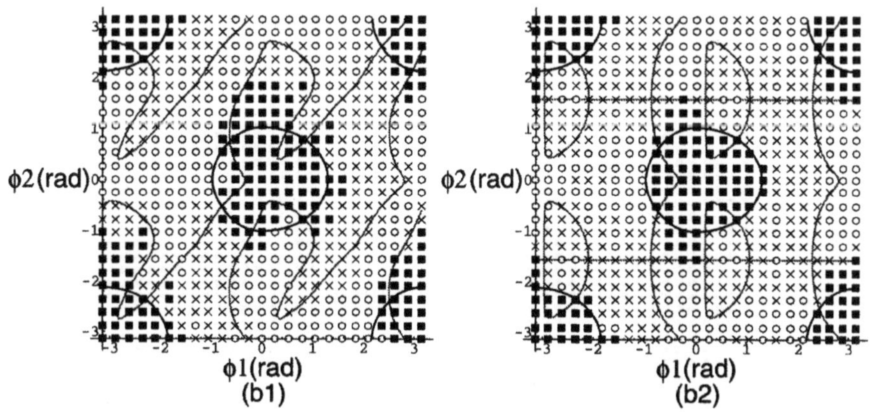

Figure 3. Simulation of the junction mappings b_1 and b_2, showing one period between $-\pi$ and $+\pi$, for the two angles ϕ_i between the dislocation lines and the intersection of the slip planes. Filled rectangles : junctions; crosses : crossed-states; empty circles : repulsive states. The predictions of the two simplified models (cf. Fig. 1) are superimposed to these results.

Out of the ten types of junctions determined from an energetic argument, one, labeled 5 in the line "self-energy" of Table 1, is split into three configurations corresponding to three different angles between the interacting slip planes (cf. the angles β_{12} associated with the matrix coefficients a_i labeled 5, 8 and 11). As shown in Fig. 1, the corresponding reaction mappings, b_1, b_2 and b_3, have identical junction lobes but different boundaries for neutral interaction force. These analytical results are now compared with those of DD simulations, which involve a rigorous treatment of the self-energy and self-interactions of the interacting segments, using the same methodology as in [8].

The result is illustrated in Fig. 3 by two reaction mappings b_1 ($\beta_{12} = \pi/3$) and b_2 ($\beta_{12} = \pi/2$) for the matrix coefficients 5 and 8.

For the mapping b_1, the elastic models predict rather well the boundaries between (attractive) crossed-states and repulsive states. Inside the lobes, one can see that the repulsive barriers can in a few cases prevent junction formation, whereas some junctions are found outside the lobes in attractive regions. For the mapping b_2, the simplified analytical approach yields only qualitative predictions. The reason is that, with slip planes making an angle $\beta_{12} = \pi/2$, the interactions between the two dislocations are weak and the results become more sensitive to other effects, especially line tension effects.

The collinear annihilation (energy case 4 in Table 1) is also split in three different cases (corresponding to $a_i = 4$, 7 and 10 and to the mappings Ω_1, Ω_2 and Ω_3). Then, as mentioned above, the mappings are essentially driven by the barrier effect. The influence of the additional angle β_{12} is rather strong, as can be checked by comparing the two simulated mappings Ω_1 ($\beta_{12} = \pi/3$) and Ω_2 ($\beta_{12} = \pi/2$) of Fig. 4.

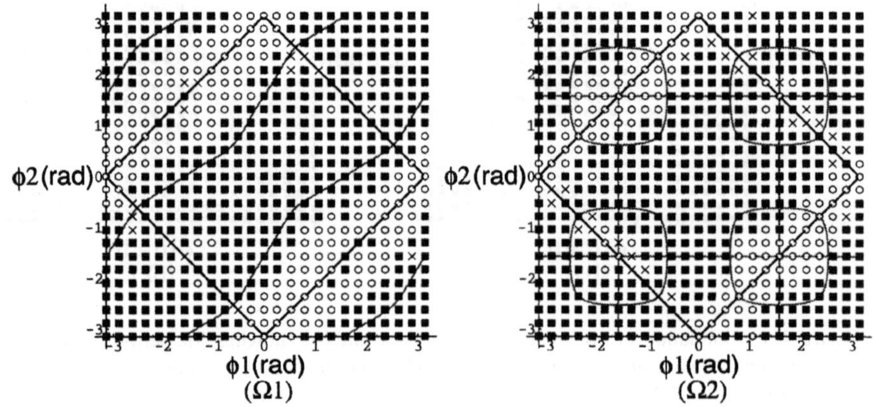

Figure 4. Simulation of the collinear anihilation mappings Ω_1 and Ω_2. Same conventions as in Fig. 3

In addition to the periodicity common to all mappings (cf. section 2), the collinear interactions have one additional symmetry, with respect to the two diagonals, due to invariance with respect to an exchange between the two angles ϕ_1 and ϕ_2 and because the Burgers vectors b_1 and b_2 are collinear (see Fig. 4). A similar symmetry, but with respect to one diagonal only, can also be found for junctions when $\beta_1 = \beta_2$, as is the case for mappings (a), (c), (g), (h) and (j) in Fig. 1. Finally, when the slip planes are perpendicular and one of the two β_i is zero (cases b_2 with $\beta_2 = 0$), the other axis ϕ_j is also a symmetry axis for the mapping. For the collinear annihilations Ω_2 (with $\beta_1 = \beta_2 = 0$) there is symmetry with respect to the two axes.

5. HIGH TEMPERATURE HARDENING

DD simulations have now reached a stage where they are able to tackle such questions as dislocation microstructure and forest hardening in bulk single crystals [10, 11]. As far as forest hardening is concerned, however, to obtain a quantitative insight into this mechanism it is necessary to take into account the multiplicity of the active slip systems and the variety of their interactions. This is formally achieved for fcc metals, as well as for bcc metals above the athermal temperature, by first defining an interaction matrix for the pair interactions between slip planes. The coefficients of this matrix are, for bcc metals, those given in Table 1-a. They are now written in the form a^{su}, as they express the hardening of system (s) by a density of forest dislocations ρ^u in system (u). In a second step, the critical stress for the onset of plastic flow in system (s), τ_c^s, is written [12]:

$$\tau_c^s = \mu b (a^{su} \rho^u)^{1/2}. \tag{3}$$

This relationship is valid only in conditions where forest interactions govern the flow stress. It does not apply to bcc crystals in the low temperature domain, i.e., in the presence of a lattice friction. The effective connection between the mesoscopic scale and the continuum approaches of plasticity provided by Eq. (3) is now established for fcc crystals. Of the six coefficients needed to describe forest hardening in fcc crystals, five were measured by DD simulations and found in good agreement with experimental results. The sixth one corresponds to the much less well-known collinear interaction. Actually, it is found to be, by far, the strongest of all and this may have several interesting consequences that are now under investigation [13].

Regarding bcc metals, the current state of affairs is less advanced. A first step was carried out in the present work by defining the symmetries of the interaction matrix. The latter can be split into three sub-matrices, one for the interactions between 1/2<111>{110} systems, one for those of 1/2<111>{112} systems and the last one for the mixed interactions between the two sets of systems. The next step, which is under way, consists in measuring the seventeen related matrix coefficients. Preliminary results on the interactions between 1/2<111>{110} systems confirm, in this case too, the very large strength of the collinear interaction as compared to other reactions. As there are three types of collinear interactions in bcc crystals, in contrast to fcc crystals where there is only one, their respective strengths remain to be checked.

6. CONCLUSION

The present study establishes the symetries of the interaction matrix between slip systems in bcc crystals, assuming that slip takes place in {110} and {112} planes. Seventeen different types of interactions were determined, one for self-hardening, one for the dipolar interaction, twelve for junction formation and three for the collinear interaction. In some cases, and notably for the collinear interaction, both elastic models and DD simulations show that rather different reaction mappings are obtained when the angle between the interacting slip planes is modified, all other geometrical parameters being fixed. This may entail different strengths for the corresponding interactions. Measuring the coefficients of the interaction matrix is now the next step leading to a physical model for the strain hardening of bcc crystals at high temperatures.

Acknowledgments

The authors wish to thank Dr. B. Devincre for many helpful discussions. RM acknowledges the support of the Centre Européen de Calcul Atomique et Moléculaire (CECAM).

References

[1] G. Saada, *Acta Metall* **8**: 841-847 (1960)

[2] J. Friedel, Dislocations (Pergamon Press, Oxford, 1964)

[3] G. Schoeck and R. Frydmann, *Phys. Stat. Sol.(b)* **53**: 661 (1972)

[4] W. Püschl, *Phys. Stat. Sol.* **90**, 181 (1985)

[5] L.K. Wickham, K. Schwarz and J.S. Stölken, *Phys. Rev. Lett.* **83**, 4574 (1999)

[6] R. Madec, B. Devincre, L.P. Kubin, *Computational Materials Science* **23**: 219-224 (2002)

[7] P. Franciosi, *Acta metall* **31**: 1331-1342 (1983)

[8] L.P. Kubin, R. Madec, B. Devincre, *in "Multiscale Phenomena in Materials", 2003 MRS Spring Meeting*, Eds. HM Zbib et al., in press

[9] J. Hirth and L. Lothe, Theory of Crystal Dislocations (Krieger, Malabar, Florida, 1982), p 123

[10] R. Madec, B. Devincre, L.P. Kubin, *Phys. Rev. Lett.* **89**:255508 (2002)

[11] R. Madec, B. Devincre, L.P. Kubin, *Scripta Materialia* **47**: 689-695 (2002)

[12] P. Franciosi, M. Berveiller and A. Zaoui, *Acta metall* **28**: 273 (1980)

[13] R. Madec, B. Devincre, L.P. Kubin, T. Hoc and D. Rodney, *submitted*

DISCRETE DISLOCATION PREDICTIONS OF SINGLE CRYSTAL FATIGUE CRACK GROWTH

V.S. Deshpande
Cambridge University, Department of Engineering,
Trumpington Street, Cambridge CB2 1PZ, UK
vsd@eng.cam.ac.uk

A. Needleman
Division of Engineering, Brown University,
Providence, RI USA.
needle@engin.brown.edu

E. Van der Giessen
University of Groningen, Department of Applied Physics,
Nyenborgh 4, 9747 AG Groningen, The Netherlands.
Giessen@phys.rug.nl

Abstract A framework for analysis of crack growth under cyclic loading conditions is discussed where plastic flow arises from the motion of large numbers of discrete dislocations and the fracture properties are embedded in a cohesive surface constitutive relation. The formulation is the same as used to analyze crack growth under monotonic loading conditions, differing only in the remote loading being a cyclic function of time. Fatigue, i.e. crack growth in cyclic loading at a driving force for which the crack would have arrested under monotonic loading, emerges in the simulations as a consequence of the evolution of internal stresses associated with the irreversibility of the dislocation motion. The predictions for the qualitative features of fatigue crack growth are in remarkable accord with experimental observations.

Keywords: Dislocations; mechanical properties, fatigue, plastic; computer simulation

H. Kitagawa and Y. Shibutani (eds.), IUTAM Symposium on Mesoscopic Dynamics of Fracture Process and Materials Strength, 79–86.

1. INTRODUCTION

The essence of fatigue crack growth is that it occurs even when the driving force for crack growth is much smaller than what is needed for the same crack to grow under monotonic loading conditions. Consider a cracked solid subject to loading corresponding to a stress intensity cycling between K_{min} and K_{max}, fatigue crack growth occurs even when K_{max} is much smaller than the value of K needed for the same crack to grow under monotonic loading conditions. Typically, there is a threshold value of $\Delta K_{\text{I}} = K_{\text{max}} - K_{\text{min}}$ below which cracks do not grow at a detectable rate. Above this threshold value, in the regime where the amount of crack growth per cycle, da/dN, is on the order of a few lattice spacings, there is a steep increase in da/dN with ΔK_{I}. For larger values of ΔK_{I}, the increase in da/dN becomes less steep and the Paris law regime [1] is entered where $da/dN \propto (\Delta K_{\text{I}})^m$, see [2].

Fatigue requires irreversibility. Fatigue cannot occur in an elastic system because the state of the system then only depends on the current value of the loading parameter and not on its history; crack growth in an elastic system either occurs in the first cycle or it does not occur at all. Dissipative mechanisms are key for fatigue. As a consequence, for crystalline metals, both the plastic flow mechanism and the process of material separation play important roles in determining the fatigue behavior.

As noted by Cleveringa *et al.* [3], dislocations play a dual role in the fracture process under monotonic loading. On the one hand, plastic flow caused by the motion of dislocations delays crack initiation and increases the resistance to crack growth. On the other hand, it is the local stress concentrations associated with discrete dislocations in the vicinity of the crack tip that leads to stress levels of the magnitude of the cohesive strength, causing the crack to propagate. This dual role is key for fatigue in crystalline metals – the dissipation from dislocation motion provides the irreversibility, while the high stresses associated with the dislocation structures that form near the crack tip precipitate crack growth.

Here, we present results from our series of analyses of crack growth under cyclic loading conditions in Deshpande *et al.* [4, 5, 6, 7]. Plastic flow arises from the motion of large numbers of discrete dislocations, which are treated as singularities in an isotropic elastic solid. The material model is independent of the presence of a crack. The fracture properties of the material are embedded in a cohesive surface constitutive relation, [8], so that crack initiation and crack growth are stress as well as deformation driven. A key aspect of the formulation is that the plastic stress versus strain response and the evolution of the dislocation structure, as well as crack initiation and growth are outcomes of the solution of the boundary value problem. Furthermore, the only distinction between an analysis of monotonic crack growth and fatigue crack growth is that in fatigue the remote loading is specified to be an oscillating function of time.

2. THEORY

A brief overview of the theoretical framework is presented; background and further descriptions are given in [4, 5, 6, 7] and references cited therein. Initially, the crystal is assumed to be free of mobile dislocations, but to contain a random distribution of dislocation sources and point obstacles. The rules for dislocation nucleation and motion are based on those proposed in [9] and use the Peach-Koehler force as the driving force. The sources mimic Frank-Read sources and generate a dislocation dipole when the magnitude of the Peach-Koehler force exceeds a critical value for a specified period of time. The obstacles, which represent small precipitates or forest dislocations, pin dislocations and release them once the Peach-Koehler force attains a specified obstacle strength. Annihilation of two dislocations with opposite Burgers vector occurs when they approach each other within a critical annihilation distance. Dislocation motion is assumed to occur only by glide with no cross slip. The magnitude of the glide velocity $v^{(k)}$ of dislocation k is taken to be linearly related to the Peach-Koehler force $f^{(k)}$ through the drag relation $f^{(k)} = Bv^{(k)}$. There is no special dislocation nucleation from the crack tip.

In [4, 5, 7], loading is prescribed in terms of displacements corresponding to the isotropic elastic mode I singular field remote from the crack tip while in [6] remote uniaxial tension is imposed. There is a single cohesive surface [8] that lies in front of the initial crack. At each time step, an increment of the remote loading (the mode I stress intensity factor $\dot{K}_I \Delta t$ for small scale yielding) is prescribed. At the current instant, the stress and strain state of the body is known, and the Peach-Koehler forces on all dislocations can be calculated. On the basis of these forces the dislocation structure is updated, which involves the motion of dislocations, the generation of new dislocations, their mutual annihilation, their pinning at obstacles, and their exit into the open crack. After this, the new stress and strain state can be determined.

The field quantities, i.e. the displacement u_i, the strain ϵ_{ij} and the stress σ_{ij} are determined using the superposition method in [10],

$$u_i = \tilde{u}_i + \hat{u}_i\,, \quad \epsilon_{ij} = \tilde{\epsilon}_{ij} + \hat{\epsilon}_{ij}\,, \quad \sigma_{ij} = \tilde{\sigma}_{ij} + \hat{\sigma}_{ij}\,. \tag{1}$$

The ($\tilde{\ }$) fields are the superposition of the singular fields of the individual dislocations in their current configuration while the ($\hat{\ }$) fields represent image fields that correct for the actual boundary conditions. The sum of the ($\tilde{\ }$) and the ($\hat{\ }$) fields in (1) gives the solution that satisfies all boundary conditions. Since the ($\hat{\ }$) fields are smooth in the region of interest, the boundary value problem for them can conveniently be solved using a finite element method.

Both reversible and irreversible cohesive traction-displacement relations are used. As the cohesive surface ahead of the crack separates, the magnitude of the traction increases, reaches a maximum and then approaches zero with increasing separation. In a vacuum, there is no oxidation of the newly formed surface and it is expected that this relation is followed in a reversible manner. When the newly formed surfaces oxidize, the cohesive relation will not be followed in a reversible manner. The effect of the formation of the oxide layer and the subsequent surface contact during unloading is modeled by specifying unloading from and reloading towards the monotonic cohesive law to occur according to a linear incremental unloading relation but with no degradation in strength with continued cycling.

A reference stress intensity factor K_0 is introduced that provides a convenient normalization for the imposed stress intensity factor. It is defined in terms of the work of separation of the cohesive surface, ϕ_n, by

$$K_0 = \sqrt{\frac{E\phi_n}{1-\nu^2}}, \tag{2}$$

Crack growth in an elastic solid with the given cohesive properties takes place at $K_I/K_0 = 1$.

3. DISCRETE DISLOCATION PREDICTIONS

In the small scale yielding calculations in [4, 5, 7], the applied stress intensity is varied between $K_{\rm min}$ and $K_{\rm max}$ with a rather high loading rate to shorten the computation time.

Fatigue threshold results from [4], summarized in Fig. 1. Crack growth under cyclic loading occurs if and only if (i) the cyclic amplitude ΔK_I exceeds a critical value $\Delta K^*_{\rm th}$, and (ii) the maximum stress intensity $K_{\rm max}$ exceeds a critical value $K^*_{\rm max}$.

With a reversible cohesive constitutive relation, which models conditions in a vacuum, this can be rationalized as follows: For sufficiently low $K_{\rm max}$, no dislocations are generated and the system is elastic. Therefore, for fatigue to occur with a reversible cohesive law, $K_{\rm max}$ must exceed some minimum $K_{\rm max}$ denoted by $K^*_{\rm max}$. For $K_{\rm max} >> K^*_{\rm max}$, interac-

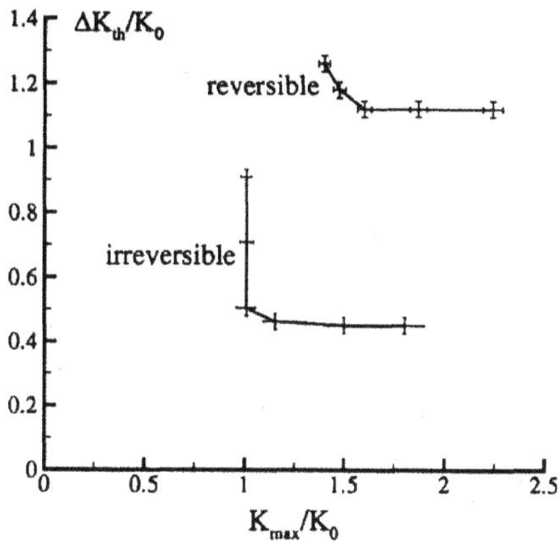

Figure 1. Discrete dislocation predictions for the variation of $\Delta K_{\rm th}$ with $K_{\rm max}$. From [4].

tions within the now dense dislocation structure act to retard dislocation motion. Accordingly, a minimum cyclic stress intensity factor range ΔK_I is needed to induce dislocation motion during unloading and reloading. Thus, in this regime, a critical $\Delta K^*_{\rm th}$ is needed. For an irreversible cohesive relation, which models conditions in an oxidizing environment, contact plays an important role [4].

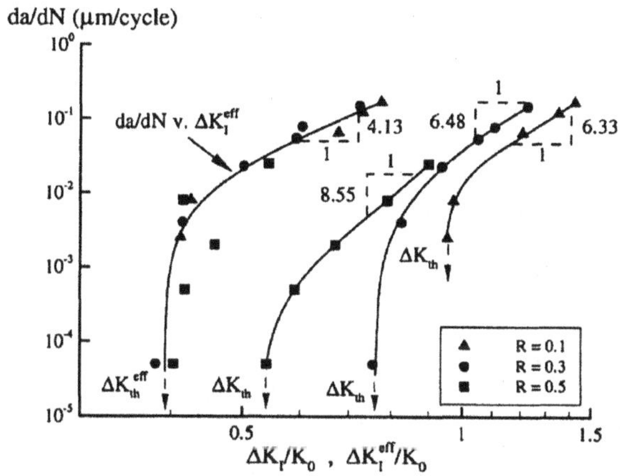

Figure 2. The cyclic crack growth rate da/dN versus $\Delta K_{\rm I}/K_0$ and $\Delta K^{\rm eff}_{\rm I}/K_0$ for an interface crack. From [5].

The form of the $\log(da/dN)$ versus $\log(\Delta K_{\rm I})$ curve seen experimentally, with a threshold and a Paris law regime, is captured in Fig. 2. The effective stress intensity range ΔK^{eff} responsible for crack growth is defined by $K_{\text{max}} - K_{\text{op}}$ where K_{op} is the stress intensity factor at which the crack faces first separate at the current location of the crack upon reloading. The effect of crack closure is more pronounced at the lower values of ΔK_{I} so that ΔK_{th} is sensitive to the load ratio $R = K_{\text{min}}/K_{\text{max}}$ and $\Delta K_{\text{th}}^{\text{eff}}$ is much less than ΔK_{th}.

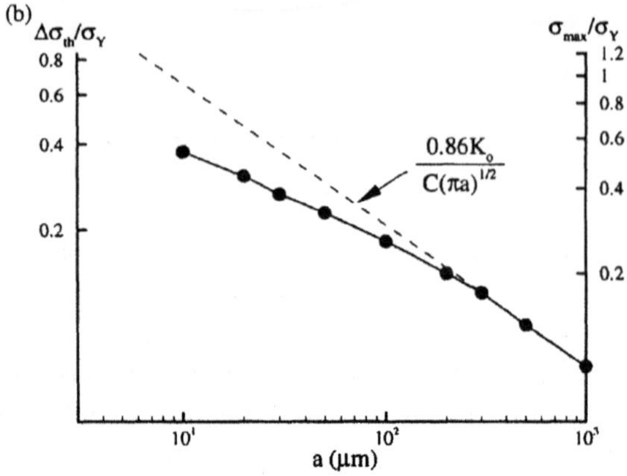

Figure 3. The fatigue threshold, $\Delta\sigma_{\text{th}}/\sigma_Y$, versus crack length a (corresponding values of $\sigma_{\text{max}}/\sigma_Y$ are shown on the right axis). For long cracks $\Delta K_{\text{th}} \approx 0.86K_0$ with deviation from ΔK-governed behavior seen for crack lengths less than 300μm. From [6].

The results of fatigue threshold calculations carried out in [6] for geometrically similar edge cracked specimens are shown in Fig. 3 which are identified by the edge crack length, a. For crack lengths less than 300μm, the deviation from ΔK-governed fatigue increases with decreasing crack size, with the fatigue threshold for smaller cracks tending to be $\Delta\sigma$ rather than ΔK-governed. A consequence is that crack growth under cyclic loading conditions occurs even when the remote applied K_{max} is less than the stress intensity K_0 at which the crack would grow in an elastic solid. However, values of $K_{\text{max}}/K_0 < 1$ are consistent with $\sigma_{\text{max}}/\sigma_Y < 1$ for small cracks. To illustrate this, neglect constants of order unity and write the ratio K_{max}/K_0 as

$$\frac{K_{\text{max}}}{K_0} \approx \left(\frac{\sigma_{\text{max}}}{\sigma_Y}\right) \frac{\sqrt{a/\delta_n}}{\sqrt{E\sigma_{\text{coh}}/\sigma_Y^2}} = \sigma_{\text{max}} \frac{\sqrt{a/\delta_n}}{\sqrt{E\sigma_{\text{coh}}}} . \tag{3}$$

Here, σ_{coh} is the cohesive strength, σ_Y is the yield strength and δ_n is a cohesive characteristic length such that the work of separation is $\propto \sigma_{\text{coh}}\delta_n$. For a 10 micron crack size, the calculations give a value of the fatigue threshold of $\sigma_{\max}/\sigma_Y \approx 0.5$. With $E/\sigma_Y \approx 10^3$, $\sigma_{\rm coh}/\sigma_Y \approx 10$, and $a/\delta_n \approx 10^4$, (3) implies $K_{\max}/K_0 \approx \sigma_{\max}/\sigma_Y < 1$. Given $\sigma_{\max}$, the value of $K_{\max}/K_0$ in (3) is independent of the flow strength σ_Y. Also, $K_{\max}/K_0$ in (3) only depends on $\sigma_{\rm coh}$ and δ_n through their product, and the work of separation is of the order of 1 J/m^2 for a wide range of materials.

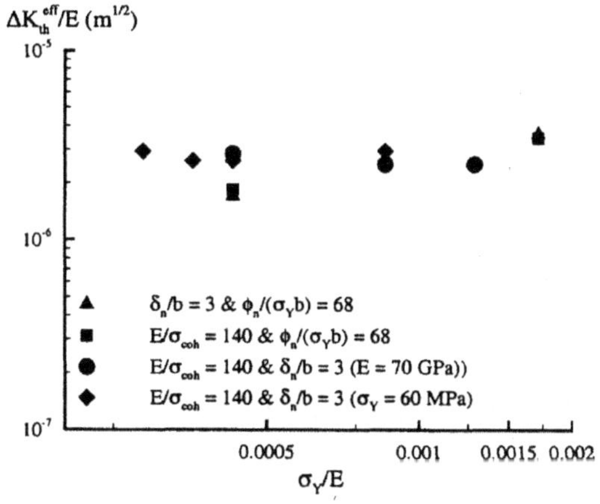

Figure 4. Discrete dislocation predictions showing that similar to experiment, $\Delta K_{\rm th}^{\rm eff}$ is relatively independent of the yield strength σ_Y and scales approximately linearly with Young's modulus E. From [7].

At least in the near-threshold and Paris law regimes, fatigue crack growth rates are relatively independent of the yield strength of the material but scale with the elastic modulus. This rather surprising observation has been borne out in experimental studies on a variety of metallic alloys. Results for $\Delta K_{\rm th}^{\rm eff}$ from [7] are shown in Fig. 4. Consistent with experimental data, the calculations show that $\Delta K_{\rm th}^{\rm eff}/E$ is rather independent of the normalized strength σ_Y/E over approximately a decade. The results in [7] show that the observed relative lack of dependence of the fatigue threshold in ductile metals on yield strength emerges from a cohesive fracture model with the stress concentration arising from near crack tip organized dislocation structures. In contrast, conventional continuum plasticity models predict that fatigue crack growth is sensitive to the value of the yield strength.

4. CONCLUSIONS

Results of plane strain analyses of crack growth under cyclic loading conditions have been discussed where plastic flow arises from the motion of large numbers of dislocations. The only difference between the analyses for monotonic and cyclic crack growth is in the specification of the applied loading.

- Crack growth occurs under cyclic loading conditions when the driving force is smaller than what is needed for the crack to grow under monotonic loading conditions.

- The results in Refs. [4, 5, 6, 7] show that the fatigue threshold, Paris law behavior, striations, the accelerated growth of short cracks as well as the observed scaling of the fatigue threshold and of fatigue crack growth rates in the near-threshold regime emerge naturally from a unified framework where plastic flow arises from the motion of large numbers of discrete dislocations and the fracture properties are embedded in a cohesive surface constitutive relation.

Acknowledgments

Support from the AFOSR MURI at Brown University on *Virtual Testing and Design of Materials: A Multiscale Approach* (AFOSR Grant F49620-99-1-0272) is gratefully acknowledged.

References

[1] Paris P.C., Gomez M.P., Anderson W.P. A rational analytic theory of fatigue. Trend Engin 1961; 13:9-14.

[2] Suresh S., *Fatigue of Materials*, Cambridge: Cambridge University Press, 1991.

[3] Cleveringa H.H.M., Van der Giessen E, Needleman A. A discrete dislocation analysis of mode I crack growth. J Mech Phys Solids 2000; 48:1133-57.

[4] Deshpande, V.S., Needleman, A., Van der Giessen, E. A discrete dislocation analysis of near-threshold fatigue crack growth. Acta Mat 2001; 49:3189-3203.

[5] Deshpande, V.S., Needleman, A., Van der Giessen, E. Discrete dislocation modeling of fatigue crack propagation. Acta Mat 2002; 50:831-46.

[6] Deshpande, V.S., Van der Giessen, E., Needleman, A. Discrete dislocation plasticity modeling of short cracks in single crystals. Acta Mat 2003; 51:1-15.

[7] Deshpande, V.S., Needleman, A., Van der Giessen, E. Scaling of discrete dislocation predictions for near-threshold fatigue crack growth. Acta Mat 2003; to be published.

[8] Needleman A. A continuum model for void nucleation by inclusion debonding. J Appl Mech 1987; 54:525-31.

[9] Kubin L.P., Canova G., Condat M., Devincre B., Pontikis V., Bréchet Y. Dislocation microstructures and plastic flow: a 3D simulation. Solid State Phenomena 1992; 23-24:455-72.

[10] Van der Giessen E., Needleman A. Discrete dislocation plasticity: a simple planar model. Modeling Simul Mater Sci Eng 1995; 3:689-735.

A CRYSTAL PLASTICITY ANALYSIS FOR ACCUMULATIONS OF GEOMETRICALLY NECESSARY DISLOCATIONS AND DIPOLES AROUND SHEAR BAND

Yoshiteru AOYAGI* and Kazuyuki SHIZAWA**
** Graduate School of Science and Technology, Keio University*
*** Department of Mechanical Engineering, Keio University*

Abstract: In this paper, generalized geometrically necessary (GN) dislocation density and GN incompatibility are newly defined by extending Kröner's dislocation density and incompatibility tensors. A new model of dislocation-crystal plasticity coupling a deformation field with a dislocation field is developed by introducing these dislocation densities into the hardening modulus matrix of crystal plasticity through the Bailey-Hirsch relation. The corrected Seeger model for the dislocation mean free path is extended to the third stage of strain hardening. Moreover, a finite element simulation is carried out for an f.c.c. single crystal under plane strain tension. It is numerically predicted that the shear band can be regarded as a subgrain.

Key words: Dislocation-crystal plasticity, Geometrically necessary dislocation density, Incompatibility, Dipole, Geometrical nonlinearity, Dislocation mean free path, Finite element method, Shear band, Subgrain.

1. INTRODUCTION

From a microscopic viewpoint, plastic deformation and work-hardening of crystals are caused by dislocation motions and dislocation accumulations, respectively. Recently, crystal plasticity studies with the dislocation information have actively been done by many researchers [1]-[5]. However, in the conventional theories using geometrically necessary (GN) dislocation density, the GN dislocation generating long-range stress fields can not express other crystal defects such as dislocation pairs and so on. A method

H. Kitagawa and Y. Shibutani (eds.), IUTAM Symposium on Mesoscopic Dynamics of Fracture Process and Materials Strength, 87–96.

adding statistically stored (SS) dislocation density to the GN dislocation density is proposed to represent these defects associated with short-range stress fields [6]. In this case, two kinds of quantities belonging to different categories, i.e., geometry and statistics are mixed in a hardening law of crystal plasticity. Therefore, when we use the GN dislocation density, it is desirable to express all defects as geometrical quantities in a uniform manner by introducing some geometrical quantity which plays a role of the dislocation density associated with short-range stress fields instead of the SS dislocation density. According to the differential geometry of crystal defects, dislocation density and incompatibility tensors correspond to the GN dislocation and the other crystal defects, respectively. Not only dislocation density but also incompatibility should be taken into account from the geometrical viewpoint of crystal defects.

In this paper, a multiscale model of crystal plasticity is proposed by considering the GN dislocation density and incompatibility from a completely geometrical standpoint. Geometrical nonlinearities of these defects are introduced into Kröner's dislocation density and incompatibility tensors in order to predict large deformations of a crystal precisely, and the generalized GN dislocation density and GN incompatibility are newly defined. A new model of dislocation-crystal plasticity coupling a deformation field with a dislocation field is developed by introducing these dislocation densities into the hardening modulus matrix of crystal plasticity through the Bailey-Hirsch relation. A transition of stress response to the third stage of the hardening curve of a single crystal is represented by increasing the dislocation mean free path after start of dynamic recovery. Furthermore, a finite element dislocation-crystal plasticity simulation for an f.c.c. single crystal under plane strain tension is carried out in order to verify the validity of this model. Distributions of GN dislocation density and GN incompatibility are visualized. It is numerically predicted that the shear band can be regarded as a subgrain.

2. DEFINITIONS OF DISLOCATION DENSITIES

2.1 Dislocation density and incompatibility

Kröner [7] defines a dislocation density tensor $\boldsymbol{\alpha}_K$ and an incompatibility tensor $\boldsymbol{\eta}_K$, such that

$$\boldsymbol{\alpha}_K \equiv -\mathrm{curl}\,\boldsymbol{\gamma} \tag{1}$$

$$\boldsymbol{\eta}_K \equiv -(\mathrm{curl}\,\boldsymbol{\alpha}_K^T)_S \tag{2}$$

where $(\,)_S$ denotes the symmetric part of a second order tensor and $\boldsymbol{\gamma}$ the slip tensor. The incompatibility tensor implies almost all crystal defects other than dislocation, i.e., dipole, vacancy and disclination. When we compare the above theory with the theory of non-Riemannian plasticity [8], $\boldsymbol{\alpha}_K$ is equivalent to torsion of differential geometrical space, and expresses line defects such as a dislocation generating a long-range stress field due to isolated dislocations accumulating at an inclusion or on a grain boundary; while, $\boldsymbol{\eta}_K$ corresponds to Riemann-Christoffel curvature, and expresses the other defects associated with the short-range stress field caused by latent obstacles such as forest dislocations. The following equations are obtained in consideration of geometrical nonlinearities originated in a large deformation of a crystal [9].

$$\boldsymbol{\alpha}_K \equiv -\mathrm{curl}\,\boldsymbol{\gamma} + 2\,\mathrm{grad}\,\boldsymbol{\gamma} \dot{\times} \boldsymbol{\gamma}_S \qquad (3)$$

$$\boldsymbol{\eta}_K \equiv -(\mathrm{curl}\,\boldsymbol{\alpha}_K^T)_S + 2(\mathrm{grad}\,\boldsymbol{\alpha}_K^T \dot{\times} \boldsymbol{\gamma}_S)_S \qquad (4)$$

The second terms in the right side of Eq. (3) and Eq. (4) are geometrical nonlinearity terms. Usually, the dislocation density is defined by the total length of dislocation line per unit volume. Although Kröner's dislocation density and incompatibility tensors express the crystal defects surely, it is not clear whether these can be directly treated as dislocation density or not. Then, in order to regard these tensors as two types of dislocation densities in the same dimension, we define new dislocation density tensors, extending Kröner's dislocation density and incompatibility tensors so that these quantities are consistent with a concept of the "Geometrically Necessary quantities" proposed by Ashby [10].

2.2 Generalized GN dislocation density

Figure 1 shows a local coordinate system (x_1, x_2, x_3) whose plane x_1x_3, axis x_1 and unit basis coincide with a slip plane, a slip direction and crystal basis $\boldsymbol{s}^{(\alpha)}$, $\boldsymbol{m}^{(\alpha)}$ and $\boldsymbol{t}^{(\alpha)} = \boldsymbol{s}^{(\alpha)} \times \boldsymbol{m}^{(\alpha)}$, respectively. Ashby [10] defines the edge and screw components of GN dislocation density as

$$\rho_s \tilde{b} = \partial \gamma / \partial x_3 \qquad (5)$$

$$\rho_e \tilde{b} = -\partial \gamma / \partial x_1 \qquad (6)$$

where γ and $\tilde{b}$ denote the slip and the magnitude of the Burgers vector. In the local coordinate system, since nonzero component of slip tensor $\boldsymbol{\gamma}$ is

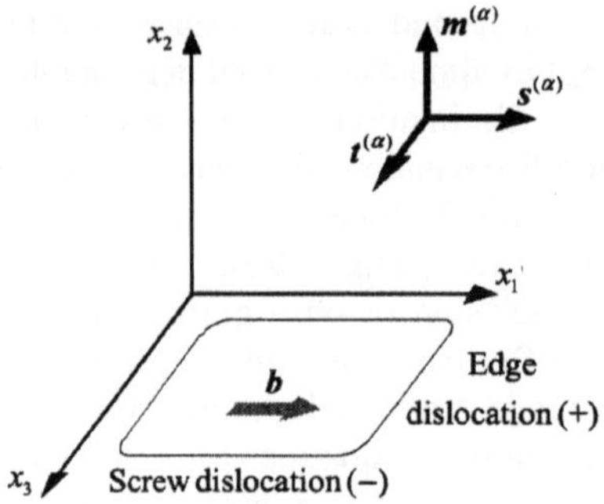

Figure 1. A local coordinate system and crystal bases

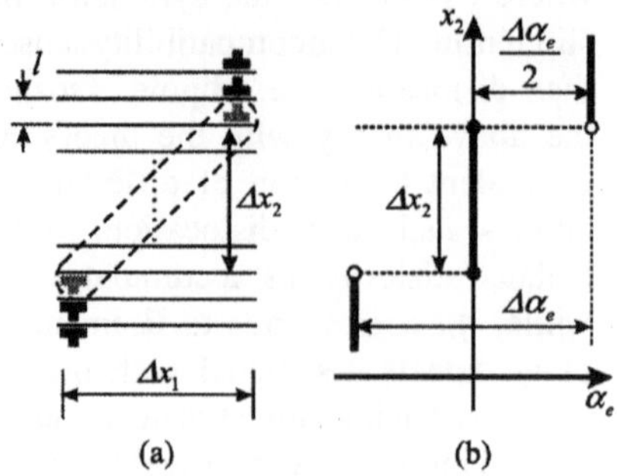

Figure 2. Distribution of GN dislocation density

only $\gamma \equiv \gamma_{12}$, two components of the dislocation density tensor remain, such that

$$\alpha_{K(I)11} = \partial\gamma / \partial x_3 \tag{7}$$

$$\alpha_{K(I)13} = -\partial\gamma / \partial x_1 \tag{8}$$

From Eqs. (5)-(8), GN dislocation density can be extended to

$$\boldsymbol{\alpha} \equiv \frac{1}{\tilde{b}} \boldsymbol{\alpha}_K \tag{9}$$

We may call $\boldsymbol{\alpha}$ the generalized GN dislocation density.

2.3 GN incompatibility

The number of edge dislocations n_e piercing the plane x_1x_2 in length Δx_1 on a slip plane is $n_e = l\,\Delta x_1\,\alpha_e$, where l and α_e are the distance of two adjacent slip planes and the density of edge dislocation, respectively. On the other hand, dislocation dipoles form a couple of positive and negative edge dislocations whose location is shown in Fig. 2(a). Therefore, the number of dipoles n_d existing in domain $\Delta x_1 \Delta x_2$ is equal to the number of dipolarized edge dislocations existing in domain $l\Delta x_1$. Noting that the dipoles are formed in the region where the sign of the edge dislocation density changes from positive to negative as in Fig. 2(b), we may write the positive edge dislocation density in the dipoles as $\Delta\alpha_e / 2$ because the difference of edge dislocation density $\Delta\alpha_e$ implies the density of a pair of dipolarized edge dislocations. Hence, the number of dipolarized positive edge dislocation n_d is expressed as

$$n_d = l\,\Delta x_1 \frac{\Delta\alpha_e}{2} \tag{10}$$

In the same way, the dipole density in domain $\Delta x_1 \Delta x_2$ is written in the following form by use of Eq. (10).

$$\alpha_d = \frac{n_d}{\Delta x_1\,\Delta x_2} = \frac{l}{2}\frac{\Delta\alpha_e}{\Delta x_2} \tag{11}$$

If we define α_d in the limit, the dipole density in infinitesimal volume element is defined by

$$\alpha_d = \lim \frac{l}{2}\frac{\Delta\alpha_e}{\Delta x_2} = \frac{l}{2}\alpha_{e,2} = \frac{l}{2}\frac{1}{\bar{b}}\alpha_{K13,2} \tag{12}$$

Component η_{K33} of the incompatibility tensor in Eq. (4) coincides with $\alpha_{K13,2}$ in Eq. (12). Therefore, the incompatibility tensor is extended to the following form.

$$\eta \equiv \frac{l}{2\bar{b}}\eta_K \tag{13}$$

We may call η the geometrically necessary incompatibility (GN incompatibility), since η exists at a local point in a crystal at which the gradient of dislocation density exists.

2.4 Component of dislocation densities for the plane strain condition

Here, we assume that slip γ is constant in direction x_3 in Fig. 1 for the plane strain condition. Since the gradient of slip in direction x_3 is always zero, the densities for a slip system in Eq. (9) and Eq. (13) are described as

$$\boldsymbol{\alpha}^{(\alpha)} = \alpha_e^{(\alpha)} \boldsymbol{s}^{(\alpha)} \otimes \boldsymbol{t}^{(\alpha)} \tag{14}$$

$$\boldsymbol{\eta}^{(\alpha)} = \eta_{de}^{(\alpha)} \boldsymbol{t}^{(\alpha)} \otimes \boldsymbol{t}^{(\alpha)} \tag{15}$$

where $(\ )^{(\alpha)}$ denotes a quantity for the slip system α. In Eq. (14), the component of edge dislocation density $\alpha_e^{(\alpha)}$ is given by

$$\alpha_e^{(\alpha)} = -\frac{1}{\tilde{b}}\gamma_{,1}^{(\alpha)} - \frac{1}{\tilde{b}}\gamma_{,2}^{(\alpha)}\gamma^{(\alpha)} \tag{16}$$

In Eq. (15), $\eta_{de}^{(\alpha)}$ denotes the pair of positive and negative edge dislocations on a different slip plane (dipole). This component is expressed by

$$\eta_{de}^{(\alpha)} = \frac{l}{2}\alpha_{e,2}^{(\alpha)} + l\alpha_{e,1}^{(\alpha)}\gamma^{(\alpha)} \tag{17}$$

Additionally, we describe the norms of the generalized GN dislocation density and GN incompatibility as

$$\rho_\alpha^{(\alpha)} = \left\|\alpha^{(\alpha)}\right\| \tag{18}$$

$$\rho_\eta^{(\alpha)} = \left\|\eta^{(\alpha)}\right\| \tag{19}$$

3. DISLOCATION-CRYSTAL PLASTICITY THEORY

A scheme of crystal plasticity theory is the same as from Peirce et al. [11]. The evolution equation of flow stress $g^{(\alpha)}$ for a crystal plasticity theory is usually adopted by

$$\dot{g}^{(\alpha)} = \sum_\beta h_{\alpha\beta}\left|\dot{\gamma}^{(\beta)}\right| \tag{20}$$

where $h_{\alpha\beta}$ is the hardening modulus representing the interaction of slip systems. In the same fashion to Ohashi [1], we differentiate Bailey-Hirsch's relation extended to multislip systems with respect to time and use the Orowan equation. Comparing Eq. (20) with the result, we find the hardening moduli $h_{\alpha\beta}$ as

$$h_{\alpha\beta} = \frac{1}{2}\frac{ac\mu\Omega_{\alpha\beta}}{L^{(\beta)}\sqrt{\rho^{(\beta)}}} \tag{21}$$

where a and c are the numerical factors of the order of 0.1 and 1, and μ is the elastic shear modulus. Matrix $\Omega_{\alpha\beta}$ expresses the dislocation interaction between different slip systems [1]. The dislocation density $\rho^{(\beta)}$ and dislocation mean free path $L^{(\beta)}$ are introduced into the hardening moduli.

We express dislocation density $\rho^{(\beta)}$ by adding a norm of the GN incompatibility $\rho_\eta^{(\alpha)}$ instead of the SS dislocation density to the initial dislocation density $\rho_0^{(\beta)}$ and a norm of the generalized GN dislocation density $\rho_\alpha^{(\alpha)}$, such that

$$\rho^{(\beta)} = \rho_0^{(\beta)} + \rho_\alpha^{(\beta)} + \rho_\eta^{(\beta)} \tag{22}$$

Note that $\rho_\alpha^{(\alpha)}$ is the first-order gradient of the slip and $\rho_\eta^{(\alpha)}$ the second-order gradient of the slip, whereas higher-order gradients more than third-order are disregarded in the present paper.

3.1 Dislocation mean free path

Models of the dislocation mean free path $L^{(\alpha)}$ are proposed variously. Ohashi [1] corrects the Seeger model in order to correspond to multislip. During single slip in the early stage of deformation, the mean free path is a constant and thus work-hardening curve is parabolic. This type of deformation arises in the first stage of hardening of a single crystal. After the onset of multislip, the mean free path decreases with increase of slip and the stress-slip curve becomes linear. This region is the second stage of the hardening curve. The transition from the first stage to the second stage is expressed by Ohashi [1]. However, this model cannot represent the three-stage hardening, namely the third stage in the large deformation is ignored. In the third stage, the dislocation mean free path should increase again due to dynamic recovery. In this paper, we attempt to extend the corrected Seeger model to the three-stage hardening model of the single crystal as

$$L^{(\alpha)} = \begin{cases} L_0^{(\alpha)} & \text{:Stage I} \\ \dfrac{\Lambda}{\gamma - (\gamma^* - \Lambda / L_0^{(\alpha)})} & \text{:Stage II} \\ \left. \begin{array}{l} L_1^{(\alpha)} + \dfrac{1}{2}(L_0^{(\alpha)} - L_1^{(\alpha)})\left\{1 - \cos\left(2\pi \dfrac{\gamma - \gamma^{**}}{\gamma_r}\right)\right\} \\ L_0^{(\alpha)} \quad (\gamma > \gamma^{**} + \gamma_r / 2) \end{array} \right\} & \text{:Stage III} \end{cases} \tag{23}$$

where $L_0^{(\alpha)}$ is the initial dislocation mean free path, γ^* the value of beginning of multislip, Λ a material constant, $L_1^{(\alpha)}$ and γ^{**} the values of $L^{(\alpha)}$ and γ when the third stage starts and γ the summation of $\left|\gamma^{(\alpha)}\right|$.

4. NUMERICAL ANALYSIS

In the present simulation, an f.c.c. single crystal of aluminium is assumed, and plane strain condition is employed. Asaro's 2-slip model is used in this analysis. The initial aspect ratio of the specimen is L/W=3, and the initial width is W=1.0mm. Material constants and numerical parameters used here are E=60.84GPa, ν =0.3, $\rho_0^{(\alpha)}$ =0.5μm^{-2}, $L_0^{(\alpha)}$ = 500μm, Λ =15μm, $\tilde{b}$ =25.6nm, τ_0 =1MPa, m=0.005, $\dot{\gamma}_0$ =0.001s^{-1}, γ^* =0, γ^{**} =0.1, γ_r =0.3, a=0.3 and c=3.

Figure 3 and Fig. 4 show the distributions of slip, GN dislocation density and GN incompatibility with the deformed meshes at u/L=0.2 where u denotes the end displacement of specimen. The shear band is observed in Fig. 3. From Figs. 3(a), (b) and (c), the values of the slip, the GN dislocation density and the GN incompatibility are high along the shear band. Figures 4(a), (b) and (c) represent the results calculated without the GN incompatibility, the geometrical nonlinearities and the dynamic recovery, respectively. We find that the shear band is not formed in these cases. The disappearance of shear band is considered to be due to the neglect of GN incompatibility. These results suggest that, in a large deformation analysis with strain localizations, it is necessary to introduce the three stage hardening into the mean free path model and the geometrical nonlinearities into the dislocation density and incompatibility.

The resolved shear stress versus slip curves are shown in Fig. 5. Since the crystal orientation inducing double slip from the initial state is used, the first stage does not appear in the hardening curve. According to Fig. 5, work-hardening begins from the second stage first, and it changes to the third stage after that. In the curve (iii) of Fig. 5, the value of dislocation density

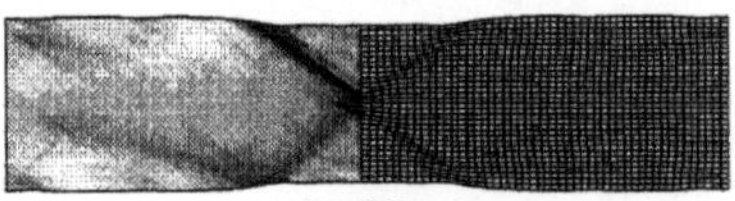

(a) Slip

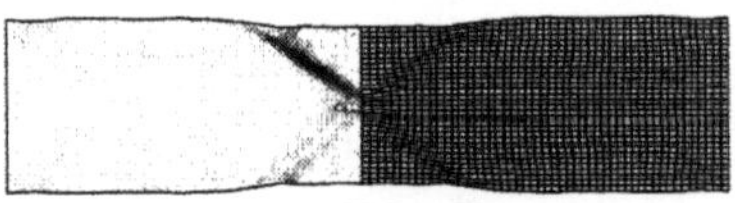

(b) GN dislocation

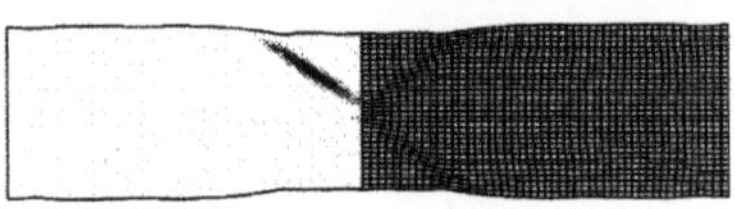

(c) GN incompatibility

Figure 3. Distribution of slip and dislocation densities

(a) Without GN incompatibility

(b) Without geometrical nonlinearities

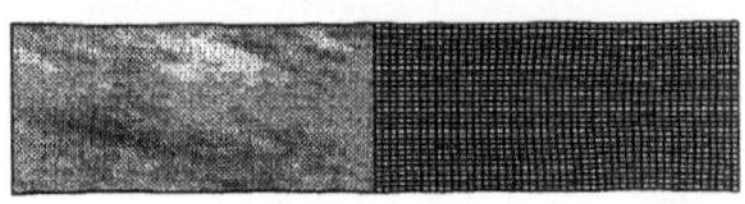

(c) Without stage III

Figure 4. Distribution of slip

becomes large and the resolved shear stress increases with the increase of slip. The other curves express a similar work-hardening although the value of slip is small. Comparing curve (i) with that of an experimental result in which the loading direction is almost the same as one of curve (i), we find that both results are well in agreement. Also, the other curves show hardening curves of the second and the third stages.

Figure 6(a) describes schematic crystal orientations, in which the line segments correspond to the slip planes. The rotation angles of the crystal orientation are about 10° around the shear band, and in other domain where the strain localization is not observed are in the range of −3° to 3°. These results predict that the small angle tile grain boundary is formed between the shear band and the domain apart from the shear band. Figure 6(b) shows the schematic accumulations of dislocations. The black and gray symbols denote the edge dislocations and the positive and negative dipoles, respectively. The positive and negative edge dislocations accumulate respectively on the upper and lower boundaries of the shear band. Therefore, the shear band can be regarded as an induced subgrain with a small angle tilt grain boundary

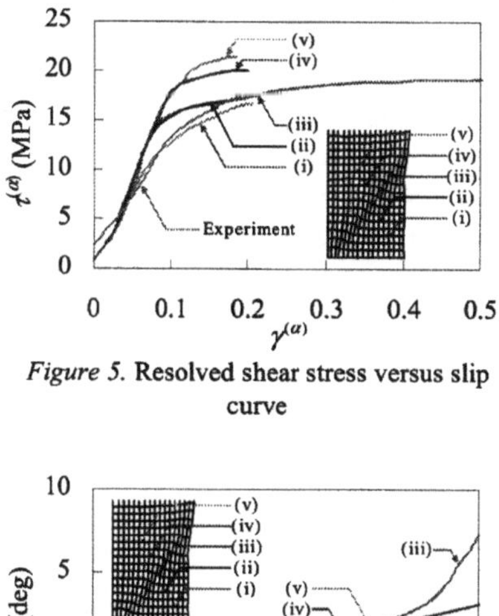

Figure 5. Resolved shear stress versus slip curve

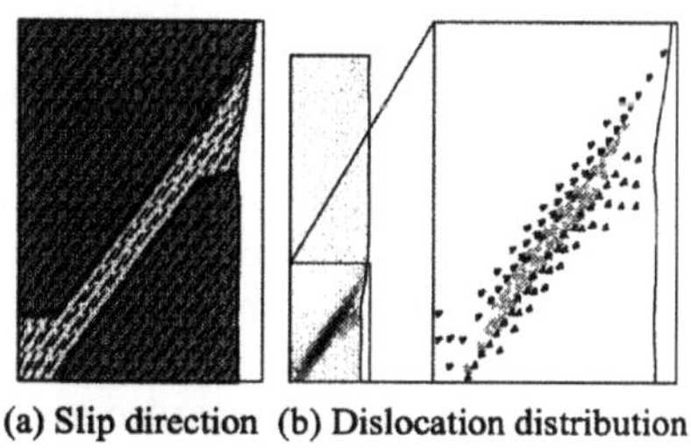

Figure 6. Small angle tilt grain boundary due to shear band formation

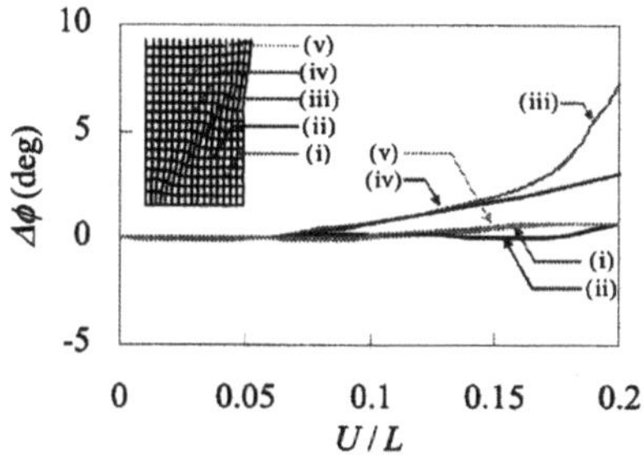

Figure 7. Angle of rotation of crystal orientation versus elongation curve

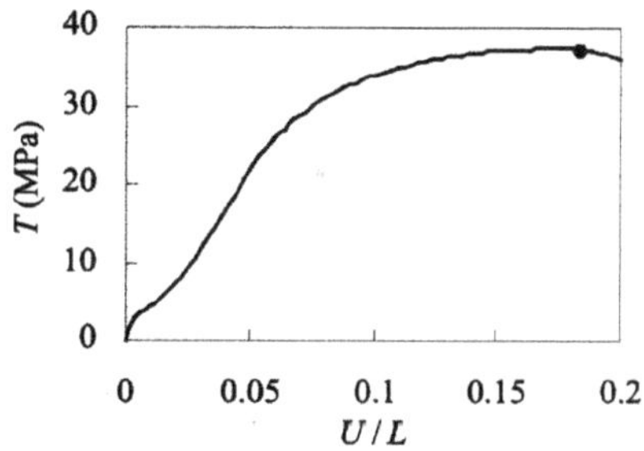

Figure 8. Load versus elongation curve

caused by large strain. This indication is also supported by the following fact. The rotation angle of crystal orientation versus elongation curves are shown in Fig. 7. The rotation of the crystal orientation at the center of the shear band (iii) is larger than the other domain. Comparing curve (iii) with Fig. 8 which shows the load versus elongation curve, the elongation percentage of the bifurcation point in Fig. 8 almost coincides with that of the point at which the rotation angle starts to increase in curve (iii) of Fig. 7.

5. CONCLUSIONS

We propose a new dislocation-crystal plasticity model by defining the generalized GN dislocation density and GN incompatibility. Finite element dislocation-crystal plasticity simulations for an f.c.c. crystal are carried out. Distributions of dislocation densities are visualized and the validity of this model is discussed. The results obtained here are summarized as follows.

(1) The GN incompatibility is defined by use of the incompatibility tensor, the distance of two adjacent slip planes and the magnitude of Burgers vector.

(2) In large deformation analyses with strain localizations, it is necessary to consider not only the GN dislocation density but also GN incompatibility, and to introduce geometrical nonlinearities into dislocation density and incompatibility. Moreover, the model of dislocation mean free path should be extended to the third stage of strain hardening.

(3) The shear band is regarded as a subgrain induced by large strain.

REFERENCES

[1] Ohashi, T., (1994), *Phil. Mag.*, **70**, p. 793.
[2] Ohashi, T., (1999), *J. Phys. IV France*, **9**, Pr9, p. 279.
[3] Zikry, M. A. and Kao, M., (1996), *J. Mech. Phys. Sol.*, **44**-11, p. 1765.
[4] Teodosiu, C., (1997), Large Plastic Deformation of Crystalline Aggregates, p. 21, Springer-Verlag.
[5] Busso, E. P., et al., (2000), *J. Mech. Phys. Sol.*, **48A**, p. 2333.
[6] Fleck, N. A., et al., (1994), *Acta Metall.*, **42**-2, p. 475.
[7] Kröner, E., (1958), Kontinuumstheorie der Versetzungen und Eigenspannungen, Springer-Verlag.
[8] Kondo, K., I(1955), II(1985), III(1962), IV(1968), *RAAG Memoirs of the Unifying Study of Basic Problems in Engineering and Physical Sciences by Means of Geometry*.
[9] Shizawa, K. and Zbib, H. M., (1999), *Int. J. Plasticity*, **15**, p. 899.
[10] Ashby, M. F., (1970), *Acta Metall.*, **21**, p. 399.
[11] Peirce, D., Asaro, R. J. and Needleman, A., (1983), *Acta Metall.*, **31**, p. 1951.

A NEW MODEL OF SCALE DEPENDENT CRYSTAL PLASTICITY ANALYSIS

Tetsuya Ohashi

Kitami Institute of Technology,
Koencho 165, Kitami, 090-8507, Japan
E-mail: ohashi@newton.mech.kitami-it.ac.jp

Abstract: Crystal plasticity analysis of slip deformation in metal microstructure enables one to calculate densities for the statistically stored (SS) and the geometrically necessary (GN) dislocations, and such densities are utilized to evaluate the critical resolved shear stresses for slip systems. In this paper, we propose a new model, where the mean free path of moving dislocations is defined as a function of the densities for SS and also, GN dislocations, and the critical resolved shear stresses for slip systems are given only by SSDs. Scale dependent characteristics of GN dislocations is transmitted to the SSDs via the mean free path and finally, to the slip resistance of slip systems. Tensile deformations of six-grained multicrystal models whose average grain diameter, $\bar{d}$, ranges from 0.1 to 500 μm, are analyzed with this new model and micro- and macroscopic aspects are examined. Plastic flow stresses increase almost linearly with $\bar{d}^{-1/2}$ within the range, $500 > \bar{d} > 1\ \mu m$, showing the Hall-Petch relation, and this grain refinement effect is gradually reduced for finer microstructures.

Key words: metal microstructure, crystal plasticity analysis, dislocations, scale dependent plasticity.

1. INTRODUCTION

Scale dependent characteristics of yield stress, plastic flow stress, or indentation hardeness of metal polycrystals are well known and they are summarized in the Hall-Petch type description:

$$\sigma = \sigma_0 + k / \sqrt{\bar{d}} ,$$

H. Kitagawa and Y. Shibutani (eds.), IUTAM Symposium on Mesoscopic Dynamics of Fracture Process and Materials Strength, 97–106.

where, σ_0 and k are considered to be material constants, while $\bar{d}$ is a representative length of material's microstructure. A lot of efforts have been made to establish theoretical framework of solid mechanics that show material scale dependency and among them, a model by Fleck *et al.* (1994) induced a wider range of discussion. In their model, the critical resolved shear stresses of slip systems are modeled by the ordinary Taylor type expression but they introduced the geometrically necessary (GN) dislocations as well as the statistically stored (SS) ones into it. That is,

$$(critical\ resolved\ shear\ stress) \propto \sqrt{\rho_S + \rho_G}$$

where, ρ_S and ρ_G denote the densities of the SS and GN dislocations, respectively. Density increment of the SS dislocations is related to the increment of plastic shear strain and the mean free path of moving dislocations, while the density of the GN ones is related to the spatial gradient of the plastic shear strain on slip systems. The critical resolved shear stress reflects its scale dependent characteristics of the GN dislocations through the above expression. However, it is still open to discussion whether the above expression is valid and sufficient (for example, Weertman, 2002, and Needleman and Gil Sevillano, 2003).

In this paper, we propose a new model of scale dependent crystal plasticity analysis, where the mean free path of moving dislocations is defined as a function of the densities for SS and also, GN dislocations, and the critical resolved shear stresses for slip systems are given only by SS dislocations. Scale dependent characteristics of GN dislocations is transmitted to the SS dislocations via the mean free path and finally, to the slip resistance of slip systems. Tensile deformations of six-grained multicrystal models whose average grain diameter, $\bar{d}$, ranges from 0.1 to 500 μm, are analyzed with this new model and micro- and macroscopic aspects are examined.

2. BASIC EQUATIONS

Slip deformation is assumed to take place on {111} crystal plane and in <110> crystal direction. The activation condition of the slip system n is given by the Schmid law;

$$P_{ij}^{(n)}\sigma_{ij} = \theta^{(n)},\ P_{ij}^{(n)}\dot{\sigma}_{ij} = \dot{\theta}^{(n)},\quad (n = 1,\cdots,12), \tag{1}$$

and,

$$P_{ij}^{(n)} = \frac{1}{2}\left(\nu_i^{(n)} b_j^{(n)} + \nu_j^{(n)} b_i^{(n)}\right), \tag{2}$$

where, σ_{ij} and $\theta^{(n)}$ denote the stress and the critical resolved shear stress on the slip system n, respectively. The slip plane normal $\nu_i^{(n)}$ and the slip direction $b_i^{(n)}$ define the Schmid tensor $P_{ij}^{(n)}$. Quantities with dot indicate increments of them. Increment of the critical resolved shear stress is written as follows;

$$\dot{\theta}^{(n)} = \sum_m h^{(nm)} \dot{\gamma}^{(m)}. \tag{3}$$

Here, $\dot{\gamma}^{(m)}$ denotes the increment of plastic shear strain on slip system m. If the deformation is small and rotation of the crystal orientation is neglected, the constitutive equation is written as follows (Hill, 1966),

$$\dot{\sigma}_{ij} = [S^e_{ijkl} + \sum_n \sum_m \{h^{(nm)}\}^{-1} P_{ij}^{(n)} P_{kl}^{(m)}]^{-1} \dot{\varepsilon}_{kl}. \tag{4}$$

S^e_{ijkl} denotes elastic compliance. Summation is made over the active slip systems.

Let us suppose that the critical resolved shear stress is given by the following equation (Ohashi, 1987, 1994);

$$\theta^{(n)} = \theta_0 + \sum_m \Omega^{(nm)} a \mu \tilde{b} \sqrt{\rho_S^{(m)}}, \tag{5}$$

where, θ_0 denotes the lattice friction term, which is, in general, very small for fcc crystals, and $\rho_S^{(m)}$ denotes the density of SS dislocations that accumulate on the slip system m. Reaction between dislocations on slip systems n and m defines the magnitude of the interaction matrix $\Omega^{(nm)}$. Diagonal terms in $\Omega^{(nm)}$ are unity and in the present study, we choose off diagonal terms to express pseudo-isotropic hardening character for every slip system[1]. μ and $\tilde{b}$ denote the elastic shear modulus and magnitude of Burgers' vector, respectively. a is a numerical factor, which is close to 0.1.

---

[1] To ensure the existence of the inverse of the hardening coefficient, which will be given in eq. (11), we also introduce a numerical "noise" of the order of 10^{-4} in $\Omega^{(nm)}$.

Increment of the SS dislocations is given as follows;

$$\dot{\rho}_S^{(n)} = c\dot{\gamma}^{(n)} \Big/ \tilde{b}L^{(n)}, \tag{6}$$

where, c is a numerical coefficient of the order of 1. $L^{(n)}$ denotes the mean free path of dislocations on slip system n and, in this paper, we use the following model for it;

$$L^{(n)} = \frac{c^*}{\sqrt{\sum_{m \neq n} (s \cdot \rho_S^{(m)} + g \cdot \|\rho_G^{(m)}\|)}}, \tag{7}$$

where, c^* is a material constant of the order of 10 - 100 (for detailed discussion, Kuhlmann-Wilsdorf, 1989) . Parameters s and g are introduced to control the analysis and $0 \le s \le 1,\ 0 \le g \le 1$.[2]

The edge and screw components of the geometrically necessary dislocations are obtained from the strain gradients (Ohashi, 1997);

$$\rho_{G,edge}^{(n)} = -\frac{1}{\tilde{b}} \frac{\partial \gamma^{(n)}}{\partial \xi^{(n)}}, \qquad \rho_{G,screw}^{(n)} = \frac{1}{\tilde{b}} \frac{\partial \gamma^{(n)}}{\partial \zeta^{(n)}}. \tag{8}$$

Here, $\xi^{(n)}$ and $\zeta^{(n)}$ denote directions parallel and perpendicular to the slip direction on the slip plane, respectively. Norm of two components defines the scalar density for the GN dislocations,

$$\|\rho_G^{(n)}\| = \sqrt{\left(\rho_{G,edge}^{(n)}\right)^2 + \left(\rho_{G,screw}^{(n)}\right)^2}. \tag{9}$$

Evaluation of the edge and screw components for the GN dislocations enables one to calculate the tangent vector $\mathbf{l}^{(n)}$ of the dislocation line segments (Ohashi, 1999);

$$\mathbf{l}^{(n)} = \frac{1}{\|\rho_G^{(n)}\|} \left(\rho_{G,screw}^{(n)} \cdot \mathbf{b}^{(n)} + \rho_{G,edge}^{(n)} \cdot \mathbf{b}^{(n)} \times \mathbf{v}^{(n)}\right). \tag{10}$$

---

2 A similar approach was made by Acharya and Beaudoin (2000). In their model, total dislocation density was given by geometrically and statistically determined densities; mean free paths for the former quantity was determined by the slip plane lattice incompatibility and the latter by density of the total forest dislocations.

Data for GN dislocations are obtained for each finite element and then, we can draw line segments of dislocations in three-dimensional space. We will draw one line segment at the center of each element. Direction of the line segment is given by eq. (10) and its length and thickness are determined by the density norm $\left\|\rho_G^{(n)}\right\|$.

The strain hardening coefficient in equation (3) is given by the following equation;

$$h^{(nm)} = \frac{1}{2} a c \mu \Omega^{(nm)} / \left[L^{(m)} \sqrt{\rho_S^{(m)}} \right], \tag{11}$$

and by substituting eq. (7) into eq. (11), we have

$$h^{(nm)} = \frac{ac\mu}{2c^*} \Omega^{(nm)} \sqrt{\frac{\sum_{k \neq m} (s \cdot \rho_S^{(k)} + g \cdot \left\|\rho_G^{(k)}\right\|)}{\rho_S^{(m)}}}. \tag{12}$$

Table 1. Euler angles for crystal grain 1-6, and their Schmid factor for the primary and critical slip systems.

grain	Euler angle κ	θ	ϕ	Schmid factor primary	critical
1	85.5	24.0	212.9	0.4968	0.4862
2	80.0	16.0	101.4	0.4881	0.4775
3	53.3	13.6	215.3	0.4632	0.4416
4	88.5	35.6	182.9	0.4665	0.4593
5	56.3	17.7	219.4	0.4747	0.4397
6	68.1	5.71	213.1	0.4426	0.4398

3. RESULTS AND DISCUSSION

3.1 Microscopic aspects of slip deformation in six-grained multicrystals

Figure 1(a) shows the six-grained multicrystal model employed in this study. Ratio of the width (w), height (h), and thickness (d) of the specimen is w:h:d=5:15:1. We made nine specimens of similar shape but with different dimensions: the width w of the specimens are 0.1, 0.2, 0.5, 1, 2.5, 5, 50, 100, and 500 μm. The average grain diameters $\bar{d}$ of the models are approximately equal to w. The models are divided into finite elements and their tensile deformations in y direction are analyzed by our crystal plasticity

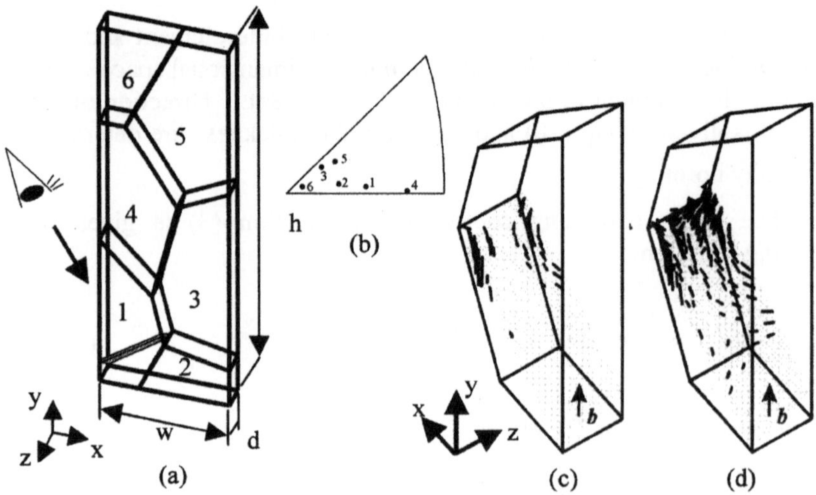

Fig.1 (a):Six-grained multicrystal model employed in this study and a thin foil region in the grain 1. The foil is cut parallel to the primary slip plane. (b): crystal orientations. (c),(d):Development of GN dislocations on the primary system when the average tensile strain is 0.049 and 0.053 %, respectively.

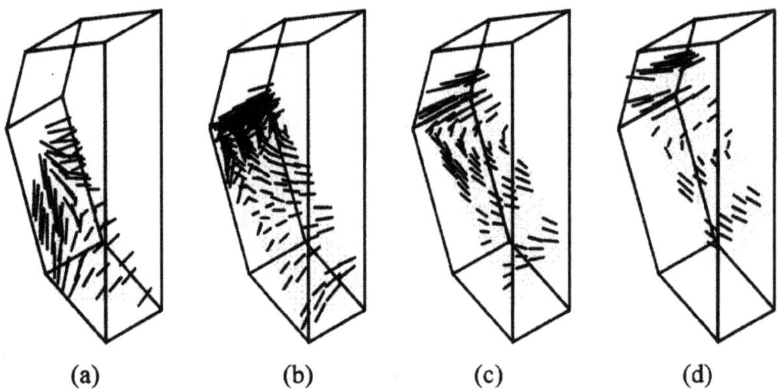

Fig.2 (a) - (d): Three-dimensional distribution of the GN dislocations on the primary system when the average tensile strain is 0.068 %.

software code *clp7*. We use a fine mesh (4864 elements in total) for detailed study of microscopic aspects of slip at small strain range, and a coarse one (480 elements) to calculate macroscopic response of the whole specimen up to 20% strain. Elastic compliance data for Cu are used for the six grains and the initial dislocation densities for twelve slip systems are supposed to be $1x10^{12}$ m^{-2} and c*=100 in all the specimens. Crystal orientations are shown in Figure 1(b) and Schmid factors for the primary and a secondary slip system are summarized in Table1.

First, we examine slip deformation in detail by scale independent (s=1,

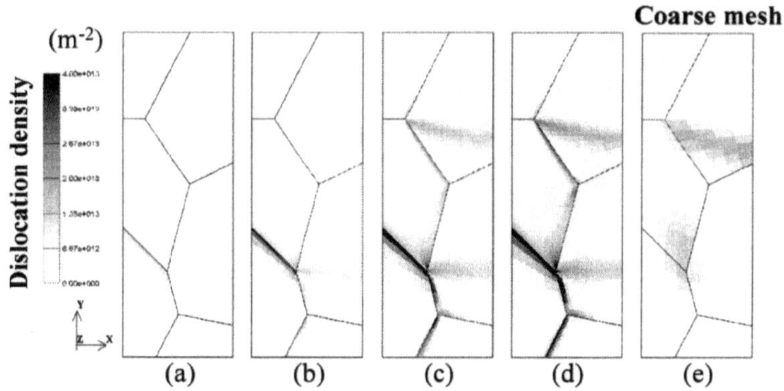

Fig.3 (a) - (d): Density distribution of the GN dislocations on the primary system when the average tensile strain is 0.08, 0.16, 0.32, and 0.4 %, respectively. The fine mesh is used. (e):Result obtained by the coarse mesh. The average tensile strain is 0.4%.

g=0) hypothesis. Figures 1(c) and (d) show development of GN dislocations in a thin foil in the grain-1 at the very beginning of plastic slip deformation. The foil is parallel to the primary system and positioned as illustrated in Figure 1(a). Viewing direction of Figs.1(c) and (d) is also given in Fig.1(a). When the average tensile strain is about 0.049%, slip deformation starts near a grain boundary triple junction. Segments of GN dislocations make up half loop shaped structures. As the deformation proceed (Figure 1(d)), the loop shaped structure grows and dislocation pile-up is observed near the triple junction. The shape of GN dislocation segments in places far apart from the triple junction is rather straight. Figures 2(a)-(d) compare the structure of GN dislocations in different positions. Dislocation pile-ups at grain boundaries are commonly observed, but other features differ gradually from place to place. Figures 3(a)-(d) show growth of accumulation of GN dislocations when the average tensile strain is less than 0.4%. Not only grain boundary pile-ups, but also a formation of thin region of higher density growing straight from some grain boundary triple junctions are observed. The latter type of structure is supposed to be formed by many body interaction of crystal grains; interaction of at least three crystal grains contribute to the formation of this type of structure (Ohashi, 1989). Fig 3(e) shows the result obtained with the coarse finite element mesh, as a comparison to Fig. 3(d). Details of grain boundary pile-ups are not clear in the result by the coarse mesh, while densities at grain interior shows to be approximately equal to the ones obtained by the fine mesh.

3.2 Scale dependent analysis

We now analyze slip deformation with the scale dependent model by setting s=g=1 in eq. (7). We use a coarse mesh, which enables us to

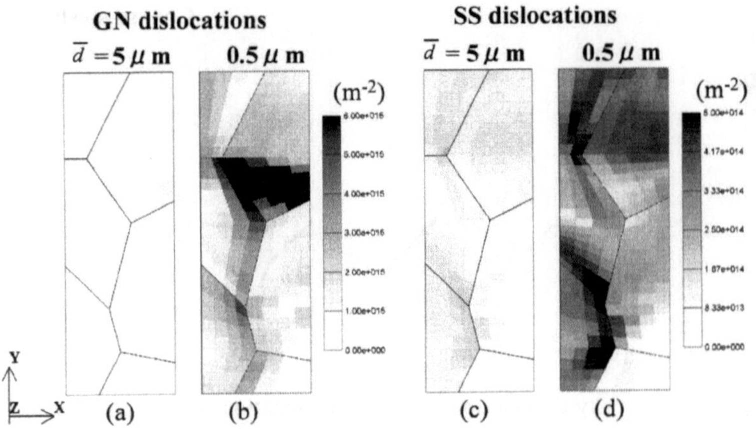

Fig.4 (a),(b): Density distribution of the GN dislocations on the primary system when the average tensile strain is about 19%. The average grain diameter is 5 and 0.5 μm, respectively. (c),(d): Density distribution of the SS dislocations on the primary system when the average tensile strain is about 19%. The average grain diameter is 5 and 0.5 μm, respectively.

analyze the deformation up to 20% tensile strain in a realistic computer time. With this coarse mesh, some details of dislocation structure will be missing, on which we made a comparison in Figures 3(d) and (e). Effects of such a fine structure of dislocations on the macroscopic response of the specimen are not fully understood and a point of interest for further study.

Figures 4(a)-(d) compare results for the GN and SS dislocations obtained for specimens with average grain sizes of 5 and 0.5 μm. Specimens are deformed until the average tensile strain is about 19%. It should be noted that color bars used in these figures are different from each other and also from the one used in Figures 3. Densities of GN dislocations in the specimen of $\bar{d}$=0.5 μm is about 10 times as large as the ones in the $\bar{d}$=5 μm specimen, which is a logical consequence of the theory that we evaluate the GN dislocations. At the same time, densities of SS dislocations in the specimen of $\bar{d}$=0.5 μm is about 5 times as large as the ones in the $\bar{d}$=5 μm specimen and this cause higher flow stress for specimen with smaller grain size.

Figure 5 shows nominal stress - strain curves for specimens with different mean grain diameters. Yield stresses for the all specimens are substantially the same, but the strain hardening is more significant for specimens with smaller grain size. Differences in strain hardening characteristics for specimens with $\bar{d}$>50 μm and scale independent models are negligibly small, while the strain hardening of the specimens with sub-micron grain diameter becomes remarkably high with the reduction of the mean grain diameter. Figure 6 shows the flow stress at 20% tensile

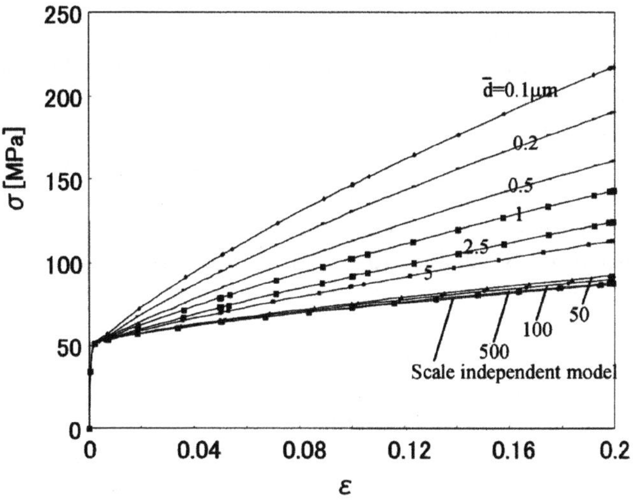

Fig.5 Nominal stress - strain response of specimens with various mean grain diameter.

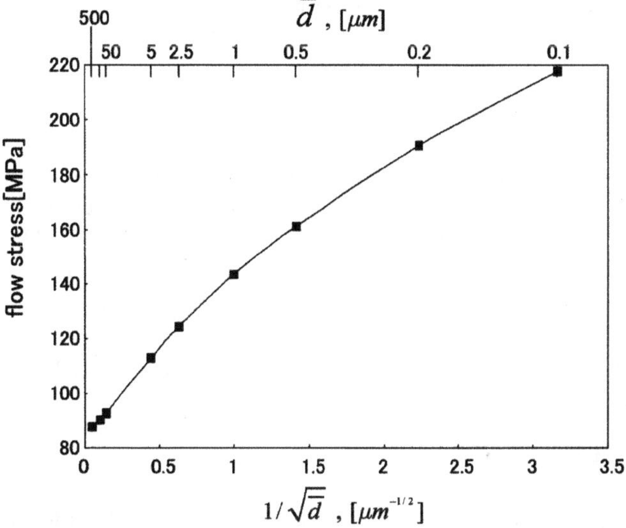

Fig.6 Hall-Petch plot of the flow stresses when the mean tensile strain is about 20%.

strain plotted against $1/\sqrt{\bar{d}}$. The flow stress increase almost linearly within the range $\bar{d}>1$ μm, showing a Hall-Petch relation, and the slope gradually decreases when $\bar{d}$ is smaller than 1 μm. If we apply the Hall-Petch relation for the flow stress data for the specimens of $\bar{d}>1$ μm, the coefficient k is calculated as about 1.84 MPa-mm$^{1/2}$ (=5.82x10^{-2} MPa-m$^{1/2}$). This is about a half of the data obtained by experiment for Al (Ono, *et al.*, 2002). Number of crystal grains in the thickness direction is unity in the present multicrystal model, and mutual constraint of their deformation is quasi two-dimensional. On the other hand, three-dimensional constraint of deformation will generally be taking place in polycrystals. This difference in mutual constraint between crystal grains will make a significant change in the strain hardening phenomenon and its scale dependency. This point should be explored in future work. Also, the parameter c* and the initial densities of SS dislocations will have effects on the strain hardening, as eq. (12) implies. Then, the effects of c* and the initial densities of SS dislocations and its non-uniform distribution on the strain hardening phenomenon are also issues of future studies. It is a well known phenomenon that the Hall-Petch relation fails for sub-micron and nanometer regions. A full understanding of its mechanism has not been obtained, although it is usually attributed to grain boundary sliding. It was shown in the present analysis that some mechanism other than grain boundary sliding can contribute to the decrease of the k value for the region $\overline{d}<1$ μm. Then, it seems to be interesting to examine how the strain hardening and the scale dependency occur in microstructures with sub-micron sizes. Scale dependent characteristics of the yield stress is another important point for further study.

4. SUMMARY

We made a new scale dependent crystal plasticity model, where not only the statistically stored but also the geometrically necessary dislocations contribute to the mean free path of moving dislocations. Obtained results showed an approximately linear relationship between the inverse of the square root of the mean grain diameter of multicrystal models and the flow stress for the range that the mean grain diameter was larger than 1 μm. For smaller grain sizes, the grain refinement effect was gradually reduced.

References

Acharya, A. and Beaudoin, A.J., 2000, J. Mech. Phys. Solids, **48**, 2213.
Fleck, N.A., *et al.*, 1994, Acta metall. mater., **42**, 475.
Hill, R. 1966, J. Mech. Phys. Sol., **14**, 95-102.
Kuhlmann-Wilsdorf, D., 1989, Mat. Sci. Eng. **A113**, 1.
Needleman, A. and Gil Sevillano, J., 2003, Scripta Mat., **48**, 109.
Ohashi, T., 1987, Trans. Japan Inst. Met. **28**, 906.
Ohashi, T., 1990, Colloque de Physique, Col. C1, **51**, c1-593.
Ohashi, T., 1994, Phil. Mag., **A70**, 793.
Ohashi, T., 1997, Phil. Mag. Lett., **75**, 51.
Ohashi, T., 1999, J. Phys. IV France, **9**, Pr9-279.
Weertman, J., 2002, Acta Mat., **50**, 673.

A MOLECULAR DYNAMICS STUDY OF MECHANICAL PROPERTIES AND FRACTURES OF AMORPHOUS METAL

Keiko Nakatani[1], Yoshihiko Sugiyama[1] and Hiroshi Kitagawa[2]

[1]*Department of Aerospace Engineering, College of Engineering, Osaka Prefecture University, 1-1 Gakuen-cho, Sakai, Osaka 599-8531, Japan*

[2]*Department of Adaptive Machine Systems, Graduate School of Engineering, Osaka University, 2-1 Yamadaoka, Suita, Osaka 565-0871, Japan*

Abstract A molecular dynamics simulation of mode I crack propagation is carried out. We propose a definition of the J integral in terms of atomistic information, i.e. site potential energy, atomic stress, gradient of displacement and so on, and call it J^a parameter. The crack tip opening displacement, the crack extension, stress distribution and number density are evaluated in the vicinity of the crack tip. The relationship between the crack growth and J^a parameter is discussed.

Keywords: Computational mechanics, Fracture mechanics, Molecular dynamics, J integral, Amorphous iron, Mode I crack, Void nucleation, Crack propagation.

1. INTRODUCTION

Amorphous metals do not possess regular crystal structures. As a result, their fracture mechanisms are considered to proceed in completely different ways from the commonly accepted mechanism for crystals – i.e., arising from microscopic phenomena, such as the breaking of bonds between atoms, and the occurrence of dislocations near the crack tip, which are observed in the fracture processes of ordinary crystalline materials (Masumoto and Maddin, 1975).

In our previous studies on amorphous metal, we have demonstrated that the mechanical properties change significantly owing to damage from deformation (Nakatani et al., 1996b; Nakatani et al., 1997; Nakatani et al., 1998). The propagation of mode I crack was considered, and the characteristic manners of the fracture of amorphous metals were investigated (Nakatani et al., 2000a). Moreover, the quantitative evaluation of fracture near the crack tip was also studied (Nakatani et al., 2000b).

In this paper, for the first-step of the quantitative evaluation of fracture near the crack tip, an atomistic parameter J^a is proposed which corresponds to the conventional J integral. J integral is a representative mechanical parameter

H. Kitagawa and Y. Shibutani (eds.), IUTAM Symposium on Mesoscopic Dynamics of Fracture Process and Materials Strength, 107–116.

commonly used as a criterion for crack propagation in continuum mechanics. It is shown that the mechanical state near the crack tip can be predicted quantitatively using J^a.

2. ANALYSIS MODEL AND METHODOLOGY

2.1 Setting up for Cracked Model

Molecular Dynamics (MD) method in which the behavior of atoms is assumed to be governed by classical mechanics based on the Newtonian equations of motion, is applied to the problem of crack propagation in an amorphous metal.

The cracked model is of a cylindrical region (radius $r = 6.99$ nm) , which consisted of 22251 Fe atoms (Nakatani et al., 1996a). The displacements corresponding to the field near a mode I crack tip are given to all atoms. A periodic boundary condition is applied along the crack front direction, z, with a periodic length of $l_z = 1.76$ nm. The initial atomic arrangement of a cracked model is illustrated in Fig. 1.

Propagation simulation is performed by letting the stress intensity factor, K_I, increase with time, where K_I is given by specifying the displacement of the boundary atoms. The initial stress intensity factor K_I^{init} is 0.75 MPa$\sqrt{m}$ and it is updated by 0.0125 MPa$\sqrt{m}$ every 1000 steps. The time increment Δt is 2.0×10^{-15} s, the total step is 150000 (30.0 ps), and then the K_I is updated 150 times from 0.75 to 2.625 MPa$\sqrt{m}$. The values of the J^a parameter, crack tip opening displacement (CTOD) and number density are evaluated for each K_I level. The values are averaged using the values for the final 500 steps for each K_I level.

2.2 J^a Parameter, CTOD, Crack Extension and Number Density

Atomistic definition for J^a parameter, CTOD, crack extension and number density are given as following. (See also (Nakatani et al., 2000b).)

2.2.1 Definition of J^a parameter.

The J integral (Rice, 1968) is defined in a continuum as a line integral along a path Γ surrounding the crack tip as follows.

$$J = \int_{\Gamma} \Big(W\delta_{1j} - \sigma_{ij}\frac{\partial u_i}{\partial x_1} \Big) n_j \, d\Gamma, \tag{1}$$

where W, σ_{ij} and u_i are the strain energy density, stress, and displacement, respectively. By introducing a function Q, of which values are 0 and 1 along Γ_1 and Γ_2 respectively, to area S^* which is enclosed by a closed path Γ^* as

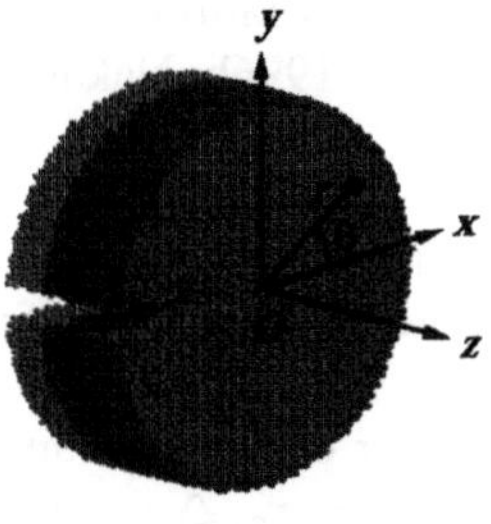

Figure 1. A schematic cracked model for analysis showing three layers.

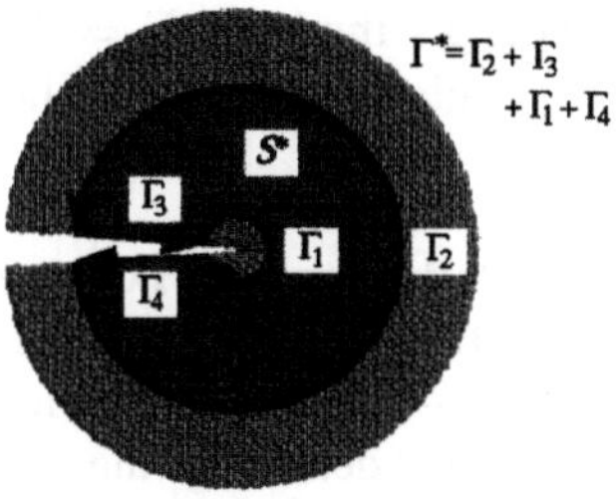

Figure 2. Summation region for J^{a} integral of an atomic model of an amorphous metal

shown in Fig. 2, another form of J integral can be considered as follows.

$$J = \iint_{S^*} \left[\left(W\delta_{1j} - \sigma_{ij}\frac{\partial u_i}{\partial x_1} \right) \right] \frac{\partial Q}{\partial x_j}\, dS. \tag{2}$$

It is originally introduced in (Li et al., 1985) and the equivalence between Eqs. (1) and (2) are proved.

To estimate J, we propose a parameter J^{a} which can correspond to an atomistic expression of Eq. (2). By assuming that W could be calculated by the change in the potential energy density $W^{(\alpha)}$, J^{a} is calculated as follows.

$$J^{\mathrm{a}} = \sum_{\alpha \mathrm{in} S^*} \left[\left(W^{\mathrm{C}(\alpha)} - W^{\mathrm{O}(\alpha)} \right) \delta_{1j} - \left(\sigma_{ij}^{\mathrm{C}(\alpha)} - \sigma_{ij}^{\mathrm{O}(\alpha)} \right) \frac{\partial u_i}{\partial x_1} \right] \frac{\partial Q}{\partial x_j} V^{(\alpha)} \frac{1}{l_z}, \tag{3}$$

where $V^{(\alpha)}$ denotes the volume occupied by the α-th atom, and $\boldsymbol{\sigma}^{(\alpha)}$ means the atomic stress at α-th atom. For the simplicity, $V^{(\alpha)}$ is assumed constant V_{a} and the other variables with superscript (α) are smoothed by taking the weighted average over the neighbouring atoms. Variables with superscript C are evaluated for current cracked model that has a carck. The variables with superscript O denote the reference states, which were introduced to remove the influence of the initial stress and the increase in free surface energy from the introduction of the crack, and to evaluate the increase in pure stress and pure strain energy, and bulk model and surface model and surface models are used for calculation (Nakatani et al., 2000b).

Within the linear elastic region and small scale yielding assumption, the value of the J integral, J_{LE} is equivalent to the rate of energy release under plane strain condition as follows.

$$J_{\mathrm{LE}} = \frac{(1-\nu^2)K_{\mathrm{I}}^2}{E}, \tag{4}$$

where Young's modulus $E = 183.0$ GPa and Poisson's ratio $\nu = 0.26$, according to uniaxial tensile simulations (Nakatani et al., 1996b; Nakatani et al., 1997).

2.2.2 Definition of CTOD and Crack Extension.

The CTOD is defined as the minimum value of the y components of relative position $(q_y^\alpha - q_y^\beta)$ among the atoms in the range $-\Delta x \leq q_y^\alpha, q_y^\beta \leq 0$. The extention Δa is defined as the minimum value of the x coordinates among the atoms in the range $-\Delta y/2 \leq q_x^\alpha \leq \Delta y/2$. Here, $\Delta x = r_c$, $\Delta y = r_c/4$. The $r_c = 0.357$ nm means the cut off distance of the interatomic potential.

2.2.3 Definition of the Number Density.

The number density is defined by $\rho(\boldsymbol{q}^\alpha)$ at a certain position, $\boldsymbol{q}^\alpha$, as follows.

$$\rho(\boldsymbol{q}^\alpha) = \sum_\beta w\left(\left|\boldsymbol{q}^\beta - \boldsymbol{q}^\alpha\right|\right) V_a \tag{5}$$

Here, the normal distribution in which the standard deviation is r_c, is adopted as weight function w. The number density is used for detecting defects – e.g., crack surfaces and voids. In this paper, if the value of $\rho(\boldsymbol{q}^\alpha)$ is smaller than 0.8, then we regard the atom as being a defect.

3. RESULTS AND DISCUSSION

3.1 Distribution of the Radial Stress

Figure 3 shows the logarithmic relationships between the stress, σ_θ, and the distance, r, from the crack tip along radial directions measured at 0, 30, 60, and 90 degrees counter-clockwise from the direction of the crack propagation. The solid lines drawn in the figures indicate the linear elastic prediction. For small $K_{\rm I}$, the relationships agree with the linear elastic prediction, except for a very narrow region near the crack tip, in which the material is damaged inelastically. On increasing $K_{\rm I}$, the region that does not agree with an elastic body spreads out, which means that the material becomes damaged.

3.2 Number Density

Figure 4 shows the atomic configuration at some $K_{\rm I}$ levels. In the figure, the color of the atoms indicates the value of the number density, and the darker the atomic color, means relatively lower density around the atom, the closer the status of that atomic region to that of a void. Relatively low density region is expanded in front of the crack tip. As the load increasing, crack is blunted and it is extended.

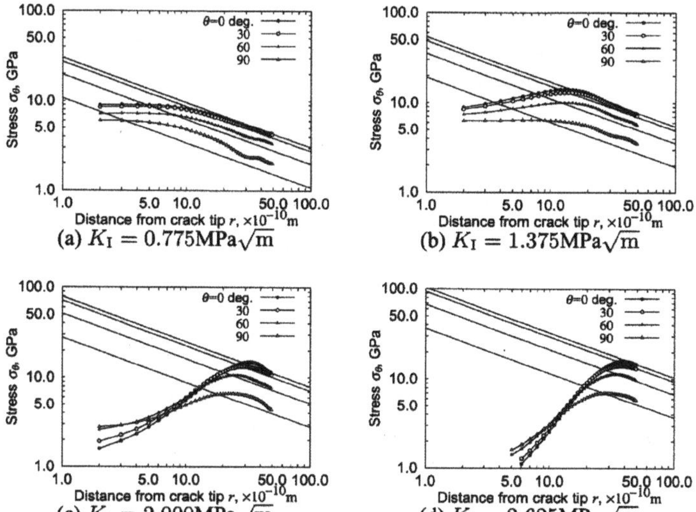

(a) $K_{\mathrm{I}} = 0.775\mathrm{MPa}\sqrt{\mathrm{m}}$ (b) $K_{\mathrm{I}} = 1.375\mathrm{MPa}\sqrt{\mathrm{m}}$

(c) $K_{\mathrm{I}} = 2.000\mathrm{MPa}\sqrt{\mathrm{m}}$ (d) $K_{\mathrm{I}} = 2.625\mathrm{MPa}\sqrt{\mathrm{m}}$

Figure 3. Distribution of radial stress near the crack tip (log plots).

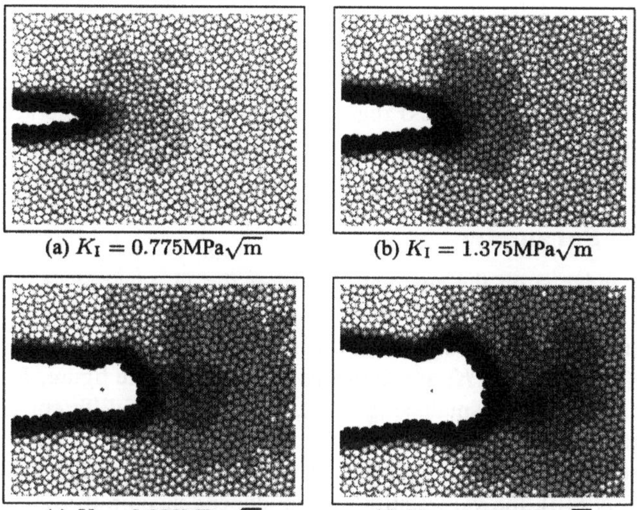

(a) $K_{\mathrm{I}} = 0.775\mathrm{MPa}\sqrt{\mathrm{m}}$ (b) $K_{\mathrm{I}} = 1.375\mathrm{MPa}\sqrt{\mathrm{m}}$

(c) $K_{\mathrm{I}} = 2.000\mathrm{MPa}\sqrt{\mathrm{m}}$ (d) $K_{\mathrm{I}} = 2.625\mathrm{MPa}\sqrt{\mathrm{m}}$

Figure 4. Number density of atoms near a mode I crack tip.

3.3 Distribution of Hydrostatic Stress near a Crack Tip

The distribution of hydrostatic stress is shown in Fig. 5. In the figure, the color of the atoms indicates the value of the hydrostatic stress. If an atom has

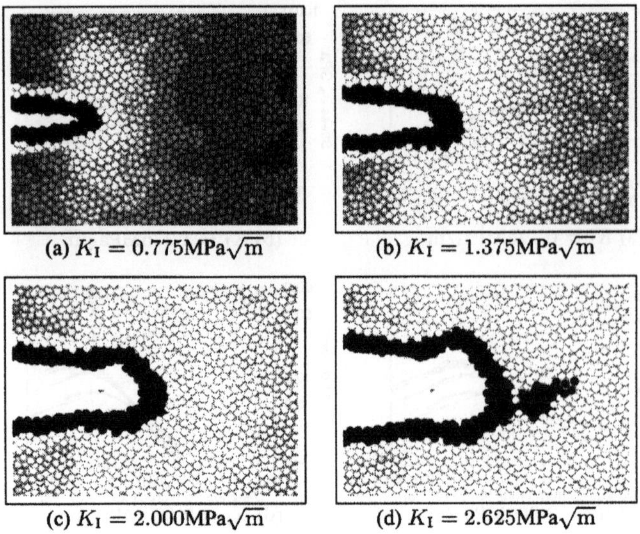

Figure 5. Distribution of hydrostatic stress and a void near a crack tip.

a high stress value, then its atomic color is brighter. Perfectly black-colored atoms are judged as being voids ($\rho < 0.8$), using the number density criterion defined previously. High stress regions appear in front of the crack tip, and they extend as the load increases. In Fig. 5(d), a region of voids is seen to appear.

3.4 J^a Parameter

The dependence on the summation region for J^a parameter is shown in Fig. 6.

The nominal J integral values based on the linear elastic prediction evaluated using Eq. (4) with given K_I are shown as solid lines in the figure. In the cases where the K_I values are small, (Fig. 6(a)), the J^a value of different summation region agrees with J_{LE}, and the dependence on the J^a parameter summation region almost vanishes. Though the J^a values become smaller than the J_{LE} values in Fig. 6(b), the values still have no dependence on the J^a parameter summation regions. However, when K_I becomes large (e.g., in Fig. 6(c)), the K_I value whose summation region approaches a crack tip (where $r_{min} = 1$ nm) shows a dependence on the summation region, and the K_I value decreases remarkably. In Fig. 6(d), the J^a value whose summation region passes through the value of $r_{min} = 2$ nm also has dependence on the summation region.

The dependence on the J^a parameter summation region of a few values is thus demonstrated, though the mechanical properties become changed near

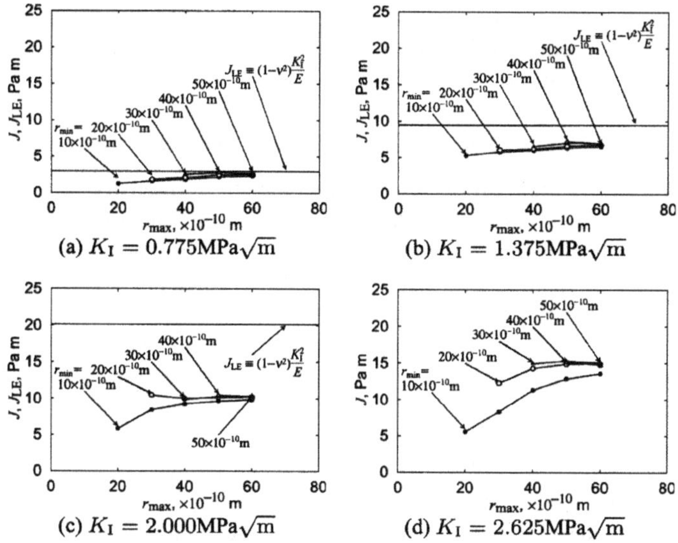

Figure 6. Dependence of J^a on the summation region.

the crack tip, where material is damaged because of the high K_I value. We can predict the mechanical status surrounding the crack tip using the J^a parameter chosen using an summation region, avoiding the damaged zone.

By comparing these results and those of Fig. 3, a remarkable decrease in the J^a parameter value can be seen in the case of a J^a parameter summation region passing through the non-linear region, where r_{min} is small. However, even in the case where the region near the crack tip is non-linear, if the J^a parameter summation region is taken to be some distance away from the crack tip, then it has little dependence on the J^a parameter summation region. The J^a parameter value converges more, owing to the larger width of the summation region, which is smaller than the nominal value predicted by Eq. (4) under large deformation. This means that the nominal value is no longer as large as the true value, because of the non-linearity caused by the deformation and damage near the crack tip.

3.5 Change in CTOD and Crack Extension

The relationship between K_I and CTOD, crack extension Δa is shown in Fig. 7. The solid curve shows a CTOD based on the linear elastic predictions deduced from Eq. (6).

Both the CTOD value and the crack growth extension Δa, increase with the value of K_I. Some steps occur in the crack growth versus K_I, and the crack growth increases discontinuously. The discrepancy between MD result and theoretical prediction is not only due to the definition but it is due to the non-elastic deformation.

3.6 Load Level Dependence of the Strain Energy

Figure 8 shows the relationship between the stress intensity factor, K_{I}, and the J integral value. The J integral value based on the linear elastic prediction of Eq. (4) is also drawn as a solid curve. The plots in black indicate that the minimum radius of the summation region r_{min}, is greater than or equal to 2 nm. The plots in open symbol indicate that the minimum radius of the summation region r_{min}, is greater than or equal to 1 nm. The error bars show the minimum and maximum values calculated for the summation regions shown in Fig. 6. In other words, if the length of the error bars is long, then the J^{a} value has a dependence on the summation region. When the value of K_{I} is small, then the J integral values obtained by the MD simulation agree with the values predicted by the linear elastic prediction. However, as K_{I} increases, the J^{a} values become smaller than the value predicted by the linear elastic prediction.

Comparing the data in Fig. 5 and Fig. 7 in detail, some characteristic loading levels concerning the data in the figure arise. According to Fig. 7, when K_{I} is nearly equal to 1.0, 1.5, 1.525, and 2.5 $\mathrm{MPa}\sqrt{\mathrm{m}}$, the crack extents, and/or the CTOD value increases with the gap, which means that the crack tip becomes blunt. At that time, the rate of increase in the value of J begins to decrease. When K_{I} is nearly equal to 1.75 $\mathrm{MPa}\sqrt{\mathrm{m}}$ in Fig. 8, the lower limit of error bars of the open symbols (for $r_{\mathrm{min}} \geq 1$ nm) suddenly decreased. From considering Figs. 3(b) and (c), the stress fields near 1 nm from the crack tip deviate from the linear elastic prediction for $K_{\mathrm{I}} \simeq 1.75\mathrm{MPa}\sqrt{\mathrm{m}}$. Moreover, the rate of increase in the J values for K_{I} around 2.5 $\mathrm{MPa}\sqrt{\mathrm{m}}$ decrease remarkably. From Fig. 5(d) for $K_{\mathrm{I}} = 2.625\mathrm{MPa}\sqrt{\mathrm{m}}$, a region consisting of atoms that have a low number density, which we consider as being a void, appears in front of the crack tip. The negative rate of increasing J value is therefore related to the void nucleation near a crack tip.

3.7 Dugdale Model Estimation for J^{a}-CTOD Relation

The relationship between J^{a} ($r_{\mathrm{min}} \geq 2$ nm) and CTOD are plotted in Fig. 9. There often seems to be expressed by Dugdale model (Dugdale, 1960) in which the CTOD value δ_{t} has a linear relationship with the J integral value as follows (Rice, 1968).

$$\sigma_{\mathrm{Y}}\delta_{\mathrm{t}} = J_{\mathrm{LE}}, \tag{6}$$

where the yield stress σ_{Y}. Estimation by Eq. (6) with choosing $\sigma_{\mathrm{Y}} = 6.5\mathrm{GPa}$ is drawn on the Fig. 9 as a solid line. The MD result and theoretical linear relation of Dugdale model has good agreement. The value $\sigma_{\mathrm{Y}} = 6.5\mathrm{GPa}$

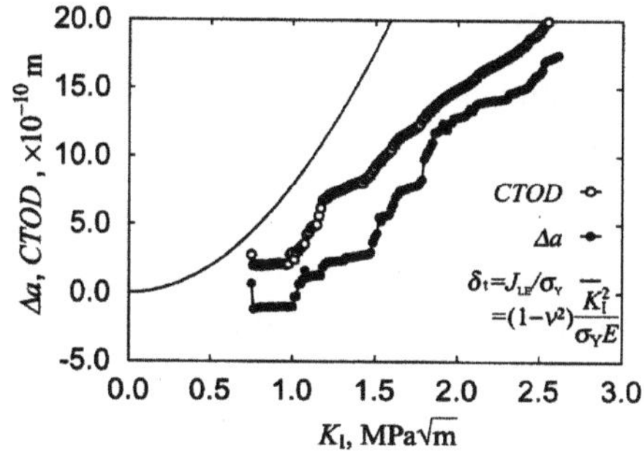

Figure 7. Crack tip opening displacement (CTOD) and crack extension (Δa)

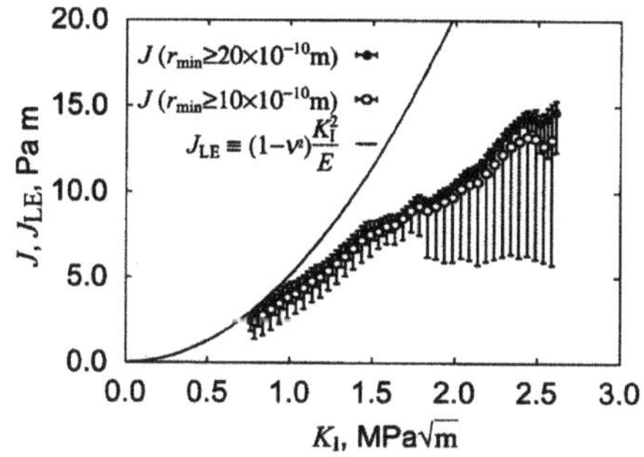

Figure 8. Plots of the J integrals. The error bars denote the maximum and minimum values of J integrals passing through various summation regions for the same K_I value.

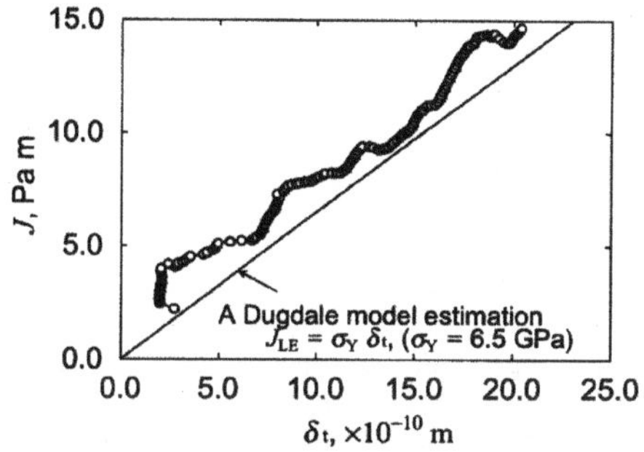

Figure 9. The Relationship between the CTOD and J integral.

is not so far from the yield stress obtained in a uniaxial tensile using same material model simulation (Nakatani et al., 1996b; Nakatani et al., 1997).

4. CONCLUSIONS

The MD method was applied to a cracked amorphous structure with mode I loading, to evaluate the mechanical fields near the crack tip actually.

The J^a parameter was proposed as a representative mechanical parameters near a crack tip for an atomistic model. By using the J^a value, the mechanical damage caused by the change of internal mechanical properties and the relationship between the crack propagation and J^a parameter were evaluated in detail.

The linear relationship between J^a and CTOD was seen and the property of crack blunting agrees well with using a Dugdale model.

Acknowledgments

The authors are grateful to Professor Peter Gumbsch for discussions on unique mechanism of fracture in amorphous and quasi-crystals. Suggestion from Professor Akihiro Nakatani with regard to the visualization of crack-void interaction is gratefully acknowledged. This work is partially supported by the Grant-in-Aid for Scientific Research through the Ministry of Education, Culture, Sports, Science and Technology, Japan.

References

Dugdale, D.S. (1960). ielding of steel sheets containing slits. *J.Mech.Phys. Solids*, 8:100 – 104.

Li, F.A., Shih, C.F., and Needleman, A. (1985). A comparison of methods for calculating energy release rates. *Eng.Frac.Mech.*, 21:405 – 421.

Masumoto, T and Maddin, R. (1975). Structural stability and mechanical properties of amorphous metals. *Mat.Sci.Eng.*, 19:1–10.

Nakatani, K., Kitagawa, H., and Nakatani, A. (1996a). Molecular dynamics simulation of the local structure and internal stress of single-component amorpheus metal. *Trans.Jpn.Soc.Mech. Eng.*, 62(595):847 – 852.

Nakatani, K., Kitagawa, H., and Nakatani, A. (1996b). Molecular dynamics study of deformation mechanism and damage in single-component amorphous. In *Proc. of 3rd. Asis-Pacific Symposimu on Advances in Engineering Plasticity and its Application (AEPA '96)*, pages 489–494.

Nakatani, K., Kitagawa, H., and Nakatani, A. (1997). Molecular dynamics simulation on the mechanical strength of one-component amorpheus metal — decreasing of elastic moduli and strength induced by change of internal structure. *J.Soc.Mat.Sci,Jpn*, 46(3):244–249.

Nakatani, K., Nakatani, A., and Kitagawa, H. (1998). *Mesoscopic Dynamics of Fracture*, chapter Molecular Dynamics Study on Fracture Mechanism of Fe-Amorphous Metal (J Integral near Mode I Crack Tip in Amorphous Metal), pages 88 – 98. Springer.

Nakatani, K., Nakatani, A., Sugiyama, Y., and Kitagawa, H. (2000a). Molecular dynamics simulation on crack growth in amorphous metal. *J.Soc.Mat.Sci.,Jpn*, 49(3):275–281.

Nakatani, K., Nakatani, A., Sugiyama, Y., and Kitagawa, H. (2000b). Molecular dynamics study on mechanical properties and fracture in amorphous metal. *AIAA J.*, 38(4):695 – 701.

Rice, J.R. (1968). A path independent integral and the approximate analysis of strain concentration by notches and cracks. *J. Appl. Mech.*, 35:379 – 386.

ELASTIC PROPERTIES OF THE SURFACES AND INTERFACES OF CRYSTAL AND AMORPHOUS SILICON

Satoshi IZUMI, Shotaro HARA, Tomohisa KUMAGAI, and Shinsuke SAKAI
University of Tokyo, JAPAN
izumi@fml.t.u-tokyo.ac.jp

Abstract We propose a method to calculate well-relaxed interface/surface properties by using order-parameters to evaluate interface/surface microstructures. Structural relaxation greatly influences the energies of the a/c interface and the surface. It is found that the interface energies are smaller than the surface energy and do not depend significantly on the crystal orientation. These findings agree very well with the experiments. The interface stress is also smaller than the surface stress and involves the scattering across a broad range.

Keywords: Molecular dynamics, Interface, Amorphous silicon, Surface

1. INTRODUCTION

Polycrystalline and amorphous silicon have a number of technological applications as thin film semiconductors. In recent years, the thickness of such films has approached several nm and much effort has been devoted to control the size and shape of the crystal grain, since grain geometry greatly affects the electrical and mechanical properties of thin films. From this viewpoint, crystal nucleation in the amorphous phase has been the subject of intensive experimental and theoretical study[1][2]. The interface energy between the amorphous and crystal phases is a key parameter for detecting the critical radius of a nucleus. Interface energy and interface stress are also key parameters for predicting the intrinsic stress of thin films. The intrinsic stress, which introduces the problems such as the delamination between film and substrate and the generation of dislocation, depends on the microstructures and growth mode at the atomic level[3][4][5].

H. Kitagawa and Y. Shibutani (eds.), IUTAM Symposium on Mesoscopic Dynamics of Fracture Process and Materials Strength, 117–127.

In this study, we propose a method to calculate such interface/surface properties by using classical molecular dynamics[6]. Unlike previous methods[7], the proposed method pay special attention to structural relaxation which has significant effects on surface/interface properties as well as on bulk properties.

2. DEFINITIONS OF INTERFACE/SURFACE ENERGY AND INTERFACE/SURFACE STRESS

2.1 Interface/surface model

In order to define the interface in a molecular dynamics system, we prepared a amorphous-crystal interface model by sticking two phases as shown in Fig. 1. Periodic boundary conditions are applied in all three directions. The films must be thick enough to prevent those two interfaces from interfering with each other. In the case of a surface model, the crystal layer is replaced by a vacuum space.

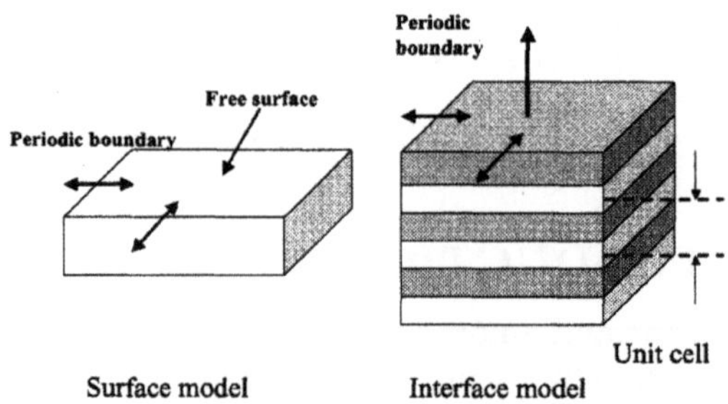

Figure 1. Schematic illustration of a/c interface and surface model

2.2 Definition of interface/surface energy

Interface/surface is the energy variation per unit area resulting from the interface/surface creation as defined by Eq. (1). E^{inter} and E^{cont} are the energy of a system that includes the interface/surface and the energy defined by the continuum approximation, which ignores the effect of interface/surface, respectively. The latter must be calculated separately from the bulk values. In the case of interface energy, the atoms in the transition region must be assigned to either amorphous phase or the crystal phase. A is the area of two interfaces/surfaces.

Here, the strain η is defined so that the cell size in the zero-strain state is equal to that of the zero-stress state under the continuum approximation.

$$\gamma^I(\eta) = \frac{1}{A(\eta)}\left(E^{inter}(\eta) - E^{cont}(\eta)\right) = \frac{E^{\gamma^I}(\eta)}{A(\eta)} \tag{1}$$

2.3 Definition of interface/surface stress

The interface/surface stress is the variation in interface/surface energy E^{γ^I} with respect to the strain η_{ij}. This variation corresponds to the stress generated by the creation of the interface/surface.

$$f_{ij}^I(\eta) = \frac{1}{A_0}\left.\frac{d(A(\eta)\gamma^I(\eta))}{d\eta_{ij}}\right|_{\eta'=0} \tag{2}$$

Applying this definition to molecular dynamics, Eq. (3) can be obtained by using E^{γ^I}.

$$\begin{aligned} f_{ij}^I(\eta) &= \frac{1}{A_0}\left.\frac{dE^{\gamma^I}}{d\eta_{ij}}\right|_{\eta'=0} \\ &= \frac{1}{A_0}\left(\left.\frac{\partial E^{inter}(\eta)}{\partial\eta_{ij}}\right|_{\eta'=0} - \left.\frac{\partial E^{cont}(\eta)}{\partial\eta_{ij}}\right|_{\eta'=0}\right) \\ &= f_{ij}^{inter}(\eta) - f_{ij}^{cont}(\eta) \end{aligned} \tag{3}$$

$f_{ij}^{cont}(\eta)$ must be calculated by the same procedure as in E^{cont}.

3. FABRICATION METHOD OF THE A/C INTERFACES AND AMORPHOUS SURFACES

The Tersoff potential[8](T3) is used for interatomic potential, which has been used in a broad range of studies of crystal and amorphous silicon and is known for its strong ability to express the physical properties of the bulk crystalline phase[9], the bulk amorphous phase[10][11], liquid phase[9][12] and surface structure[13]. Especially, unlike the Stillinger-Weber (SW) potential[14], T3 can describe the formation energy and bond length of bond defect that plays an important role in amorphization of silicon[15]. Therefore, we have chosen to use the Tersoff potential. The iso-volume and iso-thermal ensemble is used, along with a time-step of 1.08 fs.

Before a well-relaxed interface/surface model is fabricated, a well-relaxed bulk amorphous model must be prepared. Such a structure is created by the melt-quenched method, followed by annealing for 10 ns at 1600 K under zero-stress condition. After annealing, almost all point defects such as void and interstitial have disappeared. The averaged coordination number, the ratio of the 4-coordination, and bond-angle deviation become 4.05, 95.1 % and 11.7 °, respectively. Those values are in good agreement with experimental values. The details of the structure and related bulk properties of amorphous silicon are given in Ref. [11] and [16].

3.1 Amorphous-crystal interface

The amorphous model with 1000 atoms and crystal model with 600 atoms are stuck to each other in the (001) plane, as shown in Fig. 1. The in-plane cell sizes, L_x and L_y, must be set so as to minimize the misfit. The misfit strain of the a/c interface, $\varepsilon_m = (a_A - a_C)/a_C$, is about 0.005, where a_C and a_A are the lattice parameters of stress-free crystal structure and the lattice size of stress-free amorphous structure, respectively, with an equal number of atoms. In general, if the misfit is below 1 %, the effect can be ignored[17]. Therefore, we decide the cell size by using the averaged lattice size, $a_{AC} = (a_A + a_C)/2$.

Since the structure of an as-stuck interface is not sufficiently relaxed, relaxation must be advanced by thermal annealing at high temperature. We adopted an annealing temperature of 1600 K so that the temperature did not exceed the melting point (1900±50 K[18]). At 1600 K, the solid phase epitaxy (SPE) parallel to the a/c interface is observed at a slow growth rate (about 16cm/s), as shown in Fig. 2. For the analysis, we extract 30 snapshots per constant time interval. The analyses are performed after the system is cooled to 0K and optimized by using the conjugate-gradient method.

To clarify the effect of structural relaxation, a model annealed for 2 ns at 300 K is also prepared for comparison. We refer to it as the unrelaxed model. The SPE is not observed at that temperature. By using different initial conditions, three samples are prepared for each model; that is, the total of 3 samples are prepared for the unrelaxed model and the total of 90 samples (30 × 3) are prepared for the well-relaxed model. The a/c (111) interface models are prepared by using the same scheme as for the a/c (001) models.

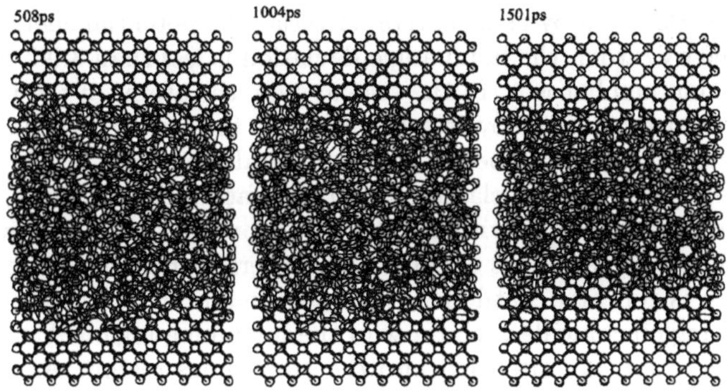

Figure 2. Snapshots of a/c (001) interface model during the solid phase epitaxy at 1600 K.

3.2 Amorphous surfaces

The surfaces are formed by removing the periodic boundary condition of the z-direction. After the surfaces are created, the systems are relaxed by the conjugated gradient method (unrelaxed surface) or by thermal annealing for 4 ns at 1200 K and 2 ns at 800 K (well-relaxed surface). This recipe is determined so that the state of the bulk does not change by annealing. Three samples are prepared.

4. METHOD TO CALCULATE A/C INTERFACE ENERGY AND STRESS

Since the interface would move due to the SPE during high-temperature annealing, the interface energy cannot be obtained simply from the energy variation before and after annealing. Therefore, a method to detect the moving interface region (transition region) is newly proposed. After that, the assignment of atoms in the interface region to both phases is conducted.

4.1 Detection method of a/c interface region

We classify each atom into three kinds of region (crystal, amorphous and interface regions) by evaluating the order parameters. The interface region is further separated into several subregions according to order parameters of the interface atoms; this separation is required for the classification. The order parameters are the potential energy per atom, E eV/atom, bond angle deviation, $\Delta\theta$ °, and averaged coordination number, N_{ave}. In the first step, the atoms of the crystal region (crystal atoms) are detected by using several criteria: the potential energy is less than -4.60 eV/atom, the bond angle deviation is less than 3.0 ° and the coordination number is 4. In the second step, the atoms of the interface regions are detected step by step according to the criteria based on bond information as described below. A non-crystal atom that bonds with at least one crystal atom is classified into interface region 1. Next, a non-crystal atom that bonds with at least one atom in region 1 (previously-defined interface region) is classified into interface region 2. The second procedure is repeated until the order parameters of the newly defined interface region coincide with those of the remaining region. Finally, the remaining atoms are classified into the amorphous region. As a result, three interface regions are established. The number of interface regions depends on the broadness of the interface.

A cross-sectional view of the classification is shown in Fig. 3 for the a/c (001) interface. In the interface region, both atoms and bonds are shown, whereas in the crystal and amorphous regions only bonds are shown. It is found that the interface regions are located on the region of transition from the ordered crystal to disordered amorphous regions, and that the width of interface region is about 6– atomic layers, which amounts to 0.8–0.9 nm.

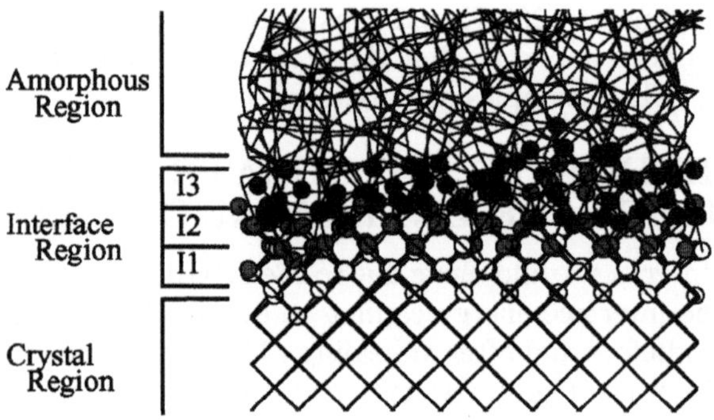

Figure 3. Cross-sectional view of the a/c (001) model after classification

4.2 Assignment of atoms to the two phases

To calculate the interface energy and interface stress, the interface atoms must be assigned between the two phases. Since this assignment includes arbitrariness, various criteria are proposed[7][19]. We assign the atoms so that no excess volume is associated with the interface[20]. Although that scheme is based on the thermodynamics of equilibrium interfaces, it is frequently used for nonequilibrium interfaces. In this system, the density of each interface region is very similar to those of the others. Therefore, this assignment is identical to the assignment in which region is divided equally between the amorphous and crystal phases, as shown in Eq. (4), where N_A, N_C and N_{int} are the number of atoms in the amorphous, crystal and interface regions, respectively and ε_A and ε_C are the energy per atom in the bulk amorphous and crystal regions, respectively.

$$E^{cont} = N_A\varepsilon_A + N_C\varepsilon_C + \frac{1}{2}N_{int}(\varepsilon_A + \varepsilon_C) \tag{4}$$

5. RESULTS AND DISCUSSIONS

5.1 Surface energy and surface stress

The surface energy γ and surface stress f (in-plane f_{11} and f_{22}) are shown in Table 1.

Table 1. Surface energy γ (J/m^2) and surface stress f (N/m) of the surfaces of crystal silicon and amorphous silicon, each with an evaluation area of 14.6 nm^2. The x-direction of the (2×1) surface model corresponds to the dimer-bonding direction.

	γ	f_{11}, f_{22}
(100) 1×1	2.27	-0.88
(100) 2×1	1.48	0.40, -1.34
a-Si(unrelaxed)	1.62±0.06	-0.50±0.19
a-Si(well-relaxed)	1.07±0.06	1.38±0.19

The surface energy of the crystal decreases due to the 2×1 reconstruction. The compressive in-plane surface stress also decreases. The well-relaxed amorphous surface has the lowest surface energy and a large tensile surface stress. These variations are caused by a large reconstruction of the amorphous surface. The coordination number and ring statistics are shown in the Tables 2 and 3, respectively. By comparing the unrelaxed amorphous surface with the well-relaxed one, we can see that annealing resulted in the disappearance of the 2-coordination N_2 and the increase of the 4-coordination N_4. Increases in the number of six- and seven-membered rings are also observed. It is thought that the strong structures of the bond network result from the surface reconstruction, and that those effects influence the surface properties greatly.

While the coordination number of a well-relaxed amorphous surface is almost the same as that of a crystal silicon surface (001)(2×1), the ring statistics show a different tendency; the amorphous surface has a larger number of five- and seven-membered rings. This reflects the difference in the bond network between amorphous and crystal surfaces.

Table 2. The ratio of the coordination number within 5.5 nm from the surface. The standard deviation of amorphous results is in the range of 0 to 2 %

	N_2	N_3	N_4	N_5	N_{ave}
(100) 1×1	22.2	0	77.8	0	3.56
(100) 2×1	0	22.2	77.8	0	3.78
a-Si(unrelaxed)	8.8	20.7	65.9	3.8	3.63
a-Si(well-relaxed)	0.0	22.6	74.8	2.5	3.79
a-Si(Bulk)	0	0.1	96.2	3.7	4.03

Table 3. Ring statistics within 5.5 nm from the surface

	R_3	R_4	R_5	R_6	R_7
(100) 1×1	0	0	0	1.33	0
(100) 2×1	0	0	0.22	1.44	0.22
a-Si(unrelaxed)	0.00	0.03	0.32	0.69	0.66
a-Si(well-relaxed)	0.00	0.03	0.40	0.81	0.80
a-Si(Bulk)	0.00	0.03	0.40	1.03	1.03

5.2 Interface energy and interface stress

The interface energy γ^I and interface stress f^I of the unrelaxed and well-relaxed models are shown in Table 4. To increase the accuracy of calculation, the averaged energies of the atoms that belong in the crystal region and bond only with the atoms of the crystal region are used as the crystal energy ε_C, and the amorphous energy ε_A is detected in the same manner. The interface stress is also calculated by the same procedure. Experimental values are shown for comparison. Our calculation shows excellent agreement with the experimental values. However, care must be taken to use our calculated values. For quantitative evaluation, ab-initio scheme such as the DFT is needed for verification. Direct comparison with the result of TBMD[7] is difficult since we adopted the different strategies for the assignment with those of Ref. [7]. However, both values are closely similar with each other.

The interface energy is lower in the well-relaxed model than in the unrelaxed model. A strong correlation can be said to exist between interface structure and interface energy. Only a small difference exists in

Table 4. Interface energy (J/m^2) and interface stress (N/m) of a/c (001) and a/c (111) interfaces. The evaluation area is 7.5 nm^2.

	γ (J/m^2)	f (N/m)
a/c(001)unrelaxed	0.96 ± 0.07	0.14 ± 0.01
a/c(001)well-relaxed	0.29 ± 0.08	0.29 ± 0.24
a/c(111)unrelaxed	0.85 ± 0.02	-0.29 ± 0.56
a/c(111)well-relaxed	0.33 ± 0.08	0.21 ± 0.31
TBMD a/c(001)[7]	0.49 ± 0.05	-
Hand-built a/c(111)[19]	0.31	-
Exp.[1]	0.23	-
Exp.[2]	0.25	-

the interface energy between a/c (001) and a/c (111). This suggests that the dependence of the a/c interface energy on the crystal orientation is very weak. The scattering of the interface stress is much broader than that of the interface energy, as is the case with surface stress. As far as we know, there are no experimental values for a/c interface stress. Therefore, a quantitative comparison is impossible. However, from the comparison with the value of metal/metal interface stress (about $\pm$ 2 N/m)[4], it can be said that the value of the a/c interface stress is comparatively smaller.

The structures of the unrelaxed and well-relaxed interface regions are then analyzed. The averaged order parameters of each region are shown in Fig. 4 in the case of the a/c (001) interface. The unrelaxed interface has higher energy and larger angle deviation, both resulting from insufficient structural relaxation. On the other hand, the distribution of the order parameters of the well-relaxed interface varies smoothly from the crystal to the amorphous regions.

Finally, in addition to the interface energy, the thermodynamic critical nucleus sizes are discussed. Piriolo et al.[1] and Kahn et al.[2] estimated them to be 0.57 nm (40 atoms) and 0.66 nm (61 atoms). Our calculation, which was about 20 atoms, does not agree with the experimental values. This difference is attributable to the difference in excess free energy Δg_{ac}. Since the critical radius amounts to several tens of atoms, it is doubtful that the continuum theory can be applied to such a small region. In order to investigate the validity of the classical nucleation theory, we have simulated the homogeneous nucleus nucleation by using molecular dynamics. Well-relaxed amorphous silicon (no initial crystal nucleus is included, unlike the case in the previous researches[21]) are annealed for 10-20 ns at 1700K. The number of atoms is 4096 and

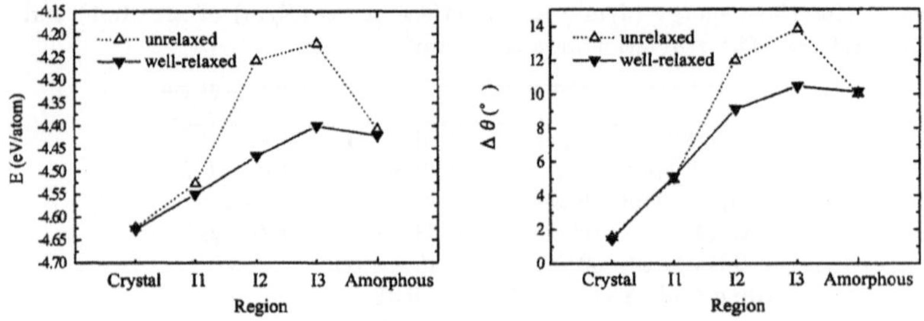

Figure 4. Averaged values of order parameters of each region; E (left) is the potential energy per atom and $\Delta\theta$ (right) is the bond angle deviation. I1, I2 and I3 indicate the number of interface regions.

periodic boundary conditions are applied. The critical radius r^* is estimated by the same criterion to calculate the interface energy. The size of the critical nucleus is estimated at 30~50 atoms, which agrees with the experimental estimation very well but disagrees with the estimation obtained by using the calculated interface energy. This disagreement is attributed to the small size of the nucleus, which is comparable to the width of the interface region ($0.8 \sim 0.9$ nm). Modified nucleation theory, including the effect of interface width, must be developed.

6. CONCLUSION

We propose a method to calculate well-relaxed interface/surface properties. Structural relaxation greatly influences the energies of the a/c interface and the surface. It is found that the interface energies are smaller than the surface energy, and that the dependence of the interface energy on the crystal orientation is small. These findings agree with the experiments very well. The critical nucleus size estimated by the calculated interface energy is smaller than that estimated by the direct MD simulation for homogeneous nucleation.

References

[1] C. Spinella, S. Lombardo, F. Piriolo, J. Appl. Phys., 48, 5383 (1998).

[2] H. Kahn, A. He, A. Heuer, Philos. Mag. A, 82, 137 (2002).

[3] R. Koch, J. Phys. Condens. Matter, 6, 9519 (1994).

[4] F. Spaepen, Acta Mater., 48, 31 (2000).

[5] J. A. Floro, S. J. Hearne, J. A. Hunter, P. Katuka, E. Chason, S. C. Seel, C. V. Thompson, J. Appl. Phys., 89, 4886 (2001).

[6] S. Hara, S. Izumi, S. Sakai, Proc. 2002 Inter. Conf. on Computational Engineering & Science, Reno, 131 (2002).

[7] N. Bernstein, M. Aziz, E. Kaxiras, Phys. Rev. B, 58, 4579 (1998).

[8] J. Tersoff, Phys. Rev. B, 37, 6691 (1988).

[9] S. Cook, P. Clancy, Phys. Rev. B, 47, 7686 (1993).

[10] M. Ishimaru, S. Munetoh, T. Motooka, Phys. Rev. B, 56, 15133 (1997).

[11] M. Ishimaru, J. Appl. Phys., 91, 686 (2002).

[12] M. Ishimaru, K. Yoshida, T. Motooka, Phys. Rev. B, 54, 7176 (1996).

[13] G. Hadjisavvas, G. Kopidakis, P. Kelires, Phys. Rev. B, 64 125413 (2001).

[14] F. Stillinger, T. Weber, Phys. Rev. B, 32, 5262 (1985).

[15] L. Marqués, L. Pelaz, J. Hernandez, J. Barbolla, G. Gilmer, Phys. Rev. B, 64, 045214 (2001).

[16] S. Izumi, S. Hara, T. Kumagai, S. Sakai, Comp. Mat. Sci, (submitted).

[17] K. Nakajima 'Mechanism of epitaxial growth', Kyoritsu, (2002) (in Japanese).

[18] L. Brambilla, L. Colombo, V. Rosato, F. Cleri, Appl. Phys. Lett., 77, 2337 (2000).

[19] F. Spaepen, Acta Metall., 26, 1167 (1978).

[20] F. Spaepen, Solid State Physics, 47, 1 (1994).

[21] J. K. Bording, J. Tafto, Phys. Rev. B, 62, 8098 (2000).

ATOMISTIC SIMULATION OF DISLOCATION GENERATION AT SURFACE STEPS IN METALS AND SILICON

S. Brochard, J. Godet, L. Pizzagalli, P. Beauchamp
Laboratoire de Métallurgie Physique, UMR 6630 Université de Poitiers-CNRS, SP2MI, BP 30179, Futuroscope Chasseneuil Cedex, France

Abstract Atomic scale simulations of a crystal with a free surface containing a surface step, submitted to a uniaxial stress have been performed using semi-empirical potentials representing metals (aluminum and copper) and silicon. In metals, different types of dislocations are nucleated for stresses well below the theoretical strength, according to the stress orientation, and the nucleation is preceded by a localization of shear confined into a single dense plane in-zone with the step. As expected, the behavior in silicon is very different from that of metals: besides complex plastic deformations appearing for large stresses, for those orientations where it has been possible to form dislocations, they have been found to glide in the shuffle set.

Keywords Dislocation nucleation; Atomistic simulations; Surface and interface

1. INTRODUCTION

In nanostructures (thin films, nanograins...), the observed presence of dislocations cannot be explained by the Frank-Read source mechanism because of the reduced dimensions and / or the absence of pre-existing dislocation to activate the source. In such materials, the interface or surface defects, such as steps, are good candidates to act as dislocation sources.

The mechanism of dislocation nucleation from a surface step being still out of reach of experiments, atomic scale simulations are expected to bring useful information. We have performed atomistic calculations of dislocation generation from a surface step under external applied stress. The first studies were conducted in metals, for which reliable interatomic potentials, well adapted for simulations at this length scale, exist. Afterwards, we studied silicon, as it is a prototype of semiconductor materials used in microelectronic devices, where the mechanisms under study are expected to occur frequently.

The main geometrical difference between the diamond-like structure of silicon and the simple f.c.c. crystals lies in the presence of two atoms per unit cell in the diamond-like structure, yielding two kinds of {111} plane sets, namely the shuffle set and the glide set [1].

H. Kitagawa and Y. Shibutani (eds.), IUTAM Symposium on Mesoscopic Dynamics of Fracture Process and Materials Strength, 129–138.

In silicon it is commonly accepted that, at least at high temperature, plastic deformation is mediated through dissociated dislocations located in glide set planes [2]. At low temperature, plastic deformation requires specific deformation conditions, as for example a confining pressure, to avoid fracture of the material. During recent experiments at low temperature perfect dislocations have been observed [3], which is consistent with gliding in the shuffle set plane, as predicted by calculations of the generalized stacking fault energies [4, 5, 6, 7]. In the study presented here, a particular attention will be paid to the type of plane, shuffle or glide, where the dislocations are nucleated in silicon, when they are.

2. MODEL

In this section, we briefly present the geometry and computational method used for this work. More detailed information can be found in references [8, 9]

2.1 Geometry – application of stress

In order to study the dislocation nucleation from a surface step, a crystal limited by a {100} free surface is constructed. In all the calculations presented here, the bulk crystal is simulated by freezing the opposite face; it has been ensured that releasing this constraint does not change the kind of dislocation nucleated, nor in a significant way the critical stress for dislocation nucleation. In the surface plane, the step lies along a <110> dense direction, intersection of a {111} glide plane and the surface. For the f.c.c. metals, this orientation determines in a unique way the resulting monoatomic step. In silicon, because of the 2×1 surface reconstruction, and depending if a single or a double step is considered, four different steps can result from this orientation [10]. The most stable step configurations (S_A and D_B) have been studied, but systematic calculations have been conducted only for D_B steps, which step height corresponds to the Burgers vector of a perfect dislocation. In this configuration, the step line is parallel to the dimerization direction (figure 1).

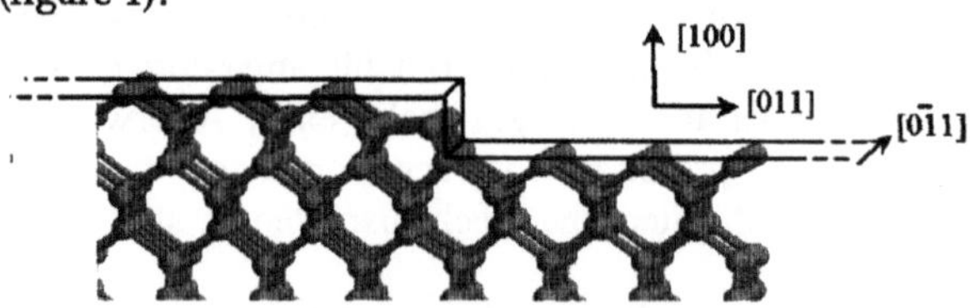

Figure -1. Orientation and geometry of the slab for silicon.

The uniaxial stress, whose orientation is contained in the surface plane, is applied through the displacements of all the atoms. These displacements are calculated using linear anisotropic elasticity (the elastic constants being determined for each potential). Both tensile and compressive stresses are studied. After the application of stress, the simulated box dimensions are maintained via two fixed buffers or through periodic boundary conditions along the direction perpendicular to the step line (in the latter case, two opposite steps are introduced on the surface). Along the step line direction, periodic boundary conditions are applied, with enough thickness to prevent spurious interaction between an atom and its image. However, the thickness along this direction, which is also the dislocation line direction when dislocations are nucleated, is too small to allow the formation of a half-loop: the nucleated dislocations are always straight. Different stress orientations have been studied. The stress direction is indicated by α (figure 2), the angle between the normal to the step line and the stress direction ($\alpha = 0°$ corresponds to a stress orientation normal to the step line).

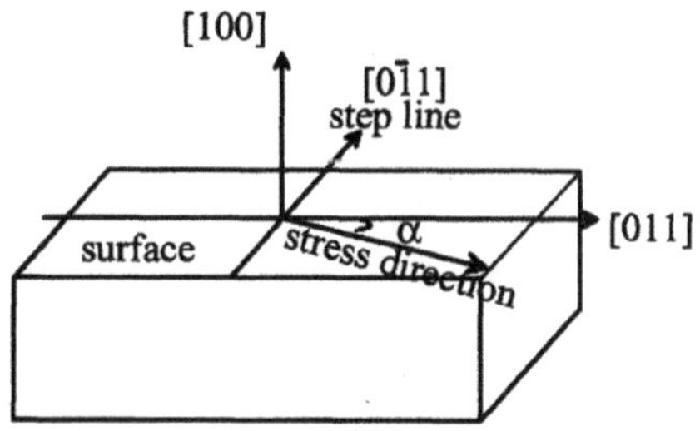

Figure -2. Geometry of the studied system.

2.2 Algorithms – potentials

For metals, simulations at 0 K were done, using a conjugate gradient algorithm for relaxation. For silicon, both relaxation at 0 K and molecular dynamics (MD) simulation [11] at finite temperature (300 K) have been performed, in order to favor dislocation nucleation. The MD simulation lasted typically 50 ps, and was then followed by a quench to minimize the energy. In all cases, the minimum energy was assumed to be reached when the mean force on each atom was less than 10^{-7} eV/Å. The interatomic interactions are derived from semi-empirical many-body potentials for the metals studied here (aluminum and copper) [12]. For silicon, we used three potentials that have proved their efficiency in different contexts, namely Stillinger-Weber potential (SW) [13], Tersoff potential (T) [14] and EDIP (Environment-Dependent Interatomic Potential) [15].

3. METALS (ALUMINUM AND COPPER)

Two different f.c.c. metals, aluminum and copper, have been studied. They differ in both their intrinsic stacking fault energy and elastic anisotropy coefficient, so that different behaviors are expected. With the potentials used, the intrinsic stacking fault energy for aluminum is 155 mJ/m^2 and the anisotropy factor is 1.07; for copper the values are 29 mJ/m^2 and 3.15. In the following, the results on metals are briefly described, emphasizing the main points for later comparison with the more complicated case of silicon. More details on metals can be found in [8, 16].

3.1 Results

The results for metals are summarized in Table 1, together with the type of leading and trailing dislocations resulting from the stacking of $\{111\}$ planes, the Schmid factors on these dislocations, and the theoretical shear strengths. The stress orientations have been chosen so that the ratio between Schmid factors on the trailing and leading dislocations is 0.5 or 2 ($\alpha = 0°$ and $\alpha = 45°$) and 1 ($\alpha = 18°$). The main results are:

(i) The dislocations are almost always formed at the step, and then glide in $\{111\}$ planes in-zone with the step.

(ii) When the nucleation event originates from the step, the stress threshold for dislocation nucleation is well below the theoretical shear strength.

(iii) Except in few cases where technical constraints can be put forward, the type of dislocation formed is well explained by the stress orientation (see the Schmid factors) and the stacking of the $\{111\}$ planes. For example, for $\alpha = 0°$, 90° Shockley partial dislocations are nucleated in traction, since they have the largest Schmid factor, and intrinsic stacking faults remain in the crystal after their formation. On the contrary, for the same orientation but in compression, the leading dislocation must be a 30° Shockley (a 90° would involve a prohibited stacking of the type ...ABCAABC...). But the Schmid factor on this partial dislocation is half the one for 90° Shockley, and no dislocation is nucleated until the theoretical strength is reached (aluminum) or a perfect 60° dislocation is nucleated for a quite large stress (copper).

(iv) Although some differences can be noted, the stress thresholds and the type of defect formed in relation with the stress orientation, are comparable in aluminum and copper.

Table -1. Dislocations nucleated in aluminum and copper with the corresponding strain and stress. The type of leading and trailing partial dislocations, as well as the corresponding Schmid factors are indicated in the first column. The value of the theoretical shear strength is given for a stress orientation normal to the step line.

	aluminum	**copper**
theoretical shear strength	13.7% (10.6 GPa)	7.3% (10.6 GPa)
$\alpha = 0°$, traction leading: 90° $(\sqrt{2}/3)$ trailing: 30° $(\sqrt{2}/6)$	two 90° Shockley partial dislocations at the step $\varepsilon = 8.4\%$ ($\sigma = 6.5$ GPa)	one 90° Shockley partial dislocation at the step $\varepsilon = 5.1\%$ ($\sigma = 7.4$ GPa)
$\alpha = 0°$, compression leading: 30° $(\sqrt{2}/6)$ trailing: 90° $(\sqrt{2}/3)$	surface and bulk nucleation $\varepsilon = 13.0\%$ ($\sigma = 10$ GPa)	one perfect 60° dislocation and one 30° Shockley partial at the step $\varepsilon = 6.6\%$ ($\sigma = 9.6$ GPa)
$\alpha = 18°$, traction leading: 90° $(9\sqrt{2}/30)$ trailing: 30° $(9\sqrt{2}/30)$	one perfect 60° dislocation at the step $\varepsilon = 9.5\%$ ($\sigma = 7$ GPa)	one 90° Shockley partial dislocation at the step $\varepsilon = 6\%$ ($\sigma = 6.5$ GPa)
$\alpha = 18°$, compression leading: 30° $(9\sqrt{2}/30)$ trailing: 90° $(9\sqrt{2}/30)$	one perfect 60° dislocation at the step $\varepsilon = 9.5\%$ ($\sigma = 7$ GPa)	one perfect 60° dislocation at the step and one 30° Shockley partial at the surface $\varepsilon = 7.7\%$ ($\sigma = 8.5$ GPa)
$\alpha = 45°$, traction leading: 90° $(\sqrt{2}/6)$ trailing: 30° $(\sqrt{2}/3)$	one perfect 60° dislocation at the step $\varepsilon = 11\%$ ($\sigma = 8$ GPa)	—
$\alpha = 45°$, compression leading: 30° $(\sqrt{2}/3)$ trailing: 90° $(\sqrt{2}/6)$	three 30° Shockley partial dislocations at the step $\varepsilon = 9.5\%$ ($\sigma = 7$ GPa)	—

3.2 Localized shear prior to dislocation nucleation

An important result of the calculations on metals is the presence, prior to any dislocation generation, of a shear localized in the {111} glide plane where the first nucleation event will occur [16]. It has been shown that this localized shear is a precursor of the fully formed dislocation, the latter appearing when the shear reaches, locally, a critical value for which the crystal becomes mechanically unstable (the theoretical shear strength in Frenkel model [17]). This shear localization has been related, via a non linear tension-shear coupling, to the local stress field originating from the step when an external stress is applied. It may be worth seeking if such a localized shear appears in silicon, as in metals.

4. SILICON

As mentioned in section 2, three different potentials were used to perform the calculations in silicon. In subsection 4.1 a comparative study of the three potentials is presented. Then, in subsection 4.2, we detail the results obtained with the SW potential, which proved to be the best suited for the problem studied here.

4.1 Comparative study of Si potentials

Preliminary tests indicated that the defects formed in strained samples were dependent on the potential used. In order to discriminate between the three potentials, their shear properties have been confronted to ab initio calculations. Two different types of tests have been conducted, the first involving homogeneous shear on {111} planes, and the second, generalized stacking fault energy surfaces ("γ-surfaces", as named by Vitek [18]) calculations on these planes. The criterion of choice for the potential is based on the description of the mechanism and energetics of bond switching which necessarily occurs at large enough strain.

The study of large homogeneous shear properties, in addition to more generally considered γ-surfaces, was partly motivated by the observation of a rather homogeneous strain field, even in the step region (absence of localized shear) in the case of silicon [9]. The results obtained with the three empirical potentials and an ab initio simulation [19] are detailed elsewhere [20]. They all agree on that, when the imposed shear is large, the mechanism of neighbor switching occurs by breaking the bonds across the shuffle set. In glide set planes, the deformation remains elastic, depending almost linearly on the internal shear stress. But the main point is that only the SW potential shows smooth energy variations and continuous internal shear stress, close to what is obtained with ab initio.

Subsequently, γ-surfaces have been computed for the three empirical potentials, and compared to first-principles results [4, 6]. Tersoff and EDIP potentials yield unstable stacking fault energy values closer to ab initio than the SW potential [20], but they show discontinuitics that arc not obtained with the SW potential, nor with ab initio techniques. This is particularly obvious on the curves derived from the γ-surfaces in the directions corresponding to a perfect 60° dislocation (<110>) in the shuffle or the glide plane or a partial Shockley dislocation (<112>) in the glide plane, where the continuity and smoothness of the ab initio curves are reproduced only by the SW potential, and not by Tersoff and EDIP potentials.

Consequently, the best potential for the problem studied here, that is formation of plastic defects under large stress, is clearly the SW potential.

4.2 Survey of the results with the SW potential

The results obtained with the SW potential are summarized in Table 2. Here again, different stress directions have been analyzed, favoring orientations for which the Schmid factor is maximum on the 90° Shockley partial dislocation ($\alpha = 0°$), on the 60° perfect dislocation ($\alpha = 22.5°$), or on the screw dislocation ($\alpha = 45°$).

Table -2. Type of dislocations nucleated in silicon and corresponding strain. The type of leading and trailing partial dislocations, as well as the corresponding Schmid factors are indicated in the first column. The value of the theoretical shear strength is given for a stress orientation normal to the step line.

	T = 0 K	T = 300 K
theoretical shear strength	32% (48 GPa)	—
$\alpha = 0°$, **traction** leading: 90° $(\sqrt{2}/3)$ trailing: 30° $(\sqrt{2}/6)$	fracture $\varepsilon = 25.1\%$	fracture $\varepsilon = 13.1\%$
$\alpha = 0°$, **compression** leading: 30° $(\sqrt{2}/6)$ trailing: 90° $(\sqrt{2}/3)$	formation of a microtwin at the step $\varepsilon = 7.8\%$	—
$\alpha = 22.5°$, **traction** leading: 90° (0.40) trailing: 30° (0.44)	one perfect 60° dislocation at the step $\varepsilon = 18.7\%$	perfect 60° dislocations at the step $\varepsilon = 12.5\%$
$\alpha = 22.5°$, **compression** leading: 30° (0.44) trailing: 90° (0.40)	plastic deformation in {111} planes $\varepsilon = 10\%$	one perfect 60° dislocation at the step $\varepsilon = 7.5\%$
$\alpha = 45°$, **traction** leading: 90° $(\sqrt{2}/6)$ trailing: 30° $(\sqrt{2}/3)$	formation of microtwins from surface and bulk $\varepsilon = 19.7\%$	—
$\alpha = 45°$, **compression** leading: 30° $(\sqrt{2}/3)$ trailing: 90° $(\sqrt{2}/6)$	one perfect 60° dislocation at the step $\varepsilon = 14.0\%$	—

It must be pointed out first that for all orientations, and at 0 K as well as at 300 K, the plastic defects (dislocations, microtwins or more complicated defects) are nucleated and glide in {111} planes of the shuffle set.

It may then be noticed that dislocation nucleation is in general much more difficult than in metals, which is consistent with the high theoretical shear strength. Another characteristic is that the defects are more easily formed in compression than in traction, in the sense of lower critical deformations. Complementary studies are needed to clarify this point. For all stress orientations, the presence of the step and temperature reduce the yield stress, but the type of defect formed is not as easily rationalized as in the case of f.c.c. metals. Except for the microtwins, which are a feature of the SW potential, the defects formed are perfect 60° dislocations.

As a rule of thumb, dislocations are not nucleated in glide set planes for the relatively low temperatures studied here, in agreement with the usual explanation that it would involve the breaking of three covalent bonds (compared to only one in the shuffle set planes). As a result, 90° Shockley partial dislocations (for $\alpha = 0°$ in traction) or 30° Shockley (for $\alpha = 45°$ in compression), which necessarily belong to a glide set plane (figure 3), are not nucleated, contrary to what happens in metals. For these stress orientations, in silicon, the applied deformation is accommodated via the propagation of a crack approximately normal to the surface for $\alpha = 0°$ in traction, and via the nucleation of a perfect 60° dislocation for $\alpha = 45°$ in compression.

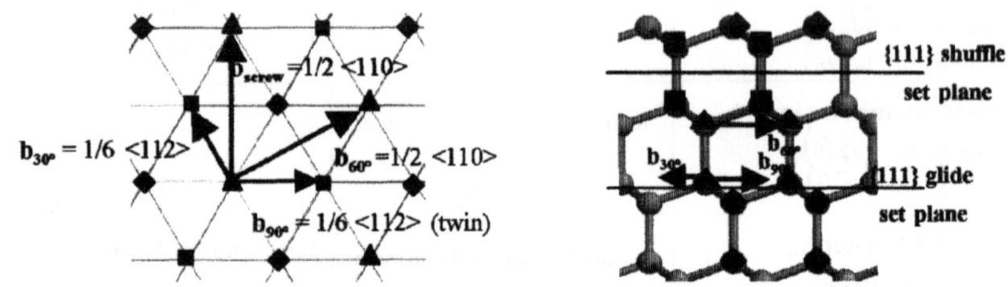

Figure -3. Diamond-like structure with the different Burgers vectors projected along <111> (left) and along <110> (right).

Regarding the two cases for which the trailing partial dislocations are favored by the Schmid factor against perfect dislocations or against the leading partial dislocations, i.e. $\alpha = 0°$ in compression and $\alpha = 45°$ in traction, a particular type of microtwin has been obtained and found to be formed by glide events occurring in the shuffle set [21]. Once again, it is the impossibility of nucleating partial dislocations which leads to the formation of this probably unphysical plastic deformation.

Finally, for $\alpha = 22.5°$, perfect 60° dislocations are nucleated (figure 4), as expected, both in traction and in compression, and more easily with temperature than without. As a matter of fact, at 0 K in compression, no dislocation forms at the step, and plastic events hardly analyzable, but clearly originating from the step in {111} planes are obtained. In traction the applied deformation must be increased up to 18.7% before the perfect dislocation forms from the step.

An important point of these calculations is that the active planes for gliding are always {111} planes of the shuffle set, a result also obtained with the other potentials (Tersoff and EDIP). The difficulty of breaking bonds seems to be a determinant factor in the process of dislocation nucleation in this covalent material.

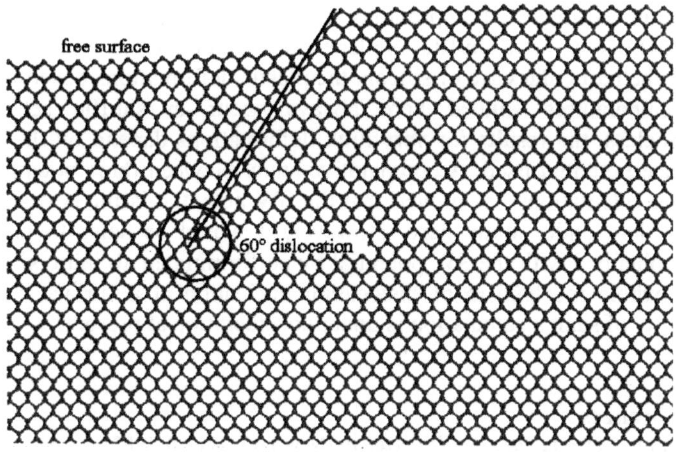

Figure -4. 60° dislocation nucleated for $\alpha = 22.5°$ and for an applied compression strain of 7.5% at 300 K. The solid line indicates the shuffle set plane where the dislocation has glided.

Furthermore, as mentioned previously, no precursory shear in the {111} glide planes in zone with the step is observed. An analysis based on a point force model has proved that the non-appearance of a localized shear is not due to the step geometry [9]. The non-appearance of the shear localization backs up once again the idea of a predominant role of the hard bond breaking in silicon.

5. CONCLUSION

Calculations of dislocation formation from an atomic surface step in a stressed crystal have been performed. Two f.c.c. metals, aluminum and copper, and a model semiconductor, silicon, have been studied. In both cases, the uniaxial stress orientation acts directly on the type of defect formed. But as expected, a different behavior is observed between metals and silicon. In silicon, dislocation formation is more difficult and requires higher stress. Perfect dislocations form in the shuffle set, and partial dislocations are never nucleated, which is quite different from what occurs in metals. In order to understand the mechanism of dislocation nucleation, the crystal structure just before nucleation has been analyzed. In metals, a localized shear in the glide plane where the first nucleation event will occur is observed, which is not the case in silicon. The localized shear observed in metals has been analyzed in terms of a tension-shear coupling [16]. Such a coupling must play a determinant role in the mechanism of dislocation nucleation, as suggested by previous studies [22], and may be different for a covalent bond across a shuffle set plane and a metallic bond.

For the particular case of silicon, the plastic deformations observed at the low temperatures studied here always occur in shuffle set planes. Nevertheless, the results for very high deformations depend on the type of empirical potential used. A comparative study between three potentials and ab initio calculations has proved that the SW potential is the best suited for the problem under study here. However, it would be safer to check the results obtained with the SW potential by performing an ab initio simulation of the whole mechanism of dislocation nucleation at a surface step. The feasibility of such a calculation is under consideration at present.

References

[1] Hirth J.P., Lothe J., *Theory of Dislocations* (New York, Wiley), 1982

[2] Alexander H., *Dislocations in Solids*, vol. 7, edited by F.R.N. Nabarro (Amsterdam: North-Holland), 1986

[3] Rabier J., Cordier P., Demenet J.L., Garem H., *Materials Science & Engineering* **A309-310** (2001) 74

[4] Kaxiras E., Duesbery M.S., *Physical Review Letters* **70** (1993) 3752

[5] Joós B., Ren Q. and Duesbery M.S., *Physical Review B* **50** (1994) 5890

[6] Juan Y.M. and Kaxiras E., *Philosophical Magazine. A* **74** (1996) 1367

[7] Miyata M. and Fujiwara T., *Physical Review B* **63** (2001) 045206

[8] Brochard S., Beauchamp P., Grilhé J., *Philosophical Magazine A* **80** (2000) 503

[9] Godet J., Pizzagalli L., Brochard S., Beauchamp P., *Scripta Materialia* **47** (2002) 481

[10] Chadi D.J., *Physical Review Letters* **59** (1987) 1691

[11] XMD code, written by Rifkin J. (URL: http://www.ims.uconn.edu/centers/simul#Software)

[12] Aslanides A., Pontikis V., *Computational Materials Science* **10** (1998) 401

[13] Stillinger F.H., Weber T.A., *Physical Review B* **31** (1985) 5262

[14] Tersoff J., *Physical Review B* **38** (1988) 9902

[15] Bazant M.Z., Kaxiras E., *Physical Review B* **56** (1997) 8542

[16] Brochard S., Beauchamp P., Grilhé J., *Physical Review B* **61** (2000) 8707

[17] Frenkel J., *Z. Physik* **37** (1926) 572

[18] Vitek V., *Philosophical Magazine* **18** (1968) 773

[19] The ABINIT code is a common project of the Université Catholique de Louvain, Corning Incorporated and other contributors (URL: http://www.abinit.org)

[20] Godet J. et al., submitted to *Journal of Physics: Condensed Matter*

[21] Godet J., Pizzagalli L. Brochard S., Beauchamp P., to be published in 12th Computational Materials Science workshop (2002) proceedings

[22] Sun Y., Beltz G.E., Rice J.R., *Materials Science & Engineering* **A170** (1993) 67

On the Plasticity and Fracture of Semiconductors

Pirouz Pirouz*, Shanling Wang*, Ming Zhang*, and Jean-Luc Demenet#
*Department of Materials Science and Engineering, Case Western Reserve University, Cleveland, OH 44106-7204, U.S.A.
#Laboratoire de Metallurgie Physique, CNRS, SP2MI, 86962 Futuroscope Cedex, France

Abstract Extending the measurement of the flow stress τ_y to low temperatures shows a transition in the deformation mode of a number of semiconductors. Instead of the $\tau_y(T)$ plot being an exponentially increasing curve with decreasing temperature, it actually consists of two branches intersecting at a critical temperature T_c. Similarly a plot of $\ln(\tau_y)$ versus $1/T$ shows two straight lines with different slopes intersecting at $1/T_c$. In all these semiconductors, it appears that the critical temperature T_c is close to the brittle-to-ductile transition temperature T_{BDT} of the crystal. In this paper, results of measurements on $\tau_y(T)$ for the wide bandgap semiconductor, 4H-SiC, are presented that show the transition point T_c. Additionally, results of 4-point bend tests, performed to directly measure the brittle-to-ductile transition temperature T_{BDT} of 4H-SiC and GaAs are shown. The results on SiC at four different strain rates confirm the identity of T_c and T_{BDT}. Based on these results, and extensive TEM investigation of the microstructure of the crystals deformed at temperatures above and below T_c, a model has been developed to explain the brittle-to-ductile transition temperature as well as deformation of semiconductors at different temperatures.

Keywords: III-V semiconductors; Other semiconductors; Mechanical properties of solids; Deformation and plasticity; Fatigue, brittleness, fracture and crack

H. Kitagawa and Y. Shibutani (eds.), IUTAM Symposium on Mesoscopic Dynamics of Fracture Process and Materials Strength, 139–152.

1. INTRODUCTION

In the first half of the last century it became clear that both plastic and fracture properties of crystals are facilitated by imperfections in the crystal. Early theoretical estimations showed that huge stresses would be required to deform a perfect crystal, and this has since been verified by experiments on nearly defect-free crystals such as whiskers. Similarly, theoretical estimates showed that the fracture of a crystal, that ultimately depends on rupturing interatomic bonds, is a nearly impossible process were it not for the existence imperfections in the crystal. The type of defects for plastic deformation and fracture of a crystal are of course very different; at moderate temperatures yielding is primarily carried forward by linear defects, i.e. dislocations, while fracture is facilitated by the existence of microscopic bulk defects, e.g. voids or microcracks.

Semiconductors are a class of materials which are primarily covalent although the compound semiconductors have a component of ionic bonding that can be quite large. In a covalent crystal, the motion of dislocations - through the elementary steps, kink nucleation and kink migration - involves breaking and re-forming interatomic bonds. Thus, because of their strong covalent/ionic bonds, semiconductors have very large lattice resistance that makes dislocation motion in them intrinsically difficult. However, kink nucleation and kink migration are both thermally activated processes and dislocation glide becomes easier at higher temperatures. Conversely, at low temperatures (say, less than half the absolute melting point), dislocation motion in semiconductors is very difficult and the stress required to move dislocations can be a large fraction of the elastic modulus *Y*, typically *Y*/100; this compares to *Y*/1000 or less for materials with a metallic bonding (e.g., Cu or Pb). The difficulty of moving dislocations in semiconductors translates to a high value of yield strength σ_y at low temperatures, of the order of a few *GPa*.

In the fracture of a material, the relevant defects are microcracks where their role is to concentrate and magnify the applied stress at the crack front to values exceeding the ideal strength of the crystal, i.e. to values required to rupture the interatomic bonds. Quantitatively, the stress concentration raises the local stress $\sigma_l(r)$ at a distance *r* ahead of a sharp crack of length *a* to [1]:

$$\sigma_l = \sigma + \sigma\sqrt{\frac{a}{2r}} \qquad \text{... (1)}$$

a value much larger than the average stress σ applied to the sample. Clearly this relation cannot hold for all *r* since, otherwise, it would go to infinity as $r \rightarrow 0$. At some distance, r_y, the local stress rises to a value reaching the yield stress σ_y of the material, giving rise to plastic flow. As expected, the value of r_y depends on the yield strength of the material, σ_y, and is given by:

$$r_y = \frac{K^2}{2\pi\sigma_y^2} \qquad \ldots (2)$$

where K is the stress intensity factor given by $K = C\sigma\sqrt{\pi a}$ with C a geometry-dependent constant with a value close to unity. A roughly hemispherical region of radius r_y ahead of the crack front may be defined as the *plastic zone*. One effect of the plastic zone is to blunt the crack tip and decrease its effectiveness as a stress concentrator. Another effect is that, depending on its size, the plastic zone shields the crack and decreases the tensile stress on it. Thus we see that the ease of bond rupture and crack propagation depends on the size and effectiveness of the plastic zone. From Eq. (2), this in turn depends on the applied stress σ and the crack length a (through the stress intensity factor K), and the yield strength of the material σ_y. In fracture mechanics, these conditions are expressed by saying that a crack can propagate in the solid only when K reaches a critical value K_c, known as the *critical stress intensity factor* (or *fracture toughness*) of the material. Microscopically, K_c is related to the rupture strength of the interatomic bonds; more macroscopically, it is related to material properties such as the elastic (Young's) modulus Y and the critical strain energy release rate G_c by $K_c = \sqrt{YG_c}$.

Plastic flow within the zone defined by r_y takes place by dislocation nucleation and motion, i.e., it depends on the yield strength of the material which is sensitively dependent on the test temperature. Thus, the effect of temperature on the fracture of the material comes through the manner in which the plastic zone affects crack propagation. At low temperatures, the yield strength of semiconductors is large and, according to equation (2), the plastic zone r_y is small, i.e. the zone is relatively ineffective in blunting and shielding the crack tip. As a result, the local stress ahead of the crack front exceeds the ideal strength of the crystal and the crack spreads rapidly between the pair of atomic planes that have the weakest bonding, a process called "cleavage". Because of the absence of local plastic deformation, this fracture surface, or "cleavage plane", is atomically flat and featureless. This is unlike the fracture surface of metals where appreciable local plastic deformation has occurred resulting in a rough surface.

In brief then, at low temperatures, semiconductors are generally brittle and fracture easily. Conversely, at higher temperatures, where dislocation nucleation and glide becomes easier, the plastic zone ahead of the crack tip becomes large and its effectiveness to reduce the local stress concentration increases; this in turn makes interatomic bond rupture ahead of the crack front and crack propagation more difficult. In terms of equation (2), one can say that starting from low temperatures, where the material is brittle, an increase in temperature decreases the yield strength of the semiconductor, thus decreasing its intrinsic lattice resistance. Dislocations can nucleate and move more easily resulting in an increase in the size of the plastic zone and causing the crack tip to become blunter. At a certain temperature, the tip is so blunt that fast cleavage can no longer occur. Thus, at high temperatures, semiconductors are generally ductile and their fracture resistance is high.

We can show the transition from brittleness to ductility graphically by considering the temperature variation of the yield stress, $\tau_y(T)$, and the fracture stress, $\sigma_F(T)$ [2]. In figure 1, the temperature dependence of yield stress is shown schematically, following the usual empirical relation for semiconductors [3-5]:

$$\tau_y(T) = A\dot{\varepsilon}^{1/n}\exp\left(\Delta H_\tau / k_B T\right) \qquad \text{... (3)}$$

In this equation. A and n are constants (different for the upper τ_{uy} and lower τ_{ly} yield stresses), and ΔH_τ is an energy parameter such that $n\Delta H_\tau$ is approximately the activation energy for dislocation glide ΔH_d. An increase in the strain rate $\dot{\varepsilon}$, say from $\dot{\varepsilon}_1$ to $\dot{\varepsilon}_2$, shifts the $\tau_y(T)$ curve to the right, i.e. to higher temperatures (Fig. 1). On the other hand, the normal stress to fracture the crystal is weakly temperature dependent and can be approximated as a constant, i.e.:

$$\sigma_n(T) \approx \sigma_F \qquad \text{.... (4)}$$

The temperature at which the $\tau_y(T)$ and $\sigma_n(T) \approx \sigma_F$ curves intersect defines the transition from brittleness to ductility [2]; this is known as the brittle-to-ductile transition temperature, T_{BDT}.

Consider now the response of a crystal to an applied stress σ_{app}. At low temperatures, the resolved normal component $\sigma_n(T)$ of the applied stress σ_{app} is less than the resolved shear stress τ_y while at higher temperatures, the situation is reversed. The simultaneous solution of equations (3) and (4) gives the intersection of the two curves at the brittle-to-ductile transition temperature and provides the following relationship [2]:

$$T_{BDT} = \frac{\Delta H_\tau}{k_B \ln\left(\dfrac{S\sigma_F}{A\dot{\varepsilon}^{1/n}}\right)} \qquad \text{... (5)}$$

In this equation, S is a geometric (Schmid) factor relating the resolved shear stress to the applied stress, $\tau = S\sigma_{app}$. From Fig. 1, it can be seen that for temperatures less than T_{BDT}, as the stress σ_{app} applied to the crystal increases, its normal component (σ_n) eventually reaches the normal stress curve for fracture $\sigma_n(T)=\sigma_F$ before its shear component (τ) reaches the yield stress curve $\tau_y(T)$. As a result, the crystal fractures and the interval $0<T_{BDT}<T$ defines the brittle regime. On the other hand, for temperatures

above T_{BDT}, the shear component (τ) of σ_{app} reaches $\tau_y(T)$ before its normal component (σ_n) reaches the normal stress curve for fracture $\sigma_n(T)$ curve and the crystal yields, i.e. the interval $T>T_{BDT}$ defines the ductile regime of the crystal.

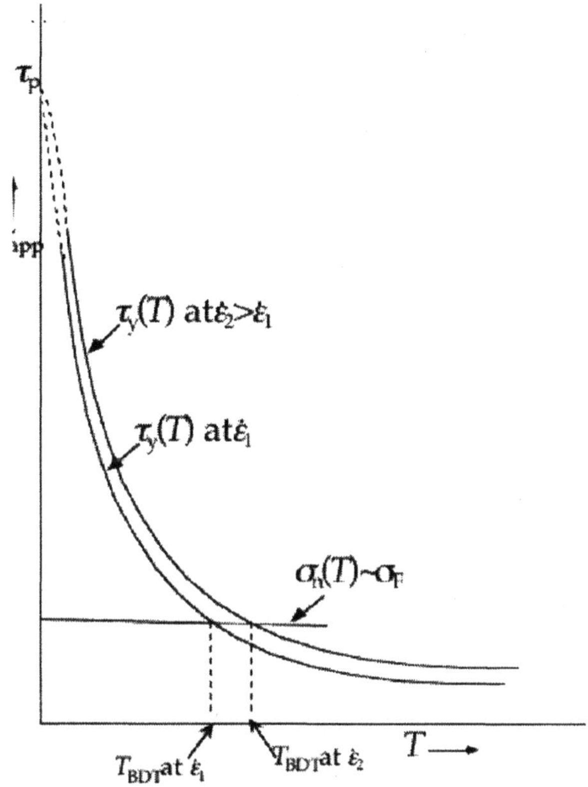

Figure 1. Temperature dependence of the yield strength $\tau_y(T)$ and normal stress to fracture $\sigma_n(T)\approx\sigma_F$ at different strain rates $\dot{\varepsilon}_1$ and $\dot{\varepsilon}_2$ ($>\dot{\varepsilon}_1$). As the strain rate increases, the yield strength curve $\tau_y(T)$ shifts to the right.

We have recently measured T_{BDT} at different strain rates $\dot{\varepsilon}$ for two materials, GaAs [6] and 4H-SiC [7]. In the remainder of this paper, we shall present a summary of the results for these two crystals (section 2). A discussion of the results is presented in section 3 together with some yield strength measurement of 4H-SiC [8-10], and a comparison of the latter with the T_{BDT} measurement in the same material.

2. EXPERIMENTAL

The BDT temperatures were measured by the 4-point bend technique – a constant $\dot{K}$ experiment - first employed by Samuels [11] in her work on silicon and more recently used by Zhang *et al.* [7] on 4H-SiC. For the measurements on undoped GaAs and 4H-SiC parallelepiped samples with dimensions 35×3×1 mm^3 (for GaAs) and 35×3×2 mm^3 (for 4H-SiC) were oriented such that the tensile stress on the cleavage plane was maximized and there was a reasonable resolved shear stress on the primary slip planes. The details of the experiments for 4H-SiC have been reported in Ref. [7] and will not be repeated here.

In the case of GaAs with a $\{110\}$ cleavage plane, the orientation of the sample was such that the 35×3 mm^2 top and bottom faces of the sample were parallel to the (001) plane, the 35×1 mm^2 side faces were parallel to the $(1\bar{1}0)$ plane, and the 3×1 mm^2 end faces were parallel to the (110) plane. In the tests, the inner and outer points of the 4-point bend jig were placed on the opposite (001) faces of the sample; the bending arm *d*, given by the difference between the outer and inner points, was 10 mm. With this geometry, the normal stress σ_{app} (in *Pa*) applied to the (110) end faces of sample is given by:

$$\sigma_{app} = 3Pd/wh^2$$

where P is the applied load (in *kg*), and w (=3 mm) and h (=1 mm) are respectively the width and thickness of the sample. Five pre-cracks were introduced in each sample by Knoop indentation prior to the deformation experiments. These pre-cracks were introduced along a line parallel to the 3 mm edge at the center of the (001) face of each sample. In each case, the indent load was 50 g and, after indentation, the sample was annealed at ~200°C to release the residual stresses. The 4-point bend jig was placed in a tube furnace and sample oxidation was prevented by flowing high purity argon gas over the sample during each experiment. The loading points in the 4-point bend jig were connected to the rod in an Instron machine. After heating up the sample to the desired temperature, the load was applied at a constant crosshead speed (i.e., a constant strain rate $\dot{\varepsilon}$) to either deform or fracture the sample.

In the brittle regime, as the load rises, the stress intensity factor K increases until it eventually reaches K_c whereby crack propagation starts from one of the pre-cracks in the sample and rapidly proceeds along the cleavage plane to break the sample in two. The crack surface (cleavage plane) in this case is flat and featureless. At higher temperatures ($T>T_{BDT}$), as the load increases, the crystal yield stress is reached and dislocations are nucleated and activated on their slip planes under the resolved shear stress to form a crack tip plastic zone which blunts the crack front and makes the crystal ductile. The crystal then continues its plastic deformation by dislocation motion and further nucleation.

In Fig. 2, the stress applied to fracture or plastically deform the sample is shown as a function of temperature for four different strain rates $\dot{\varepsilon}=1\text{x}10^{-6}$, $2\text{x}10^{-6}$, $5\text{x}10^{-6}$ and $1.5\text{x}10^{-5}\,s^{-1}$.

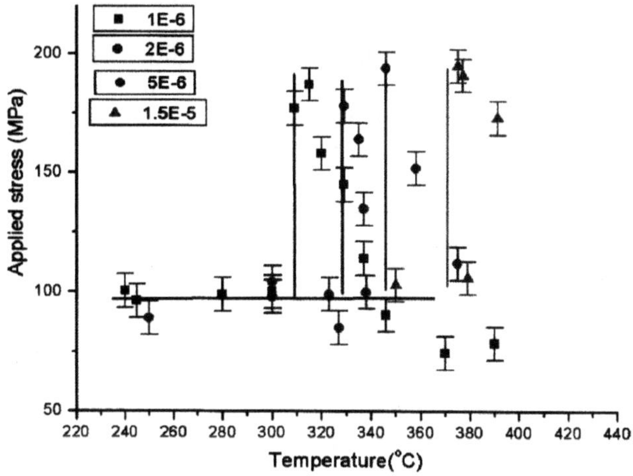

Figure 2. Temperature dependence of the applied stress σ_{app} needed to irreversibly deform the 4-point bend samples of undoped GaAs at four different strain rates. At each strain rate, the peak in σ_{app} occurs at the brittle-to-ductile temperature T_{BDT}. Note the systematic increase of T_{BDT} with the strain rate $\dot{\varepsilon}$.

In such a plot, the BDT temperature T_{BDT} at each strain rate is distinguished by a peak in the applied stress: for all temperatures T less than T_{BDT}, the sample is brittle and fractures on the cleavage plane at a nearly constant normal stress σ_{app} ($\approx\sigma_F$), whereas for all $T>T_{BDT}$, the material is ductile and deforms plastically and bends. Figure 2 shows that, in the brittle regime ($T<T_{BDT}$), irrespective of the strain rate employed, the normal tensile stress $\sigma_n=\sigma_F$ to fracture GaAs is about 100 MPa. In the ductile regime ($T>T_{BDT}$), the applied stress σ_{app} is that required to yield the crystal and, when resolved along the slip system with the highest Schmid factor S, it gives the yield stress of GaAs, $\tau_y=S\sigma_{app}$. In the present case, four slip systems, $\frac{1}{2}[01\bar{1}](111)$, $\frac{1}{2}[10\bar{1}](111)$, $\frac{1}{2}[\bar{1}0\bar{1}](\bar{1}\bar{1}1)$ and $\frac{1}{2}[0\bar{1}\bar{1}](\bar{1}\bar{1}1)$, each with S= $1/\sqrt{6}$, are equally activated. Also, as may be seen in Fig. 2, once passed the peak (i.e., at temperatures higher than T_{BDT}), σ_{app} no longer remains constant but decreases with increasing temperature; this is consistent with the decrease of the yield stress with temperature. In the ductile regime, if deformation is continued until the sample breaks, the fracture surface is rough and indicative of occurrence of appreciable localized plasticity.

According to equation (5), a plot of $\ln(\dot{\varepsilon})$ versus $1/T_{\mathrm{BDT}}$ should be a straight line with a slope $-\frac{n\Delta H_\tau}{k_\mathrm{B}} = \frac{\Delta H_\mathrm{d}}{k_\mathrm{B}}$, i.e., the slope of this plot should give the activation energy for dislocation glide. This was first suggested by St. John [12] for silicon and has since been verified by a number of other researchers [11,13-17]. In figure 3, we use the results of figure 2 to plot $\ln(\dot{\varepsilon})$ versus $1/T_{\mathrm{BDT}}$ for GaAs.

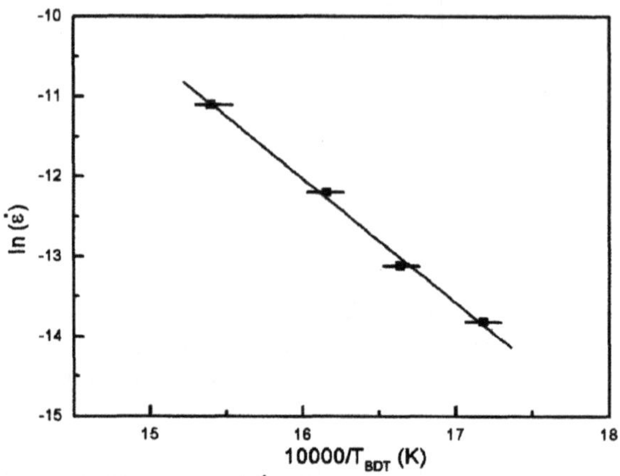

Figure 3. Plot of $\ln(\dot{\varepsilon})$ versus $1/T_{\mathrm{BDT}}$ for undoped GaAs. The slope of the straight line gives an activation enthalpy of 1.36±0.2 *eV*.

As in Si and Ge, the plot is a straight line with a slope that gives an activation enthalpy of 1.36±0.2 *eV*. A number of researchers have determined the velocity of both α [i.e., As(g)] and β [i.e., Ga(g)] dislocations in GaAs crystals with different doping concentrations [18-22]. In general there is little consistency between these results except for the agreement that α–dislocations are much faster than β-dislocations in both semi-insulating and n-type crystals while the trend reverses in p-type GaAs where β-dislocations are faster than α–dislocations. Warren [23] who used semi-insulating GaAs crystals with a dopant concentration (5.0×10^{15} cm^{-3}), probably closest to our crystals (resistivity of more than 10^7 $\Omega.cm$, corresponding to a dopant concentration of less than 10^8 cm^{-3}), determined the activation enthalpy for glide of α–dislocations to be 1.23±0.04 *eV* and for β-dislocations to be 1.35±0.02 eV. The close agreement between the value given by the slope of figure 3 (1.36±0.2 eV) and Warren's value [23] for β-dislocations (1.35±0.02 eV) may be an indication that it is the slow β dislocations that control the transition from brittleness to ductility in GaAs.

Using the same 4-point bend technique, we have also measured T_{BDT} in 4H-SiC for different strain rates [7]. The results are shown in Fig. 4 as a plot of $\ln(\dot{\varepsilon})$ versus $1/T_{BDT}$. The slope of the straight line in this figure gives an activation enthalpy of 2.47±0.2 eV. Unfortunately there are no direct measurements of dislocation velocity in any of the SiC polytypes and we are unable to compare the activation enthalpy determined from direct dislocation velocity measurements with that determined from BDT determinations.

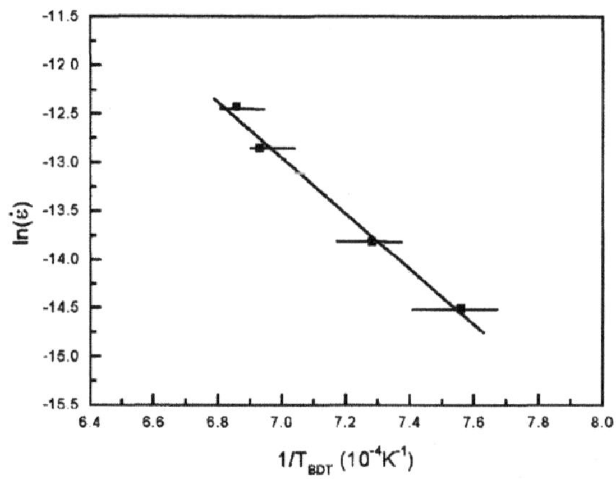

Figure 4. Plot of $\ln(\dot{\varepsilon})$ versus $1/T_{BDT}$ for 4H-SiC [7]. The slope of the straight line gives an activation enthalpy of 2.47±0.2eV.

3. DISCUSSION

According to Eq. (3), τ_y decreases exponentially with increasing temperature. Experimentally, the yield strength of many semiconductors has been measured and found to follow Eq. (3) pretty well (see, e.g. [5]). The results are usually shown as an Arrhenius plot of $\ln(\tau_y)$ versus $1/T$ at a fixed value of the strain rate. Eq. (3) shows that the slope of the resulting straight line should be proportional to the activation enthalpy for dislocation glide, $\Delta H_d = n\Delta H_\tau$. For practical reasons, most of the yield strength measurements in the literature pertain to the ductile regime of the semiconductor ($T > T_{BDT}$), where the samples undergo plastic deformation rather than fail by brittle fracture. It is possible to extend the measurements of τ_y to lower temperatures ($T < T_{BDT}$) provided brittle fracture by crack propagation is suppressed. Indeed, plastic deformation of semiconductors can be performed by indentation at very low temperatures (in their brittle regime) because the hydrostatic component of the indentation stress field suppresses fracture while its shear component drives plastic deformation. Indentation studies, however, provide only very limited quantitative information on the plasticity of a material because the indentation stress field is very complex, the resulting strains are not easily measurable, and the strain rate is not a variable. Thus, provided bulk crystals of the semiconductors are available, uniaxial compression or tension tests are much more advantageous and provide considerably more useful data. Such an experiment was first performed by Castaing *et al.* [24], who used a Griggs apparatus to deform silicon in the temperature range 300-600°C (i.e., in its brittle regime); with this apparatus, a hydrostatic pressure is superimposed during uniaxial compression of the crystal. Since then, many more tests have been performed on silicon [25-27] and a few compound semiconductors including GaAs [28-32], InP [33-35], InSb [34,35], GaP [34,35] and 4H-SiC [36]. Uniaxial deformation studies have also been conducted on 6H- and 4H-SiC without the presence of a hydrostatic pressure [8-10,37]. To succeed in the latter experiments, a few condition should generally be satisfied, e.g., presence of a low density of initial dislocations before the test, very careful alignment of sample in the deformation jig, and use of slow strain rates. Figure 5 shows the results of a series of compression tests on 4H-SiC, spanning a relatively large temperature range from the brittle to the ductile regime, at three different strain rates [10]. The data are presented in the form of Arrhenius plots as $\ln(\tau_y)$ versus $1/T$.

What is interesting in such plots is the presence of two straight lines with different slopes at each strain rate. The two lines intersect at a critical temperature T_c that shifts to a higher temperature with increasing values of $\dot{\varepsilon}$. The different slopes of the straight lines signify a change of deformation mode at T_c. Invariably, all the 4H-SiC crystals that were compressed at temperatures less than T_c contained one or more microcracks, while all crystals that were tested at higher temperatures ($T > T_c$) deformed much more readily in a ductile manner. Based on this observation, and the fact that T_c increases systematically with an increase in the strain rate, we speculated that the transition temperature T_c and the brittle-to-ductile transition temperature T_{BDT} were actually identical [2,38-40]. A comparison of T_c (from Fig. 4) and T_{BDT} (from Fig. 5) shows that there is indeed reasonably good agreement between the two transition temperatures.

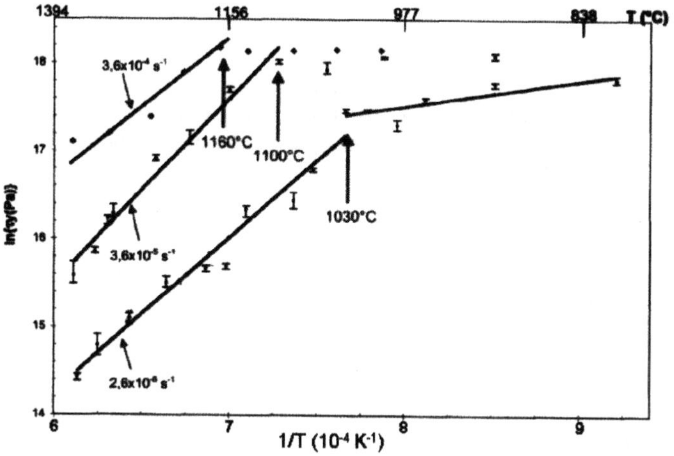

Figure 5. Temperature dependence of the yield stress of 4H-SiC at three different strain rates. At each $\dot{\varepsilon}$, the results are shown as plots of $\ln(\tau_y)$ versus $1/T$. There is a transition in the deformation mode at a critical temperature T_c where the slope of the line changes.

It should be noted that very similar curves of $\ln(\tau_y)$ versus $1/T$ have been reported by Suzuki and coworkers for four other compound semiconductors, InP, InSb, GaAs, and GaP [33-35]. We believe the existence of a critical temperature, associated with a change in the deformation mode, is actually a general feature of all tetrahedrally-coordinated semiconductors. However, in the $\tau_y(T)$ plots, or, equivalently, in the $\ln[\tau_y(1/T)]$ plots, that have been determined in the presence of a superimposed hydrostatic pressure, the transition temperature T_c is very likely shifted to lower temperatures. The reasoning for this can be seen from Fig. 1. A superimposed hydrostatic pressure counteracts the tensile stress necessary for the fracture of the crystal. This means that the $\sigma_F(T)$ line in figure 1 will be shifted to a higher value and, consequently the intersection of the $\tau_y(T)$ curve and the $\sigma_F(T)$ line will occur at a lower temperature, i.e. the BDT temperature will be shifted to the left and the sample is effectively ductile over a larger temperature span. The degree of the shift of the $\sigma_F(T)$ line and the BDT temperature will of course depend on the magnitude of the superimposed hydrostatic pressure; for the experiments in Refs. [33-35], where the hydrostatic pressure is ~1.2 *GPa*, we think the shift in T_{BDT} could be as much as a few hundred degrees.

4. CONCLUSION

The brittle-to-ductile transition temperature T_{BDT} of two semiconductors with different structures and bandgaps have been measured at different strain rates by the technique of 4-point bending of pre-cracked samples. One material has a wide bandgap (~3.2 *eV*) with a rhombohedral structure, while the other, GaAs, has a moderate bandgap (~1.4 *eV*) and a cubic sphalerite structure. In both cases, T_{BDT} is sharp and a plot of $\ln(\dot{\varepsilon})$ versus $1/T_{BDT}$ follows an Arrhenius relationship. From the slope of these plots, the activation enthalpy for glide of dislocations in undoped GaAs and 4H-SiC have been determined. The value for GaAs, 1.36±0.02 *eV*, is in good agreement with the literature value as determined by direct dislocation velocity measurements. The much higher value of T_{BDT} (~1000 versus ~300°C at a strain rate of $10^{-6}\ s^{-1}$) and larger activation enthalpy for dislocation glide (2.47 versus 1.36 *eV*) in SiC compared to GaAs reflects the different bond strengths in the two crystals. In addition, the temperature dependence of the yield strength of 4H-SiC has been measured over a large temperature range at three different strain rates. At each strain rate, the $\tau_y(T)$ plot consists of two branches that intersect at a critical temperature T_c indicating that there is a change in the mode of plastic deformation. The same transition is more clearly observed in an Arrhenius plot of $\ln(\tau_y)$ versus $1/T$. It is found that the critical temperatures T_c in the yield stress measurements of 4H-SiC correspond very closely with the brittle-to-ductile transition temperatures T_{BDT} of this material.

ACKNOWLEDGMENT

This material is based upon work supported by the National Science Foundation under Grant No. DMR-0108303 and the Basic Energy Sciences Division of the Department of Energy under Grant No. FG02-93ER45496.

REFERENCES

1. M. F. Ashby and D. Jones, R. H., *Engineering Materials 1: An Introduction to their Properties and Applications*, Vol. 1st ed. (Pergamon Press: Oxford, 1980).
2. P. Pirouz, M. Zhang, J.-L. Demenet, and H. M. Hobgood, *Yield and fracture properties of the wide band-gap semiconductor 4H-SiC*, J. Appl. Phys. **93**(6), 3279-3290 (2003).

3. H. Alexander, *Dislocations in Covalent Crystals*, in *Dislocations in Solids*, Vol. **7**, edited by F. R. N. Nabarro (Elsevier Science Publishers B.V.: Amsterdam, 1986), *pp.* 114-234.
4. A. George and J. Rabier, *Dislocations and Plasticity in Semiconductors. I - Dislocation Structures and Dynamics*, Rev. Phys. Appl. **22**, 941-966 (1987).
5. J. Rabier and A. George, *Dislocations and Plasticity in Semiconductors. II. The Relation Between Dislocation Dynamics and Plastic Deformation*, Rev. Phys. Appl. **22**, 1327-1351 (1987).
6. S. Wang, M. Zhang, J. Bradby, and P. Pirouz, *The brittle-to-ductile transition in GaAs*, To be submitted. (2003).
7. M. Zhang, H. M. Hobgood, J. L. Demenet, and P. Pirouz, *Transition from Brittle Fracture to Ductile Behavior in 4H-SiC*, J. Mater. Res. **18**(5), 1087-1095 (2003).
8. A. V. Samant, *Effect of Test Temperature and Strain-Rate on the Critical Resolved Shear Stress of Monocrystalline a-SiC*, Ph.D. Thesis, Case Western Reserve University, 1999.
9. J.-L. Demenet, M. H. Hong, and P. Pirouz, *Plastic Behavior of 4H-SiC Single Crystals Deformed at Low Strain Rates*, Scripta mater. **43**(9), 865-870 (2000).
10. J.-L. Démenet, M. Zhang, H. M. Hobgood, and P. Pirouz, *La transition fragile-ductile de 4H-SiC monocristallin*, J. Phys. IV France **106**, 53-62 (2003).
11. J. Samuels, *The Brittle to Ductile Transition in Silicon*, D. Phil. Thesis, University of Oxford, 1987.
12. C. St. John, *The brittle-to-ductile transition in pre-cleaved silicon single crystals*, Phil. Mag. **32**, 1193-1212 (1975).
13. M. Brede and P. Haasen, *The Brittle-to-Ductile Transition in Doped Silicon as a Model Substance*, Acta metall. **36**(8), 2003-2018 (1988).
14. J. Samuels and S. G. Roberts, *The brittle-ductile transition in silicon. I. Experiments*, Proc. R. Soc. Lond. A **421**, 1-23 (1989).
15. M. Brede, *The Brittle-To-Ductile Transition in Silicon*, Acta metall. mater. **41**(1), 211-228 (1993).
16. G. Michot and A. George, *Fracture and crack tip plasticity in silicon and gallium arsenide*, in International Symposium on Structure and Properties of Dislocations in Semiconductors, (Institute of Physics: Bristol, England, 1989), Inst. Phys. Conf. Ser. No. **104**, *pp.* 385-396.
17. F. C. Serbena and S. G. Roberts, *The Brittle-To-Ductile Transition in Germanium*, Acta metall. mater. **42**(7), 2505-2510 (1994).
18. V. B. Osvenskii and L. P. Kholodnyi, *Mobility of Individual Dislocations in Gallium Arsenide*, Soviet Physics - Solid State **14**(11), 2822-2825 (1973).
19. S. A. Erofeeva and Y. A. Osip'yan, *Mobility of dislocations in crystals with the sphalerite lattice*, Sov. Phys. - Solid State **15**(3), 538-540 (1973).
20. V. B. Osvenskii, L. P. Kholodnyi, and M. G. Mil'vidskii, *Influence of dopants on the velocity of dislocations in GaAs single crystals*, Sov. Phys. - Solid State **15**(3), 661-662 (1973).
21. S. K. Choi, M. Mihara, and T. Ninomiya, *Dislocation Velocities in GaAs*, Jpn. J. Appl. Phys. **16**(5), 737-745 (1977).
22. I. Yonenaga and K. Sumino, *Dynamic activity of dislocations in gallium arsenide*, J. Appl. Phys. **73**(4), 1681-1685 (1993).
23. P. D. Warren, *The Relation Between Electronic and Mechanical Properties of Non-Metals*, Ph.D. Thesis, University of Oxford, 1987.
24. J. Castaing, P. Veyssière, L. P. Kubin, and J. Rabier, *The plastic deformation of silicon between 300°C and 600°C*, Phil. Mag. A **44**(6), 1407-1413 (1981).
25. J. Rabier, P. Veyssière, and J. L. Demenet, *Plastic Deformation of Silicon at Low Temperature and the Influence of Doping*, J. Physique 44, Colloq. C4, C4-243 - C4-253 (1983).

26. J. L. Demenet, *Etude du silicium a basse et moyenne temperature sous forte contrainte*, Thèse d'etat, Université de Poitiers (UFR Sciences fondamentales et appliquees), No 457, 1987.
27. J. Rabier, P. Cordier, J.-L. Demenet, and H. Garem, *Plastic deformation of Si at low temperature under high confining pressure*, Mat. Sci. Eng. A **309-310**, 74-77. (2001).
28. J. L. Demenet, P. Boivin, and J. Rabier, *Plasticity and dislocation mobilities at low temperatures in silicon and gallium arsenide*, in International Symposium on Structure and Properties of Dislocations in Semiconductors, (Institute of Physics: Bristol, England, 1989), Inst. Phys. Conf. Ser. **104**, *pp.* 415-420.
29. J. Rabier, H. Garem, J. L. Demenet, and P. Veyssière, *Plastic Deformation of GaAs Single Crystals at Room Temperature and the Influence of Doping*, Phil. Mag. A **51**(6), L67-L72 (1985).
30. P. Boivin, J. Rabier, and H. Garem, *Plastic deformation of GaAs single crystals as a function of electronic doping. I: Medium temperatures (150-650°C)*, Phil. Mag. A **61**(4), 619-645 (1990).
31. P. Boivin, J. Rabier, and H. Garem, *Plastic deformation of GaAs single crystals as a function of electronic doping. II: Low temperatures (20-300°C)*, Phil. Mag. A **61**(4), 647-672 (1990).
32. T. Suzuki, T. Tokuoka, I. Yonenaga, and H. O. K. Kirchner, *Inverse Brittle-to-Ductile Transition in Gallium Arsenide Under Hydrostatic Pressure*, Scripta mater. **43**(7), 645-650 (2000).
33. T. Suzuki, T. Nishisako, T. Taru, and T. Yasutomi, *Plastic deformation of InP at temperatures between 77 and 500 K*, Phil. Mag. Lett. **77**(4), 173-180 (1998).
34. T. Suzuki, T. Yasutomi, T. Tokuoka, and I. Yonenaga, *Plasticity of III-V Compounds at Low Temperatures*, phys. stat. sol. (a) **171**, 47-52 (1999).
35. K. Edagawa, H. Koizumi, Y. Kamimura, and T. Suzuki, *Temperature Dependence of the Flow Stress of III-V Compounds*, Phil. Mag. A **80**(11), 2591-2608 (2000).
36. J.-L. Demenet, J. Rabier, X. Milhet, M. H. Hong, P. Pirouz, I. Stretton, and P. Cordier, *Microstructures of 4H-SiC single crystals deformed under very high stresses*, J. Phys.: Condens Matter **14**(48), 12961-12966 (2002).
37. A. V. Samant, W. L. Zhou, and P. Pirouz, *Effect of Temperature and Strain Rate on the Yield Stress of Monocrystalline 6H-SiC*, phys. stat. sol. (a) **166**, 155-169 (1998).
38. P. Pirouz, A. V. Samant, M. H. Hong, A. Moulin, and L. P. Kubin, *On temperature-dependence of deformation mechanism and the brittle-ductile transition in semiconductors*, J. Mater. Res. **14**(7), 2783-2793 (1999).
39. P. Pirouz, A. V. Samant, M. H. Hong, A. Moulin, and L. P. Kubin, *On deformation and fracture of semiconductors*, in Microscopy of Semiconducting Materials, (Institute of Physics: Bristol, England, 1999), Inst. Phys. Conf. Ser. No. **164**, *pp.* 61-66.
40. P. Pirouz, J. L. Demenet, and M. H. Hong, *On transition temperatures in plasticity and fracture of semiconductors*, Phil. Mag. A **81**(5), 1207-1227 (2001).

HVEM/AFM Studies on Crack Tip Plasticity in Si Crystals

Kenji Higashida and Masaki Tanaka
Department of Materials Science and Engineering, Kyushu University
6-10-1 Hakozaki, Higashi-ku, Fukuoka, 812-8581 JAPAN
E-mail: higasida@zaiko.kyushu-u.ac.jp

Abstract Crack tip plasticity in silicon crystals has been studied by both high voltage electron microscopy (HVEM) and atomic force microscopy (AFM). Cracks were introduced into silicon wafers at room temperature by Vickers indentation method. Specimens indented were annealed at temperatures higher than 773K to activate dislocation sources around a crack tip under the residual stress due to the indentation. In the present study, two types of plastic zones were examined: the 45° -shear-type in {001} wafers and hinge-type in {011} wafers. In AFM observations, very fine and sharp slip bands were found around crack tips in both types of plastic zones, where the step heights in the slip bands were around a few nanometers. In HVEM study, characteristic dislocation configurations have been observed in each type of plastic zone. Contrast simulations for the dislocations revealed not only their slip systems but also the signs of their Burgers vectors. The dislocation structures characterized by HVEM correspond well with the slip bands found by AFM. Those dislocations were shielding-type, which contributes to the increase of facture toughness to cause the sharp brittle-to-ductile transition of silicon crystals.

Keywords: dislocation, crack, fracture toughness, brittle-to-ductile transition, silicon

1. INTRODUCTION

Brittle-to-ductile transition (BDT) is one of the most important phenomena in the mechanical behavior of crystalline materials, which should be understood by dislocation theory. To have a comprehensive understanding of the BDT, it is essential to clarify the detailed behavior of plastic deformation around a crack tip. In highly perfect silicon crystals, a sharp brittle-to-ductile transition appears,[1] so that they have been employed as a suitable model substance to investigate the BDT process, and some significant results have been obtained by using X-ray topography, etch pits and transmission electron microscopy (TEM).[2-6] However, experimental studies on fundamental behavior of crack tip plasticity have been still limited, and

H. Kitagawa and Y. Shibutani (eds.), IUTAM Symposium on Mesoscopic Dynamics of Fracture Process and Materials Strength, 153–162.

further investigations are required to clarify the generation or multiplication process of crack tip dislocations which control the sharp BDT.

In the present study, both high voltage electron microscopy (HVEM) and atomic force microscopy (AFM) have been employed to examine the crack tip plasticity in silicon crystals. Since HVEM is applicable to observe thick specimens (thickness>10^{-6}m) in comparison with usual TEM, three dimensional aspects of crack and dislocations were easily examined by using HVEM. In addition, nano-scale fine slip bands near a crack tip was also observed by AFM. On the basis of these observations, fundamental aspects of crack tip plasticity is discussed.

2. EXPERIMENTAL

Si single crystals of {100} and {110} p-type wafers commercially available were employed. Cracks were introduced by indenting, with a Vickers hardness tester with a load of 1.96 N, on a wafer chip surface at room temperature (see Fig.1(a)). In order to introduce dislocations around the crack tip, the specimens indented were annealed at temperatures higher than 823K for a few minutes. This method enables us to seize the aspect of crack tip dislocations at the very beginning of multiplication process. For HVEM observation, a selected area around the tip of

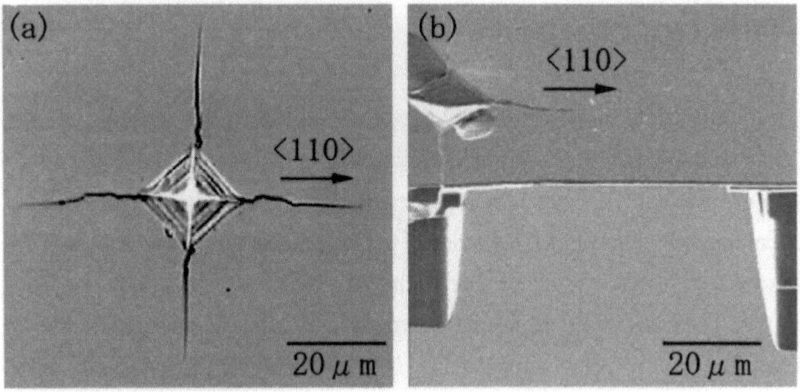

Figure 1. (a) Optical micrograph of an indent on the {100} plane. Cracks propagate from the edges of the indent along the <110> direction. (b) Image of a scanning ion microscope (SIM) taken during the FIB fabrication.

a crack was thinned by focused ion beam (FIB), as shown in Fig.1 (b). HVEM observations were carried out using JEM-1000 (JEOL) at the Research Laboratory for High Voltage Electron Microscopy, Kyushu University. Accelerating voltage was 1000 kV. In order to characterize crack tip dislocations, computer simulations of the dislocation images were carried out using a program based on the Howie-Whelan formulation.[7-8] The Burgers vector of dislocations is defined by FS/RH (finish-start/right hand) rule where the closure failure is specified in the perfect crystal.[9] According to this definition of Burgers vector **b**, the signs of screw dislocations are denoted by

$$\begin{aligned} \mathrm{sign}(\mathbf{b}.\mathbf{u}) &> 0 \quad \text{for right-handed (R-H) screw dislocations} \\ &< 0 \quad \text{for left-handed (L-H) screw dislocations} \qquad (1), \end{aligned}$$

where **u** is the dislocation line vector. Fine slip bands formed around the crack tip were observed using an atomic force microscope (AFM, SPI-400 (SII)).

3. RESULTS AND DISCUSSION

3.1 AFM Observation of Slip Bands around a Crack Tip

Figures 2 (a) and (d) show AFM images of slip bands formed around the crack tip in specimens of {001} and {011} wafers, respectively. Those slip bands were observed in specimens annealed at temperatures higher than 823K after the indentation. Without the heat treatment, no slip bands were observed, indicating that dislocations were activated around the crack tip at such high temperatures under residual stresses due to the indentations. TEM study also showed that no dislocations were generated around a crack tip before anneling.[6]

In Fig.2 (a), a crack is seen along the $[1\bar{1}0]$ direction on the (001) wafer surface. The crack plane is parallel to (110). Slip bands are observed along both the $[1\bar{1}0]$ and [110] directions being parallel to the traces of the {111} slip planes. Note here that the slip bands along $[1\bar{1}0]$ are seen more prominent than those along [110], which indicates that the plastic zone observed in Fig.2 (a) is similar to the 45° shear-type plastic zone[10] formed by mode I tensile load, as illustrated in Fig.2 (c).

Figure 2 (b) shows a cross sectional profile along the [110] direction at the position of 0.5 μm behind the crack tip in Fig.2 (a). Steps with the height of 1-2 nm are seen in the profile, corresponding to the slip bands along $[\bar{1}10]$ formed by several

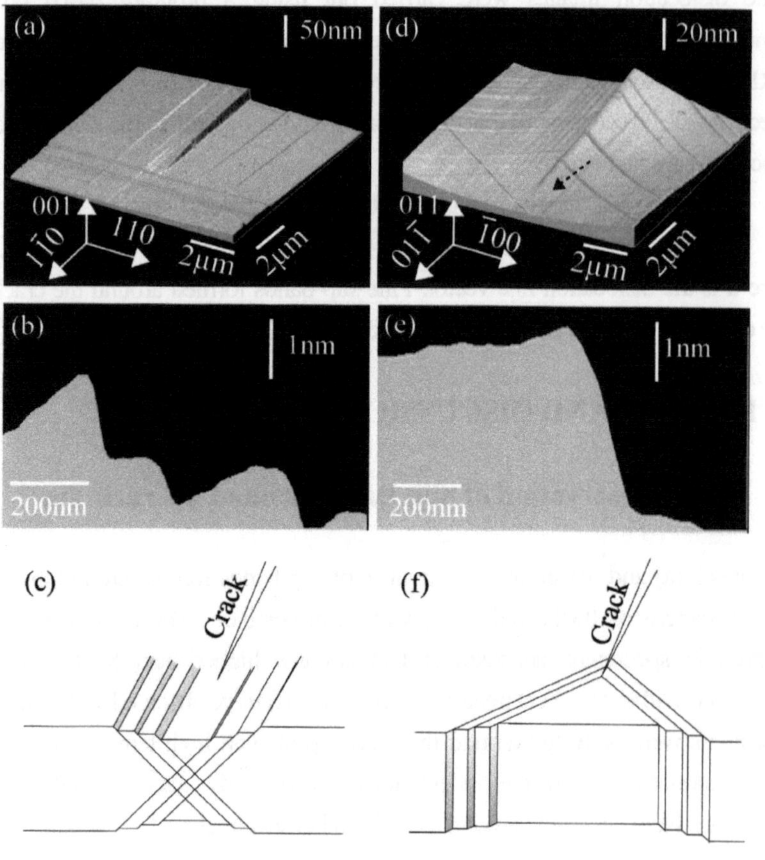

Figure 2. (a) (b) (c) 45°-shear-type plastic zone, (d)(e)(f) hinge-type plastic zone.
(a) and (d) are images of atomic force microscopy around the crack tip.
(b) and (e) are cross sectional profiles of slip bands near the cracks.
(c) and (f) are schematic drawings of a 45°-shear-type plastic zone and a hinge-type one, respectively.

dislocations gliding on the $(\bar{1}11)$ plane. Although a mode III displacement is seen along the crack, it is a minor component, and the major component of this crack is mode I opening (note the difference between the scale of vertical and horizontal directions).

Fig. 2(d) shows a crack extending along the $[01\bar{1}]$ direction on the (011) wafer surface. The crack plane is parallel to (100). Slip bands are formed with regular spacing of a few microns in both sides of the crack. The traces of the slip bands mainly observed coincide with those of the $(11\bar{1})$ or $(\bar{1}1\bar{1})$ planes, indicating that the slip bands are those formed in the hinge-type plastic zone as illustrated in Fig.2 (f). A cross sectional profile of the surface step due to one of the slip bands is shown in Fig.2 (e), which was obtained by sectioning along the dotted arrow indicated in Fig. 2(d). A step with the height of 2.7 nm is seen, which is considered to be formed by the motion of around 14 dislocations (step height due to one dislocation: 0.19 nm). In Fig.2 (d), other slip bands are also seen although the contrasts are not so sharp as those due to the hinge-type. The trace of those slip bands is parallel to that of the (111) or (111) planes, indicating that the 45° shear-type is also included as a minor component.

3.2 HVEM Observation of Crack Tip Dislocations

Figure 3(a) shows an HVEM image (bright field: BF) of a crack and dislocations observed in a 45°-shear-type plastic zone, where both the foil normal and incident beam directions are parallel to the [001] direction. The diffraction vector **g** was taken to be [220]. The crack observed is extending along the $[1\bar{1}0]$ direction. Dislocations are generated not only in front of the crack tip but also behind the tip. Since the number of the dislocations observed is very limited, the dislocation arrangement is considered to be that in the very early stage of dislocation generation process. To understand the relationship between the observed dislocations and the crack tip plasticity, we specified the slip system of the activated dislocations. First, the direction of Burgers vector and the slip plane of the dislocations were

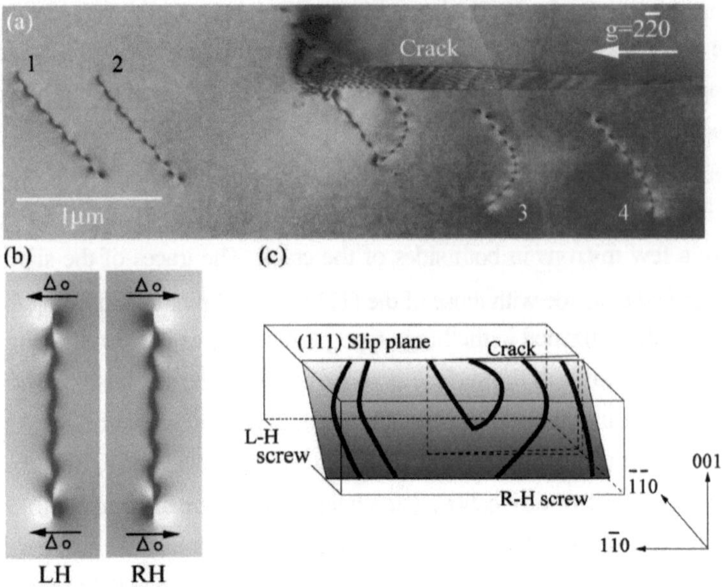

Figure3. (a) Dislocations emitted from a crack tip in a 45°-shear-type plastic zone. Bright field image, incident beam direction was <001>, g=2$\bar{2}$0.
(b) Computed images, RH and LH screw dislocations.
(c) Schematic drawing of the dislocations and the crack shown in Fig. (a).

determined by using the usual method of the invisibility criterion. As a result, the dislocations in Fig.3 (a) were found to be all lying on the (111) plane, and their Burgers vectors were almost parallel to the dislocation lines: they are almost screw dislocations. Next, to determine the signs of their Burgers vectors, contrast simulations of dislocation images were carried out.[5] Figures 3(b) shows computed bright field (BF) of the right-handed (R-H) and left-handed (L-H) screw dislocations. Each dislocation image is observed as black and white oscillating zig-zag contrasts, which corresponds to images of dislocations inclining to the foil surface. Note the black and white contrasts at both ends of each dislocation line. Here, Δ_0 is defined as the contrast vector which goes from the center of a black contrast to that of a white contrast at the point of emergence of the dislocations at the surface. When comparing the L-H with R-H, white and black contrasts appear conversely (i.e., Δ_0 of L-H is anti-parallel to that of R-H). By using this criterion, dislocations labeled 1 and 2 in front of the crack tip were determined as L-H screw dislocations, and those labeled 3 and 4 behind the tip (wake dislocations) were determined as R-H. Figure 3(c) shows a schematic illustration of the configuration, where dislocations ahead of the crack tip are L-H screw dislocations while the wake dislocations are R-H. The dislocation configuration is understood by a model where dislocation loops are emitted from a source near the crack tip under a mode I tensile load. Such dislocation behavior causes displacements in a 45° -shear-type plastic zone as illustrated in Fig.2(c).

Figure 4 shows HVEM images (BF) of a crack and dislocations observed in a hinge-type plastic zone. In Fig.4 (a), both the foil normal and the incident beam directions are parallel to [011], and diffraction vector **g** was taken to be $[02\bar{2}]$. A crack is seen in the right bottom of the figure, where the crack is extending along the $[0\bar{1}1]$ direction. In this figure, two prominent arrays of dislocations are seen almost horizontally, *i.e.*, along the $[21\bar{1}]$ direction: they are lying on the $(\bar{1}1\bar{1})$ plane which is parallel to the incident beam direction, *i.e.*, [011], so that the dislocations lying on $(\bar{1}1\bar{1})$ are observed as a row of line segments or dots. To observe the aspect of each dislocation line in the array, the specimen was tilted in the holder and the incident beam direction was taken to be [001]. Figures 4(b) and 4(c) show the images of the [001] incidence. Dislocations in each figure correspond to those encircled by the rectangles in Fig.4 (a).

Diffraction vector **g** of both figures was taken to be $[2\bar{2}0]$. As is seen in the figures, each dislocation line in the arrays tends to lie along the [011] direction, although it is curved and seems to be bowing out from the right to left. This suggests that the dislocation array was activated from a source on the crack to form a hinge-type plastic zone. To confirm this consideration, the directions and the signs of their Burgers vectors were determined by the same method as shown in Fig.3. It was found that all dislocations in the arrays had the same Burgers vector of $a/2[\bar{1}01]$. Since their dislocation lines tend to lie along [011], they are mainly 60°

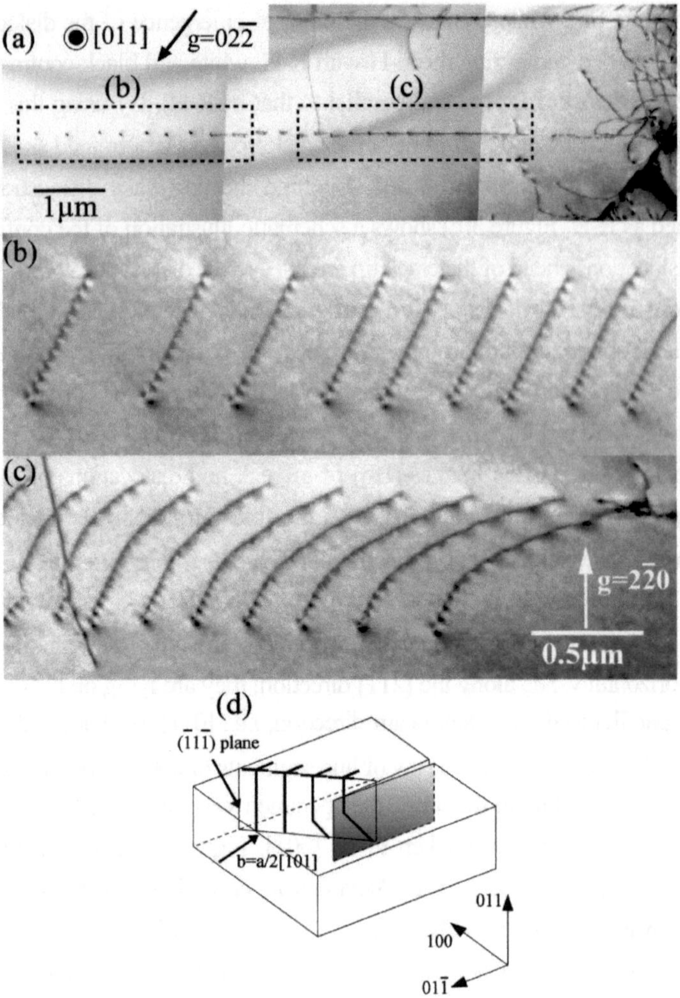

Figure 4. (a) HVEM image of dislocation arrays in a hinge-type plastic zone. Incident beam direction was parallel to [011]. Bright field image, **g**=02$\bar{2}$. (b) (c) Enlarged images of dislocations in the area surrounded by dotted rectangles in Fig.4 (a). BF image, **g**=2$\bar{2}$0. (d) Schematic drawing of the dislocation array emitted from the crack.

dislocations. The sign of their burgers vectors is shown in Fig.4 (d): extra-half planes of edge components are indicated in the figure, and the screw components are R-H. When dislocations of this type glide away from the crack, crack tip deformation occurs as shown in Fig.2 (f). Thus, the dislocation arrays observed in Figs.4 (a)-(c) exhibit the microscopic structure of slip bands in the hinge-type plastic zone.

4 CONCLUSION

Plastic zones associated with a crack under a mode I tensile load are macroscopically separated into two groups; the 45°-shear-type and the hinge-type. The former is well developed under the plain stress condition, the latter under the plane strain condition. In the present study, with the aim of clarifying the microscopic structure of such plastic zones, fine slip bands and dislocation structures near a crack tip in silicon crystals were examined by using AFM and HVEM.

In {001} wafer specimens, the plastic zone of 45°-shear-type was observed around a crack tip. Very fine slip bands with step height of several Burgers vectors were found by AFM. HVEM exhibited a dislocation structure which is formed by dislocation loops emitted from a source near the tip of a crack. In {011} wafer specimens, hinge-type plastic zones were observed, where slip bands with the step height of a few nanometers were formed with regular spacing of a few microns. In HVEM observation, sharp arrays of dislocations with the same Burgers vector were found, which correspond to the microscopic structure of slip bands in the hinge-type plastic zone

The dislocation structures observed in the present study are considered to be closely related with the increase of fracture toughness through the effect of crack tip shielding.

REFERENCES

[1] C. St. John, Phil. Mag., **32**(1975), 1193-1212.

[2] M. Brede and P. Haasen, Actta Metall., **36**(1988) 2003-2018.

[3] A. George and G. Michot, Mater. Sci. Engng., **A164**(1993) 118-134.

[4] P.B. Hirsch and S.G. Roberts, Phil.Trans. R.Soc, Lond. A, **355**(1997) 1991-2002.

[5] K. Higashida, N. Narita, M. Tanaka, T. Morikawa, Y. Miura and R. Onodera, Phi. Mag. A, **82**(2002), 3263-3273

[6] M. Tanaka, K.Higashida, T. Kishikawa and T. Morikawa, Mater. Trans., **43**(2002) 2169-2172

[7] P.B. Hirsch, A. Howie R.B. Nicholson, D.W. Pashley and M.J. Whelan: *Electron Microscopy of Thin Crystals*, (London: Butterworths, 1965)

[8] A.K. Head, P. Humble, L.M. Clarebrough, A.J. Morton and C.T. Forwood: *Computed Electron Micrographs and Defect Identification*, (Amsterdam: North Holland, 1973)

[9] J.P. Hirth and J. Lothe: *Theory of Dislocations*, (New York: McGraw-Hill, 1968) pp19-22.

[10] G. T. Hahn and A.R. Rosenfield, Acta. Metall., **13**(1965), 293-306

IRRADIATION HARDENING IN AUSTENITIC STEELS: EXPERIMENT AND SIMULATION

C. Pokor[(1)], Y. Bréchet[(2)], P. Dubuisson[(1)], J.P. Massoud[(3)] and D. Rodney[(4)]

[(1)] *CEA-DEN, SRMA, CEA-Saclay, 91191 Gif sur Yvette, France*

[(2)] *LTPCM/ENSEEG, Domaine Universitaire BP75 Saint Martin d'Hères France*

[(3)] *EDF R&D - MMC, Site des Renardières, 77818 Moret sur Loing Cedex, France*

[(4)] *GPM2/ENSPG, Domaine Universitaire BP46 Saint Martin d'Hères France*

Abstract We present a study based on Transmission Electron Microscopy and Tensile tests of irradiated austenitic steels, completed by Molecular Dynamics simulations of screw dislocations interacting with interstitial Frank loops.

Keywords: Electron Microscopic, Molecular Dynamics, Irradiation, Dislocation

1. INTRODUCTION

Austenitic steels are used as structural materials in nuclear Pressurized Water Reactor (PWR), where they are submitted to high neutron fluxes. The irradiation causes a degradation of the mechanical properties with an increase in the yield stress, a decrease in ductility and a localization of the deformation in clear bands. In order to better understand these phenomena, we performed a systematic study of the influence of temperature and pre-deformation on the microstructure and mechanical behavior of austenitic steels used in PWR, using Transmission Electron Microscopy (TEM) and post-irradiation tensile tests, in in-service temperature conditions.

This study showed that, in the present irradiation conditions, the irradiation microstructure is dominated by interstitial Frank loops. However, it offered no microscopic detail on the dislocation/loop interaction mechanisms or on the pinning strength due to the defects. This information is nevertheless required for a detailed understanding of the irradiation effect on the macroscopic behavior of the austenitic steels. For that reason, we

H. Kitagawa and Y. Shibutani (eds.), IUTAM Symposium on Mesoscopic Dynamics of Fracture Process and Materials Strength, 163–172.

have undertaken a systematic investigation at the atomic scale, by Molecular Dynamics (MD), of all the possible configurations between edge and screw dislocations interacting with Frank loops. Since interatomic potentials for austenitic steels are not available, we use Nickel as a prototypical FCC material. We present here preliminary results concerning screw dislocations.

2. EXPERIMENTAL STUDY

2.1 Experimental procedure

Three types of austenitic steels were irradiated: (1) a solution annealed 304L steel (noted ***SA304***); (2) a cold-worked 316 steel, deformed to about 15% (noted ***CW316***) and (3) a 20% cold-worked modified 316 steel containing Titanium (0.28 %wt) and Silicon (0.83 %wt) (noted ***CW316Ti***).

The SA304 and CW316 alloys are employed in PWR respectively for baffle plates and baffle bolts. They were chosen to study the influence of the initial dislocation network on the final microstructure after irradiation. The third Ti-rich material is a potential candidate for the replacement of the actual internal structural materials in PWR. The initial microstructure of the samples consisted of grains of a size of about 40 μm with, in the case of the cold-work materials, twins and a high density of dislocations organized in cells.

Three reactors were employed in order to perform irradiations at a temperatures of 375°C, representative of the maximum temperature seen by internal components, and of 330°C which is the average temperature seen by these components. The dose ranged from 0.8 dpa to 40 dpa. The latter is about half of the end-of-conception-life irradiation dose.

Two reactors used here are fast breeder reactors with high fluxes: the American EBRII which produced irradiations at 375°C up to 10 dpa and the Russian BOR60 reactor which produced irradiations at 330°C with doses up to 40 dpa. The French mixed neutron spectrum OSIRIS reactor was also used to perform irradiations at 330°C at low doses (<3.4 dpa). In the experimental conditions considered here, the temperature plays a very important role while fluxes and spectrum have little influence.

TEM observations were performed on an EM 430 Philips microscope. The faulted Frank loops were imaged using the Reciprocal Lattice Rod Technique. Cavities were quantified on bright field images out of contrast.

These characterizations provided defect size distributions, from which average defect sizes and densities were evaluated.

The mechanical behavior was tested by uniaxial tensile tests at a deformation rate of $3\ 10^{-4}\ s^{-1}$. The tests were performed post irradiation at the same temperature (300°C or 375°C). Irradiation hardening is characterized by the increase in 0.2% elastic limit: $\Delta\sigma_e = \sigma_{e\ irradiated} - \sigma_{e\ non\ irradiated}$.

2.2 Microstructure and mechanical behavior evolutions

The following general *qualitative* conclusions can be drawn from the TEM observations:

1. The irradiation microstructure is dominated by a high density of interstitial Frank loops and black dots. The Frank loops are faulted, having a a/3<111> Burgers vector and a {111} habit plane. They appear as ellipses in the TEM pictures. The black dots are defects too small to be identified with precision. From the present observations at low dose (0.8 dpa), they were identified as very small interstitial Frank loops and Frank loops seen on their side with very small apparent sizes.
2. The initial dislocation network disappears progressively as the irradiation dose is increased. At high temperature (375°C, 10 dpa), all initial dislocations have disappeared in SA304 and CW316 samples and some remain in CW316Ti samples. In CW316 samples, traces of the initial network are still visible in the form of zones where TEM contrast is different. This could be due to defect densities different in the regions where the cell walls used to be. At lower temperature (330°C), some initial dislocations remain, even for doses up to 40 dpa.
3. The twins in the cold-worked samples are unaffected by the irradiations.
4. No irradiation induced precipitation is observed.
5. No cavity is observed at 330°C and a low density of the latter is observed at 375°C, especially in SA304 specimens.

As shown in Figure 1, we extracted, for each material and irradiation condition, the average size of the Frank loops and their density as a function of temperature and dose. The main *quantitative* conclusions are:

1. A saturation is observed for doses typically above 10 dpa for all materials and temperatures. Saturation levels depend on the temperature and saturation is more pronounced for the loop size than for their density.
2. At low temperature (330°C), above 10 dpa, the diameter of the SA304 and CW316 samples is constant, close to 7 nm (Figure 1.a). The loop densities continue to increase (Figure 1.b), but slowly.
3. At high temperature (375°C), the diameter is larger (~ 12 nm) and the density lower ($30\ 10^{21}\ m^{-3}$) as compared to the values obtained at 330°C.

4. If we compare the SA304 and CW316 samples before saturation at 330°C, the former has smaller loops (~ 6 nm compared to ~ 9 nm) but in higher density (~ 60 10^{21} m^{-3} compared to ~ 30 10^{21} m^{-3}).
5. Ti-rich samples have loops with larger diameters and lower densities as compared to the other materials: at 40 dpa, the diameter is 10 nm with a density of 50 10^{21} m^{-3}, compared to 7 nm and between 60 and 80 10^{21} m^{-3} for the other materials.

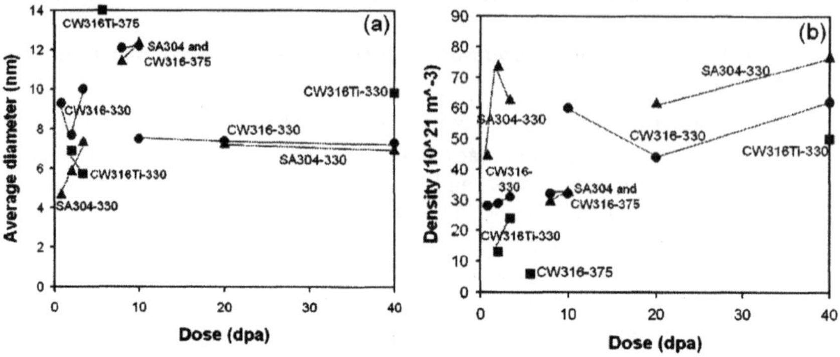

Figure 1: Evolution of the average diameter (a) and density of Frank loops (b). Errors are within 2 nm for the diameter and 10^{22} m^{-3} for the density. Triangles refer to SA304, circles to CW316, squares to CW316Ti. Irradiation temperatures are noted on the figure.

Irradiation hardening is characterized by the difference in yield stress after and before irradiation. Figure 2 presents the resulting curves obtained at 330°C with SA304 and CW316 samples. The main conclusions are:

1. In all cases, hardening increases rapidly at low doses, then saturates.
2. SA304 seems to saturate before CW316 (5 dpa compared to 10 dpa).
3. Hardening depends on temperature and is lower at 375°C than at 300°C.
4. The cold-worked 316 alloy hardens less than the solution annealed 304L alloy with saturation at about 500 MPa in the first case and 600 MPa in the second. This point is not fully understood from the present data and the influence of chemical composition and initial dislocation density can not be completely cleared

2.3 Discussion

The present results are in agreement with the literature [1]. The particular interest of the present work is that it allows to study, *on the same samples,* the effect of the temperature and pre-deformation on both the microstructure and the mechanical behavior of austenitic steels. A cluster-dynamics modeling of the microstructure evolution and a micromechanical model of the elastic limit and work hardening have been developed based on the present observations. They are detailed elsewhere [2].

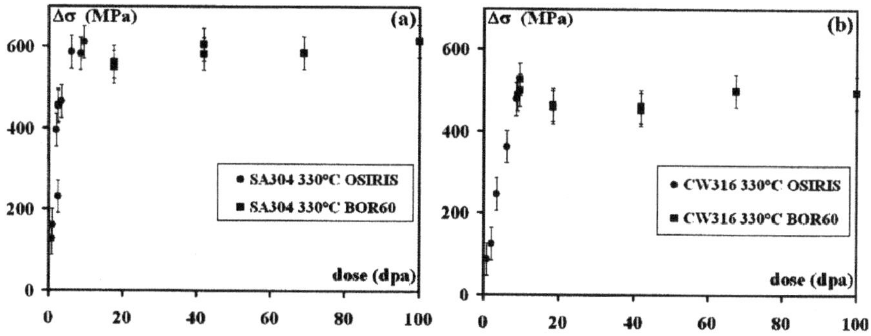

Figure 2: Increase in 0.2% elastic limit as a function of dose for the SA304 (a) and CW316 (b) alloys irradiated and mechanically tested at 330°C.

Regarding the topic of the present article, the main observation is that the *microstructure is dominated by interstitial-type defects*, namely Frank loops and black dots. This situation is very different from, for example, Copper where the vast majority of the defects observed in TEM is vacancy-type Stacking Fault Tetrahedra (SFT) [3]. This implies that the mobility of the vacancies and interstitials and the nature and strength of the sinks are different in these two materials, although, to our knowledge, there is no precise explication for this phenomenon.

Secondly, the microstructure saturates for doses typically above 10 dpa. There is a dynamicalequilibrium between the production of new defects and their incorporation in the dislocation network. The latter in turn causes dislocation climb which promotes annihilation between dislocations. Consequently, in agreement with the TEM observations, the density of loops in CW316 is smaller than in SA304 (because there are more dislocations to absorb the defects) and the initial dislocation network disappears. The fact that the loops in CW316 are larger on average than in SA304 implies that the smaller loops are absorbed in the dislocation network and not the larger ones which have had enough time to develop.

Regarding the temperature, the main conclusion is that *a higher temperature leads to larger loops with smaller densities and a significantly lower hardening.*

Ti-modified 316 samples are characterized by larger Frank loops in lower density and an irradiation hardening which evolves more slowly as compared to the two other materials, which is a definite advantage for this type of material.

3. MOLECULAR DYNAMICS SIMULATIONS

We present preliminary results concerning a screw dislocation interacting with an interstitial Frank loop that lies in a {111} plane which, either, is or is not a cross-slip plane for the dislocation.

The classical dislocation/Frank loop interaction mechanism was analyzed in the case of a vacancy-type loop [4]. The latter is unfaulted and transformed into a perfect loop when its surface is swept by a single Shockley partial. In the case of interstitial Frank loops, the situation is more complicated because their fault is made of two stacking faults in adjacent {111} planes. Their removal requires sweeping by two Shockley partials. There is still a debate [5] as to whether these partials are separated or form a single core extended over two {111} planes (the so-called D-Shockley partial). We show below that the occurrence of one or the other type of partials depends on the configuration.

3.1 Simulation technique

Figure 3 shows the simulation cell with the initial dissociated dislocation.

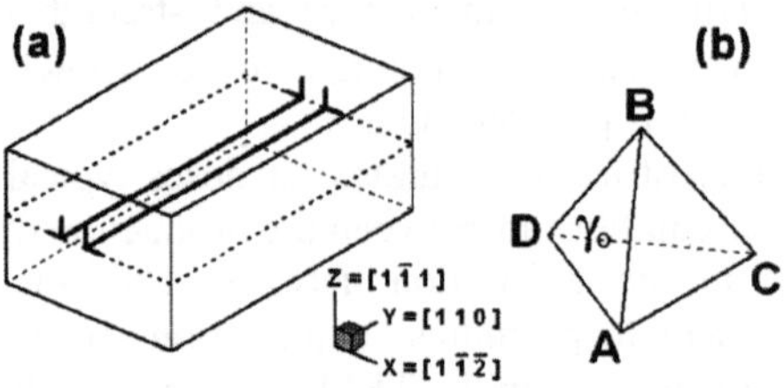

Figure 3: Simulation cell and associated Thompson tetrahedron

We use *boundary conditions that construct an infinite periodic glide plane for the dislocation.* Periodicity in the Y direction is insured by the invariance of the dislocation displacement field in this direction. In the X direction, we use periodic conditions along with a shift of +b/2 in the Y direction for atoms leaving the cell from the left-hand negative surface and reintroduced on the right-hand positive surface, and an opposite shift of –b/2 for atoms moving in the opposite direction. The reason is that the (110) planes perpendicular to the dislocation line turn around the latter with a shift close to b/2 between two opposite faces. This initial shift depends on the atomic Z coordinates; there is an initial mismatch, which however is minimized by choosing a cell with a small L_z/L_x ratio and disappears with the initial energy minimization.

In the Z direction, we apply modified free boundary conditions by allowing atomic displacements in the upper and lower Z surfaces only in the X and Y directions. A shear stress σ_{yz} may be applied by superimposing constant and opposite forces in the Y direction to the forces felt by the atoms in the upper and lower Z surfaces.

We employ the EAM potential developed by Angelo *et al.* [6] to simulate dislocations in Nickel crystals. Time is integrated using Verlet algorithm (time step of 2 10^{-15} ps). The temperature is 100 K and no control is needed because the dislocation glides over limited distances.

We consider hexagonal loops with edges aligned in <110> directions, with a diameter of 6 nm and, given the cell dimensions, a density is 80 10^{21} m^{-3}, i.e. values close to the saturation values obtained in Section 2 We consider loops lying either in an ABC cross-glide plane or in an ABD plane. Given the symmetries of the crystallography, these are the only two configurations to be considered in the case of a screw dislocation.

3.2 Loop in a cross-glide plane of the dislocation

Figure 4 shows a simulation with an applied stress of 300 MPa.

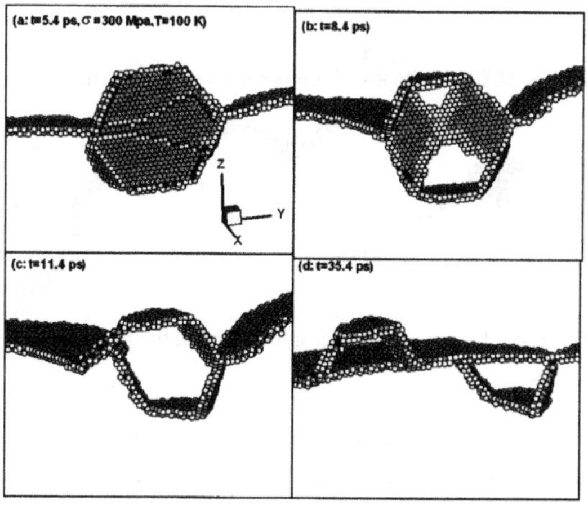

Figure 4: Case of a loop in a cross-glide plane of the screw dislocation. The atoms shown have less than 12 first neighbors in FCC position. Their color scales with this number.

The initial configuration (not shown here) was obtained using a conjugate gradient energy minimization with zero applied stress. It allowed the dislocation to dissociate and come into contact with loop surface. Application of a stress forces the section of the dislocation in contact with the loop to constrict. The dislocation cross-slips and redissociates in the plane of the loop. Atoms in the core of the two partials are visible in white in Figure 4.a. They sweep the loop surface and remove one of the two stacking faults. Thus, non-FCC atoms are visible between the partials in Figure 4.a.

Once the partials have finished sweeping the loop, two other partials are spontaneously nucleated on the edges of the loop (Figure 4.b). They remove the second stacking fault, leading to complete unfaulting (Figure 4.c). The two halves of the loop have different Burgers vectors (**CD** for the upper half and **AD** for the lower one). They are weak obstacles to the motion of the dislocation. The latter drags both halves in the direction of their respective Burgers vectors. Consequently, they get separated (Figure 4.d).

3.3 Loop *not* in a cross-glide plane of the dislocation

Figure 5 shows a simulation with an applied stress of 150 MPa.

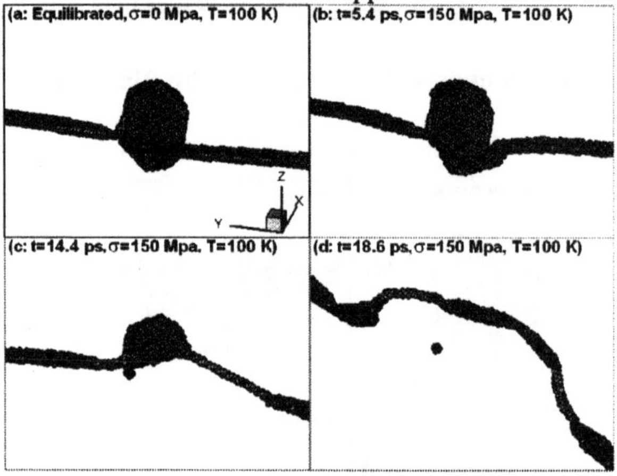

Figure 5: Case of a loop *not* in a cross-glide plane of the screw dislocation.

In the initial configuration (Figure 5.a), the dislocation is in point-like contact with the loop. Application of a stress causes the dislocation to cross-slip spontaneously from its initial ACD plane to an ABC plane where it combines an edge of the loops (Figure 5b), forming a partial dislocation with a γ**A** Burgers vector (see Figure 3 for notations). This segment is a D-Shockley partial since it is not dissociated and adjoins the interstitial fault.

The γ**A** segment is mobile and sweeps the loop surface, removing the fault. A non-dissociated segment forms (Figure 5.c). It is a a/2<110>{001} dislocation (a special type of which is the edge Lomer dislocation). The final structure (Figure 5.d) corresponds to a *helical turn* made of both dissociated a/2<110>{111} and non-dissociated a/2<110>{001} segments. It has inherited the hexagonal shape of the loop.

The helical turn is sessile because its various parts have an **AC** Burgers vector parallel to the dislocation line and can glide only in this direction. Unpinning involves an Orowan process shown in Figure 6. When the dislocation unpins , it leaves behind a perfect loop.

By using cells with different widths in the Y direction, i.e. different lengths L for the screw dislocation, we measured the strength of this interaction and found $\tau_c / \mu \approx 1.0\, b/L$. This linear relation is approximate, presumably because the size of the obstacle varies during unpinning.

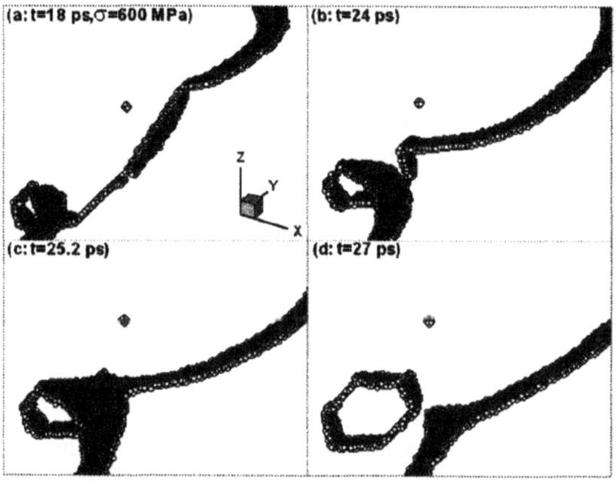

Figure 6: Unpinning reaction from a helical turn

3.4 Discussion

The present MD simulations show that atomic scale reactions between dislocations and irradiation defects can be performed with realistic sizes and densities. The boundary conditions presented here allow for the simulation of screw dislocations in an infinite glide plane without simulating a dipole and thus, with smaller simulation cells. With the corresponding edge boundary conditions [7], the influence of the dislocation character on atomic scale processes may be studied.

Both configurations considered here lead to the unfaulting of the Frank loop, in the first case with the help of two Shockley partials, in the second case, of a single D-Shockley. Therefore, *both types of partials are possible, their occurrence depending on the configuration.*

Unfaulting involves dislocation cross-slip events which occur spontaneously at low temperature. This point is important because cross-slip has been suggested as a dominant process in the formation, thickening and branching of the clear bands. The present simulations show that these cross-slip events are promoted by the *short-range core interactions between the dislocations and the Frank loops leading to local constrictions and athermal cross-slip*.

In simulations not shown here, we considered loops with edges along <121> directions, a morphology also observed in irradiated FCC materials. This case is less favourable to cross-slip because the loop edges are not in {111} planes and, indeed, we observed that the loops are not unfaulted, but sheared. Therefore *unfaulting is highly dependent on the shape of the loops.*

When the loop is in a cross-glide plane of the dislocation, the controlling reaction is loop unfaulting. Then, the half loops are easily dragged by the dislocation. By way of contrast, when the loop is not in a cross-glide plane, unpinning occurs at low stress (< 150 MPa) but the unfaulted loop, which has become an helical turn, is a strong obstacle. The unpinning stresses are high, which was expected since unpinning involves Orowan processes.

4. CONCLUSION

We have presented a TEM study of irradiated austenitic steels, completed by MD simulations. We are currently performing simulations in pure metals and model solid solutions to obtain a systematic understanding of the interactions and of their strengths. The latter will then be introduced in crystalline constitutive laws such as presented in [8] and in Dislocation Dynamics simulations to predict the macroscopic elastic limit and the dynamics of clear band formation from atomic scale information.

REFERENCES

1. P.J. Maziasz, J. Nucl. Mater. 205 (1993) 118.
2. C. Pokor, Y. Bréchet, P. Dubuisson, J.P. Massoud, A. Barbu, submitted to *J. Nucl. Mat.*
3. S.J. Zinkle and L.L. Snead, J. Nucl. Mater. 225 (1995)
4. J.L. Strudel and J. Washburn, Philos. Mag. 9 (1964) 491.
5. M. Suzuki *et al.*, Philos. Mag. A65 (1992) 1309.
6. J.E. Angelo, N.R. Moody and M.I. Baskes, Model. Simul. Mater. Sci. Eng. 3 (1995) 289.
7. D. Rodney and G. Martin, Phys. Rev. B61 (2000) 8714
8. D. Rodney, G. Martin and Y. Bréchet, Mat. Sci. Eng. A309-310 (2001) 198.

DISLOCATION GLIDE IN THE PRESENCE OF EITHER SOLUTE ATOMS OR GLISSILE LOOPS

D.J. Bacon, Yu.N. Osetsky[1], Z. Rong and K. Tapasa

Department of Engineering, The University of Liverpool, Liverpool L69 3GH, U.K.
[1]Now at: Oak Ridge National Laboratory, P.O. Box 2008, Oak Ridge, TN 37831-6158, USA.

Abstract Atomic-scale computer simulation is used to investigate obstacle effects on the dynamics of glide of an edge dislocation in two situations. In one, a dislocation in α-iron encounters copper atoms in solution and it is found that the effect on dislocation velocity under constant stress below the static Peierls stress is strong. In the other, drag of glissile interstitial loops with the same Burgers vector as the dislocation is considered for iron and copper. The drag coefficient of a loop is determined for the first time, and is shown to be related to the diffusivity of clusters of interstitials via a model of dislocation drag of discrete pinning points.

Keywords Atomic-scale modelling, dislocation dynamics, dislocation loop, drag, solute atom

1. INTRODUCTION

Plastic deformation in metals is strongly affected by the interaction of dislocations with obstacles on or near their glide planes. For example, fast-neutron irradiation creates point defects in displacement cascades and causes a substantial increase in yield stress and a reduction in ductility. These phenomena have been the subject of much research, but the mechanisms that control them are still not clear, although since a high proportion of defects form in clusters [1], dislocation-cluster interactions must be important. Molecular dynamics (MD) computer simulations have shown that self-interstitial atoms form platelets of closely-packed, parallel crowdions that are equivalent to small (nanometre-scale) interstitial dislocation loops with Burgers vector, **b**, parallel to the crowdion axis [1,2]. They can interact with

H. Kitagawa and Y. Shibutani (eds.), IUTAM Symposium on Mesoscopic Dynamics of Fracture Process and Materials Strength, 173–182.

dislocations in two ways: either as obstacles that pin dislocations by intersecting their glide plane or, being glissile, by moving to decorate dislocations. Makin [3] calculated the long-range elastic interaction between a dislocation and loop and pointed out that loops lying close to the glide plane of a dislocation line with the same **b** can be swept along by it. This should reduce line mobility, although the effect is difficult to quantify without information on the force to move a loop.

More recently, a model has been proposed, in which dislocation sources are locked by atmospheres of loops that glide to regions where they are attracted by the dislocation stress field [4]. A continuum-based dislocation dynamics method has been used to study how the critical stress for a flexible dislocation to break away from rows of loops depends on loop size, spacing and distance from the glide plane [5], and confirms the earlier estimates [4]. These treatments assume the loops do not glide with the dislocation, but this is by no means certain. Hence, in section 4 we consider loop drag effects by simulating dislocation glide in the presence of loops by atomic-scale computer simulation of Fe and Cu, two metals that offer a contrast of crystal structure and dislocation dissociation effects. This is a first step to establishing general rules on these phenomena for use in dislocation dynamics simulations.

Solute atoms also influence yielding and subsequent plastic flow. An example of interest for assessing the mechanical properties of reactor pressure vessel steels is Cu in Fe. Precipitates of Cu nucleate during operation, but growth is limited under irradiation conditions and the population of nano-scale, coherent precipitates creates substantial hardening of the Fe matrix. Dislocation-precipitate interaction has been simulated recently by MD and the atomic-scale mechanisms of strengthening analysed [6], but less is known of the effects of Cu that remains in solution. We are undertaking a modelling study of this and preliminary results are presented in section 3. As a reference for sections 3 and 4, we describe dislocation behaviour under applied stress in pure Fe and Cu in section 2.

Only an initially straight edge dislocation is considered here. The simulation method and the way in which quantitative information on stress-strain response and dislocation motion are obtained for static (T = 0K) and dynamic (T > 0K) loadings are described in [7]. The slip geometry for the Fe (BCC) and Cu (FCC) models is shown schematically in Fig. 1. Periodicity was employed in the slip direction x as well as the line direction y. The models contained between 0.3 and 1.9M atoms. Interatomic potentials from [8,9] were used for the Cu and Fe-Cu systems, respectively.

2. MOTION OF A SINGLE DISLOCATION

The variation of shear stress, τ, when an increasing shear strain is applied to a model crystal at T = 0K is plotted in Fig. 2 for both Cu and Fe. The onset

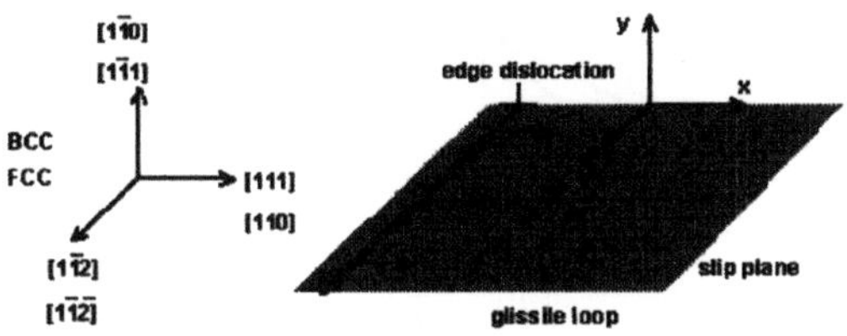

Figure 1. Geometry of glide considered for Fe (BCC) and Cu (FCC).

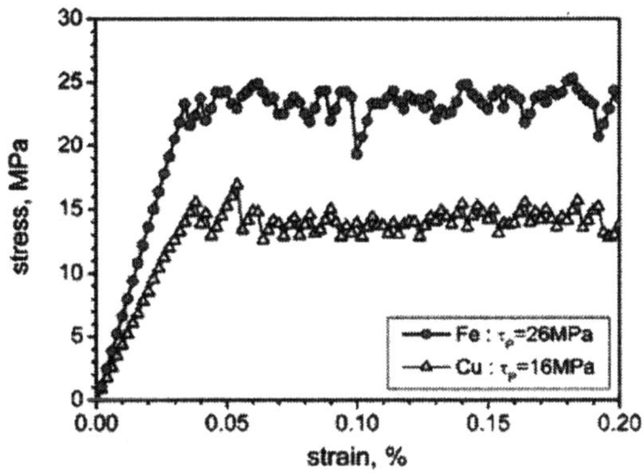

Figure 2. Stress-strain response at T = 0K of a model containing a single dislocation.

of plastic strain when the edge dislocation moves is clear: a plastic strain of 0.1% corresponds to dislocation motion of about 10-15b for the model sizes used. The critical (Peierls) stress, τ_P, required to maintain dislocation glide is seen to be about 16MPa for Cu and 26MPa for Fe. The smaller value for Cu is not surprising because the dislocation dissociates into two Shockley partials with spacing ~13a_0, where a_0 is the FCC lattice parameter.

The response of a dislocation to constant applied stress, τ, depends on temperature, but here we only consider results for T = 300K. Glide of the edge line occurs at stress levels well below τ_P. This can be backwards as well as forwards, but stress imposes a drift on this movement, and when τ rises above τ_P, the line glides uniformly. The dependence of dislocation velocity on applied shear stress is shown by the data plotted in Fig. 3(a). Although τ_P is higher for Fe than Cu, the velocity, v, at the same value of τ is actually higher. The velocity is T-dependent, increasing as T decreases, showing that glide is affected by a phonon-related drag mechanism. Defining the drag coefficient, B, in the usual way [10], i.e. $\tau b = Bv$, results in the data plotted in Fig. 3(b). B is higher for Cu than Fe, and approximately independent of τ, except for Cu at low τ where it increases strongly. The cause of this increase is unknown.

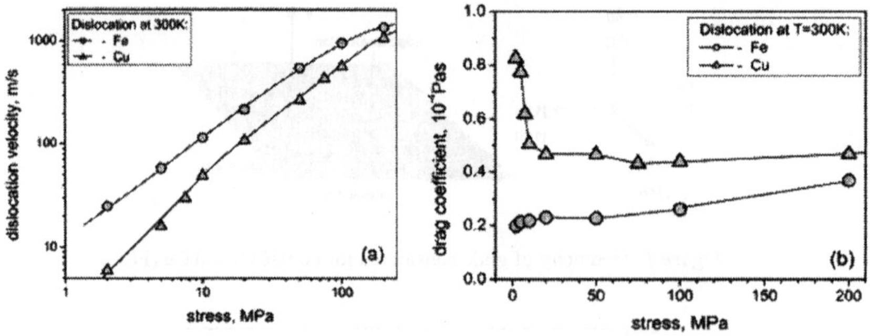

Figure 3. (a) Velocity and (b) drag coefficient versus stress for an edge dislocation in either Fe or Cu at T = 300K.

3. EFFECT OF COPPER SOLUTE IN IRON

Copper is an oversized solute in the BCC Fe matrix, with a misfit of 8.8% for the interatomic potentials used here. For solute concentrations of interest, i.e. >0.1at%, the solute spacing is such that under no applied stress, a dislocation minimises its interaction energy with the solute atoms by adopting a shape with a radius of curvature much larger than the spacing [11]. In practical terms, the radius is much larger than the periodic MD cell length, L, and so the periodicity imposes an unrealistic, almost straight line shape. We have checked that the critical value of τ for an edge dislocation to glide in a 0.5% solution at T = 0K is strongly dependent on L up to L = 22nm, i.e. 89b. Hence, for the static study (T = 0K), we have considered the effect of a single Cu atom on an edge dislocation.

Fig. 4 shows the variation along the [111] direction of the interaction energy, E_I, between the dislocation and a copper atom occupying a site in either the $(1\,\bar{1}\,0)$ atomic plane that contains the lowest atom of the dislocation extra half-plane or one of the two adjacent planes, as shown schematically in the figures. The points labelled 'atomic' are the actual values for a fully relaxed model containing an edge dislocation with **b** = 1/2[111] and a Cu atom. They are independent of the length of periodicity L for the range considered (3.5-27nm). The line labelled pΔV is an 'elastic' approximation obtained from the product of the pressure, p, at the equivalent atomic sites in a model of pure Fe times ΔV for a single substitutional copper atom in an otherwise perfect crystal, i.e. 0.088Ω. The pΔV lines exhibit the forms of repulsion/attraction (E_I > 0/0) expected from elasticity theory for the change from compression to tension between positions above and below the glide plane of an edge dislocation. The 'atomic' data for Cu in sites on the highest plane considered are consistent with this (Fig. 4(a)). The values for the plane

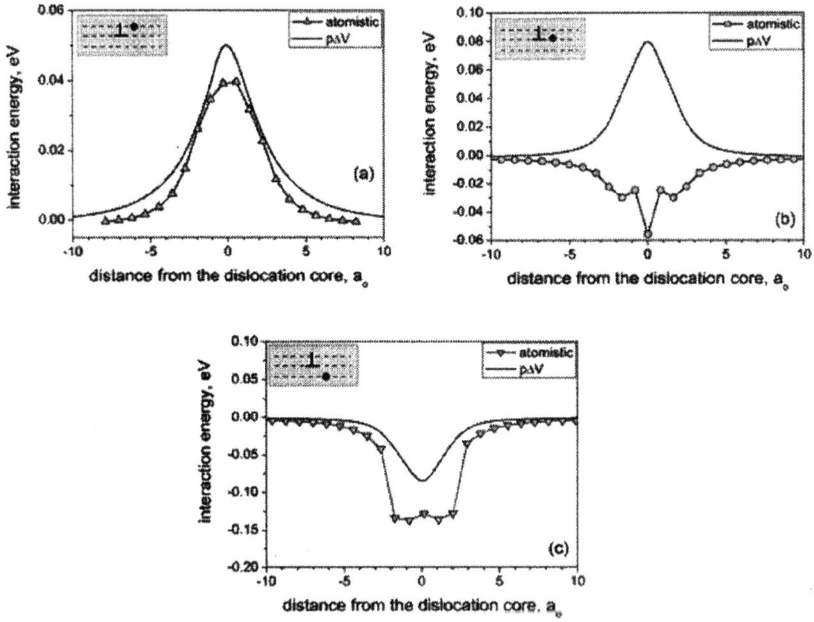

Figure 4. E_I vs x for an edge dislocation and one Cu solute atom in Fe.

immediately below the glide plane are partly consistent (Fig. 4(c)), but the minimum in E_I is so deep that the dislocation glides spontaneously over a distance of several b at T = 0K to bind strongly with the Cu atom. Surprisingly, E_I for the plane coincident with the bottom of the extra half-plane is negative (Fig. 4(b)), despite the compressive nature of the stress.

The effect of the single Cu solute in one of the three planes on the stress required to make the dislocation glide at T = 0K is shown in Fig. 5 for L = 3.5 and 11.2nm. The value of τ_P for the pure crystal is indicated by the horizontal line. The variation of τ above and below τ_P with strain reflects the attractive and repulsive variations of E_I as the dislocation moves towards and away from the solute. The maximum τ increases with decreasing L, as expected for a periodic row of localised obstacles, but not simply as L^{-1}. A reduction in L results in a decrease in line flexibility and hence in the large values of τ.

The velocity of the edge dislocation gliding under constant τ in 0.5 and 1at% solid solutions of Cu in Fe has been determined. The position of the dislocation line is plotted against time for several τ values at T = 300K in Fig. 6 for a model with L = 11.2nm, together with results for pure Fe. Velocity values for each Cu concentration are shown for each stress. It is seen that v is only weakly affected by the solute at high τ (above τ_P) but the dislocation stopped moving after a short time at 20MPa in the 1% alloy and did not move at all in either alloy when τ = 5MPa.

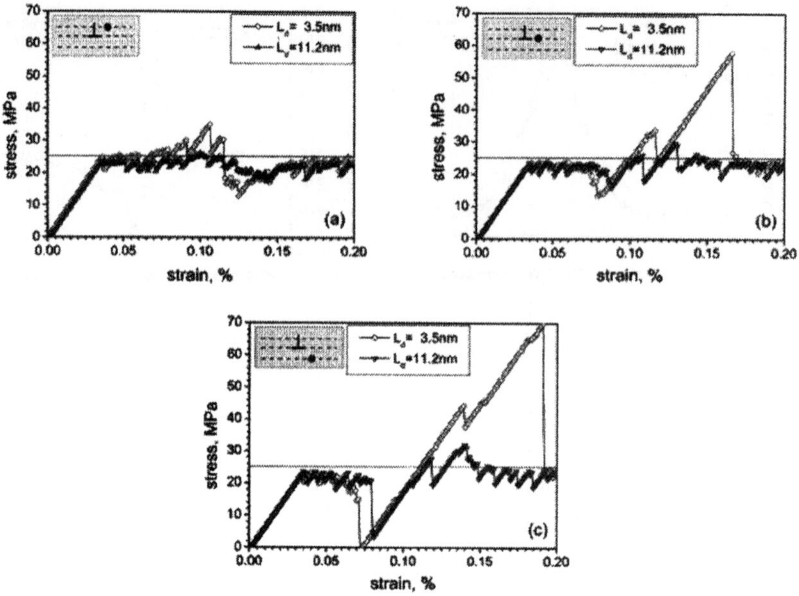

Figure 5. Stress-strain response at T = 0K of a model containing one Cu solute atom.

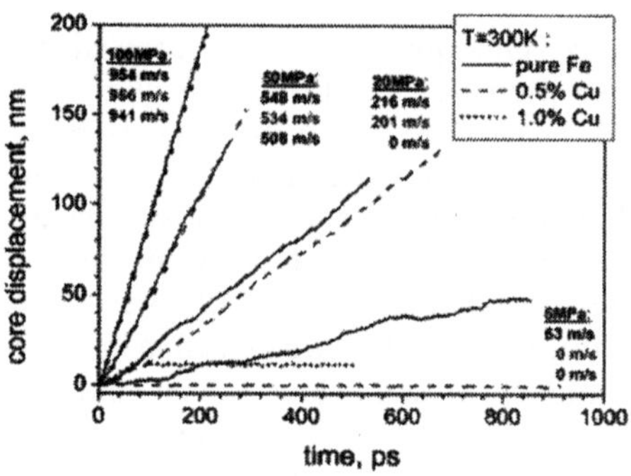

Figure 6. Distance vs time for edge dislocation glide at 300K in Fe and Fe-Cu solid solution.

4. DRAG OF A GLISSILE INTERSTITIAL LOOP

The geometry considered for the interaction and drag of an interstitial loop by an edge dislocation is shown schematically in Fig. 1. We consider here only the case when the glissile loop contains 37 interstitials and has the same perfect Burgers vector as the line. The loop is hexagonal in shape with centre a distance H below the dislocation glide plane and sides along <112>.

To understand the results more readily, the interaction energy, E_I, as a function of the distance, x, between the loop habit plane and the dislocation for H = 11, 22 or 33a_0 is plotted for Fe in Fig. 7 using two calculations. One labelled 'elastic' is simply that given by isotropic elasticity theory for an infinitesimal loop, i.e. $\sigma_{xx}bA$, where σ_{xx} is a normal stress component of the edge dislocation and A the loop area. In the elastic approximation, the loop is attracted to the positions x = ±0.58H where σ_{xx} is minimum. The data labelled 'atomic' were obtained by energy minimisation in the atomic model with periodic length L = 11nm (Fe) or 9.3nm (Cu). For H = 11a_0 the line-loop attraction is so strong that the loop glides to the energy minimum over $-40a_0 < x < 40a_0$, even in the static model. Overall, the curves indicate that a glissile loop should be dragged by a moving dislocation if the resistance to loop glide can be overcome by the maximum force exerted on it by the line, i.e. the maximum slope of the E_I vs x plot. Although E_I has longer range in the atomic calculation, because of the periodicity along the x axis, the 'atomic' and 'elastic' curves are similar in form, suggesting that it should be possible to interpret loop-dislocation forces using the elasticity analysis.

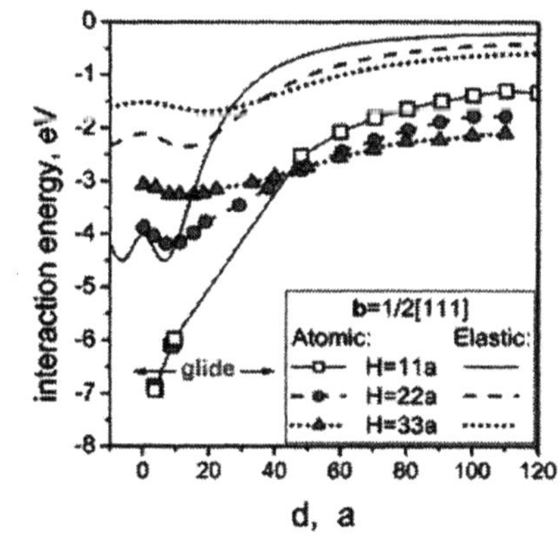

Figure 7. E_I vs distance for a loop and dislocation with the same **b** = 1/2[111].

Here, we present statics (T = 0K) and dynamics (T = 300K) results for H = 11a_0 for Fe and 11.4a_0 for Cu. The dependence of τ on strain in the statics simulation is presented in Fig 8, superimposed on the data from Fig. 2 for the dislocation alone. For the H value considered, the line moves when τ is high enough and then drags the loop in both metals. The maximum stress, τ_{max}, is 28MPa for Fe and is only slightly higher than τ_P, but the value for Cu is twice as large as τ_P, i.e. 32MPa, and is preceded by small displacements of atoms in the loop that result in small displacements of the line.

Results for the dynamics of dislocation glide at 300K in the presence of a glissile loop are presented in Fig. 9(a), together with the v vs τ data from Fig. 3(b) for the dislocation alone. The dislocation is able to drag the loop for τ up

100-150MPa in Fe and 50-75MPa in Cu, as indicated by the 'breakaway' arrows. Drag results in a reduction in v, except at small τ. (A reduction in v occurs beyond breakaway in the modelling because the moving line repeatedly interacts with the loop as a result of periodicity in the x direction.) It is striking that a loop velocity of over 800 and 200ms^{-1} is achieved prior to breakaway in Fe and Cu, respectively. The effect of the loop on the drag coefficient, B, for the dislocation is shown in Fig. 9(b). The additional contribution is less than 10μPas for Fe and about 30μPas for Cu.

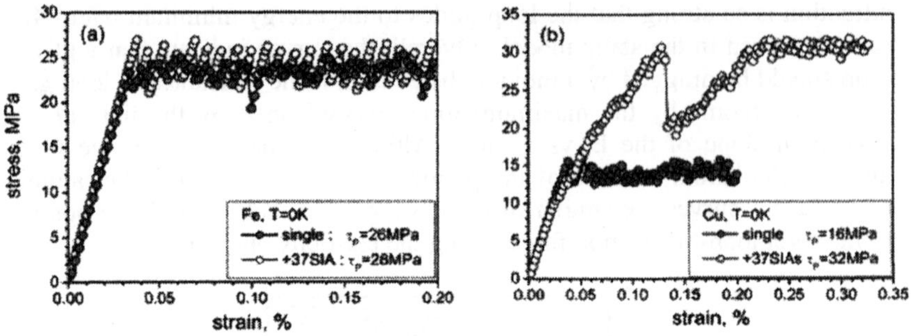

Figure 8. Stress vs strain at T = 0K of (a) Fe and (b) Cu containing a 37-interstitial loop.

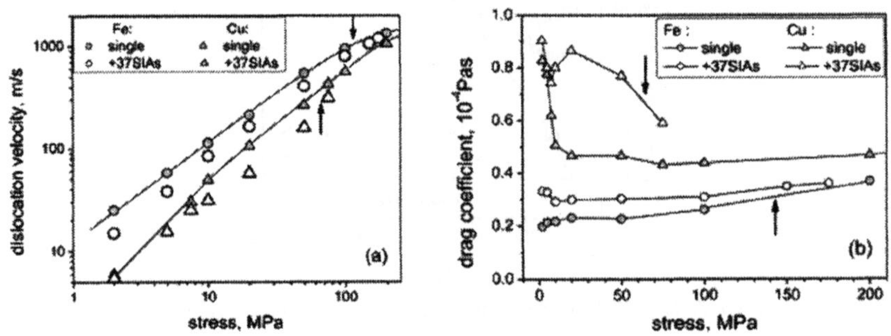

Figure 9. (a) Velocity and (b) drag coefficient for an edge dislocation in Fe and Cu at 300K with and without interaction with a loop. Arrows indicate dislocation breakaway.

5. DISCUSSION

The results demonstrate the quantitative, and sometimes unexpected, nature of the information that can be gained with the method used [7]. For example, although elasticity theory gives reasonable values for E_I between a substitutional Cu atom and an edge dislocation in Fe for solute positions on planes that are not adjacent to the glide plane, the results of Fig. 4 show that it significantly underestimates E_I for Cu for solute sites in the core just below the extra half plane (Fig. 4(c)). Furthermore, the 'atomic' data for the plane coincident with the bottom of the extra half-plane are not consistent with the pΔV data, even with regard to sign (Fig. 4(b)). This result does not seem to have a simple explanation, but presumably arises from the Fe-Cu part of the interatomic potential and may reflect the appeal for the Cu atom of sites adjacent to bond stretching in the Fe.

Another unexpected result is that the dislocation drag coefficient, B, is higher in the model of Cu than Fe, although the static Peierls stress, τ_P, is lower. It is possible that phonon scattering is enhanced by the dissociated state of the dislocation core in Cu. However, we note that in an accompanying paper by Bitzek et al. [11], MD simulation of edge dislocation glide in Ni gave B = 15μPas at 300K, i.e. about one third of the value obtained here for Cu, and τ_P = 0.15MPa, i.e. two orders of magnitude smaller than our result. This difference between the metals is unexplained. Also, it is not clear why B for Cu exhibits such a strong increase when τ falls below τ_P, in contrast to Fe. These effects require further consideration using models with different interatomic potentials.

It was seen in Fig. 8 that for a 37-interstitial cluster a distance H = 11a_0 (Fe) and 11.4a_0 (Cu) below the dislocation slip plane, the loop glides when τ reaches τ_{max} = 28MPa for Fe and 32MPa for Cu at T = 0K, compared with τ_P values of 16 and 25MPa, respectively. Although τ_{max} is different in the two cases, the force the dislocation requires to drag the loop is almost the same, i.e. $\tau_{max}bL$ ~78pN, because of the difference in L. The loop is dragged at τ_{max} because the force required to overcome its glide resistance is less than the maximum the dislocation can exert, i.e. $|dE_I/dx|_{max}$, which from the elastic line in Fig. 7 is ~0.2eV/a_0 ≅112pN.

The plots for v and B in Fig. 9 show that under constant stress at 300K, the influence of loop drag in our models is larger in Cu than Fe. To pursue this further, let B = B_{disl} + B_{loop} for a line dragging a loop, where B_{disl} is the drag coefficient for the dislocation in the absence of interaction with a loop and B_{loop} is the additional effect due to loop drag, i.e. approximately 7 and 35μPas for Fe and Cu, respectively. For a loop being dragged by a dislocation length L under stress τ, the mobility, m, is

$$m=\frac{\text{velocity}}{\text{force}}=\frac{v}{\tau bL}=\frac{1}{B_{loop}L}. \qquad (1)$$

The mobility for a diffusing defect drifting in an energy gradient is approximately D/kT, where D is the diffusivity and k is the Boltzmann constant. Hence, B_{loop} can be estimated if D is known.

The thermally-activated motion of interstitial clusters in metals has been simulated by MD in recent years and detailed information is available for Fe and Cu [2]. Clusters migrate one-dimensionally in the direction of **b** with high jump frequency, ν, because the activation energy is small (~0.02eV).

The jumps are highly correlated at low T, i.e. there is a preference for a jump to be in the same direction as the previous one, and D may be written:

$$D = f\frac{1}{2}b^2\nu, \tag{2}$$

where f is the correlation factor and b the length per elementary jump. Hence

$$B_{loop} = \frac{2kT}{fb^2\nu L}. \tag{3}$$

For a 37-interstitial cluster, f = 4 and 2.5, and ν = 0.3 and 0.2THz for Fe and Cu at 300K, respectively [2]. Eq. (3) therefore predicts that B_{loop} is approximately 11 and 26μPas for Fe and Cu, respectively.

This treatment is based on that for damping due to drag of discrete pinning points, e.g. point defects, by dislocations [12]. The finding that the estimates for B_{loop} are close to the increase in B seen in Fig. 9(b) give weight to the validity of this model for the current problem. We are currently carrying out simulations for other T, H and **b** combinations to test this. If it proves to be applicable, it should be possible to create general rules for loop drag based on only a few parameters.

ACKNOWLEDGEMENTS

Research supported by: the UK Engineering and Physical Sciences Research Council; a JREI grant of the Higher Education Funding Council for England; and the Division of Materials Sciences and Engineering and the Office of Fusion Energy Sciences, U.S. Department of Energy, under contract DE-AC05-00OR22725 with UT-Battelle, LLC. Z.R. holds an ORS award of Universities UK, and K.T. is supported by the Thai Ministry of Science.

REFERENCES

1. D.J. Bacon and Yu.N. Osetsky, International Materials Reviews, 47 (2002) 233.
2. Yu. N. Osetsky, D. J. Bacon, A. Serra, B.N. Singh and S.I. Golubov, Philos. Mag. 83 (2003) 61.
3. M.J. Makin, Phil. Mag. 10 (1964) 695.
4. H. Trinkhaus, B.N. Singh and A.J.E. Foreman, J. Nucl. Mater. 249 (1997) 91.
5. Z. Rong, V. Mohles, D.J. Bacon and Yu.N. Osetsky, in preparation.
6. Yu. N. Osetsky and D. J. Bacon, these Proceedings.
7. Yu. N. Osetsky and D. J. Bacon, Modell. Simul. in Mater. Sci. Eng. 11 (2003) 247.
8. G.J. Ackland, G. Tichy, V. Vitek, and M.W. Finnis, Philos. Mag. A, 56 (1987) 735.
9. G.J. Ackland, D.J. Bacon, A.F. Calder, and T. Harry, Philos..Mag. A 75 (1997) 713.
10. D. Hull and D.J. Bacon, *Introduction to Dislocations*, Butterworth-Heinemann, 2001.
11. E. Bitzek, D. Weygand and P. Gumbsch, these Proceedings.
12. K. Lücke and A.V. Granato, Phys. Rev. B 24 (1981) 6991.

MULTISCALE ANALYSES OF HIGH STRAIN RATE DEFORMATION: ANISOTROPIC EFFECTS

M. A. Shehadeh[a], H. M. Zbib[a,*], T. Diaz de la Rubia[b] and V. Bulatov[b]

[a] *School of mechanical and Materials Engineering, Washington State University, Pullman, WA 99163-2920, USA*

[b] *Material Science and technology Division, Chemistry and Material Science Directorate, Mail Stop L-353, Lawrence Livermore National Laboratory, 7000 East Avenue Livermore, CA 94550, USA*

Abstract The high strain-rate deformation in copper single crystals is investigated using a multiscale dislocation dynamics plasticity model. In this study, we examine the effects of crystal orientation and anisotropy on wave profiles, dislocation density histories, and dislocation microstructures. The morphologies of the relaxed configurations of dislocation microstructures show formation of deformation bands. The number and thickness of these bands are observed to be dependent on crystal orientations.

Keywords: Dislocation Dynamics, High Strain Rate, Dynamic Plasticity, Multiscale, Shock loading

1 INTRODUCTION

There are many parameters that affect the deformation process in metal single crystals in either quasi-static or high deformation rate processes. Among these parameters, crystal orientation exhibits a primary influence on the yield strength and flow stress (Driver et al., 1994, Follansbee and Gray 1991, Horstemeyer et al., 2002). Advancements in experimental capabilities over the years have improved our understanding of the dynamic response of materials. The Hopkinson bar and plate impact techniques can now be used to study the dynamic inelastic response of materials over strain rates ranging from 10^2 s^{-1} to 5×10^6 s^{-1} (Clifton, 2000, Nasser, 1992). Mogilevskii and Bushnev (1990) used the plate impact experiment to shock load copper and aluminum single crystals to pressures ranging from 50 to 100 GPa, and observed misorientation bands in both materials. The features of these bands were found to be dependent on the crystal orientation. Recently, laser based experiments have been used to study the plastic deformation in metals single crystals at different orientations. Short pulse duration (in the order of nanoseconds) was used to generate extremely strong pressure waves that

H. Kitagawa and Y. Shibutani (eds.), IUTAM Symposium on Mesoscopic Dynamics of Fracture Process and Materials Strength, 183–192.

propagate through the samples (Loveridge-Smith, 2001, Kalanter et al., 2001, Kanel et al., 2001, Meyers et al., 2001, Meyers et al., 2003).

Different dislocation microstructures were observed at different values of pressures. During shock or impact loading, plastic deformation (occurring as a result of fast moving dislocations) is often localized in narrow band like structure. These bands generally lie on the slip planes and contain a very high density of dislocations (Coffey, 1992), which arrange themselves in certain dislocation structures depending on the loading conditions. These dislocation structures form almost instantaneously up to 10^6 s^{-1} strain rate (Kuhlmann-Wilsdorf, 2001). The observed microstructures consist of dislocation cells, deformation micro bands and deformation twins (Rivas et al., 1995, Mogilevskii and Bushnev, 1990).

Deformation physics at high strain rate is a complex multiscale dynamic problem. The level of pressure and temperature involved under extreme conditions may make it difficult to address the deformation process using physical experiments. In fact, current experimental capabilities cannot address material response at pressures larger than 100 GPa. In addition the cost of full scale testing in this area of research is high and keeps escalating (Mayer, 1992). Therefore, computer simulations methodologies are used to study the dynamic deformation phenomena by attempting to bridge the length scales from atomistic to macroscopic scales. In the atomistic scale, molecular dynamic simulations are used to investigate the response of single crystals to high strain rate loading at different crystal orientations (Horstemeyer et al., 2001, 2002). Smirnova, et al. (1999) introduced a combined molecular dynamics and finite element approach to simulate the propagation of laser induced pressure in a solid. In spite of all computational advances, current atomistic simulations can model no more than billions of atoms (Clifton and Bathe, 1999).

In the microscale, discrete dislocation dynamics provides an efficient approach to investigate the collective behavior of a large number of interacting dislocations. Dislocation dynamics (*DD*) can simulate sizes much larger than the current atomistic simulation capabilities. Recently, dislocation dynamics based multiscale modeling, which couples the micro and macro length scales has emerged as an excellent numerical tool to simulate the collective behavior of dislocations in a bulk material. In this paper, we used a *multiscale dislocation dynamics plasticity* (*MDDP*) model to study the effect of crystal orientation and anisotropy on the interaction between stress waves and dislocations.

2 MULTISCALE DISLOCATION DYNAMIC PLASTICITY MODEL

The *MDDP* model is based on fundamental physical laws that govern dislocations motion and their interactions with various defects and interfaces. The model merges two length scales; the nano-microscale where plasticity is determined by explicit 3D dislocation dynamics analysis providing the material length scale, and the continuum scale where energy transport is based on basic continuum mechanics laws, i.e. linear momentum balance and energy balance:

$$div\, S = \rho\, \dot{v} \tag{1}$$

$$\rho\, C_v \dot{T} = K \nabla^2 T + S.\dot{\varepsilon}^p \tag{2}$$

where $v = \dot{u}$ is the particle velocity, $\boldsymbol{u}$, ρ, C_v and K are the displacement vector field, mass density, specific heat and thermal conductivity, respectively. Then, the strain rate tensor $\dot{\varepsilon}$ is decomposed into an elastic part $\dot{\varepsilon}^e$ and plastic part $\dot{\varepsilon}^p$, which when combined with the classical Hooke's law yields:

$$\overset{\circ}{\boldsymbol{S}} = \left[C^e\right]\left[\dot{\varepsilon} - \dot{\varepsilon}^p\right]\dot{\varepsilon}^e,\ \overset{\circ}{S} = \overset{\bullet}{S} - \omega S + S\omega,\ \omega = W - W^p \tag{3}$$

where $\overset{\circ}{\boldsymbol{S}}$ is co-rotational Cauchy stress rate, C^e is, in general, the anisotropic elastic stiffness tensor, ω is the spin of the microstructure and it is given as the difference between the material spin W and plastic spin W^p. The evaluation of the plastic strain increment is performed in the discrete dislocation dynamics component of the model, involving massive computations of dislocation-dislocation interaction, motion, multiplication, annihilation, etc. The reader is referred to various papers dedicated to the development of this model (Hirth et al., 1998, Zbib et al, 1998-2002, Rhee et al, 1998). The resulting system of equation is solved using the standard finite element (*FE*) method.

3 MDDP SIMULATIONS

The *MDDP* simulations are designed to mimic uniaxial strain loading of high strain (10^6 s^{-1}) and short pulse durations of (2 nanoseconds). As illustrated in Figure 1, a computational cell, with dimensions 2.5 μm ×2.5 μm ×25 μm is used. Velocity-controlled boundary condition was applied on the upper surface over a short period of time t^* to generate the stress waves. In this case, the applied velocity corresponds to the average strain rate over the entire domain, whereas t^* corresponds to the required pulse duration. The upper surface is then released and the simulation continues for the elastic wave to interact with the existing dislocations. In order to achieve a uniaxial strain involved in shock loading, the four sides of the block are confined so that they can displace in the loading direction only. The bottom surface is rigidly fixed. In order to isolate the effect of the reflected wave, the length of the cell (25 μm) is chosen such that once the wave front reaches the bottom surface, the value of the stresses in the region where the dislocations are located is very small so that dislocation relaxation process can take place. Since metal single crystal exhibits cubic symmetry that leads to the anisotropy in their behavior, the *FE* stiffness matrix is modified to account for this effect.

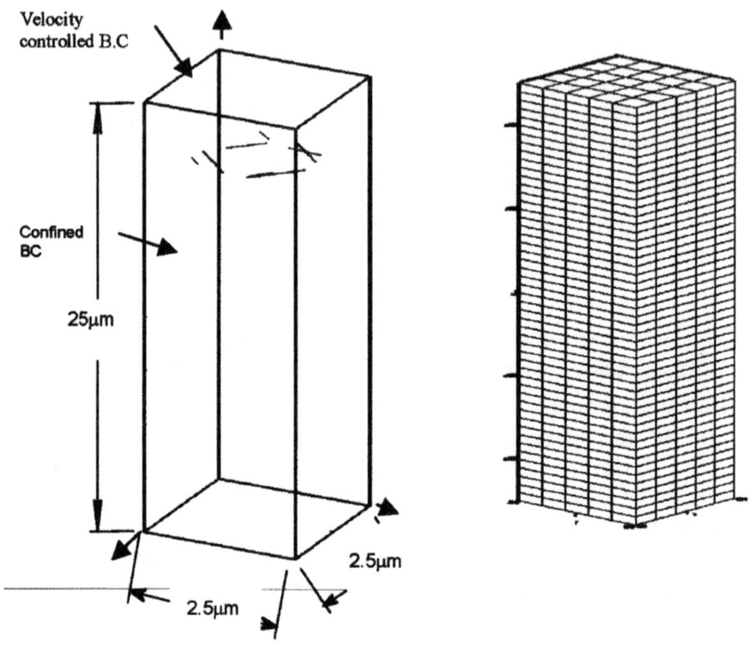

Figure 1: Setup of the *DD* simulation cell and the *FE* mesh.

Frank-Read loops randomly distributed on different slip planes are used as sources to generate dislocations as the wave interacts with these sources. The length of each source ranges from 2000b to 3000b (0.50-0.80 μm) where b is the magnitude of the Burgers vector. It is worthy mentioning that while in *FE*, the fours sides of the computational cell are constrained to displace only in the x and y direction so that confined state of stress is achieved, periodic boundary conditions are used in the *DD* part of the code in order to account for the periodicity of single crystals in infinite media.

4 RESULTS AND DISCUSSION

The effect of crystal orientation on dislocation-stress wave interaction was studied by performing simulations for three different orientations such that the number of slip systems activated in each orientation is different. The computational cell is oriented with respect to the reference frame, i.e., $(100, 010, 001)$. The orientations used for first, second and third case are $(100, 010, 001)$, $(1\bar{1}0, 11\bar{2}, 1\bar{1}1)$ and $(121, \bar{1}01, 1\bar{1}1)$, respectively. In all these three orientations, the crystal is loaded in the $[00\bar{1}]$ direction. In this section we will discuss the effect on crystal orientation and anisotropy on the dislocation density history, dislocation microstructures and the wave profiles.

4.1 Dislocation Density Histories

Figure 2 shows the dislocation density histories for the three orientations. These curves suggest the existence of three distinct regimes of interaction between dislocations and the propagating elastic waves. These regimes are: A) no interaction regime, where the wave has not yet impacted the sources, B) the interaction regime, where the wave impacts the dislocation sources leading to an avalanche in the dislocation density, and C) the relaxation regime, in which, wave surpasses the region where the dislocations are located leading to saturation in the dislocation density. In addition, these curves show that the dislocation density histories are very sensitive to crystal orientation. This indicates that the type and number of activated slip systems and their interaction differ from one orientation to another.

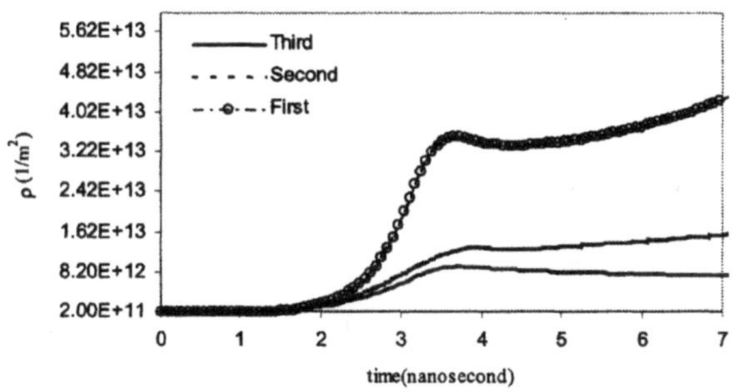

Figure 2: The effect of crystal orientations and anisotropy on the dislocation history in copper shocked to 10^6 s^{-1} for 2 nanoseconds

In FCC crystals there are 12 possible slip systems, which can contribute in the deformation process. Table 1 lists the 12 slip systems. The configurations of the slip systems before and after the load is applied are shown Figure 3a and 3b respectively.

Table 1: Slip systems in FCC metals.

Slip System	Burgers Vector	Slip Plane	Slip System	Burgers Vector	Slip Plane
1	[-1 1 0]	(-1 -1 1)	7	[1 0 -1]	(1 -1 1)
2	[-1 1 0]	(1 1 1)	8	[1 0 -1]	(1 1 1)
3	[0 1 -1]	(-1 1 1)	9	[0 -1 -1]	(1 -1 1)
4	[0 1 -1]	(1 1 1)	10	[0 -1 -1]	(-1 -1 1)
5	[-1 0 -1]	(-1 1 1)	11	[-1 -1 0]	(1 -1 1)
6	[-1 0 -1]	(-1 -1 1)	12	[-1 -1 0]	(-1 1 1)

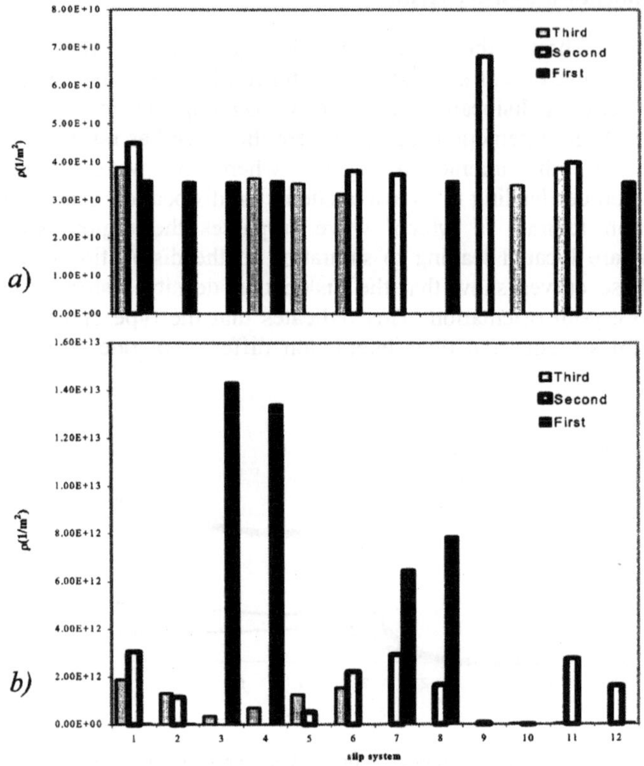

Figure 3: The configuration of dislocation density distribution on the slips systems at different orientations (a) before loading, (b) after loading.

4.2 Dislocation Microstructure

The dislocation microstructures generated by shocks depend strongly on the level of pressure carried by the stress wave (Meyers 1994). We found previously that the dislocation microstructure at strain rates on the order of 10^6 s^{-1} or higher, consists of deformation bands within which the dislocation densities are very high (Shehadeh et al., 2003). Morphologies of dislocations at different crystal orientations are illustrated in Figure 4, which shows the microstructure obtained from slices within the computational cell. It is clear that crystal orientaion does not change the main features of the micorstucture at the applied strain rate of 10^6 s^{-1} used in the present study. However, the number, thickess and the dislocation density in these these deformation bands differ from one orienatation to another.

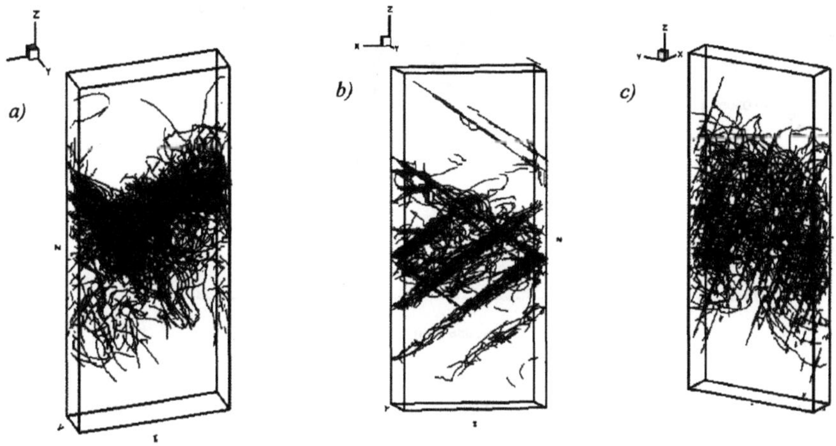

Figure 4: Dislocation microstructures at different orientations (a) first orientation (b) second orientation(c) third orientation.

Profiles of the elastic and plastic strains spatial distributions as a wave propagates in copper are shown in Figure 5a and 5b respectively. Figure 5b shows that the local plastic strain begins to evolve when the stress wave reaches the region where the dislocations are located. It is seen that the values of the plastic strain keeps on increasing until the wave surpasses that region and relaxation takes place where the dislocation density saturates and so does the plastic strain.

a)

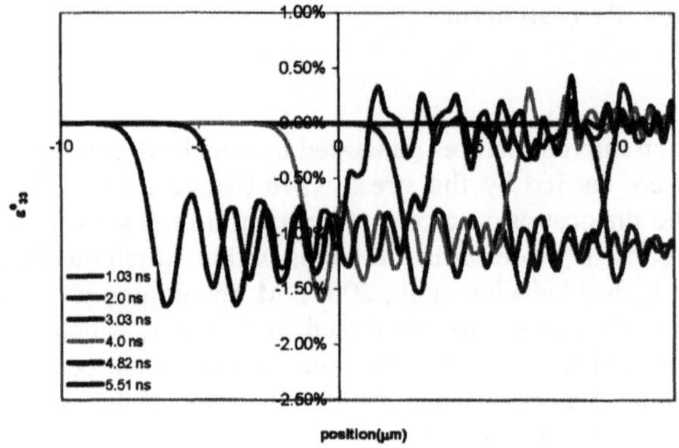

b)

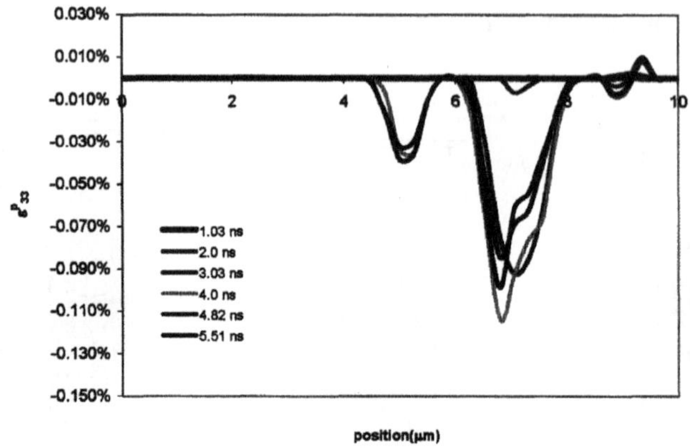

Figure 5. (a) Profiles of ε^{e}_{33} in copper single crystal and (b) profiles of ε^{P}_{33} at different times.

ACKNOWLEDGEMENTS

The support of Lawrence Livermore National Laboratory to WSU is greatly acknowledged. This work was performed, in part under the auspices of the US Department of Energy by Lawrence Livermore National Laboratory (contract W-7405-Eng-48). Fruitful discussions with Prof. John Hirth are greatly appreciated. We are thankful to Mr. Shafique Khan for reviewing the manuscript.

REFERENCES

Clifton, R.G, 2000. Response of materials under dynamic loading. International Journal of Solids and Structures, 37: 105-113.

Cilfton R. J., Bathe, N, 1999. Bridging length scales in dynamic plasticity simulations. In: Furnish, M. D., Chhbildes, L. C., Hixson, R. S., (Eds.), shock Compression of Condensed Matter, Snowbird Utah.

Coffey, C. S, 1992. Dislocation microscopic approach to shear band formation in crystalline solids during shock or impact. Shock Waves and High Strain Rate Phenomena in Materials, Meyers, M. A., Murr, L. E., Staudhammer, K. P., (Eds.), Marcel Dekker.

Driver, J. H., Juul Jensen, D., Hansen, N., 1994. Acta Mater. 42, 3105.

Follansbee, P. S., Gray, g. T., 1991. The response of single crystal and polycrystal nickel to quasistatic and shock deformation. International Journal of Plasticity. 7, 65.

Hayes, D., Hixson, R. S., McQueen, R.G, 1999. High Pressure Elastic Properties, Slid-Liquid Phase Boundary and Liquid Equation of State From Release Wave Measurements in Shock Loaded Copper. In: Furnish, M. D., Chhbildes, L. C., Hixson, R. S., (Eds.), shock Compression of Condensed Matter, 1999 June 27- July 2 in Snowbird Utah.

Hirth J. P., Zbib, H. M., Lothe, J., 1998, Forces on high velocity dislocations. Modeling Simlu. Mater. Sci. Eng. 6, 165-169.

Horstemeyer, M. F., Baskes, M. I., Godfrey, A., Hughes, D. A., 2002. A large deformation atomistic study examining crystal orientation effects on the stress strain relationship. International Journal of Plasticity. 18, 203-229.

Horstemeyer, M. F., Baskes, M. I., Plimpton, S. J., 2001. Length scale and time scale effects on the plastic flow of fcc metals. Acta Mater. 49, 4363-4374.

Kalantar, D. H. et al.2001. Laser driven high pressure, high strain-rate materials experiments In: 12th bienneial International Conference of the APS Topical Group on Shock Compression of Condensed Matter, 24-29, in Atlanta, Georgia.

Kanel, G. I., Razorenov, S. V , Baumung, K., Singer, J., 2001.Dynamic yield and tensile strength of aluminum single crystals at temperature up to the melting point. J. App. Phys. 90, 136-143.

Kuhlmann-Wilsdorf, D., 2001. Q: Dislocations structure-how far from equilibrium? A: Very close indeed. Material Science and Engineering, A315: 211-216

Loveridge-Smith, A., 2001. Anomalous elastic response of silicon to uniaxial shock compression on nanosecond time scales. Phys. Rev.Lett, 86(11), 2349-2352.

Mayer, G., 1992. New directions in research on dynamic deformation of materials. Shock Wave and High strain Rate Phenomena in Materials, M. Meyers, L. Murr, and K. Staudhammer, (Eds.) Marcel Dekker.

Meyers, M. A., 1994. Dynamic Behavior of Materials, John Wiley & Sons, INC.

Meyers, M. A.,et al., 2001. Plastic Deformation in Laser-Induced Shock Compression of Monocrystalline Copper. In: Furnish, M. D., Thadhani, N. N., Horie, Y., (Eds.), shock Compression of Condensed Matter, Atlanta, Georgia.

Meyers, M. A.,et al., 2003. Laser-induced shock compression of monocrystalline copper: characterization and analysis. Acta Materialia. 51, 1211-1228.

Mogilevskii, M. A., Bushnev L. S., 1990. Deformation structure development in Al and Cu single crystals on shock-wave loading up to 50-100 GPa. Combustion, Explosion, and Shock Waves. 26, 215-220.

Nasser, S., 1992. Dynamic deformation and failure. Shock Wave and High strain Rate Phenomena in Materials, M. Meyers, L. Murr, and K. Staudhammer, (Eds.) Marcel Dekker.

Rivas, J. M., Quinones, S. A., Murr, L. E., 1995. Hyper velocity impact cratering: microstructural characterization. Scripta Metallurgica et Materialia, 33(1), 101-107.

Shehadeh, M. A, Zbib, H. M., T. D. de la Rubia and V. Bulatov., 2003. Modeling the dynamic deformation and patterning in FCC single crystals at high strain rates: dislocation dynamic plasticity analysis. In preparation

Smirnova J. A., Zhigilei L. V., Garrison B. J., 1999, A combined molecular dynamics and finite element method technique applied to laser induced pressure wave propagation. Computer Science Communications, 118, 11-16.

Rhee, M., Zbib, H. M., Hirth, J.P., Huang, H. & de La Rubia, T.D., 1998. Models for long/short range interactions and cross slip in 3D dislocation simulation of BCC single crystals. Modeling and simulations in Maters.Sci. & Eng. 467-492.

Weertman, J., 1981. Moving dislocations in a shock front. Shock-Wave and High strain Rate Phenomena in Metals, M. Meyers, and L. Murr, (Eds.) Plenum press, NewYork,

Wright, R. N., Mikkola, D. E., LaRouche, S., 1981. Short Duration Shock Pulses as a Tool to Study the Time Dependence of Plastic Deformation." In: Meyers, M. A., Murr, L. E., (Eds.), Shock Waves and high Strain Rate Phenomena in Metals, Plenum, NewYork.

Zbib, H. M., Rhee,M.& Hirth, J.P., 1998. On plastic deformation and dynamics of 3D dislocations. Int.J.Mech.Sci. 40,113-127.

Zbib, H. M., Diaz de la Rubia, T., 2002. A multiscale model of plasticity. International Journal of Plasticity, 18, 1133-1163.

Zbib, H. M., Shehadeh, M., Khan, S.M, and Karami, G., 2003. Multiscale Dislocation Dynamics Plasticity. International Journal for Multiscale Computational Engineering. 1, 73-89.

ATOMIC-LEVEL INTERACTION OF AN EDGE DISLOCATION WITH LOCALIZED OBSTACLES IN FCC AND BCC METALS.

Yu.N.Osetsky[1,2] and D.J.Bacon[2]

[1] Computer Sciences and Mathematics Division, Oak Ridge National Laboratory, P. O. Box 2008, Oak Ridge, TN 37831-6158, USA
[2] Deptment of Engineering, The University of Liverpool, Liverpool L69 3GH UK.

Abstract Interaction between a moving dislocation and localized obstacles determines microstructure-induced hardening. The mechanisms and parameters of such interactions are necessary inputs to large scale dislocation dynamics modelling. We have developed a model to investigate these characteristics at the atomic level for dislocation-obstacle interactions under both static (T=0K) and dynamic (T>0K) conditions. We present results on hardening due to pinning of edge dislocations at obstacles such as voids, coherent precipitates and stacking fault tetrahedra in bcc-iron and fcc-copper at temperatures from 0 to 600K. It is demonstrated that atomic-scale simulation is required to determine the effects of stress, strain rate and temperature and that such effects cannot always be rationalized within continuum theory.

Keywords Atomic scale modelling, dislocation dynamics, dislocation-obstacle interaction, voids, coherent precipitates, stacking fault tetrahedra.

1. INTRODUCTION

Materials subjected to different treatments may suffer degradation of mechanical properties, e.g. hardening, loss of ductility and plastic instability. In many cases, for example under irradiation, the main reason is formation of

H. Kitagawa and Y. Shibutani (eds.), IUTAM Symposium on Mesoscopic Dynamics of Fracture Process and Materials Strength, 193–202.

localised obstacles to dislocation motion. The typical radiation defects affecting mechanical properties are defect clusters of both vacancy and interstitial types and secondary phase precipitates. Three-dimensional dislocation dynamics (3DDD), as a part of multiscale materials modelling approach, can be used to predict mechanical properties on the basis of the microstructure present but suffers from lack of understanding of short-range interactions between moving dislocations and obstacles, which makes 3DDD approximate in some cases. This gap can be filled using atomic modelling techniques at large enough scale to overlap with the continuum approach, and hence provide mutual validation and parameterization of the mechanisms considered. Another essential advantage of the atomic modelling is the possibility for direct study of thermal effects, which is impossible by other techniques. Recent achievements in atomic-scale modelling of dislocation dynamics [1-5] have demonstrated the importance of this approach. In this paper we review new modelling where a variety of obstacles, such as vacancy voids, stacking fault tetrahedra (SFTs) and coherent precipitates, to the glide of edge dislocations was studied in Cu and Fe over a range of temperatures. We focus mainly on mechanisms of dislocation-obstacle interaction, differences observed for different metals and temperature effects.

2. MODEL

The model for simulating the dynamics of the edge dislocation is described in detail in [4]. It is based on the approach of Baskes and Daw [1] using a periodic array of dislocations. An advantage of the model described in [4] is that is allows the stress-strain curve to be obtained under both statics (T=0) and dynamics (T>0) conditions. The results discussed here were obtained from bcc-Fe and fcc-Cu model crystals. Edge dislocations ½<111><112> (Fe) and ½<110><112> (Cu) were simulated in crystallites having from ~2,000,000 to ~8,000,000 mobile atoms over the temperature range from 0 to 600K. Voids or coherent copper-precipitates in Fe and voids or SFTs in Cu were created in the vicinity of the previously relaxed dislocation, the crystal was relaxed again and then loaded by applying strain or stress. Two approaches were used. Static modelling provides information on equilibrium structure under a given strain, which can be compared directly with continuum modelling of dislocation line shape for example. Dynamic modelling by molecular dynamics (MD) allows temperature effects as well as kinetic properties of moving dislocations to be investigated. The stress-strain curve can be obtained with both approaches and its dependence on strain rate studied in dynamics.

3. RESULTS

3.1 Voids and copper precipitates in iron

An example of stress-strain curves obtained for an initially straight edge dislocation overcoming a periodic row of voids (diameter D, spacing L) in Fe at T=0K is presented in fig.1. According to [4] the total process of dislocation-void interaction can be divided into four stages of dislocation motion, first before it meets the void, second as it enters the void, third as it glides over the void surface and fourth after it breaks away. These stages can be seen clearly in fig.1 for all the voids. The maximum stress for each void size gives the corresponding critical resolved shear stress (CRSS) at which the dislocation breaks away and, provided the strain is increasing, gives further plastic deformation in the fourth stage. Data on CRSS obtained for D=0.7-5nm and L=41-82nm versus harmonic mean $(D^{-1}+L^{-1})^{-1}$ are presented in fig.2 (empty circles) together with the results of continuum dislocation dynamics modelling for voids and Orowan particles from [6,7]. It can be seen that atomic scale modelling data are qualitatively similar to the result for voids in the continuum treatment, suggesting that the dependence :

$$\tau_{Void} = \frac{Gb}{2\pi L}\left[\left(D^{-1}+L^{-1}\right)^{-1}+1.52\right] \quad (1)$$

can be used for estimation of CRSS in higher level continuum modelling of void strengthening.

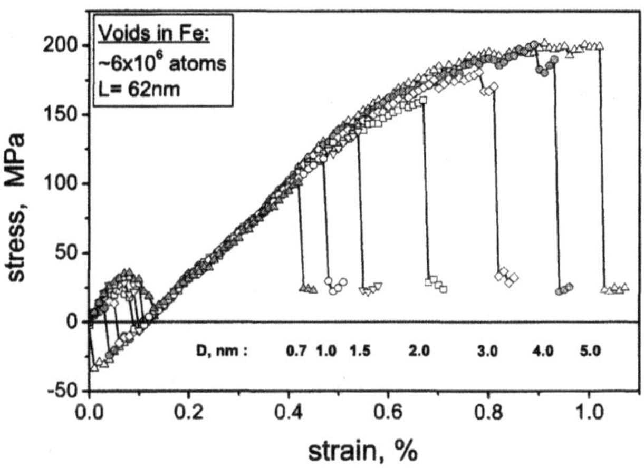

Figure 1. Stress-strain curves obtained for dislocation-void interaction by static modelling in Fe. The number of mobile atoms in the simulated crystal, void spacing, L, and void diameter, D, are indicated.

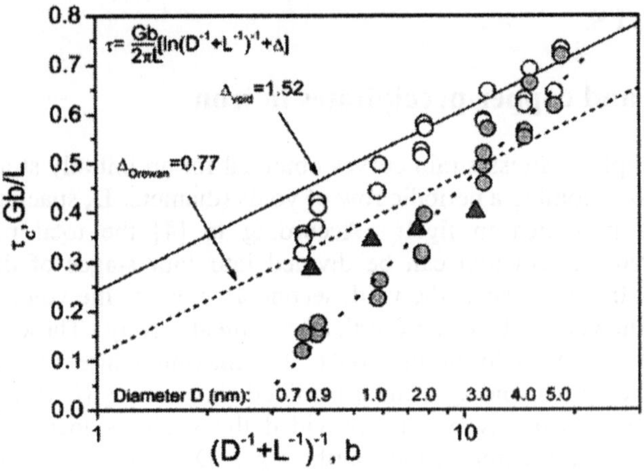

Figure 2. Critical resolved shear stress obtained for different defects as function of harmonic mean of obstacle size D and spacing L. G is the shear modulus and b is the magnitude of the Burgers vector. Open circles – voids in Fe, dark circles – Cu-precipitates in Fe, triangles – voids in Cu. Dashed and solid lines are results obtained in [6,7] by continuum modelling of an edge dislocation passing through a row of Orowan particles and voids respectively, dotted line is interpolation of precipitate results by the dependence $Gb/1.3L[\ln(D^{-1}+L^{-1})^{-1}-0.44]$.

However, the mechanism of dislocation-void interaction observed by atomic modelling in Fe has some differences from that expected from continuum modelling in [7]. The main difference is that at atomic level this interaction is accompanied by dislocation climb, to an extent that depends on the void size (see e.g. [5]). As a result of this climb, the effective size of a void decreases and it becomes a weaker obstacle for the following dislocation.

MD modelling has demonstrated that stress-strain curves are strongly temperature dependent and the CRSS decreases as T rises. An example of CRSS temperature dependence for the 2nm void is presented in fig.3 by open circles. One can see that at T=600K the CRSS drops to 60% of its zero temperature value.

The above data were obtained at strain rate $\dot{\varepsilon}=5\times10^6 s^{-1}$ at which, for the model size used, the dislocation moves with the steady state velocity $V_d \approx 15$m/s, which is a rather low velocity for atomic scale modelling. It was found that the stress at which the dislocation cuts the void depends on its velocity. As an example, consider the following data for the same 2nm void in a crystal under different $\dot{\varepsilon}$. At zero temperature the CRSS was found to be 207MPa and it drops to 179MPa at 100K. However, if simulation at T=100K is made at constant applied stress equal to 100MPa, at which $V_d \approx 900$m/s, the dislocation cuts the void. Notice, this kinetic effect depends on temperature and crystal size, e.g. higher stress should be applied at high

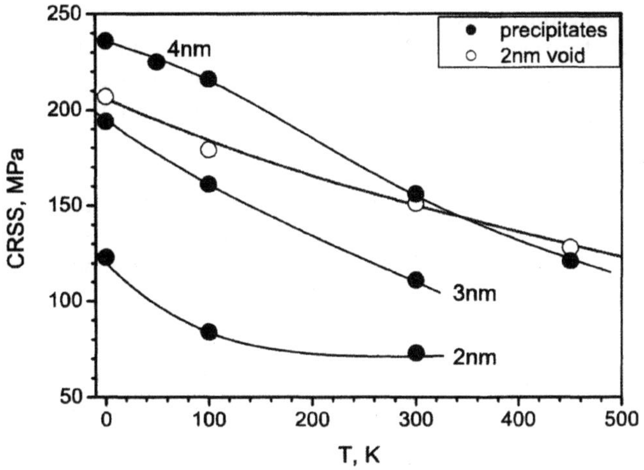

Figure 3. Temperature dependence of the CRSS due to coherent Cu-precipitates of different size and 2nm voids in Fe at applied strain rate $5x10^6 s^{-1}$.

temperature but lower stress is sufficient for the dislocation to cut the void in a larger crystal (i.e. lower dislocation density). This effect could be explained in terms of kinetic energy associated with a moving dislocation. More detailed discussion and explanation of kinetic effects can be found in [8]. We also note here that a fast dislocation climbs more than a slow one.

3.2 Copper-precipitates

Results for the CRSS at zero temperature due to a periodic row of coherent spherical Cu precipitates in Fe are presented in fig.2 by dark circles. The dependence of the CRSS on $\ln(D^{-1}+L^{-1})^{-1}$ can be approximated by a linear function (indicated by dotted line) with a slope significantly steeper than that for voids, i.e. ~1/1.3 compared with $1/2\pi$. However, the mechanism of dislocation-precipitate interaction depends on precipitate size. Thus small precipitates, D<3nm, suffer simple shear, whereas in the case of large precipitates partial transformation towards the more stable fcc structure inside precipitates and dislocation climb are observed.

It is common to attempt to describe precipitate hardening on the basis of the constant line tension approximation. For this the critical angle, φ, between dislocation segments as they break away from a precipitate are required, e.g. it is estimated from the Cu-Fe modulus difference in [9]. It is obtained accurately in our modelling and is >0 for D≤3nm and equal to zero for larger precipitates. This angle was used to estimate the corresponding stress in the line-tension treatment, and gave values up to twice the true value estimated in atomic-scale modelling (see [5]). In other words, in order to give correct stresses, the line-tension model has to use incorrect angles.

The temperature dependence of the CRSS due to Cu precipitates in Fe under a constant applied strain rate of $5x10^6s^{-1}$ is presented in fig.3 by full circles. The dependence is strong relative to that of voids. Thus, a row of 2nm voids is weaker at T=0K than 4nm precipitates, they have similar strength at T=300K and at higher temperature the voids are stronger obstacles. Another effect observed for precipitates is an increase of dislocation climb: the higher the temperature the more vacancies are left inside a precipitate.

3.3 Voids and SFTs in copper

Examples of stress-strain curves for dislocation-void interaction at T=0K in copper are presented in fig.4. It can be seen that the dependences are qualitatively different from those obtained for Fe (see fig.1). The explanation lies in the difference of the dislocation core structure in the two metals, for unlike that in Fe, the dislocation in Cu is dissociated into two Shockley partials linked by a stacking fault ribbon of width ≈13a for the edge dislocation. As a result each partial dislocation interacts with an obstacle individually. The first stress peak at yield in fig.4 for voids of each size corresponds to the stress when the leading partial breaks away, whereas the second peak is related to the same process for the trailing dislocation. The corresponding partial dislocation line shapes for D=2nm are presented in fig.5. It is interesting to note that the relative height of the two stress peaks in fig.4 depends on void size. Thus, when the dislocation cuts small voids the stress at the first peak is lower than that at the second, whereas for the

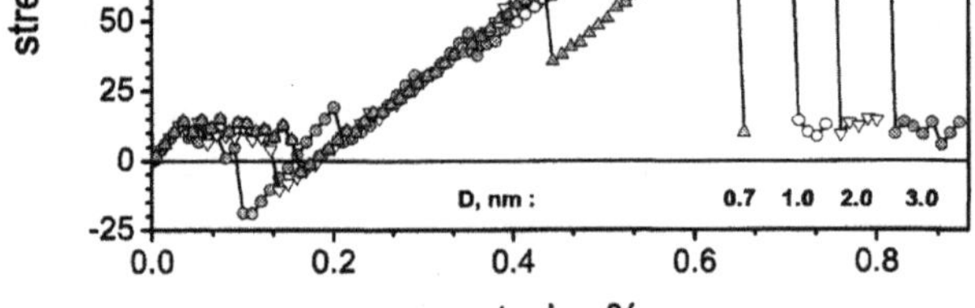

Figure 4. Stress-strain dependence for dislocation-void interaction in Cu at T=0K.

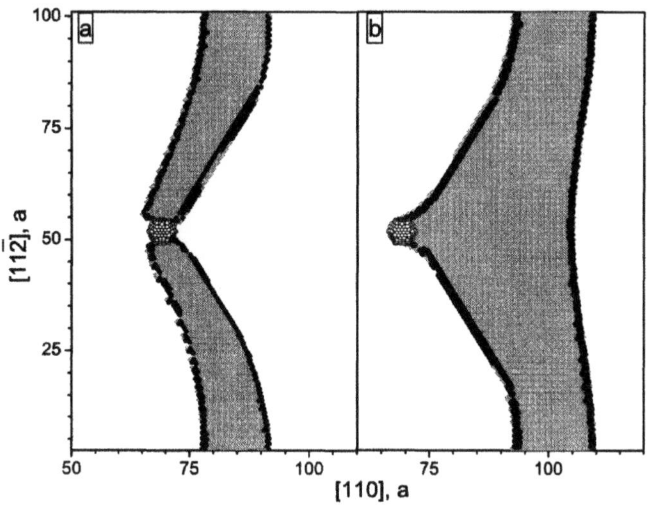

Figure 5. Critical shape of partial dislocations interacting with a 2nm void in Cu at T=0K: (a) leading partial at shear stress 115MPa; (b) trailing partial at 118MPa. The core region of partials is shown by black symbols, the stacking fault is grey and the void surface atoms are indicated by hollow spheres.

largest void studied (D=3nm) the leading partial breaks away at maximum stress. This size-dependent effect is presumably related to the pulling force the leading partial can provide for the trailing one if the area of the stacking fault increases significantly at such a high stress. The assistance of the leading partial may also explain the overall weaker strengthening due to voids in Cu in comparison with Fe, as seen in fig.2. Another feature of the dislocation-void interaction mechanism in Cu is the absence of climb and defect absorption, for only pure shear of voids has been found in all the cases studied.

Temperature effects have been studied in dynamic modelling of interaction of a gliding dislocation with 2nm (369 vacancies) and 3nm (1205 vacancies) voids and 2.5nm (45 vacancies) and 3.0nm (78 vacancies) SFTs in Cu. A significant drop in CRSS occurs when temperature is increased from 0 to 100K. Further increase of temperature gives only a weak additional decrease of CRSS, as shown by the data in fig.6. Note that the 2.5nm SFT demonstrates very similar strengthening to that of a 2nm void, despite the fact that it contains only 45 vacancies against 369 in the void. Moreover, the 2.5 nm SFT is a stronger obstacle than the 3.0nm one. A detailed analysis of dislocation-SFT interaction sheds light on this paradox. The data presented in fig.6 were obtained when the dislocation glide plain cuts an SFT through the centre of mass. In this geometry the glide plane is closer to the parallel face for a smaller SFT than a larger one. By varying the position of the glide plane, we have found that maximum strengthening occurs when a dislocation

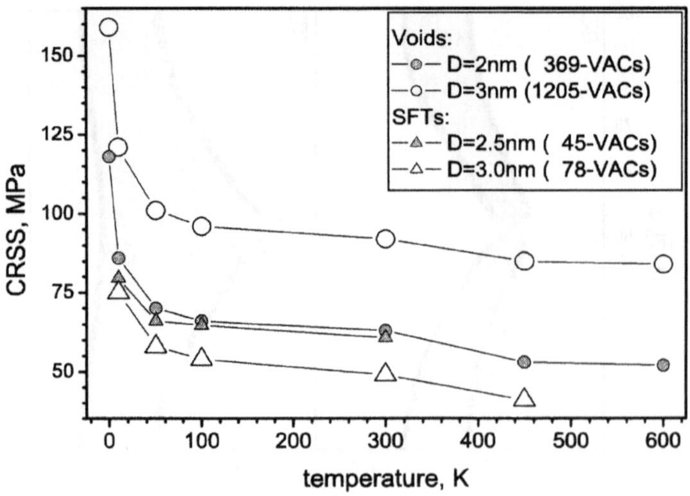

Figure 6. Temperature dependence of the CRSS of voids and SFTs in Cu obtained by molecular dynamics modelling at strain rate $5x10^6 s^{-1}$. Void diameter, SFT size and the corresponding number of vacancies is indicated.

glide plain coincides with the stacking fault of one of the faces.

The larger the distance between the face and the glide plane, the lower the CRSS. Moreover, when the larger SFT (3nm) is cut though the centre of mass at all temperatures above T=10K it recovers its regular structure via a thermally-activated propagation of a jog on the stair-rod partial dislocations of its edges. This recovery does not occur for the smaller SFT whose regular structure remained disturbed after being cut by the dislocation. Similar damage of a small SFT by a dissociated dislocation at T=0K was observed earlier [10].

4. CONCLUDING REMARKS

Atomic-scale modelling with a choice of static (T=0K) and dynamic (T>0K) conditions, together with different types of crystal loading, i.e. applied strain increment or strain rate or applied stress, allows a wide range of features of dislocation-obstacle interactions to be studied. Static modelling reveals many general features which can be treated and understood within continuum dislocation theory. An example is presented in fig.2 where stress values for voids and precipitates in Fe are close to the linear dependence of CRSS on logarithm of harmonic mean of obstacle size and spacing found earlier in simulations based on elasticity theory [6,7]. The data for voids in Cu can be described as a linear dependence in principle, but due to the dissociated character of the dislocation the mechanism is more complicated and void-size dependent: strengthening by small voids is controlled by breakaway of the trailing partial whereas that of large voids is controlled by the leading partial. Dissociation also reduces the total strength of these obstacles. Static modelling can also provide information on dislocation line shape in equilibrium at different stress levels (see e.g. [5]), and this can be used for fitting parameters required in continuum dislocation dynamics models. One of the conclusions on dislocation line shape revealed here and in [5] is that the simple constant line tension model widely used in estimations of CRSS (e.g. for hardening due to Cu precipitates in Fe [9]) provides an incorrect relationship between applied stress and line shape.

Atomic-scale details of dislocation-obstacle interactions depend strongly on dislocation core structure, as demonstrated here by comparison of Fe and Cu. The low stacking fault energy in Cu causes wide separation of the partial dislocations and leads to absence of dislocation climb in interactions with obstacles of vacancy type. The perfect edge dislocation in Fe, in contrast, may climb and absorb defects in all the cases considered.

Dynamic simulations reported here demonstrated a strong temperature effect in the stress-strain characteristics of all the obstacles studied. The nature of this is not yet fully understood. Based on the results presented, we can conclude that the temperature enhancement mechanisms may be different for different obstacles, e.g. stronger for coherent precipitates than voids in Fe (see fig.3), and for different metals, e.g. the temperature dependence of the CRSS for voids is different in Fe and Cu (see figs.3 and 6).

Kinetic effects in dislocation-obstacle interactions have been observed in all the cases treated and this seems to be important for explanation of plastic instability and creation of cleared channels in irradiated metals [11]. In fact, we have demonstrated that a dislocation moving at high speed can cut and breakaway from an obstacle at a stress significantly lower than the CRSS. Moreover, the change in structure of the obstacle due to the fast dislocation is more significant. More studies are necessary to understand these effects and to reveal mechanisms suitable for incorporation in higher level dislocation dynamics applications.

ACKNOWLEDGMENT

This research sponsored by the Division of Materials Sciences and Engineering and the Office of Fusion Energy Sciences, U.S. Department of Energy, under contract DE-AC05-00OR22725 with UT-Battelle, LLC and supported by a grant from the UK Engineering and Physical Science Research Council and a JREI grant of the Higher Education Funding Council for England.

REFERENCES

1. M.I.Baskes and M.S.Daw, Proceedings of Fourth Int. Conf. On the Effects of Hydrogen on the Behaviour of Materials (Jackson Lake Lodge, Moran, WY) eds. N.Moody and A.Thompson, The Minerals, Metals and Materials Society, Warrendale, PA, 1989.
2. D.Rodney and G.Martin, Dislocation pinning by glissile interstitial loops in nickel crystal: A molecular dynamics study. Phys.Rev.B, 2000; 61: 8714.
3. D.Rodney, G.Martin and Y.Brechet, Irradiation hardening by interstitial loops : atomistic study and micromechanical model. Materials Science and Engineering. 2001; A309-310: 198-202.
4. Yu.N.Osetsky and D.J.Bacon, An atomic-level model for studying the dynamics of edge dislocations in metals. Modelling Simul. Mater. Sci. Eng., 2002; 11: 427.
5. Yu.N.Osetsky, D.J.Bacon and V.Mohles, Atomic modelling of strengthening mechanisms due to voids and copper precipitates in alpha-iron. Philos. Mag.A, 2003; in press.
6. D.J.Bacon, U.F.Kocks and R.O.Scattergood, The effect of dislocation self-interaction on the Orowan stress, Philos. Mag., 1973; 28: 1241.
7. R.O.Scattergood and D.J.Bacon, The strengthening effects of voids. Acta Metall., 1982;30; 1665.
8. E.Bitzek, D.Weygang and P.Gumbsch, Atomistic and DD studies of inertial effects on the dynamics of dislocations. These proceedings pg.
9. K.C.Russell and L.M.Brown, A dispersed strengthening model based on a differing elastic moduli applied to the iron-copper system, Acta. Metall., 1972; 20: 969.
10. Yu.N.Osetsky, D.J.Bacon, B.N.Singh and B.Wirth, Atomistic study of interaction, accumulation and annihilation of cascade induced defect clusters, J.Nucl.Mater., 2002; 307-311: .
11. M. Victoria, N. Baluc, C. Bailat, Y. Dai, M. I. Luppo, R. Schaublin and B. N. Singh, The microstructure and associated tensile properties of irradiated fcc and bcc metals, J.Nucl.Mater., 2000; 276:114.

DEFECT NUCLEATION

Predictions through Nanoscale Experiments and Computations

Krystyn J. Van Vliet[1,4], Ju Li[5], Ting Zhu[2], Yoonjoon Choi[1], Sidney Yip[3] and Subra Suresh[1]

Departments of [1]Materials Science and Engineering, [2]Mechanical Engineering and [3]Nuclear Engineering, Massachusetts Institute of Technology, Cambridge, MA 02139

[4]Department of Surgical Research, Children's Hospital and Harvard Medical School, Boston, MA 02115

[5]Department of Materials Science and Engineering, The Ohio State University, Columbus, OH 43210

Abstract: Nanoindentation—contact-induced deformation of nanoscale material volumes—provides the potential to access experimentally the true strength of materials and the conditions for the onset of plastic deformation mechanisms. Here, motivated by experimental observations from single crystal nanoindentation, we outline the parallel development of new experimental, analytical and computational approaches regarding homogeneous defect nucleation. Together, these studies have established a semi-quantitative experimental model, validated a general energetic criterion for defect nucleation, facilitated an atomistically informed computational platform for defect analysis, and elucidated the effects of certain time and length scales on the defect nucleation events that signal the onset of plastic deformation in crystalline materials.

Key words: defect, nucleation, computational modeling, nanoindentation

H. Kitagawa and Y. Shibutani (eds.), IUTAM Symposium on Mesoscopic Dynamics of Fracture Process and Materials Strength, 203–211.

1. INTRODUCTION

Recent advances in techniques such as nanoindentation have enabled direct, experimental access to nanoscale material volumes under large mechanical strain. One important and general observation from such experiments on well-annealed face-centered cubic (fcc) metals is that, upon increasing applied load P, an instability occurs at the material surface such that the indentation depth h increases abruptly with no significant increase in P—a so-called displacement burst. Although other plausible explanations for this experimental observation exist, a hypothesis of great interest is that these displacement bursts represent discrete dislocation nucleation events, and thus signal the onset of plastic deformation in an otherwise defect-free material volume. Thus, such experiments may provide a means to probe directly the fundamental concepts of defect nucleation.

In order to test this hypothesis and formulate predictions regarding the slip character of the nucleated defects, we applied an established, two-dimensional (2D) experimental model to the specific problem of localized mechanical contact: the Bragg-Nye soap bubble raft (Gouldstone et al., 2001). As shown in Fig. 1a, this raft is a valid experimental analog to close-packed crystals insofar as the bubble interaction potential can be engineered to mimic the nearest neighbor interaction potential of fcc metals such as Cu and Ni, and can exhibit defects analogous to edge dislocations, vacancies, and grain boundaries. Nanoindentation of such initially defect-free, single crystalline rafts showed that defects indeed nucleate homogeneously beneath the indenter, at a depth z approximately equal to that predicted by 2D continuum elastic analysis ($z = 0.78a$, where a is the contact half-width).

This surprisingly accurate application of continuum elasticity in the context of a nonlinear elastic instability, coupled with the inherent restriction that load P is not a well-defined quantity in 2D (displacement-controlled) contact, prompted the development of a computational model to verify and expand on these observations (Li et al., 2002). We chose to employ atomistic computations via molecular dynamics, and replicated the site and slip character of defects observed in the 2D experimental analog (Fig. 5-1). More importantly, we realized that the pairing of these experimental and computational models of nanoindentation enabled us to test analytical models regarding defect nucleation in both 2D and 3D. Thus, we formulated an existing, formal and energetic criterion for elastic instability (Hill, 1962; Rice, 1976) such that it could be implemented via current atomistic computations. Briefly, the elastic instability required for nucleation of a defect with slip direction $\mathbf{g}$ and slip plane normal $\mathbf{n}$ occurs when the energetic Λ-criterion vanishes. Validation of this criterion was illustrated initially through exact temporal correlation of the nanoindentation load-displacement instability and the vanishing of the Λ-criterion in MD simulations of the 2D bubble raft (Van Vliet, 2003).

The parallel development of these tools has facilitated several new studies in the area of defect nucleation, particularly regarding how the onset of plastic deformation is affected by time and length scales. Below, we highlight the most recent developments that have stemmed from this collaboration.

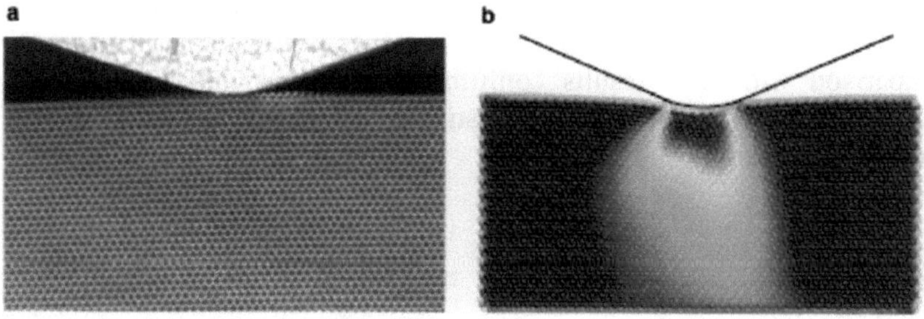

Figure 1. Pairing of (a) an experimental analog to fcc crystals, the soap bubble raft model, subject to nanoindentation; and (b) an atomistic computational approach, molecular dynamics, to quantify the load-displacement effect of defect nucleation.

2. ATOMISTICALLY INFORMED FINITE ELEMENT MODEL

As MD simulations are limited in length and time scales due to the atomistic detail inherent in such calculations, we developed in parallel a computational framework that could be informed by atomistic details such as interatomic potential and defect nucleation, yet simulate structural length scales much greater than the nanoscale volumes achievable with MD. In fact, we adapted a continuum-based finite element model (FEM) for this purpose, implementing the Cauchy-Born elasticity as a user subroutine within a commercially available software package. According to this elasticity description, deformation of a representative material volume can be considered an affine rotation of the crystallographic lattice vector $\mathbf{r}_0$:

$$\mathbf{r} = \mathbf{F}\mathbf{r}_0 \quad (1)$$

where **F** is the rotation vector and **r** is the final crystallographic lattice vector. Energy and derivative quantities of stress and elastic constants can be calculated directly for a given set of **F** and **r**, down to atomistic material volumes.

The implementation of this atomistically informed FEM is conceptually similar to the quasicontinuum method of Tadmor et al. (1996), and is in fact less general. Advantages of this approach in the present context, however, are that this is a fully continuum approach (i.e, no difficulties associated with a transition from a continuum to an atomistic description such as MD) and that this method is much simpler to design, execute and adapt, and that the Λ-criterion can be included explicitly and calculated at every time step to identify the position and slip character of defect nucleation sites. Direct comparison with MD results confirms that atomistically informed, fully continuum FEM achieves computational efficiency without sacrificing atomistic resolution (Zhu, 2003).

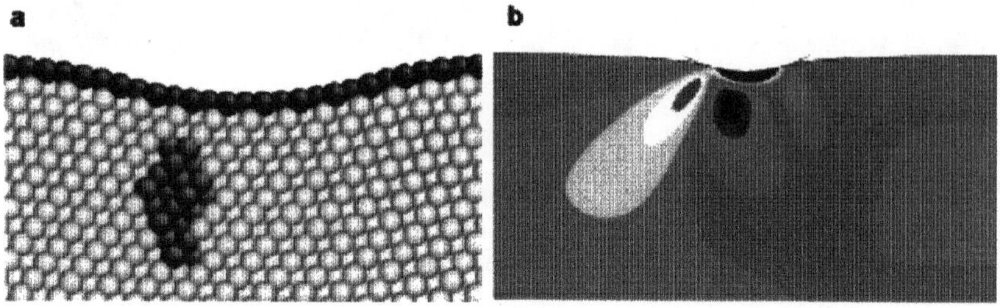

Figure 2. Comparison of (a) Molecular dynamics; and (b) Atomistically informed finite element modeling based on Cauchy-Born elasticity and including the Λ-criterion. Dark blue indicates (a) imperfect bulk coordination number; and (b) $\Lambda = 0$, or attainment of the elastic instability required for defect nucleation.

3. EFFECT OF MICROSTRUCTURE

Modification of processing methods such as inert gas condensation, severe plastic deformation, and electrodeposition have made possible the synthesis of metals with average grain size $d < 100$ nm. Since nanocrystalline metals have thus far been produced only on an experimental bases in small quantities, indentation is the most commonly employed method to assess hardness H as a function of d; trends in yield strength σ_y are usually assumed to be proportional to hardness according to the approximation $H = 3\sigma_y$ (Tabor, 1951). Note that the empirical relationship between grain size and σ_y proposed by Hall and Petch for microcrystalline materials implies increased strength and hardness as grain size decreases to the nanocrystalline regime:

$$\sigma_y = \sigma_{y0} + k_y \ d^{-1/2} \qquad (2)$$

Experimentally, however, some data indicate an opposite relationship between σ_y and d for $d < 10$ nm (e.g., Koch, 2000), a so-called inverse Hall-Petch effect. Such behavior has been ascribed to possible difficulties in producing a uniformly fine distribution of nanocrystals, uncertainties in the determination of d, and the presence of internal defects (Fougere, 1992; Masumura, 1998). Atomistic computational modeling via both MD (Yamakov, 2002; VanSwygenhoven, 2001) and conjugate gradient (energy) minimization (Schiotz, 1998) indicated that nanocrystals under uniaxial compression may indeed exhibit a decrease in σ_y with decreasing d, but the length and timescales of these simulations were admittedly disparate from those used experimentally. In an attempt to reconcile these observations, we applied the bubble raft as a model nanocrystal to evaluate whether and how grain size and grain boundaries might affect the onset of yielding under nanoindentation.

By constructing and nanoindenting rafts for which microstructural parameters including grain size d, in-plane grain boundary misorientation angle θ, and grain boundary position with respect to the indentation loading axis, we quantified the effects of certain microstructural length scales on the initiation of plastic deformation in nanocrystals. Here, plasticity onset was defined as the first observation of (heterogeneous) defect nucleation within the indented nanocrystalline raft, as parameterized by the in-plane coordinates (x, z) and contact half-width of the indenter a. This parameterization enabled us to calculate the critical pressure of indentation, which is proportional to the resolved shear stress that drives dislocation motion. We identified a transitional grain size of approximately 7 nm, greater than which σ_y increased with decreasing d and less than which σ_y decreased with decreasing d. This result was in agreement with a range of recent experimental data obtained via nanoindentation in nanocrystalline Ni. Further, we observed that this transition in d was concomitant with a transiton in deformation mode: For $d > 7$ nm, the onset of plastic deformation proceeded primarily via emission of individual dislocations from grain boundaries or grain triple junctions. For $d < 7$ nm, however, this onset proceeded primarily via migration of entire regions of a grain boundary (Van Vliet et al., 2003).

Further, as shown in Fig. 5-3, we found that although dislocation emission occurred over a wide range of θ (5° < θ < 32°), grain boundary migration occurred over a restricted range of grain boundary misorientation angles (14° < θ < 22°) that define the transition from a high angle to a low angle grain boundary. Interestingly, for a given average grain boundary misorientation angle θ_{avg} as defined over an entire nanocrystalline raft, the transition from emission to migration still occurred at a grain size *d* of approximately 7 nm. This result indicates that, at least in 2D, both θ and *d* impact the mechanisms by which plastic deformation ensues, but *d* is the controlling microstructural factor. As defect nucleation proceeded primarily at grain triple junctions, regardless of the mode of deformation or other microstructural variables, we submit that the importance of *d* is that it indicates not just grain diameter, but also the proximity of triple junctions that define regions of decreased crystalline order and increased free volume. These 2D bubble raft experiments indicate that, as triple junctions increase in number per volume and decrease in spacing, grain boundary migration is favored and resistance to plastic deformation decreases.

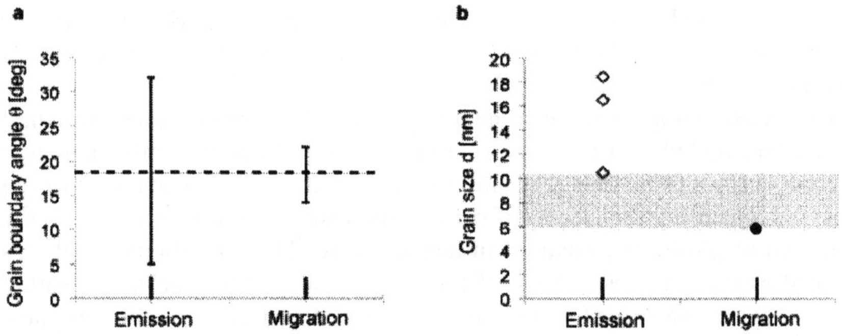

Figure 3. Effect of microstructural length scales on plastic deformation mode in nanocrystalline bubble raft nanoindentation experiments. (a) Grain boundary misorientation angle θ as a function of deformation mode; (b) Grain size *d* as a function of deformation mode, for a fixed raft-averaged misorientation angle 22° < θ_{avg} < 23°.

4. EFFECT OF MACROSTRUCTURE

The above experiments have shown that the fundamental length scales within a material can affect the strength and deformation mechanisms of that material. It is also possible that the length scales that define the structure comprising this material may also impact on defect nucleation mechanisms. One interesting example of such macrostructural considerations is the patterned metal lines utilized in the microelectronics industry as interconnect pathways among circuit components. These lines are semi-infinite in one in-plane dimension, and approach 100 nm in the perpendicular in-plane direction. In order to determine whether and how this macrostructural aspect ratio affects strength and defect nucleation, nanoindentation experiments, bubble raft experiments, finite element simulations, and molecular dynamics simulations were employed (Choi et al., 2003).

As shown in Table I, nanoindentation experiments on Al films of varying thickness indicated that the resistance to plastic deformation (i.e., were plastically "softening") decreased as film thickness increased, as quantified by the indentation depth *h* measured for a fixed indentation load *P*. The additional effect of line aspect ratio is shown clearly in Fig. 5-4, as the depth *h* also increases with decreasing line width *w* for a fixed load *P*. As the films and lines were processed on the same Si wafer, the microstructural length scales (grain size and texture) were essentially identical, and these sample sets differed chiefly in the macrostructural aspect ratio. Finite element simulations of these experiments did not corroborate such a dramatic plastic softening of the lines, indicating that a continuum analysis was insufficient to capture the effect of these macrostructural constraints on defect activity. Bubble raft experiments confirmed qualitatively that the observed plastic softening of the lines was due to the fact that dislocations nucleated within the line during nanoindentation proceeded to the free sidewalls of the line, relaxing the s tructure. In c ontrast, d islocations nucleated d uring nanoindentation of continuous films terminated motion at the film/substrate interface, presumably exerting a back stress that opposed the applied indentation stress and resulting in comparative hardening. Molecular dynamics simulations comparing film and line indentation responses confirmed the qualitative bubble raft observations (nucleated defects migrate to stress-free line sidewalls) and also the quantitative nanoindentation observations that this sidewall relaxation results in an increase in indentation depth for a given indentation load.

Table 1. Effect of macrostructural film thickness on plastic indentation depth

Film thickness [μm]	h_{avg} at P = 200 mN
1.0	88
0.7	83
0.3	70

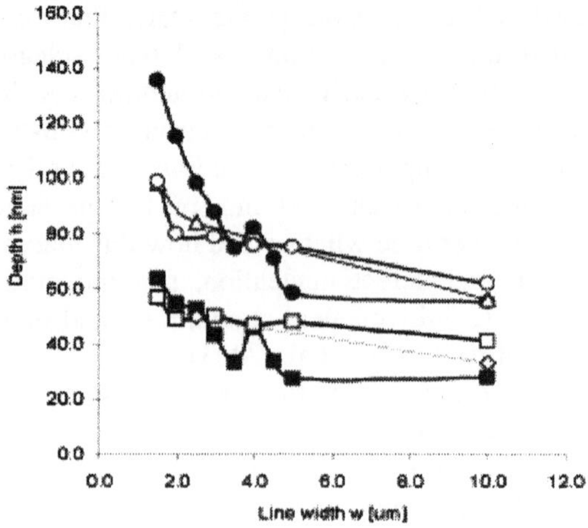

Figure 4. Nanoindentation penetration depth as a function of line width for a fixed maximum indentation load, for Al lines of 1 μm height on Si. Symbols indicate indenter manufacturer (closed, MicroMaterials, Limited, UK; open, Hysitron, Inc., USA).

ACKNOWLEDGEMENTS

KJVV, YC and SS gratefully acknowledge the Defense University Research Initiative on Nano-Technology (DURINT) on "Damage- and Failure-Resistant Nanostructured and Interfacial Materials" which is funded at the Massachusetts Institute of Technology (MIT) by the Office of Naval Research under gran N00014-01-1-0808. JL, TZ, and SY acknowledge Honda R&D Co., Ltd, AFOSR Grant No. F49260-00-10082 and NSF Grant No.DMR-9980015, and Lawrence Livermore National Laboratory under an ASCI Level 2 grant.

REFERENCES

Choi Y.J., Van Vliet K.J., Li J., Suresh S. submitted (2003).

Fougere G.E., Weertman J.R., Siegel R.W., Kim S. Scripta metal mat 1992; 26:1879

Gouldstone A., Van Vliet K.J., Suresh S. Nanoindentation: Defect nucleation in crystals. Nature 2001, 417: 656.

Hill, R. J Mech Phys Solids 1962; 10:1

Koch CC, Fedkiw P., Narayan J. NSF Partnership in Nanotechnology; 2000 January 29-30; Arlington, VA. Retrieved from:
URL

Li J., Van Vliet K.J., Zhu T., Yip S., Suresh, S. Atomistic mechanisms governing elastic limit and incipient plasticity in crystals. Nature 2002; 418:307-310.

Masumura R.A., Hazzledine P.M., Pande C.S., Acta mater 1998; 46:4527

Rice, James R. In *Theoretical and Applied Mechanics*, W.T. Koiter, ed. New York: North-Holland, 1976.

Schiotz J., DiTolla F.D., Jacobsen K.W. Nature 1998; 391:561.

Tadmor E., Ortiz M. and Phillips R. Philos Mag A 1996; 73:1529

VanSwygenhoven H., Caro A., Farkas, D. Scripta mater. 2001; 44:1513

Van Vliet K.J., Li J., Zhu T., Yip S., Suresh S. Quantifying the early stages of plasticity through nanoscale experiments. Phys Rev B 2003; 67:104105

Van Vliet K.J., Tsikata S., Suresh S. Model experiments for direct visualization of grain boundary deformation in nanocrystalline metals. App Phys Lett 2003; 83:1-4

Yamakov V., Wolf D., Phillpot S.R. and Gleiter H. Acta mater 2002; 50:61

Zhu T., Li J., Ogata S., Van Vliet K.J., Suresh S., Yip S., submitted (2003).

Formation of Prismatic Dislocation Loop of Single Crystalline Aluminum under Nanoindentation

T. Tsuru and Y. Shibutani

Dept. of Mechanical Eng. and Systems, Osaka Univ.
2-1, Yamadaoka, Suita, Osaka 565-0871, JAPAN

Abstract: Recent new finding of unstable displacement burst observed in the relation between the indent load and indent depth of nanoindentation is much possible to be related to the collective dislocation behavior. In the present paper, dislocation emission and the subsequent prismatic dislocation loop formation of a single crystalline aluminum under nano-scaled indentation are simulated by the molecular dynamics. The effect of the stress distribution due to two different types of indenters and that of the interatomic description by three different embedded atom type potentials are discussed with much emphasis. At a result, the dislocations are emitted from the severely damaged surface atomic layer due to the indenter. Afterwards, the prismatic dislocation loops, which have the same gliding direction but not on the same slip planes, are attained by both energetically unstable reaction between the shear loops emitted from the surface and the cross slip mechanism.

Key words: Nanoindentation, Dislocation Emission, Prismatic Dislocation Loop, Molecular Dynamics, Interatomic Potential, Single Crystalline Aluminum

1. INTRODUCTION

Indentation technique has been put to a wide use in order to obtain the hardness and the elastic properties of materials even at the nanoscale. Recently, it is worthy of notice that indent load and depth curves provide an opportunity to investigate the early events of plasticity (nano-plasticity). The distinctive finding of an unstable displacement burst of indent depth observed in nanoindentation [1] is thought to be linked to the dislocation emission [2] caused when maximum shear stress generated under the indenter is of the order of the theoretical shear strength. In the present paper, in order to comprehend what kinds of crystallographic incidents happen in the near contact field with the indenter, the nanoindentation process in a

H. Kitagawa and Y. Shibutani (eds.), IUTAM Symposium on Mesoscopic Dynamics of Fracture Process and Materials Strength, 213–222.

single crystalline aluminum is simulated by the molecular dynamics (MD). It has been shown that the MD simulations can provide a qualitative analysis of discrete and incipient plasticity events that are consistent with these experimental observations of the nanoindentation [3]. However, this atomistic methodology, dynamically solving the equation of motion, strongly depends on the individual movement of atoms prescribed by their intrinsic potential. Thus, three different embedded atom type potentials are employed in the present studies to elicit a distinction of the defect nucleation. The dislocation emission and the subsequent formation of the prismatic dislocation loop [4] are discussed in detail.

2. ATOMISTIC MODEL

Two atomistic models of the nanoindentation are given : 20nm wide by 20nm long by 20nm thick single crystalline aluminum containing about 500,000 atoms and 30nm wide by 30nm long by 20nm thick one containing about 1,125,000 atoms. The top and side surfaces are traction free and the bottom surface is held fixed. The fcc crystal is oriented such that the top surface is the (001) or (111) plane. Three different EAM type potentials are employed here, which have been proposed by Ackland et al. (FS) [5], Ercolessi et al. (ER) [6] and Mishin et al. (MS) [7]. The fundamental physical properties of lattice constant, cohesive energy, elastic constants and stacking fault (SF) energy, are summarized in Table 1. Generalized stacking fault (GSF) energy curves calculated by these potential are shown in Figure 1. The FS potential yields too low SF energy and the unstable SF energy doesn't appear in the ER. Totally judging, the MS is likely to be best fitting to the aluminum among them.

Two types of the indenter, the sphere punch and the Berkovich type, are atomistically modeled as the rigid carbon-based structures. The interaction between indenter and the aluminum substrate is determined by the simple Morse type potential, which parameters have been determined by averaging individual parameters of the carbon and the aluminum.

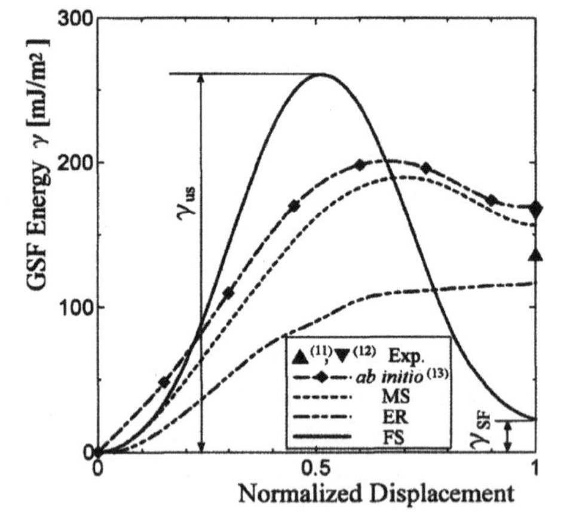

Figure 1. GSF energy curves obtained by EAMs, ab initio and experimental data.

Table 1. Physical Properties of Al predicted by three kinds of EAM potentials in comparison with experiment and *ab initio* data.

	FS	ER	MS	Exp. or *ab initio*
a_0 [Å]	4.099	4.032	4.05	4.050 [8]
E_0 [eV/atom]	-3.246	-3.36	-3.36	-3.360 [9]
C_{11} [GPa]	114	118	114	114 [10]
C_{12} [GPa]	58.9	62.2	61.5	61.9 [10]
C_{44} [GPa]	30.2	36.7	31.6	31.6 [10]
γ_{SF} [mJ/m^2]	22.6	116.8	156.6	135[11], 166[12] 169.8[13]

3. DISLOCATION BEHAVIORS

3.1 Dislocation Emission

From the overviews of dislocation emission states achieved by the molecular dynamics with the three different potentials, outstanding characteristic shapes due to the different intrinsic SF energy can be found. Figures 2 and 3 indicate the dislocation emissions under the indentations by the Berkovich punch and by the spherical punch with a tip radius of 30nm, respectively. Only the atoms with higher energy than -3.22eV (in case of FS) or -3.33eV (in cases of ER and MS) are drawn in order to detect the defect nucleation.

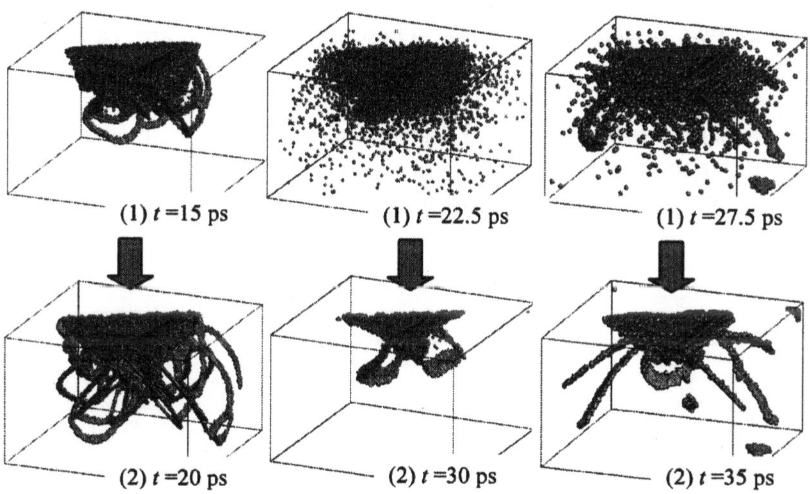

(a) FS (b) ER (c) MS

Figure 2. Dislocation emissions under the indentation by Berkovich punch.

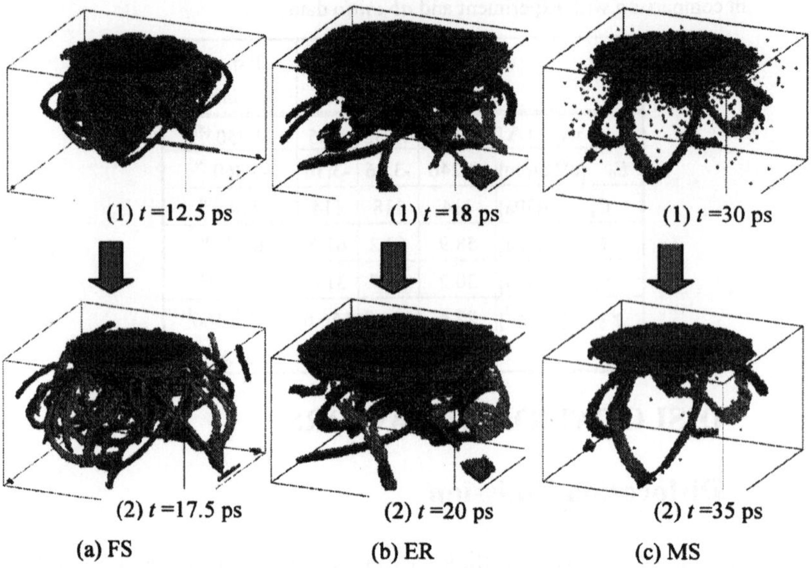

Figure 3. Dislocation emissions under the indentation by spherical punch.

A lot of emitted partial dislocations using the FS potential with the unrealistic SF energy lead the high dislocation density state. On the other hand, a few accountable perfect dislocations are observed using the ER and MS potential due to the very narrow SF width.

Figure 4 shows the indent loads obtained by the three EAM potentials and the two different indenters. These indent loads are calculated by summing the reacted forces from the Al substrate to the rigid indenter and are considerably fluctuated due to the attractive part of the Morse potential. We see that the indent load by FS potential is about four times greater than the other potentials. This suggests that the nano-plasticity triggered by defect generation is much sensitive to the potential employed.

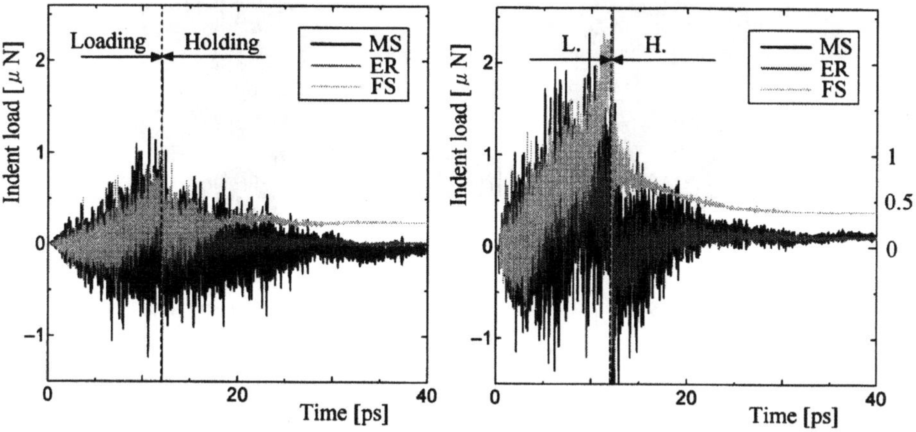

Figure 4. Time evolutions of indent loads by two kinds of punches.

3.2 Prismatic Dislocation Loops

Studies concerned to the formation of prismatic dislocation loops have been performed experimentally in the past about fifty years. Many of those remarked about the presence of a spherical or a non-spherical precipitate inside a given crystalline matrix. Several quantitative analyses for the spherical particle geometry [4] and a few studies for the non-spherical geometry [15] exist. In the present simulations, the prismatic dislocation loops are formed inside the single crystalline aluminum, and then penetrate on the side surface, leaving the rhombic steps. We show two cases of the formation of prismatic dislocation loop emitted from two kinds of surfaces: [111] as shown in Figure 5 and [001] in Figure 6, respectively.

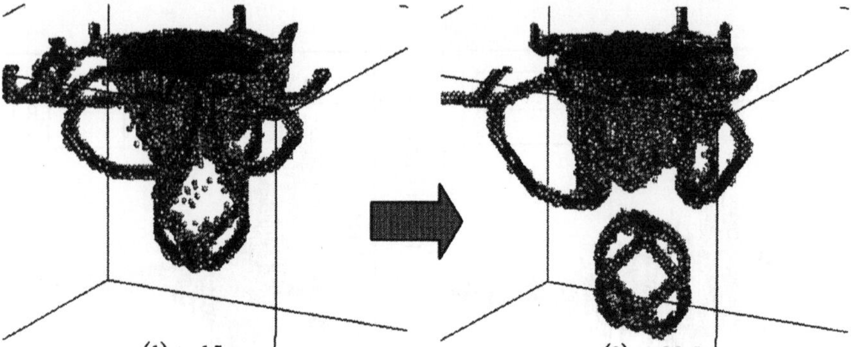

(1) t=15 ps (2) t=22.5 ps

(a) Case of FS potential

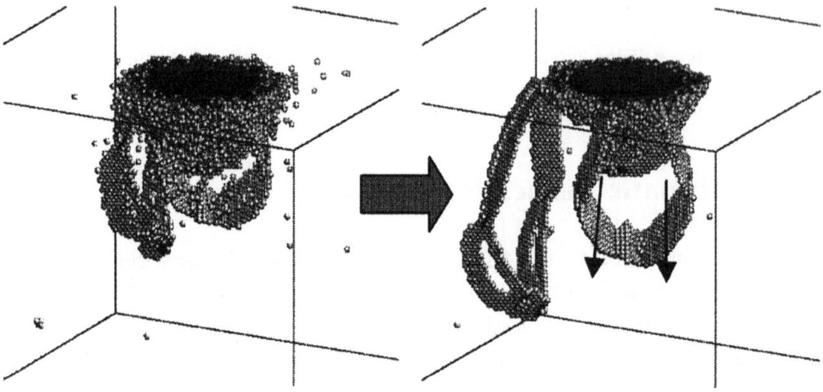

(1) t=20 ps (2) t=40 ps

(b) Case of ER potential

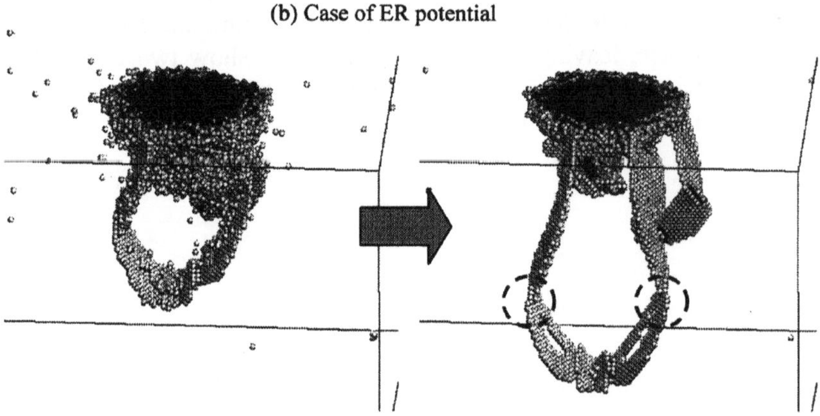

(1) t=20 ps (2) t=40 ps

(c) Case of ER potential but with the different view point from (b)

Figure 5. The formation of prismatic dislocation loop emitted from [111] surface.

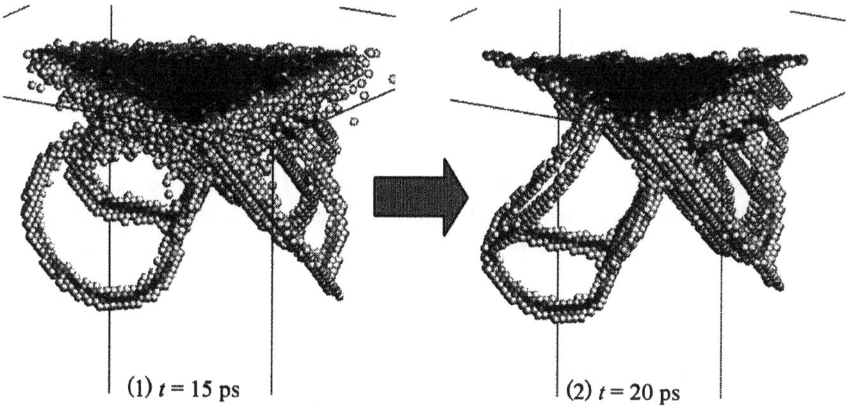

(a) Case of FS potential

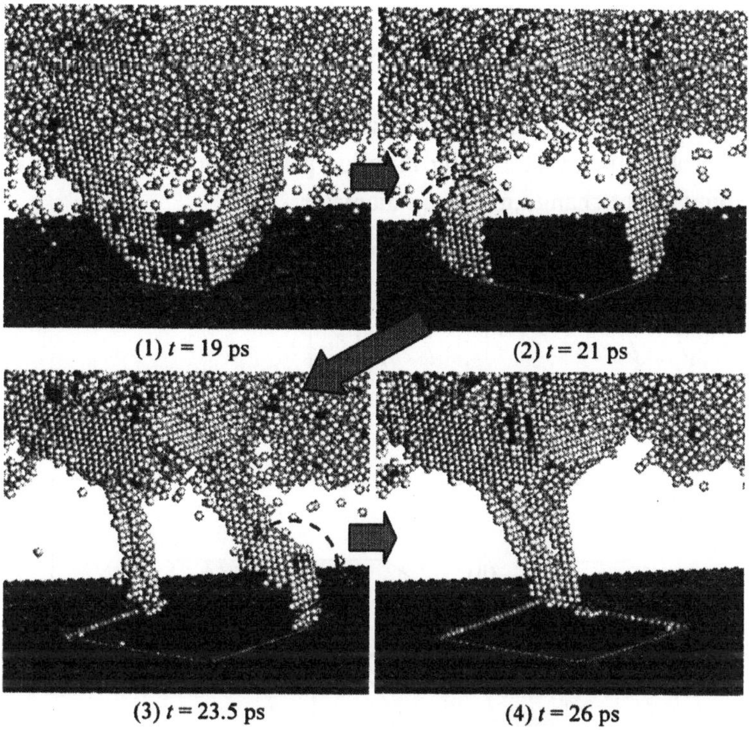

(b) Case of ER potential

Figure 6. The formation of prismatic dislocation loop and rhombic step on the side surface emitted from [001] surface.

All of the obtained prismatic dislocation loops have the same family of crystallographic orientation. Therefore, the mechanism of formation of the prismatic loops can be summarized, as shown in Figure 7. As for the first mechanism (see Figure 7(a)); (i) two shear loops emitted on the two different slip planes such as $(11\bar{1})$ and $(1\bar{1}1)$ planes unstably interact on the <011> line of intersection of these two planes [16]. (ii) The screwsides of the dislocation segment on one slip plane move to the cross-slip plane that is parallel to the other slip plane. (iii) Then, two dislocation loops on the two cross-slip planes unstably interact on <011> line again. As for the second one (see Figure 7(b)); (i) an initial shear loop emits on the slip plane such as $(\bar{1}11)$ slip plane. (ii) The screwsides on both sides of that dislocation segment move to the each cross-slip plane that is parallel to each other. (iii) Then, a dislocation loop on the cross-slip plane moves to the next cross-slip plane that is parallel to the initial slip plane. The latter mechanism is concluded to correspond to the mechanism obtained in the case of the sphere precipitate embedded in the crystalline bulk [4].

Note that the interactions of two dislocation loops under the simple nanoindentation never produce Lomer-Cottrell barriers [16] and each two dislocations on the each parallel {111} planes have opposite Burgers vector, as shown in Figure 8. Once the prismatic dislocation loop is formed, it moves through the crystal along its glide prism whose orientation is [011] direction without changing its loop diameter.

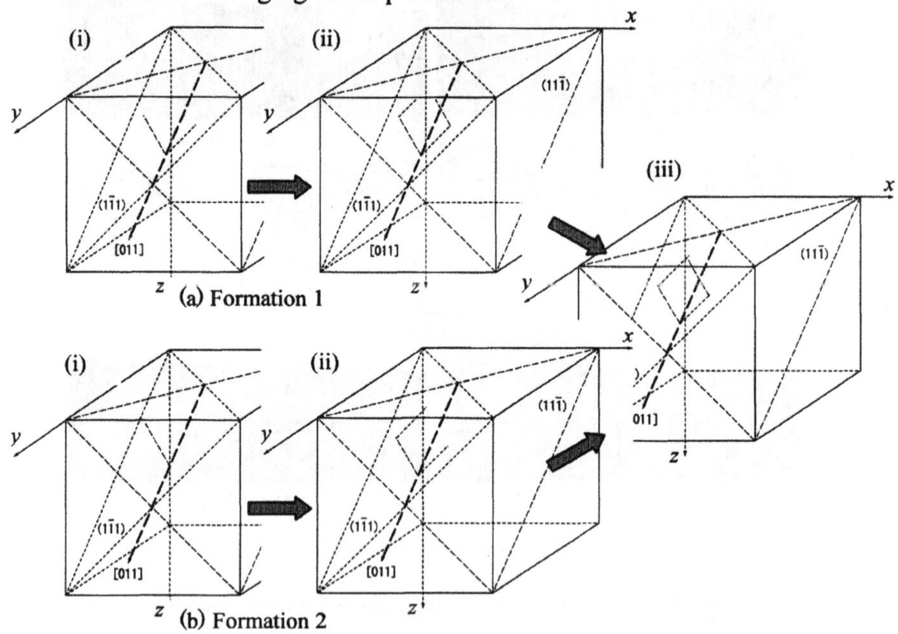

Figure 7. Two kinds of prismatic dislocation formation.

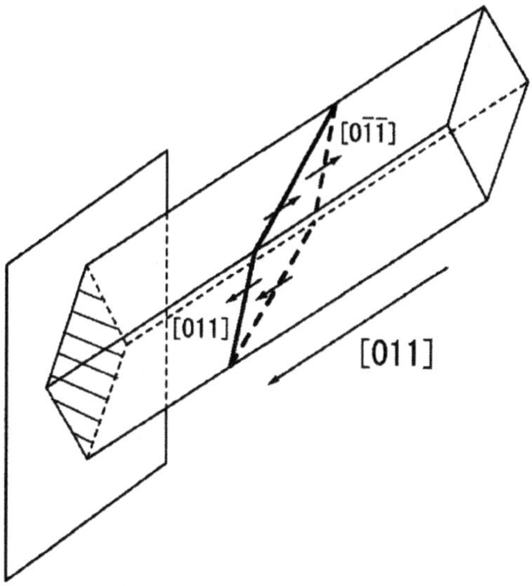

Figure 8. Simple pattern diagrams of the formation of prismatic dislocation loop.

4. CONCLUSION

We discussed on the dislocation emission and the subsequent formation of the prismatic dislocation loop under nanoindentation by the molecular dynamics simulations. All of the dislocations were emitted from the severely damaged surface due to the attractive part of the mutual interaction between the indenter and the aluminum. And the following two patterns of mechanism of the prismatic dislocation loops are observed. One is that two shear loops emitted on the two different slip planes unstably interact on the <011> line of intersection of these two planes. The screwsides of the two dislocation segments cross-slip to the plane parallel to the other slip plane, respectively. Then, the two dislocation loops on the two cross-slip planes unstably interact on <011> line again. The other is that each screwsides of an initial shear loop emitted on the (111) slip plane cross-slip to the plane parallel to each other, respectively. Then, one dislocation loop on the cross-slip plane moves to the next cross-slip plane that is parallel to the initial slip plane. The latter mechanism is concluded to correspond to the mechanism obtained in the case of the sphere precipitate embedded in the crystalline bulk.

ACKNOWLEDGEMENTS

This work was supported by the JSPS Grands-in-Aid for Scientific Research ((B)(2)15360054) and the Handai Frontier Research Center (FRC).

REFERENCES

[1] S. Suresh, T.-G. Nieh and B.W. Choi, *Scripta Mater.*, **41**-9 (1999), 951-957.

[2] Y. Shibutani and A. Koyama, submitted to JMR.

[3] J. Li, K. J. Van Vliet, T. Zhu, S. Yip and S. Suresh, *Nature*, **418** (2002), 307-310.

[4] M.F. Ashby and L. Johnson, *Phil. Mag.*, **20** (1969), 1009-1022.

[5] G.J. Ackland, G. Tichy, V. Vitek and M.W. Finnis, *Phil. Mag.*, **A56**-6 (1987), 735-756.

[6] F. Ercolessi and J.B. Adams, *Europhysics Letters*, **26**-8 (1994), 583-588.

[7] Y. Mishin and D. Farkas, *Phys. Rev.*, **B59**-5 (1997), 3393-3407.

[8] C. Kittel, *Introduction to Solid State Physics* (Wiley-Interscience, New York, 1986).

[9] *Handbook of Chemistry and Physics*, edited by R.C. Weast (CRC, Boca Raton, FL, 1984).

[10] G. Simons and H. Wang, *Single Crystal Elastic Constants and Calculated Aggregate Properties*, (1977), MIT Press.

[11] P.S. Dobson, P.J. Goodhew and R.E. Smallman, *Phil. Mag.*, **16**-9 (1967), 9-22.

[12] L.E. Murr, *Interfacial Phenomena in Meters and Alloys*, Addison-Wesley, Reading, MA (1975).

[13] S. Ogata, J. Li and S. Yip, *Science*, **298** (2002), 807-811.

[14] G. C. Weatherly, *Mater. Sci.* J, **2** (1968), 237

[15] T. Y. Tan and W. K. Tice, *Phil. Mag.*, **34**-4 (1976), 615-631.

[16] Y. Shibutani, A. Koyama and T. Tsuru, IUTAM Symposium on Multiscale Modeling and Characterization of Elastic-Inelastic Behavior of Engineering Materials, 2002-10, Morocco.

MULTISCALE MODELING OF DEFECT NUCLEATION AND REACTION: BULK TO NANOSTRUCTURES

Jinpeng Chang*, Ting Zhu*, Ju Li+, Xi Lin*, Xiao-Feng Qian*, Sidney Yip*#

*Department of Nuclear Engineering, Massachusetts Institute of Technology, Cambridge, MA 02139-4307 (USA)

+Department of Materials Science and Engineering, Ohio State University, Columbus OH 43210-1178 (USA)

#Department of Materials Science and Engineering, Massachusetts Institute of Technology, Cambridge, MA 02139-4307 (USA)

Abstract. Homogeneous nucleation of deformation twin in a metal and the reaction between a water molecule and silica are studied using classical interatomic potential and molecular-orbital theory respectively. They are examples of local (atomic-level) processes which involve the breaking and formation of chemical bonds, relevant in bulk solids as well as nanostructures.

Keywords: Multiscale modeling, defect nucleation, water-silica reaction, nanostructures

1. INTRODUCTION

We cite recent theoretical, molecular dynamics and finite-element analyses of dislocation nucleation in nanoindentation [1-3] and electronic-structure calculations of ideal shear strength [4] to suggest that atomistic simulations in which interatomic bonds are broken and formed can provide mechanistic insights common to the mechanical failure and chemical reactivity of nanostructures. The critical information is the distribution of charge densities at saddle-point configurations, particular arrangements of ions and electrons where the system is on the verge of structural instability or chemical reaction. Here two additional studies are discussed to expand on our simple notion. Homogeneous nucleation of deformation twinning [5] shows bond-strain sensitivity in the shear response of a perfect lattice. Reaction of a water molecule with a silica nanorod shows the importance of charge transfer effects. Together they motivate an electron-explicit description of mechanical and chemical behavior appropriate for bulk solids as well as nanostructures.

The aim of this contribution is to examine the common issues between the mechanics of large-strain deformation and the dynamics of

H. Kitagawa and Y. Shibutani (eds.), IUTAM Symposium on Mesoscopic Dynamics of Fracture Process and Materials Strength, 223–233.

electrochemical reactions, in the context of modeling materials structures down to nanoscale. The connection we seek lies in the observation that materials failure and reaction kinetics are traditionally regarded as different physical phenomena, yet when analyzed at the atomic level they are both governed by the valence charge densities in the system. Multiscale modeling studies in the area of mechanics of materials have largely focused on linking atomistic simulations with mesoscale or continuum methods. Usually electronic degrees of freedom are not treated explicitly, except in first-principles calculations on perfect crystals where lattice symmetry can be exploited, or on small-scale defects involving only tens of atoms. Problems of kinetics and dynamics of chemical reactions, by and large, have not attracted much attention in the modeling and simulation community. Also, there seems to be little discussion of a common perspective for understanding soft materials on the same footing as metals, semiconductors, and ceramics. In view of this situation we feel that there are particular challenges at the interface of mechanics and chemistry that can be posed through a focus on the transition from bulk to nanostructures in atomistic simulation, realizing at the same time that this would also imply a transition from hard to soft matter. Combining mechanics and chemistry, bulk and nanostructures, hard and soft materials, can all fit within the unifying framework of multiscale modeling; broad visions of such synergies have been articulated in various versions [6-8].

2. DEFORMATION TWINNING

Twinning and slip are the two primary plastic deformation mechanisms by which crystals can accommodate large strains. Deformation twins have long been identified in bcc, hcp and lower symmetry metals and alloys; more recently they have been found in fcc metals and alloys, in ordered alloys and other inter-metallic compounds, in elemental semiconductors and compounds, in non-metallic compounds such as calcite, and even in complex minerals and crystalline polymers [9]. The deformation is often characterized by very rapid formation of twinned regions, and large load drops in the stress-strain response; it operates generally at low temperatures, high stress, high strain rate, and in crystals of lower symmetry where the number of slip systems is limited. Schematically one can contrast slip and twin as lattice responses to shear where there is only one relative displacement between two adjacent layers (slip), and where such displacements occur in a stack of at least two layers (twin), see Figure 1.

Similar to most first-order lattice phase transformations, twinning is typically separated into nucleation and growth stages. A twin nucleus may be formed by the action of an applied stress in a region of near-perfect crystal (homogeneous nucleation), or formed from a pre-existing defect configuration (heterogeneous nucleation). While the latter is more commonly observed, the former can occur in highly perfect specimens, e.g., in cadmium and zinc [10], where the stresses required are an order of magnitude higher. In the present study we are concerned with the molecular dynamics simulation of homogeneous nucleation of twinning in bcc Mo using the EAM-potential of Finnis and Sinclair [11].

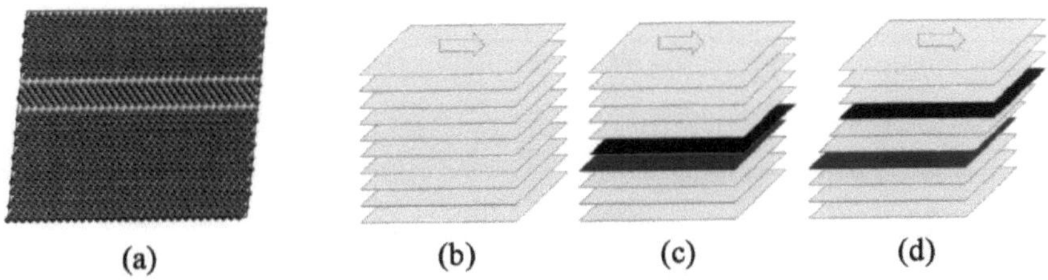

Figure 1 Atomic configuration of a twinned region in a crystal lattice enclosed by the twin boundarie, atoms are color coded according to local strain (a). Schematics depicting an undeformed stack of crystal planes prior to shear in the direction indicated by the arrow (b), deformation by slip (c), and deformation by twinning of four layers (d).

Our simulation cell with periodic boundary conditions is chosen with the X (horizontal), Y (normal to plane of paper) and Z (vertical) axes along $[111]$, $[\bar{1}2\bar{1}]$, and $[10\bar{1}]$ respectively. The corresponding dimensions are 199A, 198.6A, and 192.7A (500,000 atoms). Shear is applied at a constant rate of 3 x10^6 s^{-1} on the xy plane in the $[111]$ direction (twinning direction). At 10 K we observe homogeneous twin nucleation at a shear stress of 12.2 GPa (7.84% strain). Once nucleation sets in, a sharp decrease in strain energy and shear stress is observed. From a sequence of the instantaneous atomic configurations, shown in Figure 2, the twinned region (nucleus), delineated by color coding according to local strain and coordination, is seen to evolve into an oblate ellipsoidal shape (disk like). The disk is thickest in the middle while its edges are as thin as one layer. The defect can be described as a twinning dislocation loop with a burgers vector of $\vec{b}/3$, with $\vec{b} = (a/2)[111]$. From the MD results we estimate the velocity of the loop while the disk is expanding to be ~ 6000 m/s (sound speed is 6000-7000 m/s). Expansion in the $\{112\}$ plane is much faster than the out-of-plane growth, the fact that the former is also anisotropic causes the shape projected on the plane to be elliptical. Across the twinned region, one can verify the relative displacements are distributed in the form of (0,...,0, b/3, ..., b/3, 0, ...0), which is in accordance with Figure 1.

To interpret the simulation results we introduce a 1-D chain model to represent the essential structural characteristic of the twin defect. As shown in Figure 3, the defect is a chain of 'model atoms' specified by a set of coordinate x_i, measured on the X-axis in direction [111]. Associated with a 'model atom' i is a layer of physical atoms, the plane being perpendicular to the chain direction, with coordinates (x_i, y_i, z_i). In the chain model the only

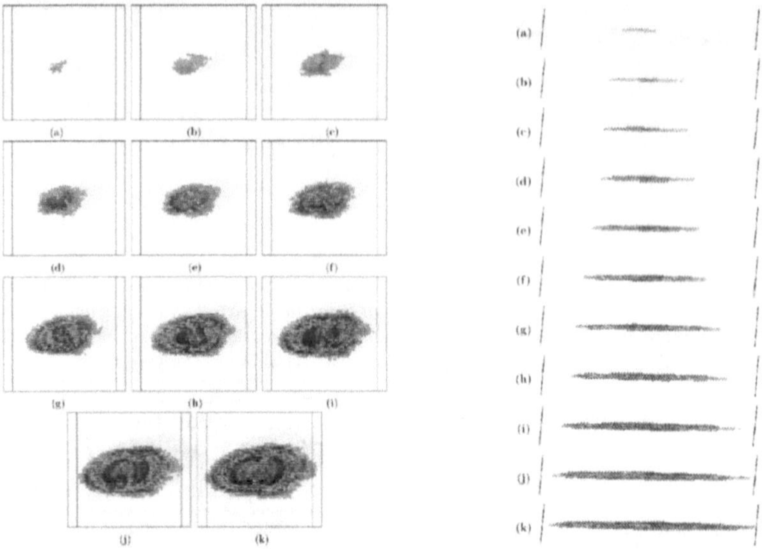

Figure 2. Growth of a disk-shape nucleus of a twinned region observed in MD simulation of shear deformation, top view (left) and side view (right).

relevant degrees of freedom are the *relative displacements* in the [111] direction *between adjacent layers*, that is, $\Delta x_i = x_{i+1} - x_i$. If the twin structure is composed of N nonzero relative displacements of 'model atoms' (N planes of atoms are displaced *successively*), we would have N *primary* degrees of freedom, Δx_i, i = 1, ..., N, specifying the defect. The other degrees of freedom, relative displacements in the Y and Z directions, Δy_i and Δz_i will be considered as *secondary*. In principle, the latter should be allowed to relax during deformation; however, one may anticipate that in a first approximation such relaxations can be neglected for the sake of computational simplicity. Whether these relaxations are actually ignored or not, our purpose is to investigate the structure-energy relation of the 1-D chain model by considering the variation of the system energy only with the primary degrees of freedom. In other words, for the system energy we would write $E = E(0,...0,\Delta x_j,...,\Delta x_{j+N},0,...0)$, with the N successively displaced planes starting at position j and ending at j+N.

The simplest case one can consider is the 1-layer shear, the rigid translation of the upper half of the lattice relative to the lower half. The energy $E(0,...,\Delta x_1,...0)$ with relaxation in the other two directions taken into

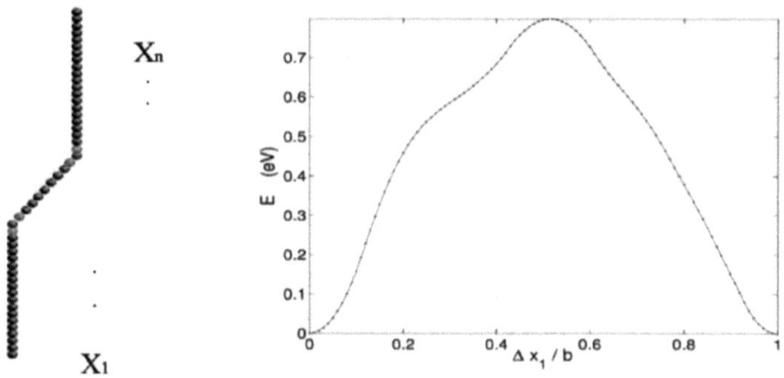

Figure 3. Schematic of the 1-D chain model of twinning (9 layers in this example). Variation of strain energy with relative displacement of the $(\bar{1}\bar{1}2)[111]$ twin in a 1-layer shear (the γ-surface) with Y and Z relaxations.

account is shown in Figure 3. This is just the conventional γ-surface energy in the [111] direction. The energy barrier is 0.82eV. Critical stress in the positive direction is 33 GPa while in the negative direction it is 29 GPa, indicating an expected asymmetry at about $\Delta x_1 = b/2$. One can discern a softening around $\Delta x_1 = b/3$, but this did not lead to a metastable state. However, the value of the relative displacement is close to the critical value for twin nucleation, as we will see in the case of 2-layer shear below. We have checked that without relaxation the energy curve peaks at a slightly higher value, 0.84eV, and a lower displacement, which are not surprising. Also, repeating the calculation for $(\bar{1}10)[111]$ shear gives a symmetric curve with a much lower barrier, 0.42eV, at b/2, and a corresponding stress of 15.13GPa, both reasonable since this is a primary slip system.

For the 2-layer shear we show the energy surface in Figure 4 where a minimum is now seen around the displacements (b/3, b/3). Under positive shear the system can either twin or slip. The energy barrier for twinning is 0.672eV with saddle point at (0.36b, 0.16b), while for slip the barrier is 0.736eV with saddle point at (0.5b, 0.09b). Under negative shear, only slip is allowed, at a barrier of 0.808eV. It is also useful to display the energy surface in the form of contour plot, see Figure 4. Knowing the energy contour one can trace out the minimum-energy path for the two deformations, twin and slip. In Figure 4 we see connections between starting configuration at the perfect lattice energy minimum (0,0) and the two ending configurations, an energy minimum corresponding to the 2-layer twin at (b/3, b/3), and another minimum corresponding to slip at (b, 0). The two paths bifurcates at (0.29b, 0.03b) before either of the saddle points is encountered. The system can either twin or slip after the bifurcation point; however, since the twinning path has a lower energy barrier, 0.672eV, than the slip path, with barrier of 0.736eV, twinning will be favored. For comparison we have also performed energy surface calculations for the $(\bar{1}10)[111]$ slip system. The only energy minima found are at (0,0), (b,0), (0,b), and (b,b), which means there is no twin minimum and the only system response to shear is slip. In this case

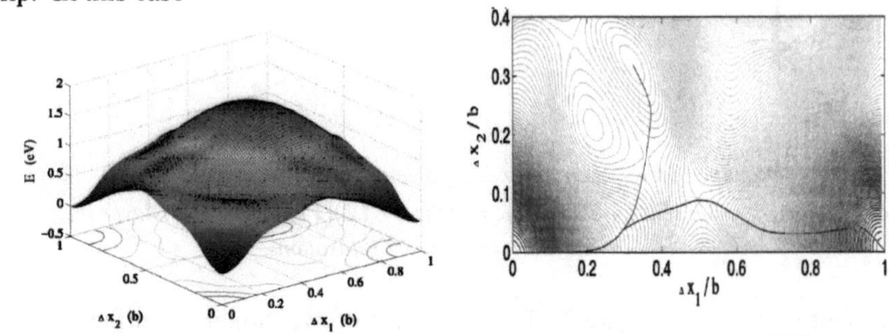

Figure 4. Strain energy surface (left) and energy contour plot (right) for the $(\bar{1}\bar{1}2)[111]$ *deformation, with Y and Z relaxations, in a 2-layer analysis [5]. An energy minimum at relative displacements of b/3 for sliding between adjacent layers confirms the existence of twin defect in the present model for bcc Mo.*

the energy surface is completely symmetric about (b/2, b/2). Comparing the energy barriers for the two $(\bar{1}10)[111]$ and $(\bar{1}\bar{1}2)[111]$ slip systems, we have 0.422eV vs. 0.736eV. This is consistent with the fact that the former is the primary slip.

3. WATER-SILICA REACTION IN A NANOROD

The ability to probe the dynamics of bond breaking and formation and the influence of local chemical environments presents an opportunity to model properties of nanostructures under conditions of their synthesis and performance. We are currently investigating the chemical reactivity of a nano-rod of SiO_2 when it comes into contact with water. By following the reactions as kinetic events involving the redistribution of charge densities, we expect to gain insight into the question of how bond strain, induced by stress or thermal environment, can affect reactivity.

The silica nanorod consists of 144 atoms (48 SiO_2 molecules) in the form of a stack of Si_6O_{18} rings capped at the two ends to eliminate any dangling bonds [12]. Using a potential description consisting of pair and Coulomb interactions [13], we have performed uniaxial tension and compression simulations on the nanorod and have observed the effects of temperature on the failure mechanism. The stress-strain responses of the nanorod, with and without a notch, as it is strained to rupture are shown in Figure 5. Also shown are the atomic configurations of the initial structure of the notched nanorod and the same rod in a highly deformed state just prior to rupture. It is notworthy that the rod thins down to a single chain of atoms before breaking. This appears to be a characteristic mode of failure that has been recently observed in gold nanojunctions by *in situ*, real time resolved high-resolution TEM [14].

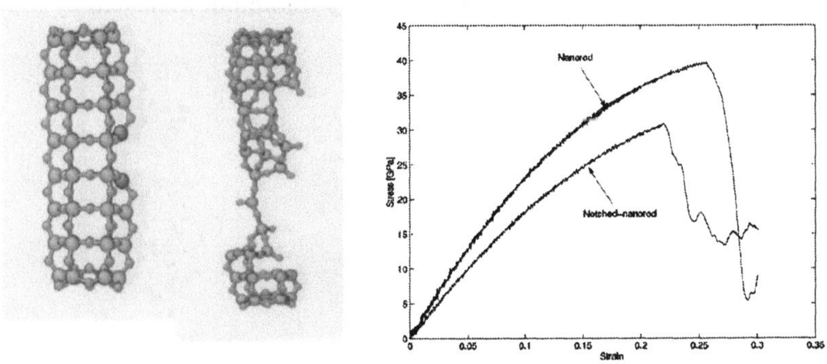

Figure 5. Tensile rupture of a quartz nanorod at 100K, atomic configurations of initial notched structure and deformed rod prior to rupture, and stress-strain responses with and without the notch.

To investigate the sensitivity to chemical reaction of a strained nanorod we consider the effect of water in contact at the surface of the rod. Our interest is probe the kinetic pathways of a water molecule reacting with a strained Si-O bond. That the presence of water has a significant effect on strength of quartz, particularly at high temperature, is widely known as the problem of 'hydrolytic weakening'. It was first proposed that the fundamental mechanism for this effect is the hydrolysis of a siloxane bond, which bridges two neighboring SiO_4 tetrahedra, to form two terminal SiOH silanol groups [15]. The silanol groups are believed to facilitate bond rupture and thus lower the strength of silica. Since the process involves bond breaking and formation, and significant effects of charge transfer, classical potential simulations will not be adequate. For initial exploration we focus on the reaction between a water molecule and a silica nanorod which is held in various stages of uniaxial tension. Interaction energies and forces are obtained by a semi-empirical molecular orbital method as coded in a general-purpose package MOPAC [16].

To identify the transition state and reaction pathways we adopt the Nudged Elastic Band method [17] for locating the saddle point configuration and the associated energy barrier. We show in Figures 6 and 7 the reaction pathway at the stress level of 16.7 GPa and the corresponding minimum energy path. These results suggest a three-step mechanism for the hydrolysis process. (1) A metastable adsorbed state is first formed (Figure 6(b)) which corresponds to a local energy minimum, point b in Figure 7. (2) A proton is transferred to the bridging O_{br} atom to form a new bond between O_{br} and H which replaces the original bridging bond between H and O_w; the saddle point configuration is shown in Figure 6(c). The maximum energy barrier on the minimum energy path occurs at c in Figure 7. (3) Rupture of the bond between bridging O_{br} and Si occurs to yield surface Si-O-H groups as shown in Figure 6(d). The system reaches a local energy minimum indicated by point d in Figure 7.

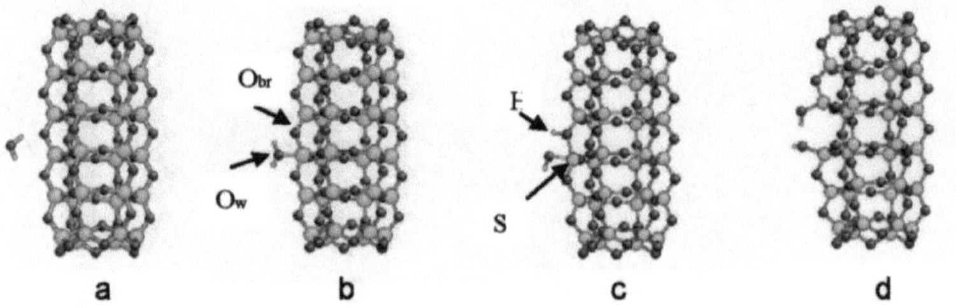

Fig. 6. Reaction pathway for a nanorod, strained at 16.7 GPa, being attacked by a water molecule.

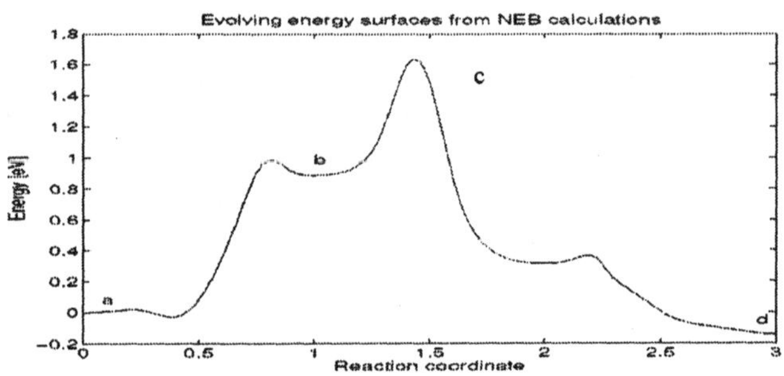

Fig. 7. Minimum energy path from Nudged Elastic Band calculation.

From the minimum energy path calculation, an activation energy barrier of 1.6 eV is obtained at the present stress level (16.7 GPa). Work is on-going to evaluate the stress dependence of this barrier.

4. MODELING FUNCTIONAL NANOSTRUCTURES

As a result of current interest in nanoscience and technology, significant opportunities are being created for understanding the physical, chemical (and biological) properties of materials structures intermediate between isolated atoms and molecules and bulk matter. Novel structures displaying unusual phenomena and functional properties have been observed on the nanoscale (a fraction of nm to 100 nm). An example is the electrical conductance across a molecular junction. Although the notion of a single organic molecule acting as an electronic rectifier had been discussed in 1974 [18], it was only in the early '90s that the prospect of building electronic circuits at the level of single molecules was recognized [19]. More recently interests in the conductivity behavior of a molecular junction have intensified considerably when it was realized that thiol-terminated conjugated oligomers in the form of a self-assembled monolayer could exhibit such transport behavior [20]. An unresolved issue is the mechanism of electron migration across a metal-molecule-metal interface, such as a junction consisting of a thiol terminated benzene ring in contact with a Au atom at each end, as shown in Figure 8. The theoretical determination of the current- voltage characteristics of this system is made all the more difficult by the contact

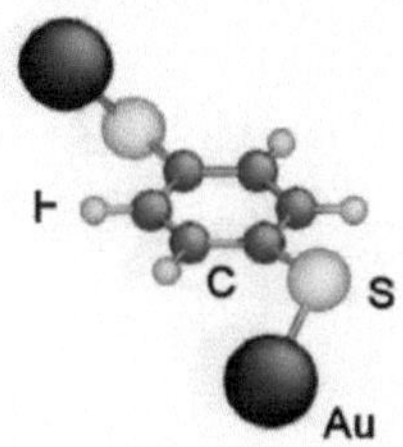

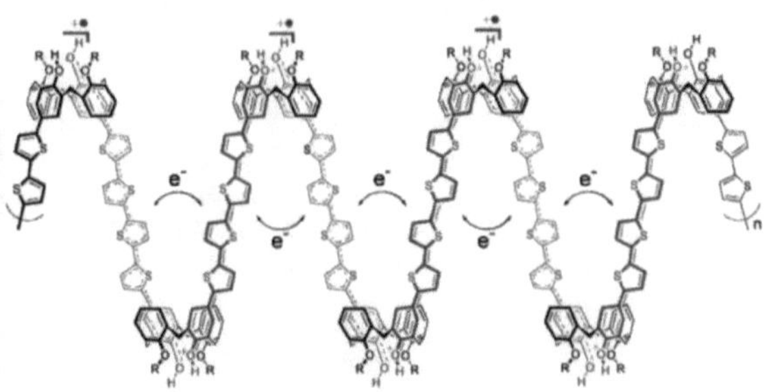

Figure 8. A molecular junction model in the form of a Au-S-(p-C_6H_4)-S-Au system is shown on the left, while on the right hinge molecules (calix[4]arene) are connected to rigid rods of quarterthiophene to form a structure that can function as a molecular actuator.

effects between metallic tips and single molecules. One approach being investigated is the Green's function scattering formalism combined with density functional theory (Kohn-Sham) Hamiltonian [21]. Another approach, which we are currently investigating, is to propagate a wave packet across the junction by numerically solving the time-dependent *Schrödinger* equation.

The molecular junction is just one example of a class of nanostructured conjugated polymers that are being synthesized and characterized for device applications. A somewhat more complicated structure is the thiophene-based conducting polymer, shown also in Figure 8, which is being considered as a candidate for molecular actuator [22]. In this case hinge molecules (cones of 4 six-member rings in various conformations) tie together rigid rods (chains of four 5-member rings, including a sulfur atom) to form a polymer backbone that can contract or expand depending on the oxidation state (the rods attract each another in the oxidized state). Since the conductivity of this system is obviously also a property of interest, the need for an appropriate method to analyze electron transport in organic nanostructures is therefore quite general. One can go even further by noting that along with conductivity, such properties as modulus, mobility, strain, and response time are also essential to the understanding and development of nanodevices.

Acknowledgements. This work was supported by the Air Force Office of Scientific Research, the Defense Advanced Research Program Agency-Office of Naval Research, Honda R&D, Inc., and the National Science Foundation (KDI and ITR). We acknowledge the collaboration of N. H. de Leeuw, R. J. Bartlett, and S. B. Trickey in the nanorod study, and of Jean-Paul Crocombette in the conductivity study.

References

1. J. Li, K. J. Van Vliet, T. Zhu, S. Yip, S. Suresh, Nature (London) **418**, 307 (2002).
2. K. J. Van Vliet, J. Li, T. Zhu, S. Yip, S. Suresh, Phys. Rev. B **67**, 104105 (2003).
3. T. Zhu, J. Li, K. J. Van Vliet, S. Suresh, S. Yip, J. Mech. Phys. Solids, accepted.
4. S. Ogata, J. Li, S. Yip, Science **298**, 807 (2002).
5. J. Chang, PhD Thesis, MIT, June 2003.
6. G. H. Campbell et al., Mater. Sci. Eng. A **251**, 1 (1998).
7. J. Computer-Aided Mater. Design (Special Issue) **6** (1999).
8. S. Yip, Nature Mater. **2**, 3 (2003).
9. J. W. Christian and S. Mahajan, in *Progress in Materials Science* (Elsevier, 1995), vol. 39, p. 1.
10. P. B. Price, Proc. Roy. Soc. A **260**, 251 (1961).
11. M. W. Finnis and J. E. Sinclair, Phil. Mag. A **50**, 45 (1984).
12. T. Zhu, J. Li, S. Yip, R. J. Bartlett, S. B. Trickey, N. H. de Leeuw, Mol. Sim, in press.
13. B. W. H. van Beest, G. J. Kramer, R. A. van Santen, Phys. Rev. Lett. **64**, 955 (1990).
14. V. Rodrigues and D. Ugarte, Phys. Rev. B **63**, 073405 (2001).
15. D. T. Griggs and J. D. Blacic, Science **147**, 292 (1965).
16.. "MOPAC 2000", J.J.P. Stewart, Fujitsu Limited, Tokyo, Japan (1999).
17. G. Henkelman, B. P. Uberuaga. H. Jonsson, J. Chem. Phys. **113**, 9901 (2000).
18. A. Aviram and M. A. Ratner, Chem. Phys. Lett. **29**, 277 (1974).

19. *Molecular Electronics: Science and Technology*, A. Aviram. ed. (American Institute of Physics: New York, 1992); *Molecular and Biomolecular Electronics*, R. R. Birge, ed. (American Chemical Society: Washington, DC, 1991).
20. M. A. Reed, C. Zhou, C. J. Muller, T. P. Burgin, J. M. Tour, Science **278**, 252 (1997), J. Chen, M. A. Reed, M. Rawlett, J. M. Tour, Science **286**, 1550 (1999), J. M. Tour, *Molecular Electronics* (World Scientific, Singapore, 2003).
21. P. A. Derosa and J. M. Seminario, J. Phys. Chem. B **105**, 471 (2001).
22. H.-H. Yu, B. Xu, T. M. Swager, J. Am. Chem. Soc. **125**, 1142 (2003); J. D. Madden, P. G. Madden, I. W. Hunter, SPIE Proc., vol 4695 (2002), p. 176.

DISCRETE-CONTINUUM MODELING OF METAL MATRIX COMPOSITES PLASTICITY

S. Groh[1], B. Devincre[1], F. Feyel[2], L. Kubin[1], A. Roos[2] and J.-L. Chaboche[2]

[1]*LEM, CNRS-ONERA, BP 72, 29 av. de la division Leclerc, 92322 Chatillon Cedex, France*

[2]*LCME, ONERA, BP 72, 29 av. de la division Leclerc, 92322 Chatillon Cedex, France*

Abstract A computing methodology is reported for modeling the plastic deformation of Metal Matrix Composites. Use is made of a discrete-continuum model based on a coupling between Dislocation Dynamics and Finite Element simulations, which provides a parameter-free quantitative description of the mechanical properties. The model and the first simulation results on MMCs are presented and briefly discussed.

Keywords: Metal Matrix Composite, Plastic deformation, Dislocation Dynamics simulation, Finite Element simulation

1. INTRODUCTION

The recent interest on materials with increasingly small characteristic length scales has clearly revealed some shortcomings in the modeling of size effects in materials plasticity. These effects, which are explicit in dislocation theory, are usually not reproduced by continuum approaches. For this reason, the prediction of the plastic properties of Metal Matrix Composites (MMCs) is sometimes considered as a bench test for theoretical or numerical models. Indeed, in MMCs the stress-strain behavior is size-dependent for reinforcement sizes and volume fractions in the micrometer range. Further, the matrix stress-strain behavior that needs to be assumed to reproduce experimental results differs from the stress-strain behavior of the unreinforced material [1–3].

As theory and experiment often do not fit very well, some authors [4–6] proposed to validate the predictions of their own theory by comparison with a Dislocation Dynamics (DD) simulation by Cleveringa et al. [7, 8], on a 2D composite material with periodic rectangular reinforcements. Such exercises

H. Kitagawa and Y. Shibutani (eds.), IUTAM Symposium on Mesoscopic Dynamics of Fracture Process and Materials Strength, 235–244.

were extremely instructive but, nevertheless, care should be exercized when extrapolating two-dimensional models to experiment.

The simulation of dislocation dynamics in 3D MMCs is a challenging problem, which goes much deeper than the usual question of CPU time limitation. One has, in addition, to define precisely the conditions for mechanical equilibrium in such complex hetero-structures (see [9] for detail). Several solutions to this problem are now potentially available [9–14] and a critical comparison of these approaches can be found in ref. [15]. In the present work, use is made of the Discrete-Continuum Model (DCM) [10, 11]. The model and its specific implementation in the case of MMCs are shortly described in part 2. Parts 3 and 4 are dedicated to a presentation of the simulation results and part 5 to concluding remarks.

2. THE DISCRETE-CONTINUUM MODEL

In essence, the DCM, is made up of an FE code (ZeBuLon), in which a DD simulation replaces the constitutive formulation for plastic properties. On the one hand, the FE code treats the boundary value problem and cares of the conditions of local equilibrium in a meshed volume element. On the other hand, the DD code cares of the topology and motion of the dislocation lines in the same volume element, hence of the plastic strain, $\underline{\varepsilon}_p$. The coupling is realized with the help of two procedures that control the traffic between the "discrete" and "continuum" codes, a homogenization procedure for the calculation of $\underline{\varepsilon}_p$ and an interpolation procedure for the calculation of the stress tensor $\underline{\sigma}$ at any point of the simulated volume (cf. [11, 15]).

2.1 Dislocation Self-Stress Fields

In agreement with Mura's Eigenstrain theory [16], the DCM can theoretically capture all details of the dislocation stress fields in isotropic or anisotropic elasticity. It is then mainly a matter of computation to design a FE mesh for computing the complex stress field of a dislocation line close to its singularity. Unfortunately, for "mass" simulations involving many dislocations, this brute force approach is numerically untractable.

For instance, considering the computational constraints discussed in part 4, the largest regular mesh that can be handled by a good workstation has $(10 \times 16 \times 6)$ quadratic cubic elements of linear length $0.137\mu m$. The mesh length may then be larger than the mean distance between dislocations and the shape function used to interpolate the stress can only reproduce the smooth variations of the dislocation stress fields far from their singularities. Indeed, the Eigenstrains associated to dislocations are homogenized in a small volume surrounding the lines, in order to remove stress singularities. For this rea-

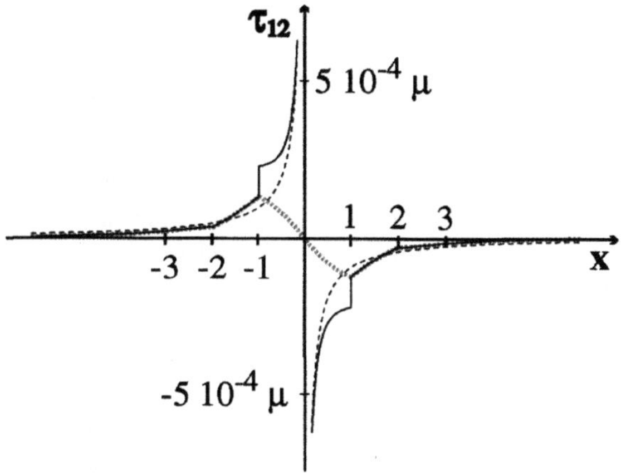

Figure 1. Isotropic shear stress field of a rigid screw dislocation calculated along a direction normal to the line. For the sake of comparison, the distances in abscissa are plotted in units of the linear dimension of the quadratic elements used in the present work. Dashed line: theoretical solution. Dotted line: solution yielded by the DCM. Continuous line: DCM solution with a short-distance correction.

son, the previous version of the DCM was restricted to problems of plastic relaxation implying no short-distance reactions between dislocations (see for instance [17]).

A simple solution is proposed to overcome this limitation. In the DD part of the DCM, the analytical (isotropic) self-stress field of a dislocation is superimposed to the stress prediction yielded by the FE code within a small volume surrounding the singularity (see Fig. 1). This volume is limited to the homogenized core region, so that the homogenization volume in the DCM is identical to the volume where the stress correction is performed. Hence, in conformity with dislocation theory, we introduce in the DCM framework the equivalent of an "elastic core surface traction" [18], which does not affect the total mechanical equilibrium (by construction the integral of the additional stress is zero), but improves the local description of dislocation interactions. In other terms, the amount of elastic energy lost in the homogenization procedure is locally restored in the DD code without affecting the consistency of the FE calculations. This correction is an essential one if one wishes to realistically reproduce the strength of dislocation reactions (cf. part 4). It must be noticed that whereas this improvement is easy to implement, it significantly increases the computational burden. This is why the small discontinuity of the stress field generated at the border of the homogenization volume (see Fig. 1) has not been removed so far. This is feasible in technical terms, but does not appears to be critical.

2.2 Initial and Boundary Conditions

The need to consider a reference cell as small as possible in order to optimize computations strongly suggests applying periodic boundary conditions (PBCs) in the case of MMCs simulations. Topological problems related to the use of these conditions, when applied with DD simulations, are discussed in [19]. Here, we restrict ourselves to the question of implementing PBCs in a FE code and to some related problems that may affect the simulation results.

As shown in Fig. 2, the simulation cell used for the study of MMCs is a parallelepipedic volume of dimension $(1.368 \times 2.188 \times 0.820)\ \mu m^3$ including four half-fibers of square cross-section at the center of the vertical faces (the edges of the fibers are rounded off in the FE mesh to eliminate local stress concentrations). Modifying the dimensions of the fibers allows obtaining various volume fractions in the simulated cell. By periodic three-dimensional replication of the simulation cell, one obtains a composite material with an hexagonal arrangement of infinitely long fibers. The displacement ($\mathbf{u}$) and strain ($\underline{\varepsilon}$) fields that verify the periodic solution in the reference cell are as follows:

$$\underline{\varepsilon}(\mathbf{u}(\mathbf{r})) = \bar{\underline{\varepsilon}} + \underline{\varepsilon}(\mathbf{u}'(\mathbf{r})) \quad \text{with} \quad \mathbf{u}(\mathbf{r}) = \bar{\underline{\varepsilon}} \cdot \mathbf{r} + \mathbf{u}'(\mathbf{r}) \tag{1}$$

where $\bar{\underline{\varepsilon}}$ is the mean deformation expected if the material is homogeneous and $\underline{\varepsilon}(\mathbf{u}')$ is a fluctuating quantity that accounts for the presence of the periodic hetherogeneities (in the present case, the fibers). This last term derives from a displacement field $\mathbf{u}'$, *i.e.*, whose values are periodically repeated at the cell boundaries in the directions of the translation vectors. Hence, the mechanical equilibrium in the simulated volume element must satisfy the following equations:

$$\langle \underline{\varepsilon}(\mathbf{u}'(\mathbf{r})) \rangle = \underline{0} \quad \text{and} \quad \langle \underline{\varepsilon} \rangle = \bar{\underline{\varepsilon}} \tag{2}$$

$$div\underline{\sigma}(\mathbf{r}) = \mathbf{0} \quad \text{and} \quad \underline{\sigma}(\mathbf{r_1}) \cdot \mathbf{n} = -\underline{\sigma}(\mathbf{r_2}) \cdot \mathbf{n}, \tag{3}$$

where $\mathbf{r_1}$ and $\mathbf{r_2}$ are opposite points at the boundary of the simulation cell and $\mathbf{n}$ is the corresponding boundary translation vector.

The main difficulty encountered when setting up proper initial conditions stems from the fulfillment of the above stress and strain conditions in the presence of a dislocation microstructure. In order for the FE code to account for the elastic fields of dislocations, the latter must imperatively be generated by a Volterra process. This implies that the DCM can only deal with closed dislocation loops. To globally satisfy the equilibrium conditions, the solution that is implemented consists in expanding the initial dipolar loops from random positions in the simulation cell (see Fig. 2-a). All the loops have same diameter, which is taken much larger than the mean distance between dislocations in order to avoid artificial screening effects. The total density is equally distributed on the twelve slip systems of the fcc structure and the total plastic shear during the Volterra process is set to zero by balancing the signs of the Burgers vectors. In the case of the unreinforced material (cf. part 3), this type of initial microstructure induces a mechanical state very close to equilibrium and only a small relaxation is observed at the beginning of the simulations. The case of the MMCs simulations is more critical and particular attention must be paid to the displacement and strain fields associated to the initial configuration (see Fig. 2-b), particularly with large volume fractions of fibers. Indeed, the fiber-matrix interfaces, which are assumed to be impenetrable barriers to dislocation glide, alter the symmetries of the initial dislocation microstructure. For this reason, it is necessary to check that the randomly generated configuration does not induce high internal stresses that could artificially modify the dynamics. In what follows, we only consider initial microstructures with an initial internal stress sufficiently small not to induce an artificial asymmetry between tension and compression tests. It is worth noting that it could be interesting in some cases to use alternative initial microstructures, for instance for reproducing the residual stresses generated during the processing of MMCs [1].

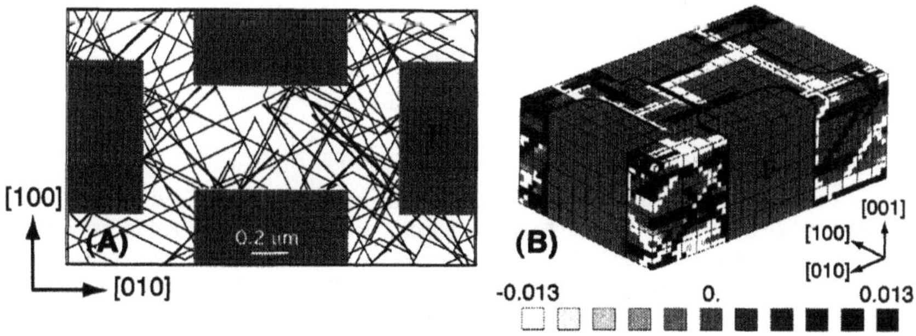

Figure 2. Initial conditions for the MMCs problem. A) A random distribution of dipolar dislocation loops generated by Volterra processes in a simulation cell. Whatever the loop positions, the dislocation lines cannot penetrate the fibers (denoted F). B) Surface mapping of the component ε_{33} of the initial strain field (the mean value is close to zero, as can be seen from the gray scale). Notice that special elements are used at the fiber's edges to avoid local stress concentrations.

Finally, whereas conventional DD simulations with PBCs make use of a cut-off distance for the stress field computations [20], the dislocation dynamics part of the DCM accounts for the totality of the periodic fields. The two types of simulations should then yield slightly different results when long-range elastic effects come into play.

3. VALIDATION TESTS

In order to validate the short-distance stress correction discussed in Sec. 2.1, simulated tensile tests have been carried out on copper crystals. The flow stress of pure fcc metals being controlled by dislocation reactions [20], such tests should be critical ones. An equivalent simulation, not reproduced here, were performed on an aluminum crystal, to determine the plastic properties of unreinforced matrix in Al_2O_3/Al MMCs. The results are very similar in both cases.

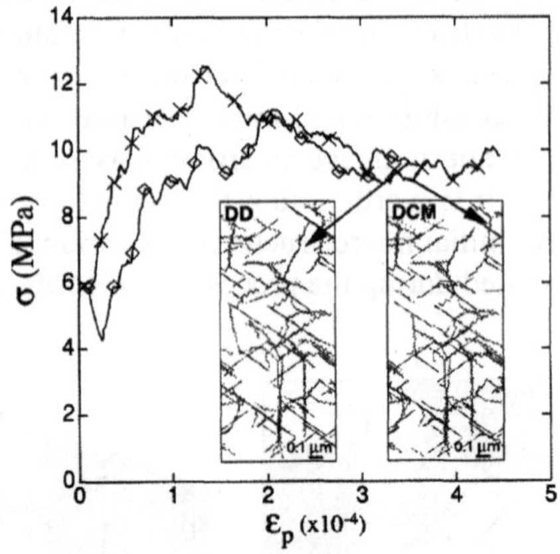

Figure 3. Tensile [100] stress-strain curves for a copper crystal, as obtained by DD (◇) and DCM (×) simulations with periodic boundary conditions. In both cases, the dimension of the reference cell is about $(10\mu m)^3$, the initial dislocation density is $10^{12}\ m^{-2}$ and the simulated specimen is deformed with a total imposed strain rate of $20\ s^{-1}$. The insets show thin foils extracted from the two simulations. One can check that the dislocation microstructures and the positions of the junctions at the end of the two simulations are nearly identical.

Figure 3 shows the results of DD and DCM simulations on a copper crystal with same loading conditions. In both cases the yield stress is exactly the one experimentally measured for a dislocation density of $10^{12}\ m^{-2}$ [20]. As a result of the small differences between PBCs in the two simulations (cf. Sec. 2.2), the two stress-strain curves slightly differ at low strains. This, actually, results from the truncation of the long-range stresses in the DD simulations. Beyond the yield stress, these differences vanish and the very good agreement between the two computations is interpreted as follows. In fcc crystals, long range interactions do not significantly contribute to the flow stress. The latter is controlled by the dislocation line tension and the numerous dislocation reactions taking place at the intersections between active slip planes.

This well-known result is illustrated in Fig. 3 by two thin foils extracted from the two simulated stress-strain curves at the same strain value. The same junctions are found at the same places in the two simulated microstructures and one has to look closely to find differences in the detail. From this result, we conclude that the local stress correction discussed in Sec. 2-1 is efficient and allows reproducing in the DCM the short-range interactions of dislocations, i.e., essentially the zipping and unzipping of junctions and dipoles. In addition, we verify that the energetic and dynamic aspects of the DCM are now consistent with their equivalents in DD simulations.

4. LONGITUDINAL TENSILE TEST IN Al_2O_3/Al

4.1 Simulation Conditions

In the present study, an Al matrix with infinitely long Al_2O_3 fibers is considered as a generic model for MMCs with long reinforcements. The mechanical properties for this composite are well documented in the literature [3], which allows performing direct comparisons with experiment. To be consistent with the dislocation microstructures generated during the processing of such MMCs, the initial dislocation density is set, in all cases, to the rather large value of $\rho_0 \approx 0.7\ 10^{14}\ m^{-2}$. Three different volume fractions of fibers are investigated, 5%, 20% and 45%, with respective fiber cross-sections of $(0.274 \times 0.274)\ \mu m^2$, $(0.547 \times 0.547)\ \mu m^2$ and $(0.828 \times 0.828)\ \mu m^2$. The distances between the centers of the fibers is constant and equal 1.29 μm, in order to check the possible occurrence of size effects. As most of the existing analyses of experimental results make use of isotropic elasticity, two isotropic matrices of elastic constants are used in the FE part of the DCM. The Young's moduli and Poisson's ratio for the two phases are, respectively, E_{Al} = 71.3 GPa, $E_{Al_2O_3}$ = 373 GPa, $\nu_{Al} = 0.347$ and $\nu_{Al_2O_3} = 0.235$.

4.2 Results

In a first step, the simulation results are compared with the simple rule of mixtures, which assimilates the material to a composition in series of two elastic phases stressed in uniaxial tension. Hence, along the tensile axis:

$$\sigma = E\varepsilon \quad \text{and} \quad E = E_{Al_2O_3}V + E_{Al}(1 - V) \tag{4}$$

where V is the volume fraction of fibers.

The composite behavior, as reproduced by the simulation, is in very good agreement with experiment on the same material [1, 3]. As shown in Fig. 4, the composite material deformed in the longitudinal direction and with a large volume fraction of fibers deforms quasi-elastically. The rule of mixtures then provides a reasonable prediction of the stress-strain dependency. A smooth deviation from the purely elastic prediction is nevertheless recorded, even at very small strains. Such behavior, viz. the absence of an initial elastic stage, is commonly observed experimentally in Al_2O_3/Al composites. In the present case, the Al matrix is soft, as it has the mechanical properties of a pure single crystal; thus, it starts plastically deforming at rather low applied stresses. As expected, this phenomenon manifests itself all the more as the volume fraction of fibers decreases.

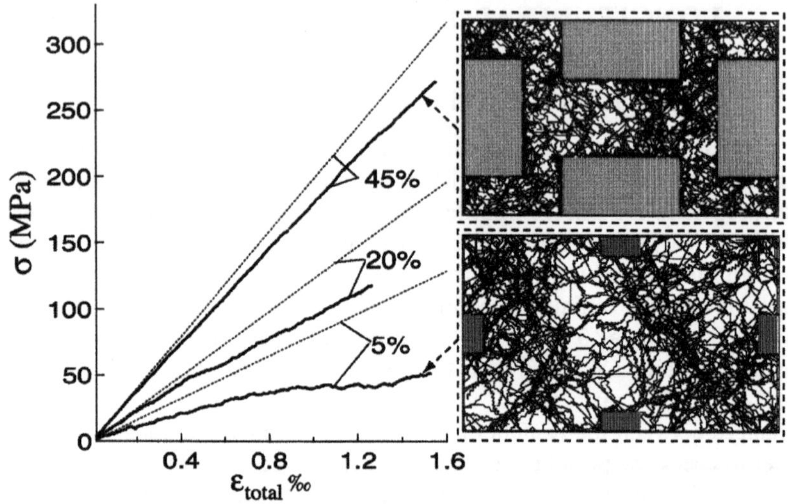

Figure 4. Longitudinal tensile stress-strain curves for Al_2O_3/Al MMCs with different fiber dimensions in a simulation cell of constant size. The volume fraction of fibers is V = 5%, 20% and 45%. The dotted lines refer to elastic predictions from the rule of mixtures. The two simulated microstructures show [001] views of the composites parallel to the fiber axis at a strain $\varepsilon = 0.15\%$. The dislocation density is about $1.35\ 10^{14}\ m^{-2}$ for $V = 5\%$ and $1.7\ 10^{14}\ m^{-2}$ for $V = 45\%$. Notice the early departure from elastic behavior for the smallest volume fraction.

When plastic deformation proceeds in the matrix, the dislocation density rapidly increases and a microstructure is formed, which contains many junctions (Fig. 4). The observed persistence of junctions at all the investigated strains is an indirect proof that the stresses developed in the matrix are still compatible with a mechanism of forest hardening. This result is again in agreement with experimental observation [1]. Nevertheless, as illustrated by Fig. 4, the differences in microstructure arrangement and dislocation density between the two volume fractions $V = 5\%$ and $V = 45\%$ strongly suggest that there is an additional hardening process. By lack of space, the analysis of this effect, which is actually size-dependent, is postponed to a future publication where the in-situ properties of the Al matrix will be investigated in full detail.

Finally, a careful observation of the simulation output reveals the progressive accumulation of a large dislocation density at the interfaces between the fibers and the matrix. In contrast with 2D simulations results, however, no pile-up of dislocation loops is formed around the fibers. This result can be explained in simple terms. In a 3D MMC structure, the dislocations can bow out between the fibers and by-pass then by an Orowan process. The critical stress for this process is inversely proportional to the distance between fibers, which is at the origin of a size effect. For topological reasons this process cannot be accounted for in two dimensions. Moreover, one should note that in 3D the mean free-path of the dislocations is large and a large amount of plastic strain can be produced by each dislocation loop. As a consequence, the sources are much less active in 3D than in 2D. This is why in multislip conditions, but in single slip conditions too, the probability for finding at the interfaces dislocations emitted in the same plane by a dislocation source is rather low.

5. CONCLUDING REMARKS

An improvement to the DCM is proposed and tested, which gives access to "massive" simulations of dislocation dynamics accounting for complex boundary value problems. The comparison between conventional DD simulations and the DCM in the case of pure fcc single crystals shows that the interactions of the dislocations at short distances, and especially their reactions, are now quantitatively reproduced.

The first original calculations of this improved version of the DCM have been dedicated to the study of a MMC. It was checked that the DCM, which is a parameter-free simulation, reproduces well the tensile properties of Al_2O_3/Al composites in the longitudinal direction. A simple comparison of 2D and 3D simulation results makes it clear that the hardening processes involved in MMCs are, in essence, three-dimensional.

This first study allows considering many future developments, which, globally, aim at discriminating between dislocation or metallurgical effects and mechanical effects through a coupled analysis of the simulated microstructures and of the stress and displacement fields. A detailed in situ investigation of the strain hardening of the aluminum matrix is under way. Since, however, the stress-strain behavior of Al_2O_3/Al composites deformed along the fiber direction is mainly elastic, size effects can more conveniently be examined by straining along the transverse direction. Lastly, experimental studies suggest the existence of a significant Baushinger effect. Its examination by the DCM could be of interest with reference to kinematic hardening in MMCs.

References

[1] J.A. Isaacs and A. Mortensen. *Metall. Trans. A*, 23A:1207-1219, 1992.

[2] C.-W. Nan and D.R. Clarke. *Acta mater.*, 44:3801-3811, 1996.

[3] P. Bystricky, H. Bjerregard, and A. Mortensen. *Metall. and Mater. Trans. A*, 30A:1843-1866, 1999.

[4] J.L. Bassani, A. Needleman, and E. Van der Giessen. *Int. J. Sol. Struct.*, 38:833-853, 2001.

[5] S. Yefimov, I. Groma, and E. Van der Giessen. *J. Phys. IV France*, 11:103-110, 2001.

[6] E. Bittencourt, A. Needleman, M.E. Gurtin, and E. Van der Giessen. *J. Mech. Phys. Solids*, 51:281-310, 2003.

[7] H.H.M. Cleveringa, E. van der Giessen, and A Needleman. *Acta mater.*, 45:3163-3179, 1997.

[8] H. H. M. Cleveringa, E. van der Giessen, and A Needleman. *Phil. Mag. A*, 79:893-920, 1999.

[9] E. van der Giessen and A Needleman. *Model. Simul. Mater. Sci. Eng.*, 3:689-735, 1995.

[10] C. Lemarchand, B. Devincre, L. P. Kubin, and J. L. Chaboche. In V.V. Bulatov et al., eds., *Multiscale Modelling of Materials*, volume 538, pp. 63-68. MRS, Warrendale, PA, 1999.

[11] C. Lemarchand, B. Devincre, and L.P. Kubin. *J. Mech. Phys. Solids*, 49:1969, 2001.

[12] D. Weygand, E. Van der Giessen, and A. Needleman. *Mat. Sci. Eng.*, A309-310:420, 2001.

[13] H. Yasin, H.M. Zbib, and M.A. Khaleel. *Mat. Sci. Eng.*, A309-310: pp. 294-299, 2001.

[14] C.S. Shin, M.C. Fivel, and K.H. Oh. In S. Forest et al., editor, *Scale Transitions from Atomistics to Continuum Plasticity*, vol. 11, p. 27, 2001.

[15] B. Devincre, A. Roos, and S. Groh. In A.. Finel et al., eds., *Thermodynamics, Microstructures and Plasticity*, NATO SCIENCE SERIES: II: Mathematics, Physics and Chemistry: Vol. 108, Kluwer Academic Publishers, The Netherlands, 2003.

[16] T. Mura. *Micromechanics of defects in Solids*. Kluwer Academic Publishers, Dordrecht, Netherlands, 1993.

[17] S. Groh, B. Devincre, L.P. Kubin, A. Roos, F. Feyel, and J.-L. Chaboche. *Phil. Mag. Lett.*, 83:303-313, 2003.

[18] R. Bullough and A.J.E. Foreman. *Phil. Mag.*, 9:315-329, 1964.

[19] R. Madec, B. Devincre, and L.P. Kubin. *This volume.*

[20] R. Madec, B. Devincre, and L.P. Kubin. *Phys. Rev. Lett.*, 89: 255508, 2002.

COMPUTATIONAL EVALUATION OF MICRO- TO MACROSCOPIC DEFORMATION BEHAVIOR OF AMORPHOUS POLYMER WITH SLIGHTLY HETEROGENEOUS DISTRIBUTION OF INITIAL SHEAR STRENGTH

Yoshihiro Tomita[1] and Makoto Uchida[2]

[1), 2)] Graduate Shool of Science and Technology, Kobe University
Nada Kobe, Japan 657-8501
E-mail: tomita@mech.kobe-u.ac.jp

Abstract: We investigate the characteristic deformation behavior of an amorphous polymer with a heterogeneous distribution of the initial shear strength (ISS). The deformation behavior of polymers under macroscopically uniform tension, and of plane strain polymer unit cell with a cylindrical void under combined straining were investigated by computational simulation with the nonaffine molecular chain network model. The results revealed the onset, evolution and interaction of microscopic shear bands emanating from slightly weak points, and the percolation of new shear bands. The micro- to macroscopic isotropy of the deformation response has been verified. Although the existence of distribution of ISS substantially affects the deformation, the effects of distribution patterns and standard deviation of ISS on the deformation were small. The interaction between the heterogeneity of ISS and voids substantially affects such micro- to macroscopic deformation behavior as onset and propagation of shear bands, mean stress distribution, and macroscopic stress vs strain relationship.

Key words: Amorphous Polymer, Distribution of Initial Shear Strength, Microscopic Shear Band, Void Containing Polymer, Molecular Chain Network Theory, Computational Simulation

H. Kitagawa and Y. Shibutani (eds.), IUTAM Symposium on Mesoscopic Dynamics of Fracture Process and Materials Strength, 245–254.

1. INTRODUCTION

The deformation behavior of polymeric materials under tension is very different from that of metallic materials. The plastic flow in an amorphous polymer due to the onset and growth of shear bands is initiated at a stress level lower than the macroscopic yielding point [1]. Such shear bands in an amorphous polymer are oriented along a direction very close to the direction of maximum shear stress. Subsequently, many shear bands form with the increase of deformation. Beyond the macroscopic yield point, shear bands are transformed into the macroscopically manifested neck and its propagation under the essentially steady state as often observed in experiments. These typical deformation behaviors are closely related to the heterogeneity of the microstructure of the amorphous polymer, nevertheless, the discussions associated with this evidence are few [2-4] and many issues are still unclear.

In this paper, by employing a nonaffine molecular chain network model [5] and finite element simulation [6] we numerically specify the characteristic deformation behavior of amorphous polymers with a slightly heterogeneous chain distribution which was replaced by the heterogeneous initial shear strength (ISS). The isotropy of micro- to macroscopic responses of a unit cell, the effect of distribution patterns and standard deviation of ISS on the deformations are investigated. The effect of the interaction between the heterogeneity of ISS and voids on such micro- to macroscopic deformation behavior as onset and propagation of shear bands, mean stress, and stress vs. strain relationship is discussed.

2. CONSTITUTIVE EQUATION

The complete constitutive equation for a polymer employed in this investigation is given in references [5]. Here, we provide a brief explanation of the constitutive equation. The total strain rate is assumed to be decomposed into the elastic strain rate and the plastic strain rate. The elastic strain rate is expressed by Hooke's law and the plastic strain rate is modeled using a nonaffine eight-chain model [5]. The final constitutive equation that relates the rate of Kirchhoff stress $\dot{S}_{ij}$ to strain rate $\dot{\varepsilon}_{kl}$ becomes

$$\dot{S}_{ij} = L_{ijkl}\dot{\varepsilon}_{kl} - P'_{ij},\ L_{ijkl} = D^e_{ijkl} - F_{ijkl},\ F_{ijkl} = \frac{1}{2}\left(\sigma_{ik}\delta_{jl} + \sigma_{il}\delta_{jk} + \sigma_{jl}\delta_{ik} + \sigma_{jk}\delta_{il}\right), \tag{1}$$

$$P'_{ij} = D^e_{ijkl}\frac{\dot{\gamma}^P}{\sqrt{2}\tilde{\tau}}\hat{\sigma}'_{kl},\ \tilde{\tau} = \left(\hat{\sigma}'_{ij}\hat{\sigma}'_{ij}/2\right)^{1/2},\ \hat{\sigma}_{ij} = \sigma_{ij} - B_{ij},$$

where D^e_{ijkl} is the elastic stiffness tensor and σ_{ij} is the Cauchy stress. The shear strain rate $\dot{\gamma}^p$ in Eq. (1) is given as [7]

$$\dot{\gamma}^p = \dot{\gamma}_0 \exp\left[\left(-\frac{A\tilde{s}}{T}\right)\left\{1-\left(\frac{\tilde{\tau}}{\tilde{s}}\right)^{5/6}\right\}\right], \tag{2}$$

where $\dot{\gamma}_0$ and A are constants, T is the absolute temperature, $\tilde{\tau}$ is the applied shear stress, $\tilde{s} = s + ap$ indicates shear strength [8], s is the shear strength which changes, with plastic strain, from the athermal shear strength $s_0 = 0.077\mu/(1-\nu)$ to a stable value s_{ss}, p is the pressure, a is a pressure-dependent coefficient, μ is the elastic shear modulus and ν is Poisson's ratio. Since s depends on the temperature and strain rate, the evolution equation of s can be expressed as $\dot{s} = h\{1-(s/s_{ss})\}\dot{\gamma}^p$, where h is the rate of resistance with respect to plastic strain. Furthermore, B_{ij} in Eq. (1) is the back-stress tensor and the principal components are expressed by employing the eight-chain model [9], as

$$B_i = \frac{1}{3}C^R\sqrt{N}\frac{V_i^2-\lambda^2}{\lambda}L^{-1}\left(\frac{\lambda}{\sqrt{N}}\right),\ L(x) = \coth x - \frac{1}{x},\ \lambda^2 = \frac{1}{3}(V_1^2+V_2^2+V_3^2), \tag{3}$$

where V_i is the principal plastic stretch, N is the average number of segments in a single chain, $C^R = nkT$ is a constant, n is the number of chains per unit volume, k is Boltzmann's constant, and L is the Langevin function. In the nonaffine eight-chain model [5], the change in the number of entangled points, in other words, the average number of segments N, may change depending on the distortion ξ which represents the local deformation of a polymeric material [5]. The simplest expression of the number of entangle points is $m = m_r \exp\{-c(1-\xi)\}$ with $\xi = 1$ in the reference state and m_r is the number of entangled points at reference state and c is a material constant [5].

3. COMPUTATIONAL MODEL

Here, we evaluate the detailed characteristics of micro- to macroscopic deformation of an amorphous polymer with heterogeneously distributed molecular chains, which is suggested by careful observation [10, 11] using an Atomic Force Microscope (AMF) under macroscopically uniform tension and shearing, and large scale Molecular Dynamic (MD) simulation [12]. Nevertheless, the information associated with the concrete distributions of chain density or ISS is not available, therefore, we restrict our investigations to the effect of the distribution of ISS on the micro- to macroscopic deformation behavior of amorphous polymer, and will employ the normal distribution of ISS.

Figure 1(a) shows the plane strain computational unit cell model for macroscopically uniform tension in which the heterogeneous distribution of ISS, s_0, is assumed. Figure 1(c) shows the normal distribution of s_0 with a mean value s_{0m} and three different standard deviations, $0.05s_{0m}$ (ND1), $0.10s_{0m}$ (ND2) and $0.15s_{0m}$ (ND3). The distribution of s_0 is specified, such that, depending on the total number of finite elements in a unit cell, a specific s_0 value is allocated to a square element and the number of elements with specific s_0 values exhibits a normal distribution, as shown in Fig. 1(c). Next, the effect of distribution pattern of ISS is considered. Figure 1(d) shows β distribution with mean value s_{0m} and the locations of the peaks of the distribution are $-0.2L$ (BD1), $0.0L$ (BD2) and $0.2L$ (BD3) from the mean value as depicted in Fig. 1(d). L is the distribution width corresponding to that of ND2 of Fig. 1(c). Furthermore, a distribution with two peaks as shown in Fig. 1(e) is considered. The distribution is established by employing two β distributions with different peak distances; $0.0L$ (DP0), $0.2L$ (DP1), $0.4L$ (DP2) and $0.6L$ (DP3).

Figure 1(b) indicates the computational model for a plane strain unit cell with a cylindrical void, which is an approximate model of an amorphous polymer containing cavitated rubbery particles [13, 14]. The macroscopically homogeneous deformations shown in Figs. 1 (a) and (b) are applied. In all cases, shear-free conditions are applied to the surface of the boundary.

Moreover, we define the average strain rates $\dot{E}_1$, $\dot{E}_2$ or stress rates $\dot{\Sigma}_1$, $\dot{\Sigma}_2$ with respect to the coordinate directions x_1, x_2, respectively, and macroscopic equivalent stress and strain are defined as $\Sigma_e = (3\Sigma_i'\Sigma_i'/2)^{1/2}$ and $E_e = (2E_i'E_i'/3)^{1/2}$. For the model in Fig.1(a), a macroscopically homogeneous strain rate $\dot{E}_e = \dot{\varepsilon}_0 = 10^{-5}/s$ is applied, whereas it is prescribed as $\dot{E}_1 = \chi\dot{E}_2$, $\dot{E}_2 = \dot{\varepsilon}_0$ for the cases of Fig. 1(b). Here, we introduce a macroscopic strain triaxiality parameter, $T_p = (E_1 + E_2)/3E_e$. $T_p = 0, 0.5$ and 1.0 correspond to $\chi = -1.0, 0.0$ and 1.0, respectively, which represent the pure shear, horizontally plane strain and biaxial strain conditions. Two different volume fractions of voids, $f_0 = 0.05$ and 0.20, are considered. Since strain rate is sufficiently low, here we disregard the heat generation due to irreversible work. The material parameters for the polymer employed are $E_m/s_{0m} = 23.7$, $s_{ss}/s_{0m} = 0.79$, $h/s_{0m} = 5.15$, $As_{0m}/T_0 = 78.6$, $\alpha = 0.08$, $\dot{\gamma}_0 = 2.0\times10^{15}/s$, $s_{0m} = 97MPa$, $T_0 = 296K$, $m_0 = 7.83\times10^{26}$ and $c = 0.33$ [5], which are the modified versions of those for the affine eight-chain model [9]. Standard deviation (S.D.) for the normal distribution of ISS is specified as $0.15s_{0m}$ for the cases unless noted otherwise.

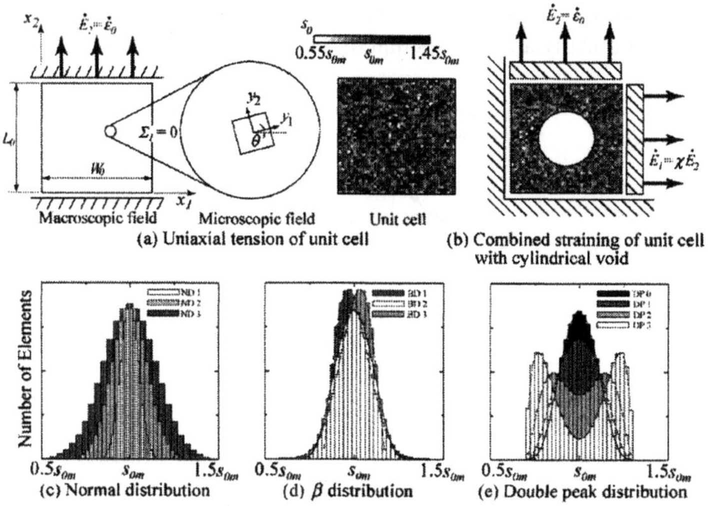

Figure 1 Computatinal Models

4. RESULTS AND DISCUSSION

4.1 Uniaxial Tension of Unit Cell

We will first discuss the effect of the heterogeneous distribution of ISS as shown in Fig. 1(a) on micro- to macroscopic deformation behavior under uniform plane strain tension with different directions. Figures 2(a) and (b) indicate the macroscopic equivalent stress-strain relation for different directional tensions with angle θ defined in the inserted figure and the equivalent plastic strain rate distribution at the corresponding macroscopic equivalent strain in Fig. 2(a), respectively. Both figures suggest the micro- and macroscopically isotropic responses. For comparison, the result for the homogeneous distribution of s_0 is also indicated in (a). The effects of heterogeneity of s_0 is substantial in the early stage of deformation. The yield due to heterogeneous distribution of s_0 causes the nonlinear response prior to the macroscopic yield and suppresses 20% of the corresponding macroscopic yield stress in the homogeneous case. Furthermore, continuous yielding at different positions of the polymer results the rather moderate increase of average stress with deformation as compared with the homogeneous case. However, the critical stretch is assumed to be identical in this investigation, therefore, the resistance of the deformation in the later stage asymptotically approaches that of the homogeneous case.

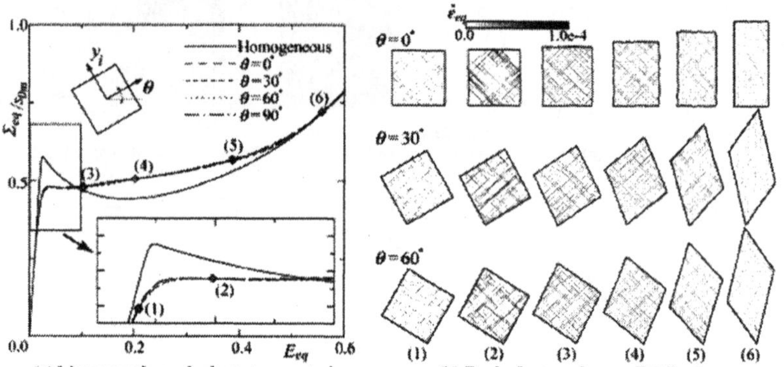

(a) Macroscopic equivalent stress - strain (b) Equivalent strain rate distribution

Figure 2 Macroscopically Differnt Directional Uniform Tension

Figure 2(b) indicates that the localized microscopic shear bands which connect the microscopically weak regions, appear at almost 45 degrees with respect to the tension direction. The macroscopic yield occurs at (2) on stress-strain curve in Fig. 2(a) where the microscopic shear bands cross the unit cell, which is responsible for lowering the yield stress as compared with the homogeneous case. Since the shear strength becomes s_{ss} in the region with a lower value of s_0, further softening may not occur. As a result, the strain rate in the initially formed shear bands possesses the maximum value at the macroscopic yielding point and it decreases with orientation hardening. The softening occurring in the subsequently generated microscopic shear band is compensated with hardening in the previously appearing shear band. Therefore, the stress vs. strain relation for the unit cell in Fig. 2(a) exhibits no softening. With additional macroscopic deformation, the percolation, intensification and propagation of shear bands along the direction normal to the shear bands are observed, accompanied by the rotation of shear bands. Subsequently, the increase of the number of shear bands in which thc maximum stretch attains the critical value, results in a significant increase of the resistance of deformation. The corresponding strain rate distribution is seen in (6) of Fig. 2(b) which indicates almost uniform deformation. Thus, the unit cell model developed shows isotropic response in both the microscopic and macroscopic deformation processes.

Furthermore, the micro- to macroscopic deformation behavior evaluated with different magnitude of the standard deviation and the profile of the distribution of ISS as indicated in Figs. 1(c), (d) and (e) suggested that their effect on the microscopic as well as macroscopic deformation is fairly small. Therefore, the investigation based on the normal distribution with a specific standard deviation of heterogeneity of ISS provides a general overview of the deformation behaviors of amorphous polymers.

4.2 Interaction between the heterogeneous distribution of ISS and voids

In order to clarify the effect of heterogeneity of ISS on the micro- to macroscopic deformation behavior of soft rubber blended polymers, the problems depicted in Fig. 1(b) are investigated by the computational model presented in section 3. A typical rubber-blended amorphous polymer contains rubber particles which are spherical with μ m length scale and are dispersed throughout the matrix. The mechanical characteristics of the blended material strongly depends on the initiation and propagation of the shear band in the matrix polymer and the cavitations of rubber particles, which are substantially affected by the volume fraction and size of the rubber particles. Although the distribution of the rubber particles is somewhat random, we assume that it is periodic and focus our attention on the essential feature of the effect of initial heterogeneous distribution of ISS on the micro- to macroscopic deformation behavior of polymers with voids. Related studies on homogeneous polymers can be seen in [13, 14]

Figure 6 indicates (a) macroscopic true stress-strain relations and (b) equivalent plastic strain rate distributions at different macroscopic strains depicted in (a) for the void containing unit cell under plane strain deformation as shown in Fig.1(b). For comparison, a homogeneous case of ISS is also indicated. The macroscopic response is sensitive to the macroscopic strain triaxiality T_p. Decreasing T_p causes a reduction in the maximum macroscopic stress and in the stiffness in the early stage of deformation. The heterogeneous distribution of ISS promotes a nonlinear response in the very early deformation stage and suppresses the maximum stress substantially, as compared with the cases in section 4.1. Due to heterogeneity of ISS, the resistance to deformation of the material does not exhibit a significant drop that can be observed in the polymer with a homogeneous ISS. This is attributable to the contribution of heterogeneity of ISS to the promotion of orientation hardening successively in the ligament area. It has been also assured that the increasing of the volume fraction of void f_0 causes the reduction of macroscopic yield stress and stiffness.

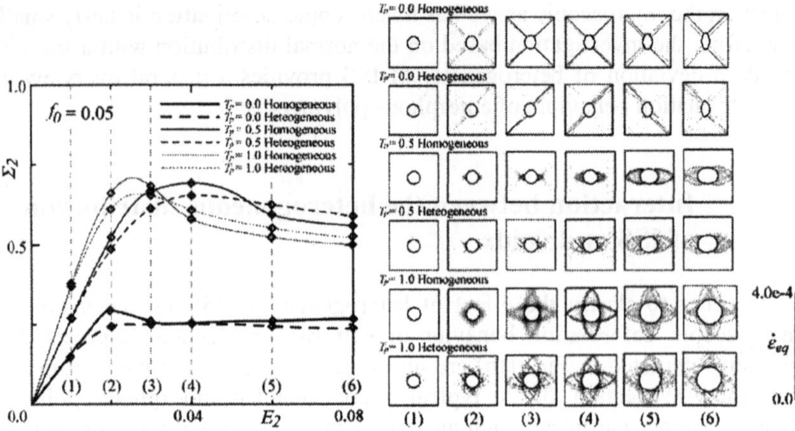

(a) Macroscopic true stress - strain (b) Equivalent strain rate distribution

Figure 3 Deformation of Unit Cell Containing Cylindrical Voids Under Combinated Strain

For the case of heterogeneous ISS, microscopic shear bands emanating from the weak points appear from the early stage of deformation, as in the case of macroscopically homogeneous deformation, which causes the nonlinear behavior in stress-strain relationships, and subsequently, continuous growth of microscopic shear bands. Their propagation and percolation cause the unified strong shear bands that connect the voids. Due to the random distribution of ISS, unified shear bands exhibit nonsymmetric with respect to the symmetric axis of the unit cell. Upon the onset of such unified shear bands, macroscopic stress-strain relationships shows a local maximum and then the stress gradually decreases with further deformation. This is attributable to the continuous onset and propagation of the shear bands due to the heterogeneous distribution of ISS.

To predict the onset of crazing, which affects the ductility and toughness of the polymer, the mean stress, particularly positive mean stress, is an essential parameter. Therefore, we will discuss the local distribution of the mean stress in the unit cell. Figure 4 shows (a) mean stress distribution and location of the maximum mean stress point at different deformation stages (1) to (6) in Fig. 3(a), (b) mean stress distribution along the void at deformation stage (6) and (c) mean stress and equivalent strain rate distribution near the surface of the void as indicated in (a). The onset and propagation of micro shear bands caused by the heterogeneous distribution of ISS promotes the occurrence of high maximum mean stress in the early stage of deformation. The effect of strain triaxiality on the local mean stress is remarkable. High triaxiality causes the high mean stress. Initially, a high positive mean stress appears near the cross points of microscopic shear bands. As the deformation proceeds, the location of the high mean stress point moves to the surface of the void where the stretch approaches a critical value.

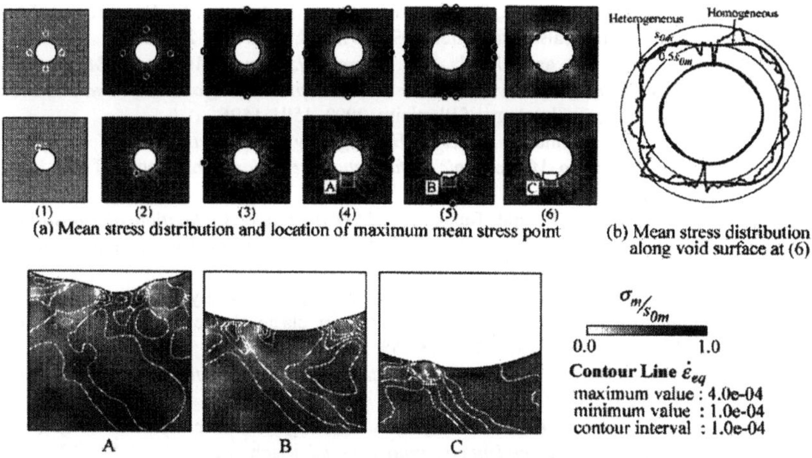

(a) Mean stress distribution and location of maximum mean stress point

(b) Mean stress distribution along void surface at (6)

(c) Mean stress and equivalent strain rate distribution at A to C in (a)

Figure 4 Mean Stress and Equivalent Strain Rate Distribution ($f_0 = 0.05,\ T_p = 1.0$)

Figure. 4(b) shows very complicated distribution of mean stress along the void surface. For the heterogeneous ISS case, the maximum mean stress value is 20% higher than that for the homogeneous case. It should be also noted that the mean stress possesses an anomalously lower value. The latter can be explained by examining the mean stress distribution near the void surface in Fig. 4(c). As can be seen in the deformation of the plane strain block with a free surface [15], the tensile strain applied to the unit cell promotes the onset and propagation of microscopic shear bands beneath the surface of voids, that causes the undulation on the void surface as can be clearly observed in Fig. 4(c). The mean stress value is very low at the convex part of the void surface. The effect of such undulation on the onset of cavitation in rubber blended polymer will be the subject of future work.

ACKNOWLEDGMENT

Financial support from the Ministry of Education, Culture, Sports, Science and Technology of Japan is gratefully acknowledged.

REFERENCES

[1] Courtney, T. H., *Mechanical Behavior of Materials*, 1990, McGraw-Hill.

[2] Boyce, M. C., and Chui, C., *Effect of heterogeneities and localization of polymer deformation and recovery*, T., De Borst, R and Van der Giessen, E., Eds. Proc. IUTAM Symposium on Materials Instabilities, 1998, Wiley & Sons Ltd., 269-285.

[3] Tomita, Y., Lu, W., and Uchida, M., *Characterization of Micro- to Macroscopic Response of Amorphous Polymers under Macroscopically Uniform Deformation*, Proc. ICTP 7, Advanced Technology of Plasticity, 2002, 1591-1596.

[4] Uchida, M., and Tomita,Y., *Effect of Heterogenity of Microstructure of Amorphous Polymer on Micro- to Macroscopic Deformation Behavior*, submitted to Trans JSME (in Japanese)

[5] Tomita, Y., Adachi, T., and Tanaka, S., *Modelling and Application of Constitutive Equation for Glassy Polymer Based on Nonaffine Network Theory*, Eur. J. Mech. A/Solids, 1997, **16**, 745-755.

[6] Kitagawa, H and Tomita, Y., An incremental finite element analysis of two dimensional large strain and large displacement problems for elasto-plastic materials, Theoretical and Applied Mechanics, 1971, **21**, 243-255.

[7] Argon, A. S., *A theory for the low-temperature plastic deformation of glassy polymers*, Phil. Mag. 1973, **28**, 839-865.

[8] Boyce, M. C., Parks, D. M. and Argon, A. S., *Large inelastic deformation of glassy polymers, Part I: rate dependent constitutive model*, Mech. Mater., 1988. **7**, 15-33.

[9] Arruda, E. M. and Boyce, M. C., *A three-dimensional constitutive model for large stretch behavior of rubber materials*, J. Mech. Phys. Solids, 1993, **41**, 389-412.

[10] Kashu,Y., Adachi, T. and Tomita, Y., *AFM observation of microscopic behavior of glassy polymer with application to understanding of macroscopic behavior*, Eds. Abe, T and Tsuta, T, Proc. AEPA'96 Pergamon, 1996, 501-505.

[11] Adachi, T., Tomita, Y., and Kashu, Y., *AFM observation of microscopic behavior of glassy polymer under macroscopic tension*, Trans. JSME, 1998, **64A**, 758-764. (in Japanese)

[12] Itoh, T., Yashiro, K., and Tomita, Y., *Molecular dynamic study on deformation of molecular chains in amorphous polymer*, JSMS 7th Symposium on Molecular Dynamics, 2002, 50-55.

[13] Steenbrink, A.C., Van der Giessen, E., Wu, P.D., *Studies on growth of voids in amorphous glassy polymers*, J. Mat. Sci., **33**, 1998, 3163-3175.

[14] Tomita, Y., and Lu, W., *Computational Characterization of Micro- to Macroscopic Mechanical Behavior of Polymers Containing Second-Phase Particles*, Int. J. Damage Mech., **11**, 2002, 129-149.

[15] Tomita, Y., and Uchida, M., *Characterization of Micro- to Macroscopic Deformation Behavior of Amorphous Polymer*, Key Engineering Materials, **233**, 2003, 673-678.

COMPUTATIONAL MODELING AND CHARACTERIZATION OF MATERIALS WITH PERIODIC MICROSTRUCTURE USING ASYMPTOTIC HOMOGENIZATION METHOD

YOSHIKAZU HIGA and HIROSHI KITAGAWA
Graduate School of Engineering, Osaka University
2-1 Yamada-oka Suita, Osaka 565–0871 JAPAN
higa@ams.eng.osaka-u.ac.jp, kitagawa@ams.eng.osaka-u.ac.jp

YOSHIHIRO TOMITA
Graduate School of Science and Technology, Kobe University
1-1 Rokkodai-cho Nada-ku Kobe, Hyogo 657–8501 JAPAN
tomita@mech.kobe-u.ac.jp

Abstract In order to clarify the micro- to macroscopic characteristic feature of two-phase single-crystalline materials, an asymptotic homogenization method has been developed for materials obeying the constitutive equation based on the dislocation-density-dependent crystal plasticity theory. We focus our attention on nickel-based superalloy that is well used in practice and in which fine and hard (gamma-prime phase: γ'-phase) precipitates are embedded in a soft (gamma phase: γ-phase) matrix. Assuming that the cuboidal γ'-phase precipitates with a periodical distribution in the γ-phase matrix, a unit cell model consisting of two phases is established. Then, a series of computa-tional simulations adopting a constant γ'-precipitate size has been performed for the alloy with the different volume fractions of γ'-precipitates associated with the width of the γ-channel and the crystallographic orientations of the two phases. Results show that the effects of the γ'-precipitate volume fraction and the applied loading direction relative to the crystallographic orientation on the macroscopic deformation behavior are considerable as compared with the strain-gradient-independent model.

Keywords: Nickel-based Superalloy, Asymptotic Homogenization Method, Strain-Gradient Crystal Plasticity Theory, Characteristic Length Scale

H. Kitagawa and Y. Shibutani (eds.), IUTAM Symposium on Mesoscopic Dynamics of Fracture Process and Materials Strength, 255–268.

1. INTRODUCTION

Nickel-based single-crystal superalloys have excellent high-temperature strength and oxidation resistance and are therefore widely used for blades in aircraft gas turbines and for components of nuclear reactors[1]. The microstructure of these alloys consists of two phases: a high volume fraction of coherently precipitated γ' cubes ($L1_2$) separated by thin channels of face-centered-cubic (f.c.c.) γ-matrix, as shown in Fig.1[2]. In order to take into account the microscopic deformation mechanisms associated with the dislocation motions, several analyses of the deformation behavior in these alloys, using the finite-element method, have been reported in the past. Fedelich[3,4] developed the constitutive model by introducing the microscopic mechanical features and investigated the overall properties under cyclic loading. His method, however, cannot be used to obtain the local and global deformation behavior simultaneously or to estimate the precipitate size effects on two-phase single crystals. On the other hand, Ohashi[5] developed the constitutive model based on the crystal plasticity theory, which contains the first gradient of shear strain derived from the geometrically necessary dislocations[6] and investigated the dislocation line profile in the γ-matrix. Busso et al.[7] and Meissonnier et al.[8] focused on the effects of precipitate size on the alloy; i.e. the effect of the relative length scale d/l, where d is the width of the γ-phase channel and l is the γ'-precipitate size, on the macroscopic deformation behavior. However these works are restricted to the [001] loading direction because their boundary conditions are limited to a fixed loading direction with respect to the crystallographic orientation; Therefore, the macroscopic deformation behavior has not been clarified.

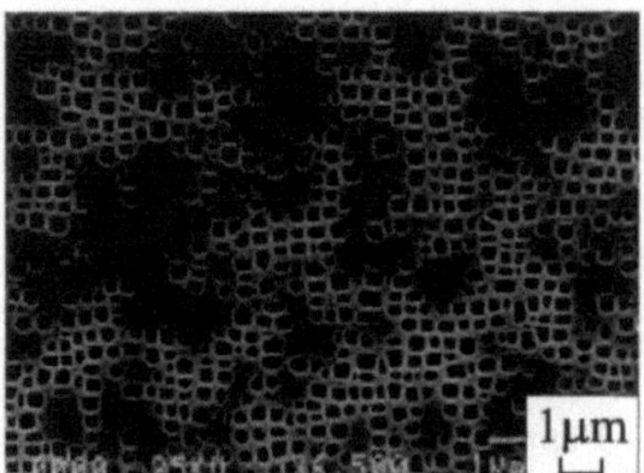

Figure 1. SEM image of the typical Ni-based superalloy morphology. The γ' precipitates appear in dark while γ channels are white[2].

In this study, we develop an asymptotic homogenization method for materials obeying the constitutive equation based on the dislocation-density-dependent crystal plasticity theory and the finite-element method. Then, a series of computational simulations for an alloy with characteristic length scales are carried out to clarify the effect of different morphological parameters, such as volume fraction of precipitates and arbitrary crystallographic orientations of two phases, on the micro- to macroscopic deformation behavior.

2. CONSTITUTIVE EQUATION BASED ON THE STRAIN-GRADIENT CRYSTAL PLASTICITY THEORY

The strain- and strain-rate dependent crystal plasticity constitutive framework used in this study is based on the formulation proposed by Peirce et al.[9], which is expressed as

$$\overset{\nabla}{S}_{ij} = D^{e}_{ijkl} d_{kl} - \sum_{\alpha} R^{(\alpha)}_{ij} \dot{\gamma}^{(\alpha)}, \; R^{(\alpha)}_{ij} \overset{\text{def.}}{=} D^{e}_{ijkl} P^{(\alpha)}_{kl} + W^{(\alpha)}_{ik} \sigma_{kj} - \sigma_{ik} W^{(\alpha)}_{kj} \quad (1)$$

where $\overset{\nabla}{\boldsymbol{S}}$ is the Jaumann rate of Kirchhoff stress, $\boldsymbol{D}^e$ the tensor of elastic moduli, $\boldsymbol{d}$ the strain rate and $\boldsymbol{\sigma}$ the Cauchy stress tensor. The slip plane normal $\boldsymbol{s}^{(\alpha)}$ and the slip direction $\boldsymbol{m}^{(\alpha)}$ define the symmetric and asymmetric parts of the Schmid tensor, $\boldsymbol{P}^{(\alpha)}$ and $\boldsymbol{W}^{(\alpha)}$, respectively.

$$P^{(\alpha)}_{ij} \overset{\text{def.}}{=} \frac{1}{2}\left(s^{(\alpha)}_i m^{(\alpha)}_j + m^{(\alpha)}_i s^{(\alpha)}_j\right), \; W^{(\alpha)}_{ij} \overset{\text{def.}}{=} \frac{1}{2}\left(s^{(\alpha)}_i m^{(\alpha)}_j - m^{(\alpha)}_i s^{(\alpha)}_j\right). \quad (2)$$

The relationship between the shear strain rate $\dot{\gamma}^{(\alpha)}$ and the critical resolved shear stress $\tau^{(\alpha)}$ on the slip system (α) is assumed to be given by the power law[10,11]

$$\dot{\gamma}^{(\alpha)} = \dot{\gamma}^{(\alpha)}_{\text{ref}} \operatorname{sgn}\left(\tau^{(\alpha)}\right) \left|\tau^{(\alpha)} / g^{(\alpha)}\right|^{1/\kappa} \quad (3)$$

where $\dot{\gamma}^{(\alpha)}$ is the reference rate of shearing and κ is the strain rate sensitivity parameter. $g^{(\alpha)}$ is the current value of the reference resolved shear stress. We assume that the reference-resolved shear stress is a Bailey-Hirsch type function[12] given by

$$g^{(\alpha)} = \tau^{(\alpha)}_0 + a\mu\tilde{b} \sum_{\beta} \omega^{\alpha\beta} \sqrt{\rho^{(\beta)}_{\text{T}}} \quad (4)$$

where $\tau^{(\alpha)}_0$ indicates a material constant such as yield stress. The second term represents the effect of accumulated dislocations in crystal. $\rho^{(\alpha)}_{\text{T}}$

denotes the total density of dislocations that accumulates on the slip system (α). $\omega^{\alpha\beta}$ is a matrix that takes into account various types of dislocation interactions[13]. $\tilde{b}$, m and a denote the magnitude of the Burgers vector, the elastic shear modulus and a material constant, respectively. The total dislocation density on the slip system (α) is given by

$$\rho_{\mathrm{T}}^{(\alpha)} = \rho_{\mathrm{S}}^{(\alpha)} + \left\| \rho_{\mathrm{G}}^{(\alpha)} \right\| \tag{5}$$

where $\rho_{\mathrm{S}}^{(\alpha)}$ and $\rho_{\mathrm{G}}^{(\alpha)}$ denote the densities of the statistically stored dislocations (SSDs) and the geometrically necessary dislocations (GNDs), respectively.

The incremental form of the SSDs is given by

$$\dot{\rho}_{S}^{(\alpha)} = \frac{c}{\tilde{b}L^{(\alpha)}} \dot{\gamma}^{(\alpha)} \tag{6}$$

where c is a material constant and $L^{(\alpha)}$ is the dislocation mean free path on the slip system (α). In this study, we consider the dislocation-dislocation interaction and introduce a power-law-type description of $L^{(\alpha)}$[14]:

$$L^{(\alpha)} = L_{0}^{(\alpha)} \left(\rho_{\mathrm{T}}^{(\alpha)} / \rho_{0}^{(\alpha)} \right)^{-n} , \tag{7}$$

where $L_{0}^{(\alpha)}$ and $\rho_{0}^{(\alpha)}$ are the reference value of the mean free path and the initial dislocation density, respectively. n is the material constant. The evolution of the density of SSDs during plastic slip is evaluated by integrating Eq.(6). On the other hand, the density of GNDs[5] is estimated by

$$\left.\begin{aligned} \rho_{\mathrm{G,edge}}^{(\alpha)} &= -\frac{1}{\tilde{b}} s_{i}^{(\alpha)} \gamma_{,i}^{(\alpha)} \\ \rho_{\mathrm{G,screw}}^{(\alpha)} &= \frac{1}{\tilde{b}} \mathrm{e}_{ijk} s_{j}^{(\alpha)} m_{k}^{(\alpha)} \gamma_{,i}^{(\alpha)} \end{aligned}\right\} \tag{8}$$

where $\rho_{\mathrm{G,edge}}^{(\alpha)}$ and $\rho_{\mathrm{G,screw}}^{(\alpha)}$ denote the edge and screw components of the GNDs, respectively. e_{ijk} is the permutation tensor and $\gamma_{,i}^{(\alpha)}$ denotes the gradient of shear strain on the slip system (α). The norm of two components defines the scalar density for the GNDs,

$$\left\| \rho_{\mathrm{G}}^{(\alpha)} \right\| = \sqrt{\left(\rho_{\mathrm{G,edge}}^{(\alpha)} \right)^{2} + \left(\rho_{\mathrm{G,screw}}^{(\alpha)} \right)^{2}} . \tag{9}$$

The evolution equation $\dot{g}^{(\alpha)}$ obtained by the time derivative of Eq.(4) is taken to be

$$\dot{g}^{(\alpha)} = \sum_{\beta} h^{\alpha\beta} \left\| \dot{\gamma}^{(\beta)} \right\| , \tag{10}$$

where $h^{\alpha\beta}$ is a matrix of the hardening moduli associated with the local strain hardening state. Diagonal components of this matrix correspond to self-hardening, while off-diagonal components describe cross-hardening effects. From Eqs.(4)-(6), (8)-(9) and (10) the hardening moduli are derived as follows:

$$h^{\alpha\beta} = \frac{\frac{1}{2} a c \mu \omega^{\alpha\beta}}{L^{(\beta)} \left(\rho_{\mathrm{S}}^{(\beta)} + \left\| \rho_{\mathrm{G}}^{(\beta)} \right\| \right)^{1/2}} . \tag{11}$$

The evolution of the density of SSDs and GNDs is affected by shear strain and shear strain gradient as shown in Eqs.(6) and (8), respectively[6]. Hence, the strain gradients are considered to be internal variables, as shown by Eq.(11). Therefore we can practically estimate the dependence of the strain gradient, in other words, the characteristic length scale, on the macroscopic phenomena without introducing higher order stresses or additional boundary conditions. In the present work, in order to decrease the interval of time steps to achieve stable computation, the algorithm of the tangent modulus method[15] is adopted for estimating the shear strain increment. The constitutive relations are summarized as follows:

$$\overset{\triangledown}{S}_{ij} = C_{ijkl} d_{kl} - \sum_{\alpha} R_{ij}^{(\alpha)} \dot{f}^{(\alpha)} , \tag{12}$$

where

$$\left.\begin{aligned}
C_{ijkl} &= D_{ijkl}^{e} - \sum_{\alpha} R_{ij}^{(\alpha)} F_{kl}^{(\alpha)} , F_{ij}^{(\alpha)} = \sum_{\beta} \left(N^{\alpha\beta} \right)^{-1} Q_{ij}^{(\beta)} , \\
Q_{ij}^{(\alpha)} &= \frac{\theta \Delta t \dot{\gamma}_t^{(\alpha)}}{m \tau^{(\alpha)}} R_{ij}^{(\alpha)} , \Delta\gamma^{(\alpha)} = \left(\dot{f}^{(\alpha)} + F_{ij}^{(\alpha)} d_{ij} \right) \Delta t , \\
N^{\alpha\beta} &= \delta^{\alpha\beta} + \frac{\theta \Delta t \dot{\gamma}_t^{(\alpha)}}{m} \left[\frac{R_{ij}^{(\alpha)} P_{ij}^{(\alpha)}}{\tau^{(\alpha)}} + \operatorname{sgn} \left(\tau^{(\beta)} \right) \frac{h^{\alpha\beta}}{g^{(\alpha)}} \right] , \\
\dot{f}^{(\alpha)} &= \sum_{\beta} \left(N^{\alpha\beta} \right)^{-1} \dot{\gamma}_t^{\beta} .
\end{aligned}\right\} \tag{13}$$

3. HOMOGENIZATION METHOD FOR STRAIN- AND STRAIN-RATE-DEPENDENT CRYSTALLINE MATERIALS

In order to estimate the macro/microscopic deformation behavior of Ni-based superalloy with periodically distributed γ'-precipitates, as

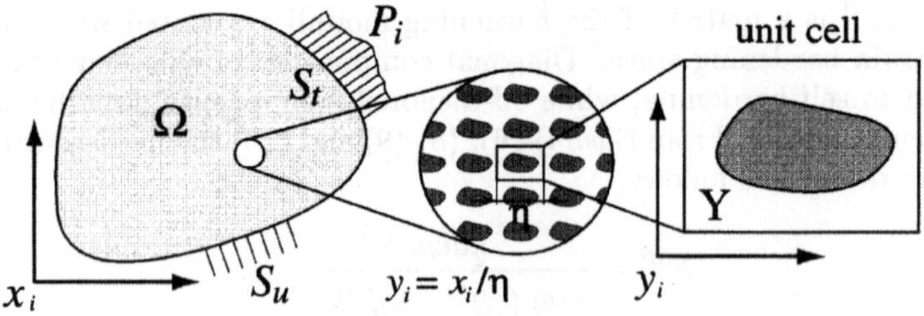

Figure 2. Global and local problems with two spatial scales.

shown in Fig.1, the formulation of an asymptotic homogenization method [16] for strain- and strain-rate-dependent crystalline materials is employed[17,18] along the same lines as indicated for the viscoplastic case[19].

Consider the problem shown in Fig.2, with domain Ω and boundary S subjected to surface force P_i on S_t and prescribed velocity on S_u. The body is formed by the spatial repetition of a base cell Y made of different materials. Assuming that the base cell is very small, of the order η, compared with the dimensions of the entire body, the global coordinate is x_i for the entire body, and the local coordinate is y_i related to the single base cell, then $y_i = x_i/\eta$. Similar to the assumption used in the case of linear elastic materials[16], the velocity and its gradients are supposed to be expressed as an asymptotic expansion with respect to parameter η. By substituting these into the virtual work principle of the updated Lagrangian formulation[20], rearranging at the same order η and taking the limit of $\eta \to 0$, we can arrive at the virtual work principle for a macroscopic body:

$$\int_\Omega \left[L^H_{ijkl} \dot{E}^0_{kl}(\boldsymbol{v}) - P^H_{ij} + \sigma^H_{ij} + \tau^H_{ijkl} \frac{\partial v^0_k}{\partial x_l} \right] \frac{\partial \delta v_i}{\partial x_j} d\Omega = \int_{S_t} \dot{P}_i \delta v_i dS \tag{14}$$

$$\left.\begin{aligned} L^H_{ijkl} &= \frac{1}{|Y|} \int_Y \left[C_{ijkl} - C_{ijpq} \frac{1}{2} \left(\frac{\partial \chi^{kl}_p}{\partial y_q} + \frac{\partial \chi^{kl}_q}{\partial y_p} \right) \right] dY \, , \\ P^H_{ij} &= \frac{1}{|Y|} \int_Y \left[\sum_\alpha R^{(\alpha)}_{ij} \dot{f}^{(\alpha)} - C_{ijkl} \frac{1}{2} \left(\frac{\partial \phi_k}{\partial y_l} + \frac{\partial \phi_l}{\partial y_k} \right) \right] dY \, , \\ \sigma^H_{ij} &= \frac{1}{|Y|} \int_Y \sigma_{mj} \frac{\partial \phi_i}{\partial y_m} dY \, , \; \tau^H_{ijkl} = \frac{1}{|Y|} \int_Y \left(\sigma_{lj} \delta_{ki} - \sigma_{mj} \frac{\partial \chi^{kl}_i}{\partial y_m} \right) dY \, . \end{aligned}\right\} \tag{15}$$

$$\dot{\varepsilon}^0_{ij}(\boldsymbol{x},\boldsymbol{y})=\dot{E}^0_{ij}(\boldsymbol{v})-\frac{1}{2}\left(\frac{\partial\chi^{kl}_i}{\partial y_j}+\frac{\partial\chi^{kl}_j}{\partial y_i}\right)\dot{E}^0_{kl}(\boldsymbol{v})+\frac{1}{2}\left(\frac{\partial\phi_i}{\partial y_j}+\frac{\partial\phi_j}{\partial y_i}\right) \tag{16}$$

$$\overset{\nabla}{S}{}^0_{ij}(\boldsymbol{x},\boldsymbol{y}) = C_{ijkl}\dot{\varepsilon}^0_{kl}(\boldsymbol{x},\boldsymbol{y}) - \sum_\alpha R^{(\alpha)}_{ij}\dot{f}^{(\alpha)}\,. \tag{17}$$

Equations (14)–(17) are the governing equations for the macroscopic and microscopic scales, respectively. The notation $\dot{\boldsymbol{E}}^0$ indicates the macroscopic strain rate and χ^{kl} and ϕ are characteristic deformation functions which satisfy the so-called Y-periodic and are determined by

$$\begin{aligned}\int_Y\left[C_{ijpq}\frac{1}{2}\left(\frac{\partial\chi^{kl}_p}{\partial y_q}+\frac{\partial\chi^{kl}_q}{\partial y_p}\right)+\sigma_{qj}\delta_{pi}\frac{\partial\chi^{kl}_p}{\partial y_q}\right]\frac{\partial\delta v_i}{\partial y_j}dY\\ =\int_Y\left(C_{ijkl}+\sigma_{lj}\delta_{ki}\right)\frac{\partial\delta v_i}{\partial y_j}dY\,, \qquad \chi^{kl}\cdots Y\text{–periodic}\end{aligned} \tag{18}$$

$$\begin{aligned}\int_Y\left[C_{ijkl}\frac{1}{2}\left(\frac{\partial\phi_k}{\partial y_l}+\frac{\partial\phi_l}{\partial y_k}\right)+\sigma_{mj}\frac{\partial\phi_i}{\partial y_m}\right]\frac{\partial\delta v_i}{\partial y_j}dY\\ =\int_Y\sum_\alpha R^{(\alpha)}_{ij}\dot{f}^{(\alpha)}\frac{\partial\delta v_i}{\partial y_j}dY\,. \qquad \phi\ \cdots Y\text{–periodic}\end{aligned} \tag{19}$$

Thus, characteristic functions χ^{kl} and ϕ for the unit cell depend solely on the material characteristics and configuration of microstructure of the unit cell, which are, in turn, obtained without interacting stress and strain of the macroscale. On the other hand, macroscopic governing equation (14) can be solved independently because the macroscopic characteristic deformation functions indicated in Eqs.(15) are identified through Eqs.(18), (19). The details have been discussed in refs.[17,18].

4. COMPUTATIONAL MODEL

Figure 3 shows the computational model for the Ni-based superalloy to be analyzed. Gray and white regions correspond to the γ-matrix and γ'-precipitate, respectively. A cube-shaped γ'-precipitate is periodically embedded in the softer γ-matrix and its size is considered to be constant at 0.5mm. Here, the geometry of the unit cell shown in Fig.3(c) is fully described by two parameters: the cuboidal γ'-precipitate size, l, and the γ-channel width, w. Therefore, the γ'-precipitate volume fraction, V_f, is related to the ratio w/l. The entire unit cell is divided into 864 eight-nodes cubic elements. The crystal orientations [100], [010] and [001] of the γ and γ' phases correspond to the y_1, y_2 and y_3 axes, respectively. The γ and γ' phases are presumed to be elasto-viscoplastic and elastic materials, respectively. The elastic properties estimated on the basis of the experimental results for [001] loading are employed[2,17]. Thereby, the elastic orthotropic parameters for the precipitate and the matrix are $C_{\gamma 11}$=210, $C_{\gamma 12}$=140 and $C_{\gamma 44}$=92, and $C_{\gamma' 11}$=230, $C_{\gamma' 12}$=155 and $C_{\gamma' 44}$=98GPa, respectively. The viscoplastic properties associated with the evolution of dislocation densities are given by choosing the following set of parameters:

$$\dot{\gamma}_{\text{ref}}^{(\alpha)}=10^{-3}\text{s}^{-1},\ \kappa=0.05,\ \tau_0^{(\alpha)}=255.5\text{MPa},\ a=0.2,\ \mu=73\text{GPa},$$
$$\tilde{b}=0.2492\text{nm},\ c=10^{-3},\ L_0^{(\alpha)}=10^{-3}\text{mm},\ \rho_0^{(\alpha)}=10^9\text{m}^{-2},\ n=2/3.$$

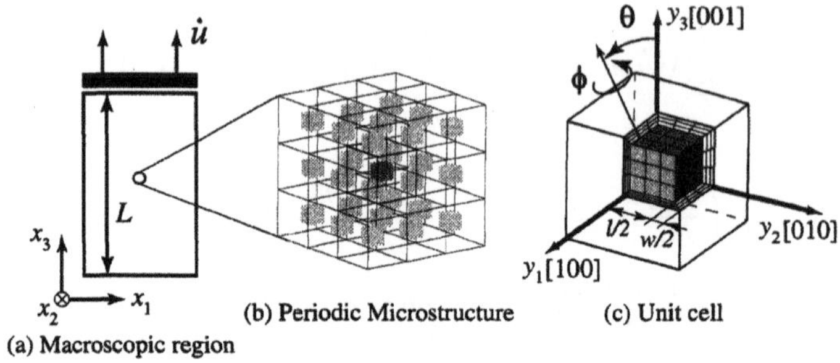

Figure 3. Computational model for Ni-based superalloy.

Table 1. {111}⟨110⟩ slip systems.

1	$(11\bar{1})[101]$	4	$(\bar{1}\bar{1}\bar{1})[\bar{1}01]$	7	$(\bar{1}11)[01\bar{1}]$	10	$(1\bar{1}1)[0\bar{1}\bar{1}]$
2	$(11\bar{1})[011]$	5	$(\bar{1}\bar{1}\bar{1})[0\bar{1}1]$	8	$(\bar{1}11)[\bar{1}0\bar{1}]$	11	$(1\bar{1}1)[10\bar{1}]$
3	$(11\bar{1})[\bar{1}10]$	6	$(\bar{1}\bar{1}\bar{1})[1\bar{1}0]$	9	$(\bar{1}11)[\bar{1}\bar{1}0]$	12	$(1\bar{1}1)[110]$

Then, the twelve (octahedral) slip systems listed in Table 1 are considered. The components of the dislocation interaction matrix $\omega^{\alpha\beta}$, as defined in Fig.4, are also introduced where the diagonal component is unity and the off-diagonal ones are C=1.0, H=1.1 and G=L=K=1.15.

The boundary conditions on the macroscopic scale are, as indicated in Fig.3(a), such that the top and bottom surfaces are shear free with a

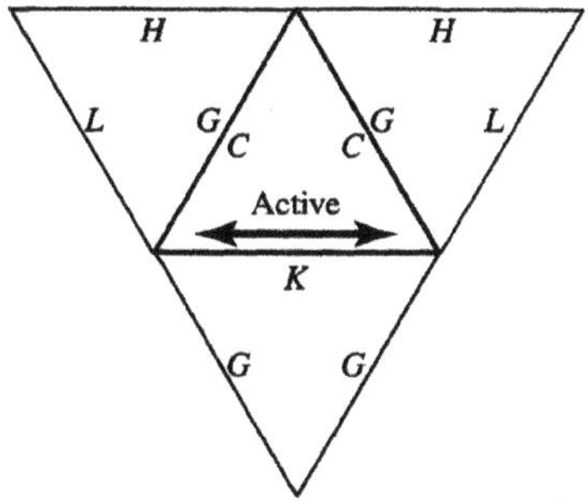

Figure 4. Indices of dislocation interaction matrix $\omega^{\alpha\beta}$ on Thompson's tetrahedron.

constant velocity, $\dot{u}/L=10^{-3}\text{s}^{-1}$, whereas the other surfaces are assumed to be stress free. In order to investigate the deformation behavior under the condition of arbitrary applied stress, the direction of the unit cell is characterized by the angles θ and ϕ corresponding to Euler angles[21] around the y_2-axis and y_3-axis, respectively.

5. RESULTS AND DISCUSSION

Figure 5 shows some results for the homogenized elastic constant $\boldsymbol{E}^{\text{H}}/\boldsymbol{E}^{\text{H}}_{[001]}$. Here $\boldsymbol{E}^{\text{H}}_{[001]}$ indicates the homogenized elastic constant corresponding to $\theta=\phi=0^{\circ}$. In the case of constant θ, the value of $\boldsymbol{E}^{\text{H}}$ is minimum at $\theta=0^{\circ}$ and also becomes stiffer with increasing angle θ. The maximum value of $\boldsymbol{E}^{\text{H}}$ obtained was at $(\theta,\phi)=(54.7^{\circ},45^{\circ})$, namely, the [111] loading direction. This result is consistent with experimentally observed evidence[22]. On the other hand, it could be seen that the effect of different volume fraction of the γ'-precipitate on the macroscopic elastic properties is not remarkable.

Figure 6 shows the average stress $\tilde{\sigma}$ versus average strain u/L at the end of the macroscopic region that undergoes uniform deformation shown in Fig.3(a). An obvious increase of deformation resistance is seen when the volume fraction of the γ'-precipitate is small, i.e., the γ-channel width is narrow. Furthermore, the results indicate a higher strength as compared with the strain-gradient-independent model[18]. With regard to the [110] loading direction, the effect of the different volume fraction of the γ'-precipitate on the macroscopic deformation resistance is considerable. To clarify such a characteristic macroscopic deformation resistance for different directional loading, further investigation associated with the microscopic deformation was performed. In the case of

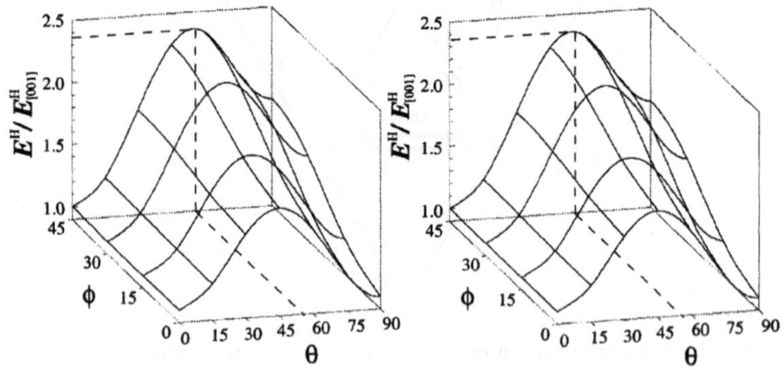

(a) $V_f = 50vol.\%$ $[w/l = 0.260]$ (b) $V_f = 70vol.\%$ $[w/l = 0.128]$

Figure 5. Variation of homogenized elastic constants $\boldsymbol{E}^{\mathrm{H}}$ with θ and ϕ. (a) and (b) correspond to the volume fraction of γ'-precipitates of 50 and $70vol.\%$, respectively.

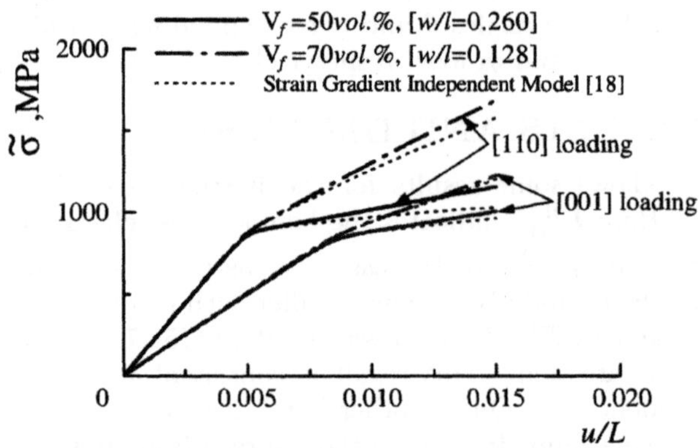

Figure 6. Effects of the different γ'-precipitate volume fraction and applied loading direction on average stress-strain relationship.

[001] loading, eight slip systems, 2, 3, 5, 6, 8, 9, 11 and 12 listed in Table 1, are markedly activated, which is identical to the case of the γ-matrix material without γ'-precipitate[17]. On the other hand, in the case of [110] loading, four slip systems, 5, 6, 11 and 12, are predominantly activated, which contributes to the absorption of the deformation in the microscopic unit cell and results in the high strain gradient that causes the higher resistance to macroscopic deformation, as shown in Fig.6.

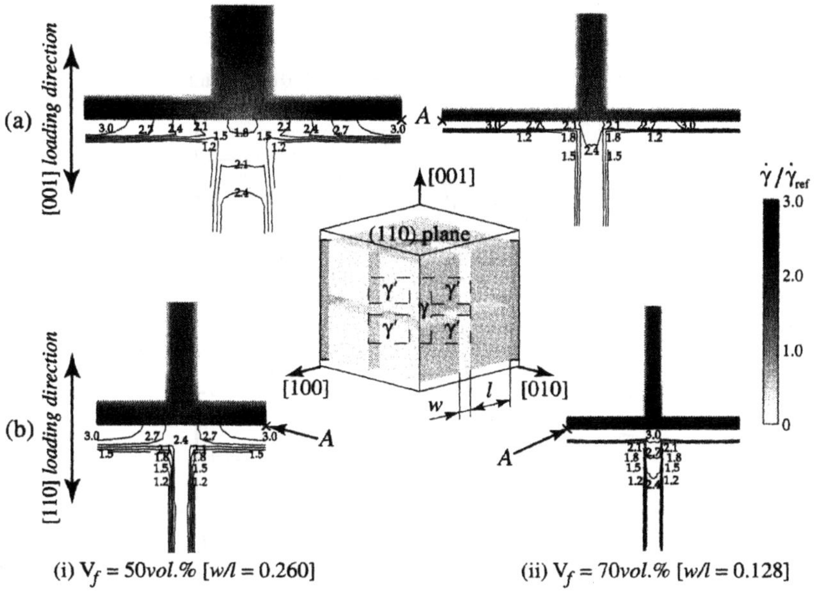

Figure 7. The accumulated plastic shear strain rate normalized by reference shear strain rate $\dot{\gamma}/\dot{\gamma}_{ref}$ distributions in γ-channel for (a) [001] and (b) [110] loading directions and different volume fractions of γ'-precipitates (i) V_f=50*vol.*% [w/l=0.260] and (ii) V_f=70*vol.*% [w/l=0.128] over cross-sectional plane (110) at u/L=0.010.

Figure 7 indicates the accumulated plastic shear strain rate normalized by reference shear strain rate $\dot{\gamma}/\dot{\gamma}_{\text{ref}}$ $(= \sum_{\alpha} \dot{\gamma}^{(\alpha)} / \sum_{\alpha} \dot{\gamma}_{\text{ref}}^{(\alpha)})$ distributions over cross-sectional plane (110) at u/L=0.010. The effect of the applied loading direction and the volume fraction of γ'-precipitates on the shear strain rate distribution is considerable. It can be observed that the high-strain-rate regions concentrate in the vertical (with respect to the applied loading direction) γ-channel, and the high gradient of $\dot{\gamma}/\dot{\gamma}_{\text{ref}}$ associated with the presence of GNDs across the channel arises appears in the vicinity of the γ/γ' interface.

Finally, we will discuss the effect of the variation of dislocation densities related to the microscopic hardening state on the macroscopic deformation behavior of the alloy. The accumulated SSDs and GNDs densities normalized by the total dislocation density are presented in Fig.8, where these dislocation densities are plotted versus average strain u/L at the nodal point 'A' shown in Fig.7. Note that ρ_i (i=T, S, G) is the

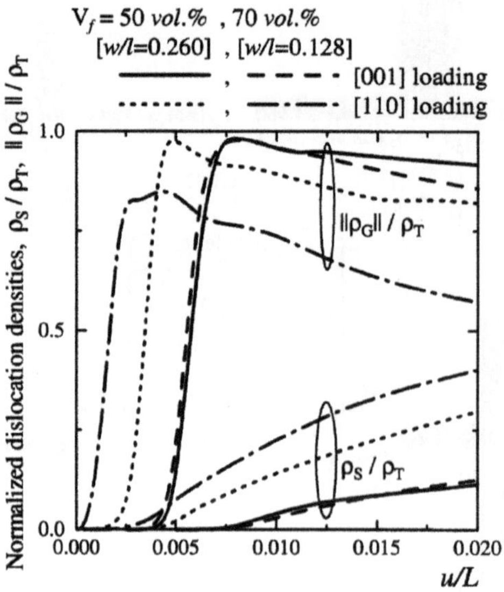

Figure 8. The accumulated SSDs and GNDs at nodal point 'A' in Fig.7 versus average strain u/L.

accumulated sum of slip systems. In this computation, the dislocation mean free path expressed by Eq.(8) is very small since the total dislocation density becomes larger with increasing GNDs density. Therefore, the macroscopic deformation resistance dependent on the local strain hardening state becomes higher with increasing γ'-precipitate volume fraction associated with the γ-channel width. With regard to the case of [110] loading, however, the dislocation mean free path is smaller than that in the case of [001] loading because the density of the SSDs, as well as of the GNDs, contributed to the local strain hardening state, as shown in Fig.6. Hence, the macroscopic characteristic feature of the alloy is high deformation resistance.

6. CONCLUSION

In this study, to elucidate the micro- to macroscopic characteristic feature of Ni-based superalloy with γ'-precipitates, an asymptotic homogenization method employing the constitutive equation based on the strain-gradient crystal plasticity theory was developed. Assuming that the cuboidal γ'-phase precipitates with periodical distribution in the

γ-phase matrix, a unit cell model consisting of two phases was established. Then, a series of computational simulations with a constant γ'-precipitate size was performed for alloys with different volume fractions of γ'-precipitate with respect to the width of the γ-channel and the crystallographic orientations of two phases. Results show that the effect of the γ'-precipitate volume fraction and applied loading direction relative to the crystallographic orientation on the macroscopic deformation behavior is significant compared with the case of the strain-gradient-independent model.

The proposed method can be used for the evaluation of the effects of not only an arbitrary applied loading direction relative to the crystallographic orientation, but also microstructural characteristic length scales, on the macroscopic deformation behavior of two-phase single-crystalline materials. Hence, the present computational strategy may provide useful data for the development of high-performance, high-strength superalloy.

Acknowledgment

Financial support from the Handai Frontier Research Center (FRC) and the Ministry of Education of Japan is gratefully acknowledged.

References

[1] Pope, D. P. and Ezz, S. S. (1984), Mechanical properties of Ni_3Al and nickel-based alloys with high volume fraction of γ', *Int. Metals Rev.*, **29**-3, 136–167.

[2] Higa, Y. (2000), *A study of high-performed and high-strengthened materials with microstructure*, Doctoral dissertation, Kobe University., (in Japanese).

[3] Fedelich, B. (1999), A microstructure based constitutive model for the mechanical behavior at high temperatures of nickel-base single crystal superalloys. *Comput. Mater. Sci.*, **16**, 248–258.

[4] Fedelich, B. (2002), A microstructure model for the monotonic and the cyclic mechanical behavior of single crystals of superalloys at high temperatures. *Int. J. Plasticity*, **18**, 1–49.

[5] Ohashi, T. (1997), Finite-element analysis of plastic slip and evolution of geometrically necessary dislocations in fcc crystals. *Phil Mag. Letters*, **75**-2, 51–57.

[6] Ashby, M. F. (1970), The deformation of plastically non-homogeneous alloys. *Phil. Mag.*, **21**, 399–424.

[7] Busso, E. P., Meissonnier, F. T., O'Dowd, N. P. and Nouailhas, D. (1998), Length scale effects on the geometric softening of precipitated single crystals., *J. de Physique IV*, **8**, 55–61.

[8] Meissonnier, F. T., Busso, E. P. and O'Dowd, N. P. (2001), Finite element implementation of a generalised non-local rate-dependent crystallographic formulation for finite strains. *Int. J. Plasticity*, **17**, 601–640.

[9] Peirce, D., Asaro, R. J. and Needleman, A. (1983), Material rate dependence and localized deformation in crystalline solids. *Acta metall.*, **31**, 1951–1976.

[10] Hutchinson, J. W. (1976), Bounds and self-consistent estimates for creep of polycrystalline materials. *Proc. R. Soc. Lond.*, **A384**, 101–127.

[11] Pan, J. and Rice, J. R. (1983), Rate sensitivity of plastic flow and implications for yield surface vertices. *Int. J. Solids Structs.*, **19**, 973–978.

[12] Ohashi, T. (1987), Computer simulation of non-uniform multiple slip in face centered cubic bicrystals, *Trans. Japan Inst. Metals*, **28**, 906–915.

[13] Franciosi, P., Berveiller, M. and Zaoui, A. (1980), Latent hardening in copper and aluminium single crystals. *Acta metall.*, **28**, 273–283.

[14] Higa, Y., Sawada, Y. and Tomita, Y. (2003), Computational simulation of characteristic length dependent deformation behavior of polycrystalline metals. *Trans. JSME Ser. A*, **69**-679, 523–529. (in Japanese)

[15] Peirce, D., Shih, C. F. and Needleman, A. (1984), A tangent modulus method for rate dependent solids. *Comput. Struct.*, **18**-5, 875–887.

[16] Guedes, J. M. and Kikuchi, N. (1990), Preprocessing and postprocessing for materials based on the homogenization method with adaptive finite element methods, *Comput. Meth. Appl. Mech. Engng.*, **83**, 143–198.

[17] Higa, Y. and Tomita, Y. (1999), Computational prediction of mechanical properties of nickel-based superalloy with gamma prime phase precipitates., In: Ellyin, F. and Provan, J. W. (Eds.), Proc. of 8th Int. Conference on Mech. Behaviour of Materials, *Adv. Mat. & Modeling Mech. Behaviour*, Vol.III, 1095–1099.

[18] Higa, Y. and Tomita, Y. (2000), Computational modeling of deformation behavior of nickel-based superalloy with gamma-prime-phase precipitates., In: Atluri, S. N. and Brust, F. W. (Eds.), Proc. of Int. Confer. Computational Engineering Sciences 2000 (ICES2k), *Adv. in Comput. Engng. Sci.*, Vol.II, 1116–1121., Tech Sci. Press.

[19] Wu, X. and Ohno, N. (1999), A homogenization theory for inelastic behavior of materials with periodic internal structures., In: Bruhns, O. T. and Stein, E. (Eds.), IUTAM Symposium on Micro- and Macrostructural Aspects of Thermoplasticity. Kluwer Academic Pub., 187–196.

[20] Kitagawa, H., Seguchi, Y. and Tomita, Y. (1972), An incremental theory of large strain and large displacement problems and its finite element application., *Ing. Arch.*, **41**, 213–224.

[21] Wei, Y. and Lee, W. B. (1993), *Mesoplasticity and its Applications.*, Springer-Verlag.

[22] Tien, J. K. and Copley, S. M. (1971), The effect of orientation and sense of applied uniaxial stress on the morphology of coherent gamma prime precipitates in stress annealed nickel-based superalloy crystals., *Met. Trans.*, **2**, 543–553.

CRITICAL BEHAVIOUR NEAR THE CRACK / DISLOCATION DEPINNING THRESHOLD

Igor L Maksimov
Nizhny Novgorod University, Department of Theoretical Physics
Faculty of Physics 23 Gagarin Ave. University, Nizhny Novgorod 603000, Russia
ilmaks@phys.unn.ru

Abstract The crack / dislocation depinning problem is studied theoretically on the basis of the exactly solvable Frenkel-Kontorova-type models of the crystal containing microcrack or dislocation. By employing a self-consistent phonon approach the system Gibbs-free energy G is calculated for a finite temperature T at a fixed external force / stress P. It is found that the Gibbs energy $G(T, P, u)$, as a function of the crack-opening/dislocation-core displacement u, suffers a fold-type catastrophe, characterized by the loss of saddle-point configuration at the depinning line $T_c(P)$. A remarkable scaling behavior of the Gibbs-free energy barrier $\Delta G(T, P) \propto [1 - T/T_c]^q$, $(q \simeq 1.5)$ has been discovered for both systems, characterized by the Morse-type atomic interaction potential. A nontrivial set of critical indices, corresponding to the Landau-type mean-field approach is obtained.

Keywords: crack, dislocation, depinning, critical behaviour, critical indices

1. INTRODUCTION

Depinning problem of topological defects (TD) in solid state systems has deserved at present a significant attention. Quite many efforts were made in order to determine the conditions of the so-called crack trapping (or pinning) by the lattice discreteness [1] or to find the dislocation mobility threshold [2]. To predict the conditions at which the crack or individual dislocation depins (i.e. advances or retreats) being driven from the equilibrium by external force and / or thermal fluctuations would be important both from fundamental and applications viewpoint. In majority of previous research this problem is analysed for the over-threshold conditions, at which TD moves in a barrier-less regime. Meanwhile, the study of the system response *below* the depinning threshold (DT) may provide valuable information on the specifics of the topological defect depinning mechanism in nanocrystals.

2. MODELS

Let us describe briefly main characteristics of two models under study, which are generically close to the well-known Frenkel - Kontorova (FK) model.

H. Kitagawa and Y. Shibutani (eds.), IUTAM Symposium on Mesoscopic Dynamics of Fracture Process and Materials Strength, 269–278.

2.1 Tensile Crack: An Atomistic Model

Consider one-dimensional system with a crack-like structure shown in Fig. 1. It consists of a semi-infinite chain of point atoms 'bonded' longitudinally by

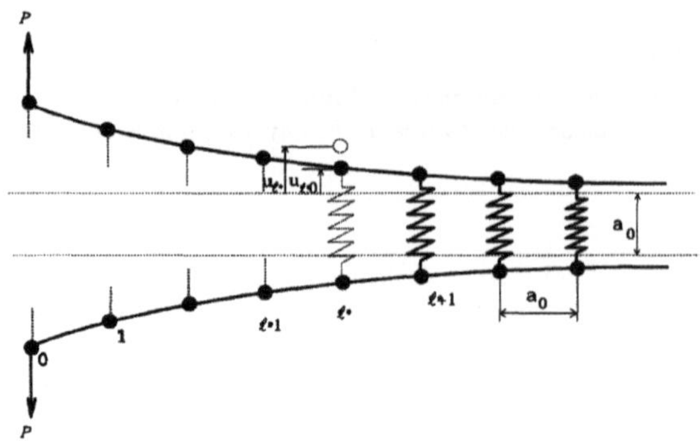

Figure 1. One-dimensional model of a crack

linear bendable elements and transversely (ahead of the crack tip) by linear stretchable elements, with the spacing a_0 and spring constants α and β, respectively. The only element assumed to be in the nonlinear force range is the l-th stretchable bond at the crack tip. Opening forces P are localized at the free ends.

The Hamiltonian of the system may be written as a function of the atoms momentum p_j and of transversal displacement u_j with respect to the chain $(j = 0, 1, \cdots \infty)$

$$H = \left(\frac{1}{2m} \sum_{j=1}^{\infty} p_j^2 \right) \times 2 + U_T(u_l). \tag{1}$$

Total potential energy of the system reads

$$\begin{aligned} U_T &= U_S + U_B(u_l) + 2\beta \sum_{j=l+1}^{\infty} u_j^2 \\ &+ \alpha \sum_{j=1}^{\infty} (u_{j+1} - 2u_j + u_{j-1})^2 - 2Pu_0. \end{aligned} \tag{2}$$

In the expression (2) we include:

1 surface energy $U_S \simeq 2a_0\gamma_S l$, γ_S being the crack-surface tension, supplemented by the contribution of the nonlinear (l-th) bond $U_B(u_l)$; here $U_B(u_j)$ is the standard atomic interaction potential energy,

2 strain energy of stretchable and flexural bonds (third and forth terms in eqn.(2)),
3 potential energy of the external loading system, described by the last term in eqn.(2).

A relatively simple, quadratic in displacement u_j $(j \neq l)$ makes the model (1) analytically tractable, as it will be seen below. With the help of the equilibrium-displacement set $u_{j,0}$, which is expressed through u_l in accordance with the exact zero-temperature solution (see e.g. [3]), one may reduce the total zero-temperature potential energy U_T to an explicit function of the crack-tip opening displacement u_l which is a key variable in our analysis

$$U_T(u_l) = U_B(u_l) + \left[U_S + \beta(\kappa - 1)u_l^2 - 2P(1 + l/\kappa)u_l\right]. \tag{3}$$

The function $U_B(u)$ is characterized by the curvature change from positive (at $u < u_m$) to negative (at $u > u_m$, u_m being the inflexion point of the function $U_B(u)$), which results in the existence of a pronounced maximum in the restoring force-displacement dependence. The analysis of the expression (3) shows that in the force range $P_- < P < P_+$ the function $U_T(u_l)$ possesses two minima separated by one maximum, representing the apparent barrier for the crack advance. Such a bistabilty manifests the existence of the nontrivial phenomenon coined as the crack trapping by the discrete lattice (crack-lattice trapping, CLT) [1]. Simple calculation shows that the barrier height as a function of force P when approaching the upper stability limit $P_c \equiv P_+$ is of power-law type:

$$\Delta U(P) \propto [1 - P/P_c]^{1.5}.$$

It is worth noting that the above power exponent does not depend on the specific shape of the function $U_B(u)$.

2.2 Edge Dislocation

According to the Frenkel - Kontorova model the energy of a dislocation in a field of external force F is written in the form $U = \sum_{j=1}[U_0(x_j) - Fx_j]$ representing a superposition of periodic and elastic components [4]

$$U_0(x_j) = U_0[1 - cos(2\pi x_j/a_0)] + 0.5Q(x_{j+1} - x_j - a_0)^2, \tag{4}$$

here U_0 is the magnitude of the periodic component and Q being an effective stiffness constant. As it is known, two cases are possible, depending on the force F magnitude:

i) Force is not extremely large ($F < F_c$), so that potential energy profile possesses a series of local minima in which a dislocation may be trapped. In this case two regimes of a dislocation propagation are possible: either thermally-activated jumping over the barrier or quantum-mechanical tunneling through the barrier.

ii) The value of force exceeds the critical value $F = F_c$, so that local minima of potential energy vanish. In this case dislocations become free to move as a result of the action of this force thus exhibiting the viscous-drag propagation regime.

At $T = 0$ the equilibrium positions of the dislocation core $x_l^{(\pm)}$, satisfying equation $f = \sin(2\pi x_j/a_0) - 2\pi\gamma\ (x_{j+1} - 2x_j + x_{j+1})/a_0$ in the limit of weak elasticity ($\gamma = Qa^2/(2\pi^2 U_0 \ll 1$; γ being nondimensional elastic constant) are expressed as follows:

$$x_l^{(+)} \approx 0.5a_0 \sin^{-1} f/\pi \quad ; \quad x_l^{(-)} \approx 0.5a_0[\pi - \sin^{-1} f]/\pi\ .$$

Here we introduced a dimensionless force variable $f = a_0F/(2\pi U_0)$, and for simplicity ignored the elastic term corrections. The barrier separating stable and saddle-point configuration takes the form

$$\Delta U(f) = U_0[f(2\sin^{-1} f - \pi) + 2(1 - f^2)^{0.5}]\ .$$

When approaching the limit $f = 1$, corresponding to the zero-temperature critical force at which dislocation depins, the barrier height decreases as $\Delta U \propto [1 - f]^{1.5}$.

Simple comparison of both systems (crack and dislocation) leads to the conclusion that the height of the topological-defect pinning barrier near the depinning threshold decreases in accord with a universal power-law dependence, characterized by the unique power exponent. This circumstance reveals a universal feature of both system response to the force/stress driven depinning transition. Moreover, in virtue of low depinning barrier thermal fluctuations will affect not only the depinning dynamics but the depinning threshold itself, as well. To describe the latter effect quantitatively hereafter we resort to the finite temperature analysis.

3. FINITE TEMPERATURE ANAYSIS

To extend the zero-temperature scheme to the case $T \neq 0$ one should perform statistical averaging with the help of the Gibbs canonical probability function $W\{H\} = Z^{-1} \exp(-H/k_BT)$ where the canonical partition function

$$Z = \int\int \exp(-H/k_BT) \prod_{j=1}^{\infty} dx_j dp_j \tag{5}$$

should be calculated employing the non-local hamiltonian H of the specific system. Unfortunately, such a nonlocality prevents obtaining closed-form results, which makes problem practically untractable. To avoid this circumstance we employ a self-consistent phonon approach suggested by Matsubara [5]. Specifically, we employ an Einstein-type hamiltonian H^0 which is characterized by a quadratic energy dependence upon the displacement $w_j = x_j - x_{j,0}$

from equilibrium atom position $x_{j,0}$ and permits an analytical study of both nonlinear and nonlocal systems.

3.1 Trial Gibbs Free Energy

Being unable to calculate exact Gibbs free energy G of the original system we should resort to a trial one, $G^{\dagger} = G^0 + \left\langle H - H^0 \right\rangle_0$, here the brackets $\langle \ \rangle_0$ mean the statistical average over canonical ensemble described by the Einstein hamiltonian H^0 . In what follows we demonstrate analytical properties of $G^{\dagger}$ of specific systems at finite temperatures.

3.1.1 Crack.

For the case of a crack we employ an Einstein-type hamiltonian H^0 in the form:

$$H^0_{crack} = \left(\frac{1}{2m} \sum_{j=1}^{\infty} p_j^2 \right) \times 2 + +2 \sum_{j=1}^{l-1} \alpha_j w_j^2 + 2 \sum_{j=l}^{\infty} \beta_j w_j^2, \tag{6}$$

where parameters α_j and β_j describe the effective stiffness coefficients of the Einstein oscillators which are characterized by a quadratic energy dependence upon the displacement $w_j = x_j - x_{j,0}$ of the j-th atom from its equilibrium position $x_{j,0}$. After performing standard calculations with the help of the Gaussian-type distribution $W^0 \equiv W\{H^0_{crack}\}$ in accord with the variational method (for details see e.g. [3, 6]) we get the trial Gibbs-free energy $G^{\dagger}_{crack}$. Minimization of $G^{\dagger}_{crack}$ with respect to α_j and β_j brings simple relations:

$$\begin{aligned} \alpha_j &\simeq 3\alpha \quad (1 \le j \le l-1), \\ \beta_j &\simeq \beta + 3\alpha \ \ (l+1 \le j) \end{aligned} \tag{7}$$

for $j \neq l$, and nonlinear integral equation to find nonlinear stiffness β_l

$$\beta_l + \int_{-\infty}^{\infty} q^2 U(q) \exp\left[iq(2u_{l,0} + a_0)\right] \exp\left[-\frac{q^2 kT}{2\beta_l}\right] dq = 3\alpha. \tag{8}$$

After numerically solving Eq. (8), one finds the function $\beta_l(u_l)$, and finally the expression for the $G^{\dagger}_{crack}$ in the form

$$G^{\dagger}_{crack} \approx G^0 + 2l\gamma_S(T,P)\, a_0 + \langle U_B\{u_l\}\rangle_0\,, \tag{9}$$

where second term represents the crack surface energy (see [3]). From the condition $\partial G^{\dagger}_{crack}/\partial u_l = 0$ one finds both $G^{\dagger\, min}_{crack}$ and $G^{\dagger\, max}_{crack}$ and thus the depinning barrier height $\Delta G_{crack} = G^{\dagger\, max}_{crack} - G^{\dagger\, min}_{crack}$.

While solving the minimization equations above we discovered that physically meaningful (i.e. positive) solution for β_l exists only within a certain temperature (or force) interval $T < T_c(P)$ (or $P < P_c(T)$) ; for $T > T_c(P)$ (or $P > P_c(T)$) no solution is available for a fixed-length crack configuration (for concreteness, in our calculations we selected $l = 20$). Physically $T_c(P)$ means the lattice softening temperature beyond which system containing a larger crack-like defect with $l > 20$, is energetically more favourable. A remarkable scaling behavior of $\Delta G_{crack}(T, P)$

$$\Delta G_{crack}(T, P) \propto [T_c(P) - T]^q \propto [P_c(T) - P]^q \,, \quad (q \simeq 1.5) \tag{10}$$

has been discovered for the tensile crack near the depinning threshold (see also [6]).

3.1.2 Dislocation.

The expression for the H^0_{disl} in the case of a FK model is more simple, and is characterized by a set of arbitrary elastic constants k_j

$$H^0_{disl} = \sum_{j=1}^{\infty} \left[p_j^2/(2m) + 0.5 k_j w_j^2 \right] \tag{11}$$

to be defined from variational procedure. Trial Gibbs energy of a dislocation $G^{\dagger}_{disl}$ takes the form

$$\begin{aligned} G^{\dagger}_{disl} &= G^0_1 + U_0[1 - cos(2\pi x_l/a)] \exp[-2\pi^2 k_B T/(a^2 k_l)] \\ &\quad +0.5Q[x_l^2 + k_B T/k_l] - F x_l + 0.5 k_B T \ln k_l, \end{aligned} \tag{12}$$

where $x = x_l$ defines the dislocation-core location and k_l denotes effective stiffness. From the minimization conditions $\partial G^{\dagger}_{disl}/\partial k_l = 0$ and $\partial G^{\dagger}_{disl}/\partial x_l = 0$ one finds both the dislocation depinning barrier height $\Delta G_{disl} = G^{\dagger\, max}_{disl} - G^{\dagger\, min}_{disl}$ and the depinning threshold.

For the most simple case $Q = 0$ (negligible interatomic stiffness) the dependence of a reduced force $f_c(\tau)$ is described parametrically as

$$\tau = \cos^3 \xi \exp(-\cos^2 \xi) \; ; \; f = \sin \xi \exp(-\cos^2 \xi) \tag{13}$$

by means of a non-dimensional parameter ξ $(0 \le \xi \le \pi/2)$; here $\tau = k_B T/(2U_0)$. Asymptotical expressions for the function $f_c(\tau)$ are as follows: $f_c(\tau) \approx 1 - (\tau)^{2/3}$, for $\tau \to 0$, and $f_c(\tau) \approx [2\tau_c(\tau_c - \tau)]^{1/2}$, for $\tau \to \tau_c$, where $\tau_c = \exp(-1) \approx 0.37$.

For the case $Q \neq 0$ analytical results are not available; nevertheless the series of curves $f_c(\tau)$, obtained numerically for different values of a non-dimensional parameter $\gamma = Qa^2/(2\pi^2 U_0)$ is presented in Fig. 2.

In the reduced units $\tilde{f} = f/f_c(0, \gamma)$, $\tilde{\tau} = \tau/\tau_c(0, \gamma)$ all curves coincide in one universal dependence $\tilde{f}(\tilde{\tau})$ shown in Fig. 3.

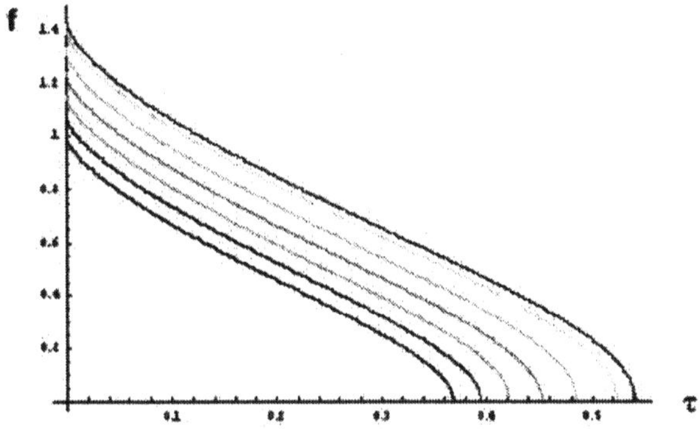

Figure 2. Temperature dependence of a dislocation mobility threshold; 1- $\gamma = 0$; 2 - $\gamma = 0.05$; 3 - $\gamma = 0.1$; 4 - $\gamma = 0.15$; 5 - $\gamma = 0.2$; 6 - $\gamma = 0.25$; 7 - $\gamma = 0.27$;

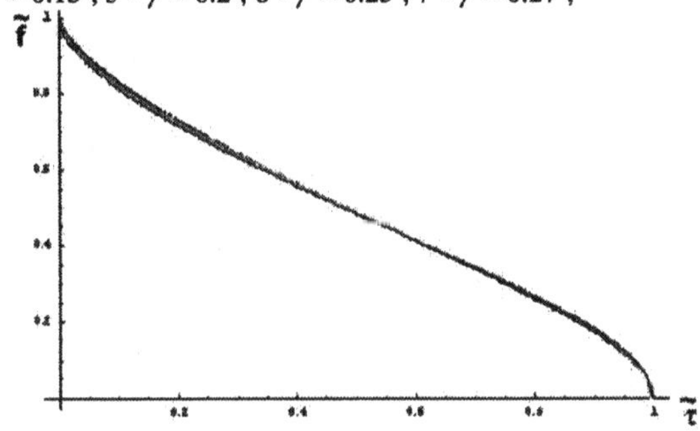

Figure 3. Universal temperature dependence of a dislocation mobility threshold

As regards barrier behaviour we find the following asymptotical expressions for ΔG_{disl}

$$\Delta G_{disl}(f, \tau) \propto [f_c(\tau) - f]^{1.5}. \tag{14}$$

4. DISCUSSION

The results obtained above demonstrate several important features of both systems response to the effect of external force / temperature:

i) the change of the Gibbs-free energy (GFE) profile at the depinning threshold,

ii) the existence of the universal power exponent $q = 1.5$ (see Eq.(10) for the crack case and Eq.(14) for the dislocation case) governing force / temperature dependence of the depinning barrier.

Let discuss peculiarities of the GFE in more detail. To be specific, hereafter we refer to the crack problem mainly.

As it was revealed earlier, below the depinning line (DL) ($T < T_c(P)$) the GFE $G(P, T, u)$ as a function of the displacement u possesses two minima separated by the barrier. On the contrary, above DL ($T > T_c(P)$) the barrier disappears, which ultimately drives the system from metastable state to a more stable one. Thus, the topography of the Gibbs-free energy $G(P, T, u)$ in the temperature(force)- displacement space suffers a dramatic change, which, according to the Arnold-Thom scheme [7], belongs to the fold-type catastrophe class.

The universal behaviour of the Gibbs-free energy near DL determines quite nontrivial system response to both temperature and force. For example, Figs. 4 and 5 demonstrate the behaviour of the mechanical "order parameter" defined as the width of the equilibrium displacement region $\Delta u(T, P) = u_+ - u_-$, which is constructed according to the conventional Maxwell suggestion, between the lowest u_- and the highest u_+ equilibrium displacement of the nonlinear bond. Square-root scaling of $\Delta u(T, P)$: $\Delta u(T, P) \propto [1 - T/T_c]^{0.5}$ (see Fig. 4), and $\Delta u(T, P) \propto [1 - P/P_c]^{0.5}$ (see Fig. 5), revealed by means of data fitting near depinning threshold, suggests the validity of the Landau-type mean-field relations. However, the differential compliance (DC) of the trapped crack, being second derivative of the GFE over force P, *diverges* near the depinning threshold (see Fig. 6):

$$DC \equiv \mid \partial \Delta u(T, P)/\partial P \mid \propto [1 - P/P_c]^{-0.5}. \tag{15}$$

Similar behaviour, albeit with respect to temperature, demonstrates heat capacity c_P:

$$c_P(T) \equiv -T\partial^2 G(T, P)/\partial T^2 \propto [1 - T/T_c]^{-0.5}. \tag{16}$$

The occurrence of a square-root singularity of both response functions (15) and (16), signifies extremely high system sensitivity to external force / temperature near the depinning threshold. Such a diverging susceptibility is a generical feature of the so-called structural transitions during which the system configuration suffers a dramatical change. Indeed, when $P > P_c(T)$ ($T > T_c(P)$) crack jumps forward by one lattice unit in accordance with the analysis of Section 2.1. The latter event represents an abrupt change of the crack length and, consequently, of the crack structure. Similar conclusions may be formulated with respect to the dislocation response characteristics near depinning threshold, as well.

The reported above results demonstrate universal behavior of both the order parameters (equilibrium displacement, equilibrium entropy) and the response functions (such as heat capacity and differential compliance) of two apparently different systems (crack, dislocation) near depinning line. Employing standard

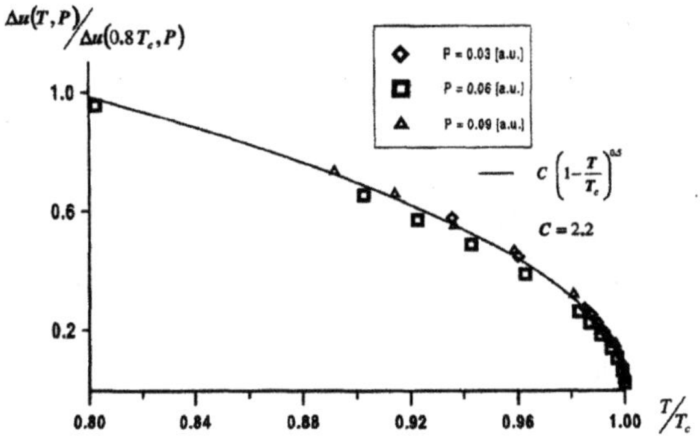

Figure 4. Temperature-dependent mechanical "order parameter"

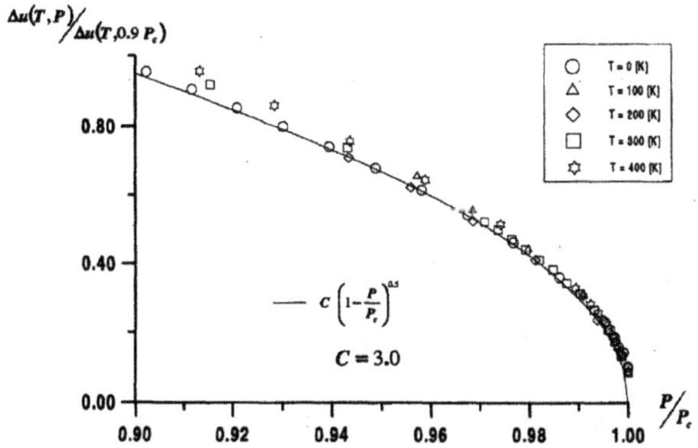

Figure 5. Force-dependent mechanical "order parameter"

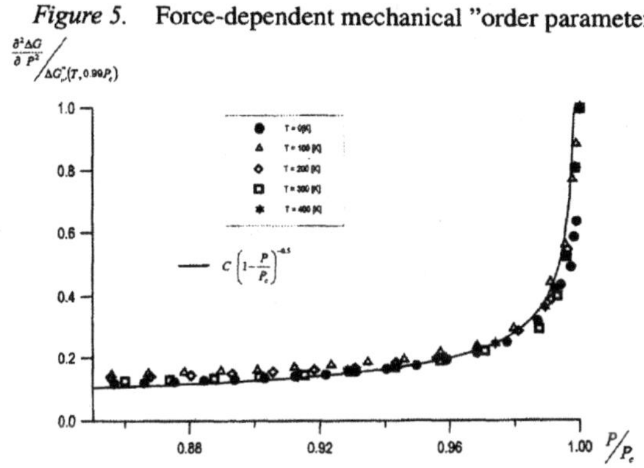

Figure 6. Divergency of differential compliance

identification scheme effective in the equilibrium phase transition theory [8] one derives a nontrivial set of critical indices

$$\alpha = \beta = \gamma = \epsilon = 0.5 \; ; \; \delta = 2, \tag{17}$$

which describes critical behaviour of an individual topological defect near depinning threshold. We expect that this set of critical indices is generical for a wide class of different physical systems, which are distinguished by the presence of a definite process-initiation (depinning) threshold. Among the candidate processes we could name, for example, the dislocation emission from the crack tip or the indentation dislocation emergence.

Acknowledgments

Author is obliged to Prof. H. Kitagawa and Prof. A. Nakatani for helpful discussions and hospitality during his stay at Osaka University. I wish to thank Mr. K. Kitamura and Mr. A. Telezhnikov for assistance in numerical calculations. This work was in part supported by the Russian Foundation for Basic Research (Grant N 02-02-16526).

References

[1] Thomson, R.M Physics of fracture. In Seitz, F., Turnbull, D., editors, *Solid State Physics*, Vol. 39, pages 1 – 129, Academic Press, New York, 1986.

[2] Nabarro, F.R.N. Theory of Crystal Dislocations. Oxford University Press, New York, 1967.

[3] Maksimov, I. L., K. Kitamura, and K. Nishioka Temperature-dependent crack surface tension. *Phil. Mag. Lett.*, 81:547, 2001.

[4] Hirth, J. P., and J. Lothe Theory of Dislocations. Wiley, New York, 1982.

[5] Matsubara, T., Self-consistent Einstein method and surface Debye temperature. *J. Phys. Soc. Japan*, 40:603, 1976.

[6] Kitamura, K., I. L. Maksimov, and K. Nishioka Temperature effect on crack-lattice trapping. *Phil. Mag. Lett.*, 75: 343, 1997.

[7] Gilmore, R. Catastrophe Theory for Scientists and Engineers. Wiley, New York, 1981.

[8] Stanley, H. E. Introduction to Phase Transitions and Critical Phenomena. Clarendon Press, Oxford, 1971.

TOPOLOGICAL SOLITON DYNAMICS IN MEDIA WITH MICROSCOPIC ROTATIONS

Sergey V. Dmitriev*, Nobuhiro Yoshikawa** and Aleksey A. Vasiliev***

*National Institute of Materials Science, Tsukuba, Japan
**Institute of Industrial Science, The University of Tokyo, Tokyo, Japan
***Department of Mathematical Modeling, Tver State University, Tver, Russia

Abstract. Microscopic rotations can play a very important role in the molecular crystals or in some dielectric crystals where atoms are joined in comparatively rigid clusters and the inter-cluster interactions are comparatively weak. The range of unusual phenomena observed in such materials is fascinating, from incommensurate phase to negative Poisson ratio. Here we focus on the topological soliton dynamics in such materials and describe a number of effects, such as the trigger off and propagation of autowaves, soliton multiplication in the metastable media, soliton collisions and annihilations. All these phenomena are addressed in frame of the elastically hinged molecule (EHM) model having microscopic particles with rotational degrees of freedom. For this model we construct various continuum approximations. New important features of the continuum models for media with microscopic rotations are the increase in number of equations of motion due to the increase in number of degrees of freedom and the appearance of the higher gradient terms in the resulting equations. We also construct the multi-field continuum approximation capable of description not only long but also short waves.

Keywords. Finite size particles, microscopic rotations, discrete model, continuum model, topological solitons.

1. INTRODUCTION

The present paper is concerned with a class of crystalline materials for which, in many practical problems, the constituting microscopic units (atomic clusters) cannot be reduced to point-wise particles, e.g., molecular crystals, liquid crystals, some dielectric crystals like silica polymorphs (SiO_2), etc. In the simplest models for such crystals, the rigid finite size microscopic particles have not only translational but also rotational degrees of freedom

H. Kitagawa and Y. Shibutani (eds.), IUTAM Symposium on Mesoscopic Dynamics of Fracture Process and Materials Strength, 279–288.

and they interact via central forces and torques. We focus on the peculiarities of topological soliton dynamics in such materials.

First successful attempt to describe some basic properties of a dislocation in frame of 1D model has been done by Frenkel and Kontorova (see the review [1]). Their model describes a chain of point-wise particles elastically coupled to their neighbours and subjected to a periodic on-site potential. In the continuum limit, the Frenkel-Kontorova model reduces to the exactly integrable sine-Gordon equation for which the exact solutions of the form of topological solitons (kinks) and dynamical solitons (breathers) are known. It is true that some important effects, e.g., Peierls-Nabarro potential, excitation of soliton internal modes, energy exchange in soliton collisions, and radiation are lost in the continuum approximation, but the exact solutions give a very useful hint to the dislocation dynamics in the discrete model [1]. There exist a number of similar discrete models, the one closely related to the present discussion is the discrete φ^4 model with the next-nearest interaction, which can also be presented in the form of the DIFFOUR model [2]. The second-neighbour bonding gives rise to the high-order gradient terms in the continuum analogue of this model. These terms are introduced in the phenomenological gradient elasticity which brings the singularity-free dislocation solutions [3]. To study the soliton dynamics in crystals with finite size particles, the elastically hinged molecule (EHM) model has been proposed by the present authors [4,5]. In its simplest 1D version, the EHM model was shown to be mathematically identical to the φ^4 model with the next-nearest interaction and thus, to the DIFFOUR model. In continuum approximations constructed for the EHM model the higher gradient terms also appear though the physical meaning of these terms is different. It appears that the EHM model is closely related to the Cosserat-type generalized continua.

The 1D discrete model we consider here has been extensively studied by Slot and Janssen [2], Hlinka, Orihara, and Ishibashi [6], and others (see our recent review [7]). In these works, the constant amplitude approximation (CAA) has been employed to obtain the solitary wave solutions propagating in different commensurate periodic phases. Later we have demonstrated the importance of the so-called multi-field approximations [8,9] in constructing the soliton solutions [5,10].

The goal of this paper is to demonstrate that the media with microscopic rotations, simulated here by the EHM model, shows a number of new interesting effects in soliton dynamics.

In Section 2, the 1D EHM model is described and then, the properties of the topological solitons and the autowave dynamics are discussed in Sec. 3. Section 4 concludes the paper.

2. EHM MODEL OF CRYSTAL

The 1D EHM model (see Fig. 1) has the following equations of motion [4]

$$\ddot{u}_n + F(u_{n-2} - 4u_{n-1} + 6u_n - 4u_{n+1} + u_{n+2}) \\ + P(u_{n-1} - 2u_n + u_{n+1}) + u_n + u_n^3 = 0. \tag{1}$$

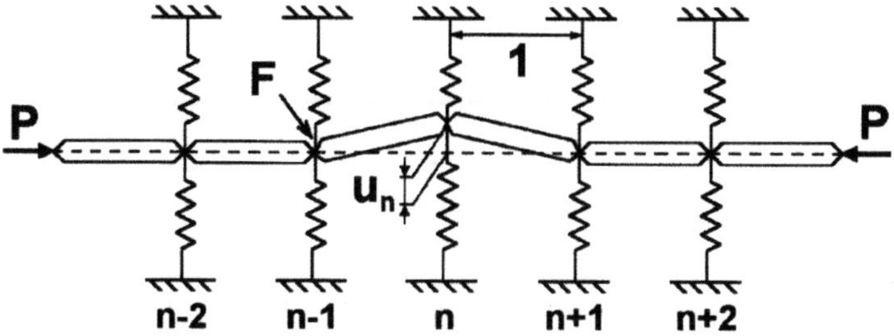

Fig. 1. One-dimensional EHM model Rigid particles of unit length and unit mass are connected in a chain by the elastic hinges with the elastic constant F and subjected to the compressive force P. The fourth-order polynomial background potential is shown by the vertical nonlinear springs. u_n is the transversal displacement of nth hinge.

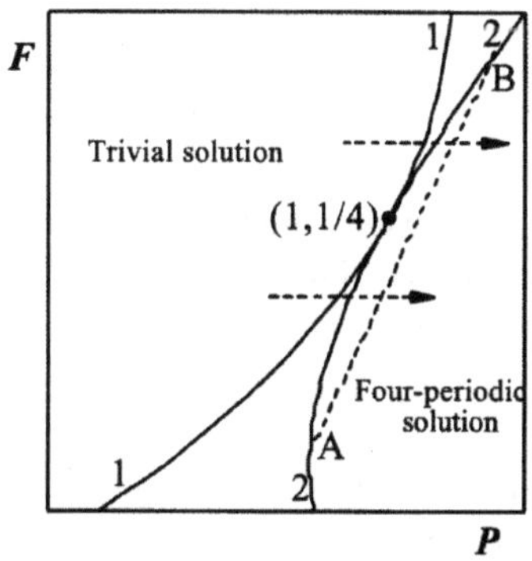

Figure 2. Phase diagram of EHM model in the vicinity of the point $(P,F)=(1,1/4)$ (schematic). To the left of curve 1 ($4F>P^2$) the trivial solution is stable. To the right of curve 2 defined by Eq. (3) the four-periodic solution is stable.

The model describes a chain of rigid particles of unit length and unit mass connected to each other by the elastic hinges with the elastic constant F and subjected to the compressive force P, acting along the chain. Hinges are in the fourth-order polynomial background potential. Each hinge has one degree of freedom, transversal displacement u_n, assuming that $u_n \ll 1$. The role of the elastic hinges is to keep the chain as a straight line because, if the angle between axis of two neighboring molecules is nor zero, the hinge produces a moment which tends to decrease the angle. On the other hand, the external force P plays a destructive role. The competition between F and P gives rise to modulational instability in the model. The background potential is to simulate the interaction of the considered molecule chain with the rest part of the crystal.

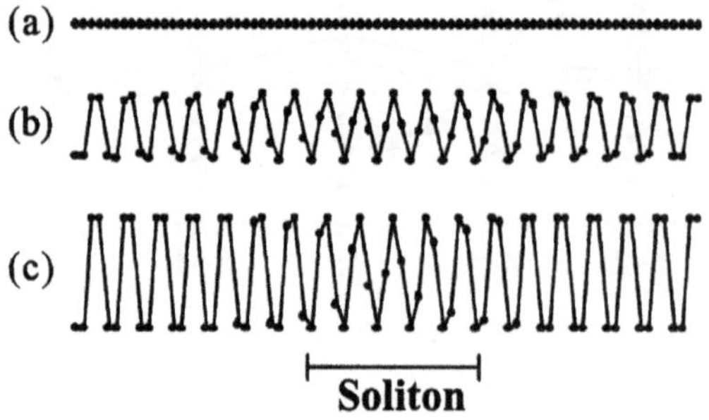

Figure 3. Transformation of the structure upon increase of P at a constant F close to 1/4. (a) External force P is small and the trivial solution is stable. (b) Modulational instability with the wave vector close to 1/4. Amplitude of the modulation wave (one period is shown) increases with increase in P. (c) For a large P, the periodic set of topological solitons separating the domains of four-periodic structure is formed.

Equation (1) supports various modulated structures [10]. Here, for the sake of illustration, we discuss the topological solitons in the four-periodic structure,

$$u_n = A\cos[\pi(n+m)/2+\pi/4], \qquad A=\sqrt{4P-8F-2}, \tag{2}$$

where $m=\{0,1,2,3\}$.

The four-periodic solution Eq. (2) is stable when

$$144F^2+3P^2-72FP+24F<0. \tag{3}$$

Phase diagram of the EHM model is schematically shown in Fig. 2. To the left of curve 1 ($4F>P^2$) the trivial solution is stable. To the right of curve 2, defined by Eq. (3), the four-periodic solution is stable. The curve 1 is tangent to the curve 2 at the point $(P,F)=(1,1/4)$. If one starts from the region of stability of the trivial solution and moves along a line $F=const$ to the right then the structural transformations presented in Fig. 3 will be observed. If F is exactly equal to 1/4, the trivial solution will be transformed into the four-periodic one as the result of modulational instability. In a general situation, when F is close but not equal to 1/4 (dashed arrows in Fig. 2), the modulated phase with the wave-vector $k\sim1/4$ will be formed upon crossing curve 1 (see Fig. 3b) since there is a gap between stability regions of the trivial and four-periodic solutions. The amplitude of the modulated structure increases with increase in P and, upon crossing curve 2, the domains of the four-periodic structure separated by the spatially localized topological solitons appear (Fig. 3c). The larger is P the narrower are the solitons. The chain of structural transformations presented in Fig. 3 simulates the phase transition into incommensurate phase [2,4,5,10].

3. TOPOLOGICAL SOLITONS IN EHM MODEL

With the help of CAA, the problem of the modulated four-periodic structure can be reduced to the sine-Gorgon equation with respect to the slowly changing phase [10]. Then the soliton depicted in Fig. 3c can be expressed in the following form

$$u_n(t) = A\cos\left\{\frac{\pi}{2}(n+m) + \frac{\pi}{4} + \arctan\exp\left[S(n - x_0 - ct)\right]\right\}, \qquad (4)$$

where $m=\{0,1,2,3\}$, c is the velocity of the soliton, and the inverse width is

$$S = \pm\frac{A}{\sqrt{4F - c^2}}. \qquad (5)$$

Note that S can be positive or negative. Positive S corresponds to an extended soliton (kink) and negative S to a compressed one (antikink). Kink is obtained by introducing one extra node (hinge) in the ideal four-periodic structure while antikink by removal of a node. Integer parameter m specifies the choice of a node to add/remove from a period.

Potential energy of the soliton is

$$U = \pm\frac{A^4}{8S} \pm \left[\pi(4F - P) + 2FS\right]\frac{A^2}{4}. \qquad (6)$$

From Eq. (4) it is clear that the soliton solution is just a phase kink. An unusual point is that the potential energy of the soliton can be positive or negative depending on P, F, and on the sign of S. This is the consequence of the fact that the four-periodic structure carrying the soliton can be metastable. Negative energy soliton can be regarded as a nucleus of a lower energy structure in the four-periodic structure. This point is important for the discussion in below.

There exists an alternative approach to describe the soliton in the four-periodic structure. Introducing $v_n = u_{4n} = -u_{4n+2}$ and $w_n = u_{4n+1} = -u_{4n+3}$, the following equations of motion for the two nearest hinges can be derived from Eq. (1)

$$\begin{aligned} &\ddot{v}_n - F(v_{n-2} - 4w_{n-1} - 6v_n + 4w_{n+1} + v_{n+2}) \\ &- P(w_{n-1} + 2v_n - w_{n+1}) + v_n + v_n^3 = 0, \end{aligned} \qquad (7)$$

$$\begin{aligned} &\ddot{w}_n - F(w_{n-2} + 4v_{n-1} - 6w_n - 4v_{n+1} + w_{n+2}) \\ &+ P(v_{n-1} - 2w_n - v_{n+1}) + w_n + w_n^3 = 0. \end{aligned} \qquad (8)$$

If v_n and w_n vary slowly with n, one can consider the two unknown fields, $v(x,t)$, $w(x,t)$, and write the following multi-field continuum approximation

$$3v_{tt}+(P-4F)w_{xxx}-12Fv_{xx}+6(P-4F)w_x \\ +3(1-2P+4F)v+3v^3=0, \tag{9}$$

$$3w_{tt}-(P-4F)v_{xxx}-12Fw_{xx}-6(P-4F)v_x \\ +3(1-2P+4F)w+3w^3=0. \tag{10}$$

If the terms v_{xxx} and w_{xxx} are neglected in Eq. (9) and Eq. (10) then the following multi-field soliton solution can be derived [5]

$$u_{4n}=\pm\frac{A}{\sqrt{2}}-\frac{S}{A\sqrt{2}}(P-4F)\cosh^{-2}\left[\frac{S}{2}(n+x_0-ct)\right],$$
$$u_{4n+1}=\frac{A}{\sqrt{2}}\tanh\left[\frac{S}{2}(n+x_0-ct)\right], \tag{11}$$
$$u_{4n+2}=-u_{4n}, \qquad u_{4n+3}=-u_{4n+1}.$$

There exist eight different solitons. To describe the four kinks, it is necessary to choose sign (+) in the first line of Eq. (11) and to shift the indices by unity sequentially. Four antikinks correspond to sing (-). In the solution Eq. (4) and Eq. (5), a particular soliton is chosen by m and particular sign of S.

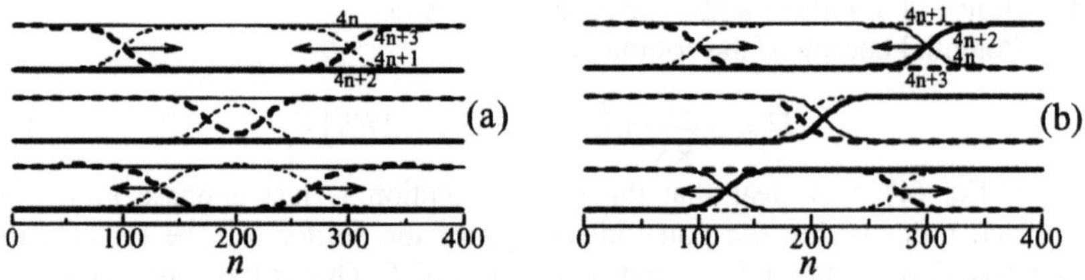

Figure 4. Collision of solitons moving in the same (a) and different (b) sublattices. Since $P=4F$, odd nodes do not interact with even nodes. In (a), colliding solitons can annihilate in principle but in (b) they cannot.

One can see that, compared to the sine-Gordon kinks, the topological solitons in systems supporting periodic structures show a greater variety.

Different properties of the solitons become very clear in their collisions. In Fig. 4, for example, we compare collisions between solitons moving in the same (a) and different (b) sublattices. In this and in the following figures, nodes with numbers $4n$ and $4n+2$ are connected by thin and thick solid lines while nodes $4n+1$ and $4n+3$ by thin and thick dashed lines, respectively. We set $F=0.251$ and $P=4F$. Note that at $P=4F$, Eqs. (7), (8) and consequently Eqs. (9), (10) become independent, that is odd and even hinges move independently. In Fig. 4a, collision is similar to the kink-antikink collision in sine-Gordon equation, but in Fig. 4b, solitons pass through each other even without phase shift. If $P?4F$ then some interaction between solitons mowing in different sublattices appears but still there is a qualitative difference compared to the case of solitons moving in the same sublattices because in the former case solitons cannot annihilate in principle but in the latter case they can.

In Fig. 5 we show how the shape of compressed soliton (antikink) changes if one moves along the line $P=2F+0.502$, schematically shown by the dashed line AB in Fig. 2. In (a), parameters P and F are chosen to be close to the point B, in (b) $P=4F$, i.e., somewhere in the middle between points A and B, and in (c), they are close to the point A. One can see that the shape of the soliton changes from the convex to the concave. For the extended solitons (kink) the evolution of the shape is opposite, i.e., near the point B it is concave and near the point A convex. It is worth noting that CAA solution Eq. (4) cannot reproduce the concave shape of the soliton and in this case the multi-field solution Eq. (11) gives a much better result [10]. The terms v_{xxx} and w_{xxx} in Eqs. (9), (10) were neglected in deriving the soliton profile Eq. (11), but they become important for convex solitons.

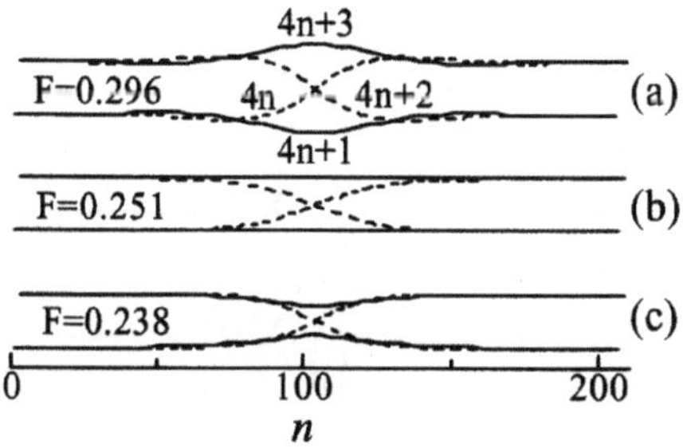

Figure 5. Displacements of hinges near compressive soliton (antikink) at different points of the dashed line AB, $P=2F+0.502$, shown in Fig. 2. (a) Structure close to the point B, (b) structure at P=4F, i.e., somewhere in the middle between points A and B, and (c) structure close to the point A.

Note that we cannot approach the points A and B very closely because the carrying four-periodic phase itself becomes unstable at these points. The solitons play a role of imperfections and their instability takes place before the collapse of the ideal four-periodic structure.

In Fig. 6 and Fig. 7, we present the transformation of unstable concave and convex solitons respectively. The results were obtained by numerical integration of Eq. (1) with the initial conditions specified from Eq. (11). We also introduced a viscosity term to damp the radiation emitted in course of structure transformations. Parameters here are chosen in a way that the four-periodic structure is still stable but the solitons are not.

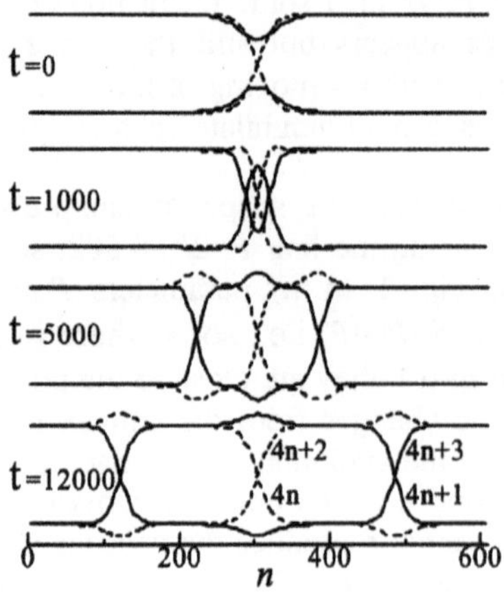

Figure 6. Transformation of an unstable concave soliton into three stable convex solitons.

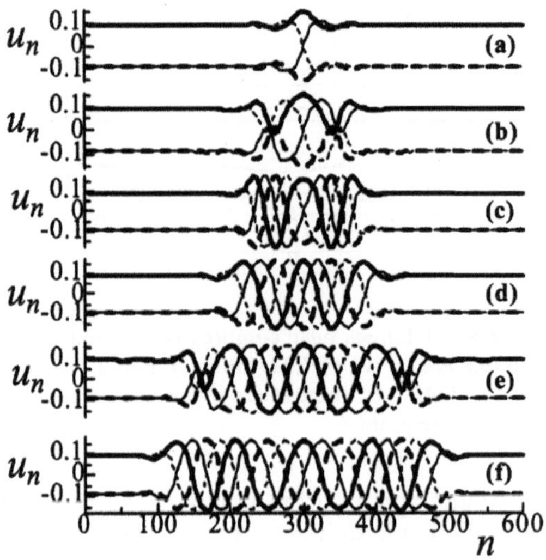

Figure 7. Toy model for the martensitic transformation. An unstable convex soliton triggers off a pair of autowaves propagating in opposite directions and transforming the metastable four-periodic structure into a lower energy structure.

The concave soliton in Fig. 6 transforms into three convex mutually repulsive solitons. Even more interesting is the structural transformation due to the instability of a convex soliton in Fig. 7. Here energy of the soliton is negative which means that the four-periodic structure is metastable and a soliton plays a role of nucleus of a lower energy phase. When soliton becomes unstable, it triggers off a pair of autowaves propagating in opposite directions and transforming the metastable four-periodic structure into a lower energy structure. The autowaves move with a constant velocity in the presence of the viscosity term and the driving force for the motion is the energy released in transformation of the four-period structure into a lower energy structure. This model can be used as a toy model for the martensitic transformation.

4. DISCUSSION AND CONCLUSIONS

In this paper, the peculiarities of structure and dynamics of topological solitons in crystalline media with microscopic particles having rotational degrees of freedom were discussed in frame of the one-dimensional EHM model.

Finite size particles with rotational degrees of freedom may favour the appearance of modulated structures and the latter may support a variety of topological defects with a rich phenomenology. Our simple analysis has revealed the following new effects.

- In the four-periodic structure, the topological solitons may propagate in the same sublattice or in different sublattices. In the former case they can annihilate when colliding while in the letter case they cannot annihilate in principle.
- There exists a line in the space of model parameters, $P=4F$, where the crystal splits into two embedded sub-crystals with no interaction between them. Any dynamics of one of the sub-crystal has no effect on the other one.
- An unstable soliton with positive energy (convex soliton in our example) can generate a few stable solitons with opposite topological charge.
- An unstable soliton with negative energy (concave soliton in our example) can initiate a dynamic structural transformation by means of propagating autowaves. This transformation simulates the martensitic phase transition.
- The EHM model reproduces incommensurate structural transformations.

Continuum analogues for the model with rotational degrees of freedom contain the high gradient terms due to the existence of extra degrees of freedom. Similar terms are introduced in the phenomenological gradient theories or they can appear under the assumption of non-locality.

For the sake of brevity and simplicity, here we restricted ourselves to the one-dimensional problem and only to the four-periodic structure. Treatment of the two-dimensional variants of the EHM model can be found in [11,12]. The CAA soliton solutions for various modulated structures have been given in [10,13]. Propagation of autowaves in two- and three-periodic structures has also been simulated [5,14].

REFERENCES

1. Braun O. M., Kivshar Yu. S. Nonlinear dynamics of the Frenkel-Kontorova model. Phys. Rep. 1998; 306:1-108.
2. Slot J. J. M., Janssen T. Dynamics of kinks in modulated crystals. Physica D 1988; 32:27-71.
3. Aifantis E. C. Update on a class of gradient theories. Mech. Mater. 2003; 35: 259-280.
4. Dmitriev S. V., Abe K., Shigenari T. One-dimensional crystal model for incommensurate phase. I. Small displacement limit. J. Phys. Soc. Jpn 1996; 65:3938-3944.
5. Dmitriev S. V., Shigenari T., Vasiliev A. A., Abe K. Dynamics of domain walls in an incommensurate phase near the lock-in transition: One-dimensional crystal model. Phys. Rev. B 1997; 55:8155-8164.
6. Hlinka J., Orihara H., Ishibashi Y. Rippled Commensurate Phases in DIFFOUR Model: Continuum Approximation. J. Phys. Soc. Jpn 1998; 67:3488-3492.
7. Dmitriev S. V., Vasiliev A. A., Yoshikawa N. "Microscopic rotational degrees of freedom in solid state physics." In *Recent Research Developments in Physics*, (in press).
8. Vasiliev A. A. Multiple-field approach in modelling stability of a cylindrical shell stiffened with frames at external pressure. Moscow University Mechanics Bulletin 1994; 49,4:71-74.
9. Vasiliev A. A. Continual modelling of two-row finite discrete system deformation with regard for boundary effects. Moscow University Mechanics Bulletin 1996; 51,5:44-46.
10. Dmitriev S. V., Abe K., Shigenari T. Domain wall solutions for EHM model of crystal: structures with period multiple of four. Physica D 2000; 147:122-134.
11. Vasiliev A. A., Dmitriev S. V., Ishibashi Y., Shigenari T. Elastic properties of a two-dimensional model of crystal containing particles with rotational degrees of freedom. Phys. Rev. B 2002; 65:94101-94107.
12. Dmitriev S. V., Shigenari T., Abe K. Mechanisms of transition between 1*q* and 2*q* incommensurate phases in a two-dimensional crystal model. Phys. Rev. B 1998; 58:2513-2522.
13. Shigenari T., Dmitriev S. V., Vasiliev A. A., Abe K., Domain wall solutions for EHM model of crystal J. Phys. Soc. Jpn 1999; 68:117-125.
14. Shigenari T., Vasiliev A. A., Dmitriev S. V., Abe K., Domain walls in one-dimensional 3-periodic structure. Ferroelectrics 1997; 203:335-347.

Dependence of Ductile and Brittle Response on Initial Energy Dissipation Mechanism at Crack Tip

From the Atomic Scale to the Mesoscale

Jeffrey W. Kysar
Department of Mechanical Engineering, Columbia University, New York, NY 10027, USA

Abstract: The criteria which determine the initial energy dissipation mechanism that is activated at or near a crack tip are derived. The possible mechanisms considered are cleavage, crack tip dislocation nucleation and also Frank-Read source activation near the tip. The criteria can be succinctly expressed in graphical form. It is suggested that a change in the initial energy dissipation mechanism may determine the conditions under which the brittle to ductile transition occurs. The criteria compare favourably to several experiments in the literature which address the competition between the various energy dissipation mechanisms.

Keywords: Fracture, dislocation, crack tip dislocation nucleation, Frank-Read dislocation source, ductile to brittle transition

1. INTRODUCTION

The brittle to ductile transition of a crystalline material that contains a crack is typically defined as an abrupt change in the overall capacity of the material to dissipate energy by means of irreversible deformation processes. It is thought that the transition is triggered by a change, at or near the crack tip, from one type of irreversible deformation mechanism to another. Traditionally two distinct competitions between deformation mechanisms have been considered in an attempt to explain the brittle to ductile transition. One approach invokes the competition between dislocation nucleation from

H. Kitagawa and Y. Shibutani (eds.), IUTAM Symposium on Mesoscopic Dynamics of Fracture Process and Materials Strength, 289–300.

a crack tip and cleavage failure. The second approach assumes that the mobility of dislocations plays the key role, which sets up a competition between cleavage failure and plastic deformation due to pre-existing dislocations in the material near a crack tip.

A recent analysis (Kysar, 2003) has reconciled the two approaches by considering the energetic competitions among the following three energy dissipation mechanisms: cleavage, dislocation nucleation from a crack tip, and dislocation nucleation from a Frank-Read source near a crack tip (which requires dislocation mobility). The result of the analysis is a set of three criteria involving dimensionless parameters which determine whether cleavage, crack tip dislocation nucleation or Frank-Read dislocation nucleation is the initial energy dissipation mechanism that is activated at or near a crack tip upon the commencement loading. Thus, the brittle-ductile behaviour of a material can be grouped into three classifications. If dislocation nucleation from a Frank-Read source is predicted to occur at a lower load level than either cleavage or crack tip dislocation, the material is said to possess *extrinsic ductility*. The condition for this to occur is stated succinctly in the form of two ranges of the dimensionless parameter $b\rho_{disl}^{1/2}$ (where b is the Burgers vector and ρ_{disl} is the dislocation density). One of the ranges addresses the condition for which Frank-Read dislocation nucleation is predicted to occur prior to cleavage failure, and the other range addresses the condition for which Frank-Read dislocation nucleation is predicted to occur prior to crack tip dislocation nucleation. If $b\rho_{disl}^{1/2}$ is smaller than the upper bound of each range, Frank-Read dislocation nucleation is expected to occur at a lower load level than either cleavage or crack tip dislocation nucleation. If $b\rho_{disl}^{1/2}$ exceeds the lower bound of each range, the pre-existing dislocation that comprises the Frank-Read source is ensured to be mobile, which is a necessary condition for Frank-Read dislocation nucleation. The Peierls stress, τ_p, which is the minimum stress necessary to move a dislocation through an otherwise perfect crystal lattice at absolute zero plays an important role in establishing the lower bound. Thus a material for which $b\rho_{disl}^{1/2}$ lies within both ranges is predicted to exhibit extrinsic ductility, irrespective of the competition between cleavage and crack tip dislocation nucleation.

If the material does not possess extrinsic ductility, the competition between cleavage and crack tip dislocation nucleation must be addressed. This has been the subject of many studies, most notably by Rice and Thomson (1974) and Rice (1992) in which the dimensionless quantity $\gamma_s/\mu b$, or similarly, γ_s/γ_{us}, is used to predict which of the two mechanisms occurs at a lower load level (γ_s is surface free energy, μ is elastic shear modulus, and γ_{us} is unstable stacking energy). If cleavage occurs at a lower load level than crack tip dislocation nucleation, the material is said to possess intrinsic brittleness. If crack tip dislocation nucleation occurs prior to cleavage, the material is intrinsically ductile.

The conditions for extrinsic ductility in Kysar (2003) assume that a Frank-Read (FR) dislocation source exists in a crystalline material a characteristic distance away from a crack tip. The derivation equates the Peach-Koehler (PK) force on the pre-existing dislocation of the Frank-Read source with the PK force necessary to activate the source. As in Rice and Thomson (1974), the PK force on the pre-existing dislocation includes contributions from the crack tip stress field as well as the image force due to the presence of the free surface. The PK force necessary to activate the FR source is determined by the increase in dislocation line energy as the pre-existing dislocation bows out past its impediments (e.g. McClintock and Argon, 1966). The result of equating the two forces is an estimate of the crack tip loading at which FR sources are expected to be activated. This crack tip loading can then be compared against the crack tip loading required for cleavage and also for crack tip dislocation nucleation. Lattice friction is accounted for by requiring that the PK force on the pre-existing dislocation be larger than the Peierls stress, which ensures that the dislocation is mobile.

Strictly speaking, this method of formulating the problem is valid only when the lattice friction contribution is either much less or much greater than the contribution due to the increase in elastic line energy of the pre-existing dislocation (this statement will be defined more precisely in what follows). When this condition does not hold, it is necessary to sum the two contributions in order to determine the critical PK force at which the source is expected to be activated.

In the present paper, the analysis to estimate the crack tip loading at which FR dislocation source activation near a crack tip occurs is modified to account for the case where the contributions from Peierls stress and increase of elastic line energy are approximately equal. The resulting predictions are amenable to graphical representation, which in the appropriate limits reduces to the aforementioned ranges of the dimensionless parameter $b\rho_{disl}^{1/2}$.

2. ENERGETIC COMPETITIONS

The goal of what follows is to determine the initial energy dissipation mechanism activated at or near a crack tip upon the commencement of mechanical loading. The energy dissipation mechanisms considered are: cleavage, crack tip dislocation nucleation, as well as dislocation nucleation from a Frank-Read source near a crack tip. The strategy to deduce the criteria is to determine the applied energy release rate at which the three energy dissipation mechanisms are activated. Then the competition between any two of the energy dissipation mechanisms can be expressed in terms of the dimensionless ratio of the respective critical energy release rates.

2.1 Energy Dissipation Mechanisms

The fracture criterion most commonly used for materials in which a significant amount of plastic deformation occurs is $G = 2\gamma_s + \gamma_p$, where G is the applied energy release rate available to effect fracture, γ_s is the free energy of the newly created surface, and γ_p is the energy dissipated through various irreversible processes in the near crack tip region. The magnitude of γ_p often exceeds γ_s by orders of magnitude, nevertheless it is known that $\gamma_p = \gamma_p(\gamma_s)$, so the energetic contribution of the newly created surfaces can not be neglected.

We consider in a dimensional analysis the effect that other variables have on γ_p. It is assumed that an atomically sharp crack exists in an elastic-plastic material within which plastic deformation occurs via the motion and creations of dislocations on discrete slip planes and in discrete directions at a critical resolved shear stress. In addition to γ_s, other important variables include: μ the elastic shear modulus, μ_{ij} the so-called Schmid factor that contains information about orientations of the plastic slip systems relative to the crack tip, and σ_{max}, the maximum theoretical tensile stress that the material can support in the absence of any defect. Yield stress, σ_y, is also a critical parameter. However since σ_y is a phenomenological and ill-defined quantity, we instead appeal to the physics-based variables which determine yield stress: b the Burgers vector, ρ_{disl} the density of mobile dislocations, and τ_p the Peierls stress. Performing a straightforward dimensional analysis yields

$$\frac{\gamma_p}{\mu b} = F\left[\frac{\gamma_s}{\mu b}, b\rho_{disl}^{1/2}, \frac{\tau_p}{\mu}, \frac{\sigma_{max}}{\mu}, \mu_{ij}\right]. \tag{1}$$

The physical significance of each of these terms is discussed in Kysar (2003). It should be noted that the unstable stacking energy, γ_{us}, introduced by Rice (1992) scales as μb, so that the dimensionless parameter $\gamma_s / \mu b$ in

Eq.(1) can be expressed equivalently as γ_s/γ_{us}. Given the nature of dimensional analysis, it is expected that the criteria for the various energy dissipation mechanisms will consist of only these dimensionless parameters.

2.2 ENERGY RELEASE RATE FOR FRANK-READ DISLOCATION NUCLEATION

The applied crack tip energy release rate to activate the Frank-Read (FR) sources nearest the tip is now derived. The derivation is elementary and necessarily approximate, but nevertheless suffices to obtain, at least, an order of magnitude estimate of the critical applied load. It is well-known that the resolved shear stress necessary to activate a FR dislocation source, τ_{FR}, scales as $\tau_{FR} = \alpha\mu b/L$, where L is considered to be the mean distance between impediments to motion of the pre-existing dislocation and α is a dimensionless constant with magnitude $1/4 < \alpha < 10$, depending upon the density of pre-existing dislocations (see Beltz, *et al.*, 1996, for a succinct discussion of appropriate choice of α). In pure single crystals, the distance between impediments scales as $L \approx \rho_{disl}^{-1/2}$, which corresponds to the mean distance between dislocations. Thus, $\tau_{FR} = \alpha\mu b\rho_{disl}^{1/2}$, which is dimensionally consistent with the overall macroscopic yield stress of face-centred cubic single crystals (Taylor, 1934). This expression accounts only for the contribution of elastic energy due to increase in dislocation line length as the dislocation bows past the impediments, which is appropriate for materials with a very small Peierls stress, τ_p. However for classes of materials for which the Peierls stress is not sufficiently small, the effect of the Peierls stress must be taken into account. Under such conditions, it is appropriate to express the critical resolved shear stress for FR source activation as the sum of the contributions from both the increase in dislocation line length as well as the Peierls stress, which yields

$$\tau_{FR} = \frac{\alpha\mu b}{L} + \tau_p . \qquad (2)$$

It is not possible to define *a priori* the precise position of FR sources near the crack tip. Nevertheless, it is well-accepted that the distance between dislocations, and hence between FR dislocation sources, scales as $\rho_{disl}^{-1/2}$. Therefore we assume that a FR dislocation source exists at a radial distance $r \approx \rho_{disl}^{-1/2}$ from a crack tip. The source consists of a pre-existing edge dislocation that spans two impediments and is parallel to the crack front. The spacing between the two impediments also scales as $L \approx \rho_{disl}^{-1/2}$. We assume the material to be fully annealed so that the Burgers vectors of all other surrounding pre-existing dislocations do not have any preferred orientation. Thus the net stress due to all other surrounding dislocations at any particular position is, on average, zero. Likewise the net image stress on a dislocation at any particular position due to all surrounding dislocations is, on average, zero.

Therefore the dislocation of the FR source interacts only with the stress field of the crack tip and with its own image force. The interactions can be quantified in terms of the Peach-Koehler (PK) force, f, defined as $f = \tau b$, where τ is the resolved shear stress on the dislocation slip plane in the direction parallel to the Burgers vector. The FR source is activated once the sum of the PK forces from the crack tip and image force equals the force necessary to overcome $\tau_{FR} b$ from Eq.(2). Finally for simplicity we assume that the dislocation exists on a slip plane which intersects the crack tip (this assumption is not necessary, and does not affect the order of magnitude of the result). Rice and Thomson (1974) discussed the PK forces on a dislocation due to a crack tip and its image force under the same conditions assumed here. Setting the sum of the PK force from the crack tip and from the image force equal to $\tau_{FR} b$ yields

$$\frac{K_I b}{\sqrt{8\pi r}} - \frac{\mu b^2}{4\pi(1-\nu)r} = \frac{\alpha\mu b^2}{L} + \tau_p b \tag{3}$$

where K_I is the Mode-I stress intensity factor, ν is Poisson's ratio and r is the radial position of the dislocation relative to the crack tip. The first term on the left side of Eq.(3) is from the singular stress field of the crack tip and the second term represents the image force. It should be noted that the trigonometric functions normally associated with the crack tip term are approximated as unity, which does not affect the order of magnitude of the result. Substituting $r \approx \rho_{disl}^{-1/2}$ and $L \approx \rho_{disl}^{-1/2}$ and solving for K_I yields

$$K_I^{FR} = \alpha_k \mu b \rho_{disl}^{1/4} \left[1 + \frac{\sqrt{8\pi}}{\alpha_k} \left(\frac{\tau_p}{\mu} \right) \left(\frac{1}{b \rho_{disl}^{1/2}} \right) \right] \tag{4}$$

where K_I^{FR} is the order of magnitude of the applied energy release rate in pure Mode I at which the FR source is expected to be activated, and $\alpha_k = \sqrt{8\pi}\{\alpha + 1/[4\pi(1-\nu)]\}$. The term containing ν in α_k is due to the image force term and will be about an order of magnitude smaller than α, which indicates that the image force plays a minor role in determining the activation criterion for the FR sources. Nevertheless, its contribution is maintained in what follows for completeness.

It is straightforward to express the condition for FR source activation near a crack tip in terms of applied energy release rate, G, through Irwin's relationship, $G=(1-\nu)K_I^2/2\mu$, to yield

$$G_{FR}=\frac{(1-\nu)\alpha_k^2}{2}\mu b^2\rho_{disl}^{1/2}\left[1+\frac{\sqrt{8\pi}}{\alpha_k}\left(\frac{\tau_p}{\mu}\right)\left(\frac{1}{b\rho_{disl}^{1/2}}\right)\right]^2. \tag{5}$$

It is now possible to state with precision the conditions under which K_I^{FR} and G_{FR} are dominated by either elastic line energy effects or by Peierls stress. A cursory glance at Eq.(4) and Eq.(5) shows that the FR activation criterion is dominated by elastic line energy effects if $\tau_p/\mu << b\rho_{disl}^{1/2}$. On the other hand, if $\tau_p/\mu >> b\rho_{disl}^{1/2}$, the activation criterion is dominated by the Peierls stress. It will become evident that the interplay between τ_p/μ and $b\rho_{disl}$ plays a major role in determining the transition from one energy dissipation mechanism to another.

2.3 COMPETITIONS TO DETERMINE INITIAL ENERGY DISSIPATION MECHANISM

Now that we have obtained a quantitative criterion for the activation of the FR sources nearest a crack tip, we can address the competition between FR source activation and both cleavage and crack tip dislocation nucleation. The competitions will be expressed in terms of various ratios of the activation criteria of the three energy dissipation mechanisms under consideration. To that end, it is known that, G_{CL}, the applied energy release rate at which cleavage occurs is $G_{CL}=2\gamma_s$. Similarly, Rice (1992) showed that the applied energy release rate at which crack tip dislocation nucleation, G_{CT}, occurs is $G_{CT}=\beta\gamma_{us}$, where β is a dimensionless constant which contains the crystallographic details of the slip system of an incipient dislocation relative to the orientation of the crack in the crystal, and is thus related to μ_{ij}.

If the condition $G_{FR}/G_{CT}<1$ holds, energy dissipation via FR source activation is expected to occur at a lower applied load level than that necessary for crack tip dislocation nucleation. From Eq.(5), this can be expressed as

$$\frac{G_{FR}}{G_{CT}}=\frac{(1-\nu)\alpha_k^2}{2\beta}\left(\frac{\mu b}{\gamma_{us}}\right)\left(b\rho_{disl}^{1/2}\right)\left[1+\frac{\sqrt{8\pi}}{\alpha_k}\left(\frac{\tau_p}{\mu}\right)\left(\frac{1}{b\rho_{disl}^{1/2}}\right)\right]^2<1. \tag{6}$$

Likewise, if the condition $G_{FR}/G_{CL}<1$ holds, energy dissipation via FR source activation is expected to occur at a lower applied load level than that necessary for cleavage, which can be expressed as

$$\frac{G_{FR}}{G_{CL}}=\frac{(1-\nu)\alpha_k^2}{4}\left(\frac{\mu b}{\gamma_s}\right)\left(b\rho_{disl}^{1/2}\right)\left[1+\frac{\sqrt{8\pi}}{\alpha_k}\left(\frac{\tau_p}{\mu}\right)\left(\frac{1}{b\rho_{disl}^{1/2}}\right)\right]^2<1. \tag{7}$$

It is evident that if both Eq.(6) and Eq.(7) are satisfied, FR sources are expected to be activated prior to both cleavage and crack tip dislocation nucleation, *irrespective of the competition between cleavage and crack tip dislocation nucleation.* It should be emphasized that G_{FR} is an order of magnitude approximation, however as discussed in Kysar (2003), this type of approximate criterion will often suffice to successfully predict the outcome of the energetic competitions.

3. DISCUSSION

The criteria for extrinsic ductility are amenable to graphical interpretation. There are only three dimensionless parameters in Eq.(6) and Eq.(7): $\gamma_s/\mu b$, $b\rho_{disl}^{1/2}$ and τ_p/μ. Two of these, $b\rho_{disl}^{1/2}$ and τ_p/μ, range over many orders of magnitude as materials and deformation states vary. The other variable, $\gamma_s/\mu b$, changes by about one order of magnitude, at most, among different classes of materials. This suggests that it is advantageous to plot Eq.(6) and Eq.(7) in $b\rho_{disl}$ vs. τ_p/μ space.

The two criteria are plotted in Fig.(1) as heavy lines. Since Eq.(6) and Eq.(7) are identical algebraic forms, both lines share the same general features. For the sake of argument, we can assign the solid heavy line to represent the competition between FR source activation and crack tip dislocation nucleation, $G_{FR}/G_{CT}<1$, embodied in Eq.(6). The dashed heavy line then represents the competition between FR source activation and cleavage, $G_{FR}/G_{CL}<1$, in Eq.(7).

We now consider the properties of the heavy lines. If the values of $b\rho_{disl}^{1/2}$ and τ_p/μ for a particular material lie below the dashed line, the inequality in Eq.(7) is satisfied so that FR source activation occurs prior to cleavage. In addition, since the dashed line itself is below the solid line, the inequality in Eq.(6) is also satisfied, so that FR source activation also occurs before crack tip dislocation nucleation. Therefore, FR source activation is expected to be the initial energy dissipation mechanism, rendering the material extrinsically ductile (Kysar, 2003).

On the other hand, what happens when the values of $b\rho_{disl}^{1/2}$ and τ_p/μ for a particular material lie between the dashed line and the solid line? In that case, we would expect cleavage to occur prior to FR source activation, yet FR source activation would be expected to occur prior to crack tip dislocation nucleation. Therefore, cleavage would unambiguously be expected as the initial energy dissipation mechanism at or near the crack tip, so that the material would be intrinsically brittle.

Finally, what is predicted when the values of $b\rho_{disl}^{1/2}$ and τ_p/μ for a particular material lie above both the dashed line and also the solid line? Under those conditions, crack tip dislocation nucleation as well as cleavage would be expected to occur prior to FR source action. Thus the energetic competition reduces to that between crack tip dislocation and cleavage, G_{CL}/G_{CT}, which can be expressed by taking the ratio of Eq.(6) to Eq.(7) to yield the condition for an intrinsically brittle material (Rice, 1992)

$$\frac{G_{CL}}{G_{CT}} = \frac{2\gamma_s}{\beta\gamma_{us}} < 1. \tag{8}$$

It is readily observed that this condition holds if the dashed line in Fig.(1) lies below the solid line. Such a material, too, is predicted to be intrinsically brittle.

To summarize, if the combination of $b\rho_{disl}^{1/2}$ and τ_p/μ lies below the dashed line, the material is predicted to be extrinsically ductile. If it lies anywhere above the dashed line, the material is predicted to be intrinsically brittle. This naturally suggests a transition in overall brittleness and ductility, which would be triggered as $b\rho_{disl}^{1/2}$ and τ_p/μ for a given material cross over the dashed line,.

We now consider the case in which the dashed line of Fig.(1) would lie above the solid line. This would occur whenever the inequality in Eq.(8) is reversed. If the combination of $b\rho_{disl}^{1/2}$ and τ_p/μ for a material would lie below the solid (and interior) line, the material would again be extrinsically ductile. If it would lie anywhere above the solid (and interior) line, the material would be intrinsically ductile.

The heavy lines in Fig.(1) have two asymptotes each. It can be shown that that the left asymptote of each curve corresponds to the lower limit of the respective range of the variable $b\rho_{disl}^{1/2}$ in the criteria for extrinsic ductility by Kysar (2003), which can be derived by assuming that $\tau_p/\mu >> b\rho_{disl}^{1/2}$ in each of Eq.(6) and Eq.(7). Likewise, the right asymptote corresponds to the respective upper range of the variable $b\rho_{disl}^{1/2}$, which is strictly valid for $\tau_p/\mu << b\rho_{disl}^{1/2}$. Therefore, the extrinsic ductility criteria of Kysar (2003) are represented by the thin dashed and thin solid asymptotes in Fig.(1), which have an identical graphical interpretation as the more general criteria presented herein. As one would expect, the two sets of criteria agree quite closely except for the case where τ_p/μ and $b\rho_{disl}^{1/2}$ are approximately equal.

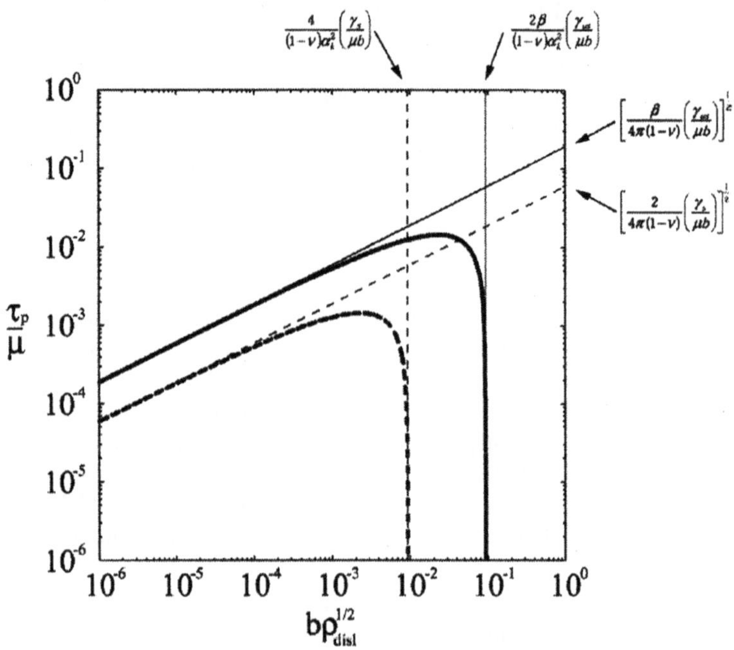

Figure 1: Initial Energy Dissipation Mechanism Map

The theory predicts the conditions under which a transition in initial energy dissipation mechanism occurs. It is suggested that this transition serves as a trigger for the overall ductile to brittle transition. Experiments by Michot and Loyola de Oliveira (2001), Argon and Gally (2001), Gumbsch, *et al.* (1998), as well as Qiao and Argon (2003) quantify the transitions between the three energy dissipation mechanisms in materials such as Si, W, and Fe-3wt.%Si. A common theme among the experiments is that cleavage is the initial energy dissipation mechanism at low temperatures. However as the temperature increases, a transition to crack tip dislocation nucleation occurs accompanied by a moderate increase in the overall fracture toughness. Eventually at an even higher temperature, dislocation motion away from the crack tip is reported, which signals the transition to FR source activation as the energy dissipation mechanism. This transition is accompanied by a very significant increase in fracture toughness.

The progression in the transition of energy dissipation mechanisms can be explained qualitatively by the criteria derived herein. The initial transition, from intrinsic brittle to intrinsic ductile behaviour, is governed (Rice, 1992) by the relationship in Eq.(8). This transition is sensitive to temperature because of thermal activation of crack tip dislocation nucleation. The subsequent transition to FR source activation occurs as the temperature increases further, which causes the *effective* Peierls stress to decrease. Once it reaches a critical level defined by Eq.(6) and Eq.(7), the transition to FR source activation occurs, which is known experimentally to lead to a dramatic increase of fracture toughness.

The elementary model used to derive the criteria has many deficiencies and there is much room for its improvement. However it is worthwhile to keep in mind that the model is intended to address only the transitions in energy dissipation mechanisms which apparently act as a trigger for the overall ductile to brittle transition. The detailed behaviour of the material once the transitions occur can not be predicted by the present model, and should be topic of future research.

4. CONCLUSIONS

The competitions between various energy dissipation mechanisms at or near a crack tip are considered. A set of criteria is derived which predicts the conditions under which Frank-Read sources near a crack tip are the initial energy dissipation mechanism. Such a material is said to be extrinsically ductile. If the criteria are not satisfied, the material is either intrinsically brittle or intrinsically ductile. The various criteria are amenable to straightforward graphical interpretation. Comparison with experiment suggests that this set of criteria can be used to address the conditions which determine the transition from a ductile to a brittle behaviour.

ACKNOWLEDGEMENTS

This work was supported by the National Science Foundation under the Faculty Early Career Development Program with grant CMS-0134226.

REFERENCES

Argon, A. S., and Gally, B. J. (2001) *Scripta Mater.* **45**, 1287-1294.
Beltz, G.E., Rice, J.R., Shih, C.F., Xia, L. (1996) *Acta Mater.* **44**, 3943-3954.
Gumbsch, P., *et al.* (1998) *Science*, **282**, 1293-1295.
Kysar, J. W. (2003) *J. Mech. Phys. Solids,* **51**, 795-824.
McClintock, F. A., and Argon, A. S. (1966). *Mechanical Behavior of Materials*, Addison-Wesley, Reading, Massachusetts.
Michot, G., and Loyola de Oliveira, M. A. (2001) *Mater. Trans. JIM*, **42**, 14-19.
Rice, J. R. (1992) *J. Mech. Phys. Solids*, **40**, 239-271.
Rice, J. R., and Thomson, R. (1974) *Phil. Mag.*, **29**, 73-97.
Qiao, Y., and Argon, A. S. (2003) *Mech. Mat.*, **35**, 903-912.
Taylor, G. I. (1934) *Proc. R. Soc. Lond. A*, **145**, 362-387.

A COHESIVE ZONE MODEL AND INTERFACIAL CRACK PROBLEMS

Kikuo KISHIMOTO and Masaki OMIYA
Department of Mechanical and Intelligent Systems Engineering, Tokyo Institute of Technology, 2-12-1, O-okayama, Meguro-ku, Tokyo 152-8552, Japan

Abstract: Based on the internal variable theory of thermodynamics, a continuum interface constitutive model relating the interface traction with interface separation is developed. The interface damage variable is introduced, and the evaluation equation is derived to characterize the degradation of interface rigidity with interface debonding. The cohesive law is formulated by taking into account of the anisotropy of interface debonding. The present constitutive model is applied to the interfacial crack problems.Finite element computation is performed in which the cohesive model is embedded along the line extending ahead of the crack tip. The critical stress intensity factors at the crack initiation and propagation are evaluated for a wide range of bimaterial constant and the fracture boundary curves are obtained. Numerical simulation on peeling test is also presented.

Key words: Cohesive Zone Model, Interface Crack, Fracture Criterion, Peel Test

1. INTRODUCTION

Recently, bimaterial systems and composite materials are widely used. Typical examples include fiber reinforced materials, adhesive joints, microelectronic devices and so on. The performance of these systems strongly depends on the strength of weak interfaces. Therefore, it is important to evaluate the interface strength precisely. In homogeneous materials, the fracture criterion under tensile loading can be well characterized by a material property, the fracture toughness. However in dissimilar materials, it is necessary to specify the loading phase angle,

H. Kitagawa and Y. Shibutani (eds.), IUTAM Symposium on Mesoscopic Dynamics of Fracture Process and Materials Strength, 301–310.

because the tensile and in-plane shear is inherently coupled at the tip of an interface crack. The interface does not have a unique fracture energy, but instead, the toughness of the interface depends on the loading phase angle (Hutchinson and Suo[1]).

In order to investigate the material parameters which govern the fracture phenomenon of an interface, it is important to model an interface itself. Needleman[2] introduced an interface potential that specified the dependence of the interface traction upon the interface separation. Tvergaard and Hutchinson[3] used a traction-separation law to model the fracture process ahead of the crack tip and they computed the crack growth resistance in homogeneous solids and along interfaces. Ma and Kishimoto[4] proposed the interface model based on the internal variable theory of thermodynamics and the continuum constitutive relation between interface traction and interface separation has been developed. Chaboche et al.[5] extended Tvergaard's model (Tvergaard[6]) to keep the continuity and monotonicity in the tangential stiffness degradation and introduced the Coulomb friction criterion.

The intent of this paper is to investigate the mixed mode fracture toughness of an interface by numerical simulation and obtain the mixed mode fracture criteria for the interface crack. A cohesive zone model can give the interfacial mechanical properties and makes it possible to describe the fracture process on the interface. The present work adopts the cohesive zone model proposed by Ma and Kishimoto[4] with minor modifications[7]. This cohesive zone model is embedded at the crack tip of the interface and the propagation of the interface crack is simulated by finite element analyses. Interface fracture criteria based on the stress intensity factors and the fracture toughness are discussed. To show the application of this cohesive zone model, a numerical simulations of peel test is also presented.

2. COHESIVE ZONE MODEL

The mechanical properties of an interface are complicated and exhibit inhomogeneity, since there may be many microdefects, cracks and various phases exist in fracture process zone(Figure 1(a)). From the idea of continuum mechanics, the mechanical response of the interface can be equivalently modeled by distributed nonlinear springs as shown in Figure 1(b). The displacement of an interface can be scaled by the components, U_t, U_n and U_b. We use $u_t = U_t / \delta_t$, $u_n = U_n / \delta_n$ and $u_b = U_b / \delta_b$ as normalized scales for interface deformation, where δ_t, δ_n and δ_b are the critical interfacial separation for each direction.

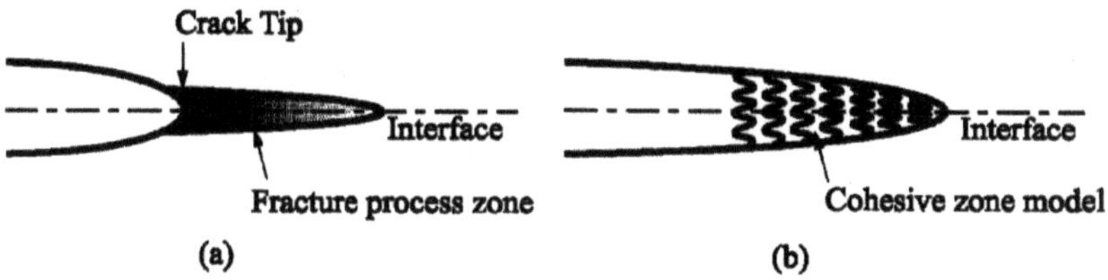

Figure 1: The representation of an interface crack tip and the equivalent spring model.

An equivalent interface separation is defined by:

$$\lambda = \sqrt{\left(\frac{U_t}{\delta_t}\right)^2 + \left(\frac{U_n}{\delta_n}\right)^2 + \left(\frac{U_b}{\delta_b}\right)^2} = \sqrt{u_t^2 + u_n^2 + u_b^2} \, . \tag{1}$$

Here, normal displacement of interface is confined to $U_n \geq 0$. The components of the interfacial traction can be related to the interface separations by the interfacial rigidity, in the normal and tangential directions as follows:

$$P_t = K_t u_t, \; P_n = K_n u_n, \; P_b = K_b u_b \, , \tag{2}$$

where K_t, K_n and K_b are the interfacial rigidities for each direction. Since the interfacial rigidities decrease with interface separation, we introduce three damage parameters, D_t, D_n and D_b , to characterize the degradation of the normal and tangential interfacial rigidities, respectively. Without the occurrence of damage, the interfacial rigidities take their initial values, K_t^0, K_n^0 and K_b^0, respectively. Restrictions on the interface damage processes are developed based on standard methods of irreversible thermodynamics. The interfacial separation energy density as the Helmholtz free energy, which satisfies the thermodynamic restrictions and the interfacial potential of dissipation due to the damage evolution are assumed as in Ma and Kishimoto[4]. In this paper, interface is assumed to be monotonically loaded, that is, no unloading and cyclic loading occurred. Then, the damage parameters can be related to the interfacial separation as,

$$\left.\begin{aligned} D_t &= \beta_t\left(\lambda - \lambda_t^0\right) \text{-} \left(\lambda^2 - \left(\lambda_t^0\right)^2\right) \\ D_n &= \beta_n\left(\lambda - \lambda_n^0\right) \text{-} \left(\lambda^2 - \left(\lambda_n^0\right)^2\right) \\ D_b &= \beta_b\left(\lambda - \lambda_b^0\right) \text{-} \left(\lambda^2 - \left(\lambda_b^0\right)^2\right) \end{aligned}\right\} , \tag{4}$$

where λ_t^0, λ_n^0 and λ_b^0 are the damage thresholds of the interfacial rigidities in pure normal or tangential separation. From Equations (2)-(4), the constitutive equation of the interface can be described as,

$$\left.\begin{aligned} P_t &= K_t^0\left(1-\beta_t\left(\lambda-\lambda_t^0\right)+\left(\lambda^2-\left(\lambda_t^0\right)^2\right)\right)u_t \\ P_n &= K_n^0\left(1-\beta_n\left(\lambda-\lambda_n^0\right)+\left(\lambda^2-\left(\lambda_n^0\right)^2\right)\right)u_n \\ P_b &= K_b^0\left(1-\beta_b\left(\lambda-\lambda_b^0\right)+\left(\lambda^2-\left(\lambda_b^0\right)^2\right)\right)u_b \end{aligned}\right\}. \tag{5}$$

When the bearing capacity of the interface is lost at a critical value $\lambda = \lambda^c$, the interface parameters become,

$$\beta_t = \frac{1+\left(\lambda^c\right)^2-\left(\lambda_t^0\right)^2}{\lambda^c-\lambda_t^0},\ \beta_n = \frac{1+\left(\lambda^c\right)^2-\left(\lambda_n^0\right)^2}{\lambda^c-\lambda_n^0},\ \beta_b = \frac{1+\left(\lambda^c\right)^2-\left(\lambda_b^0\right)^2}{\lambda^c-\lambda_b^0}. \tag{6}$$

When $\lambda^c = 1$ and $\lambda_n^0 = \lambda_t^0 = \lambda_b^0 = 0$, the interface parameters are $\beta_t = \beta_n = \beta_b = 2$. In this special case, when in-plane deformation is considered, the explicit decohesion energy can be expressed as,

$$\Gamma(\phi) = \Gamma_0\left\{\frac{1+\dfrac{K_n^0}{K_t^0}\tan^2\phi}{1+\dfrac{\delta_n^2}{\delta_t^2}\left\{\dfrac{K_n^0}{K_t^0}\right\}^2\tan^2\phi}\right\}, \tag{7}$$

where ϕ is the phase angle which is defined by the ratio of the normal and tangential stress at the crack tip ($\phi = \tan^{-1}\left(\sigma_{12}/\sigma_{22}\right)$) and Γ_0 is the decohesion energy in pure normal interfacial separation,

$$\Gamma_0 = \frac{1}{12}K_n^0\delta_n^2. \tag{8}$$

3. FRACTURE CRITERIA OF INTERFACE CRACK

3.1 Computational model

The interface crack propagation is simulated for several material pairs under various mixed mode loadings by the Finite Element Method. For the Finite Element Analysis, ABAQUS(Ver.5.6) was used. Figure 2 shows the configuration of the compact tension shear(CTS) specimen for the mixed mode fracture test[8]. Here, P is the resultant force and α is the loading direction. The interface crack is introduced along the interface and it's length is a=0.5W. The specimen is attached by a special jig that makes it possible to change the loading direction. By changing the loading direction, several kinds of mixed mode loadings can be applied to the interface crack tip. In the calculation model, 1164 two dimensional eight-node isoparametric elements and 3625 nodes are used. The width of the smallest elements(Δ) are $\Delta / W = 1.25 \times 10^{-2}$. Young's modulus of each material is denoted as E_1 and E_2 and Poisson's ration is $\nu_1 = \nu_2 = 0.3$. The ratio between E_1 and E_2 is changed variously. The interface constitutive relation is embedded at the crack tip. In this paper, we assume that the interfacial debonding induces identical degradation in the normal and tangential directions. The isotropic degradation model given by $\beta = \beta_t = \beta_n = \beta_b$ is used under plane stress conditions.

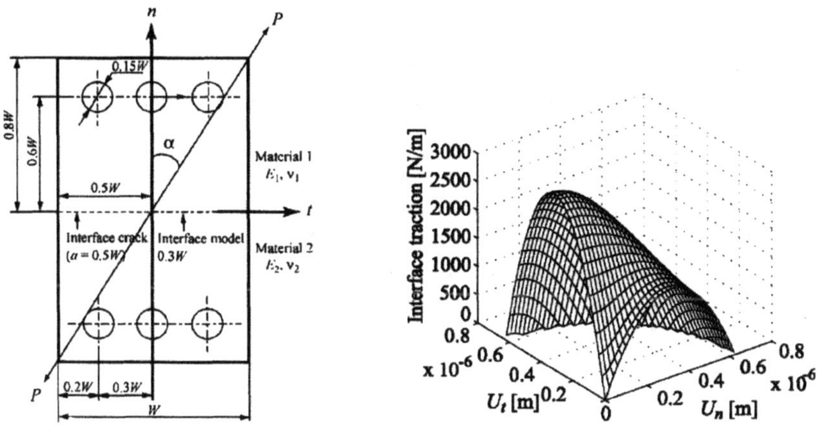

Figure 2: Numerical model and the mechanical properties of the interface cohesive zone model ($\lambda_n^0 = \lambda_t^0 = 0, \delta_n = \delta_t = 5\Delta, 2K_n^0 = K_t^0$).

The critical interface separation of each direction is also identical ($\delta_t = \delta_n$) and only the initial rigidities in Equation (5) are changed. The variation of interface traction with interface separation is presented in Figure 2 for the orthotropic interface model$\left(2K_n^0 = K_t^0\right)$.

The initiation of interfacial crack growth can be defined by using the interface constitutive relation in Equation (5). When the interface separation comes to a critical value, the interface traction becomes zero and crack growth initiates. In this paper, we assume that crack initiation occurs at $\lambda^c = 1$. At that time, the *J*-integral is calculated and this value will be termed as interface fracture. In order to know the mixed mode condition at an interface crack tip, the stress intensity factors are computed by the M_1-integral method.

3.2 Numerical results

In the computation, the initial rigidity of tangential direction is assume to be twice as that of normal direction ($2K_n^0 = K_t^0$) as shown in Figure 2. Figure 3 represents the fracture boundary curves for the orthotropic interface model. The characteristic length of the interface stress intensity factors is assumed to be $L = 2a$. When $E_1/E_2 = 1$, the shape of fracture boundary curve is elliptical and K_1 and K_2 correspond to the stress intensity factors for mode I and mode II, respectively. The fracture toughness of a shear (mode II) type fracture is about 1.5 times that of the opening (mode I) type fracture. However, in the case of $E_1/E_2 = 10,100$, the fracture boundary curves rotate around the origin and K_1 and K_2 do not correspond to the stress intensity factors for each fracture mode. This is due to the definition of the characteristic length included in the interface stress intensity factors and will be discussed in the next section.

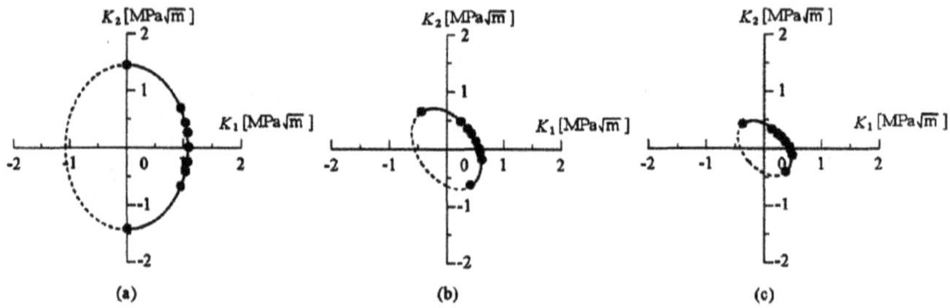

Figure 3: Fracture boundary curves for the orthotropic interface model ($2K_n^0 = K_t^0$, $L = 2a$): (a) $E_1/E_2 = 1$, (b) $E_1/E_2 = 10$ (c) $E_1/E_2 = 100$.

3.3 Interfacial fracture criterion

Stress intensity factors for an interface crack depend on the characteristic length. When this length is changed from L to L', the stress intensity factors are also changed as,

$$\begin{Bmatrix} K_1' \\ K_2' \end{Bmatrix} = \begin{bmatrix} \cos\theta & -\sin\theta \\ \sin\theta & \cos\theta \end{bmatrix} \begin{Bmatrix} K_1 \\ K_2 \end{Bmatrix}, \tag{9}$$

$$\theta = \varepsilon \ln\left(\frac{L'}{L}\right). \tag{10}$$

In Figures 3(b)(c), the fracture boundary curves are rotated around the origin when Young's moduli of the two materials are different. This is due to the definition of the characteristic length. By using Equations (9) and (10) with $L' = 1.8\times10^{-4}\ W$, which is considered to be the length of the fracture process zone, the fracture boundary curves are transformed. Then, K_1 and K_2 correspond to the stress intensity factors for each fracture mode. By normalizing K_1 and K_2 with the critical stress intensity factors of each fracture mode($K_{1CL'}$, $K_{2CL'}$), the intrinsic fracture boundary curve can be obtained as shown in Figure 4. Therefore, the fracture criterion based on the stress intensity factors could be expressed as,

$$\left(\frac{K_1}{K_{1CL'}}\right)^2+\left(\frac{K_2}{K_{2CL'}}\right)^2=1. \tag{11}$$

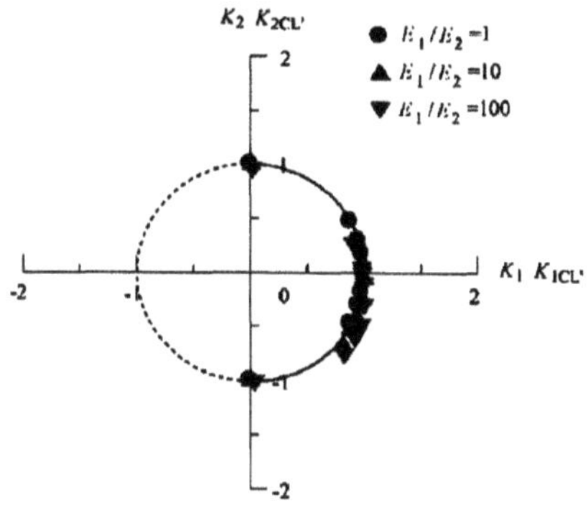

Figure 4: Intrinsic fracture boundary curve for orthotropic interface model ($2K_n^0 = K_t^0$, $L' = 18\mu$m).

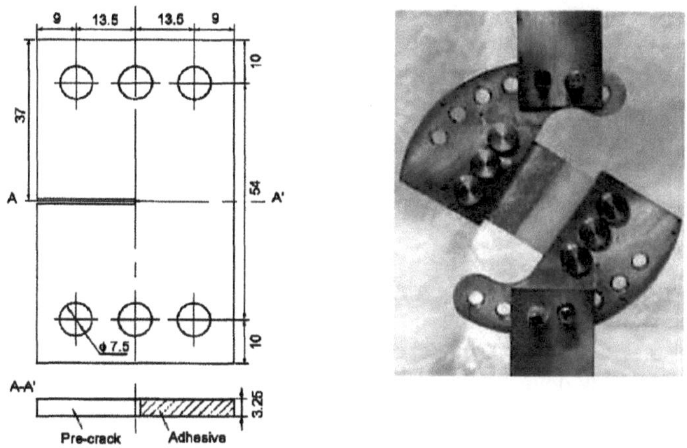

Figure 5: The schematic representations of specimen and testing apparatus.

In order to investigate the applicability of Equation(11) for mixed mode fracture criteria of an interface, mixed fracture tests of adhesively bond materials were carried out. Figure 5 shows schematic representations of the specimen and testing apparatus. One material used in this test is Aluminum and the others are Steel or ABS resin. These materials are bonded by an epoxy adhesive. In the experiment, the fracture loads and loading directions can be obtained. From these experimental results, the stress intensity factors were obtained by finite element analyses.

Figure 6 shows the fracture boundary curves for these two cases. When we take $2a$ as the characteristic length, the fracture boundary curves turn around the origin as shown in Figure 6(a). On the other hand, by choosing the characteristic length as the order of plastic zone size on the interface[9], the stress intensity factors are corresponding to each fracture mode and the fracture boundaries become as shown in Figure 6(b). It is found that Equation(11) well describes the fracture initiation of interface cracks.

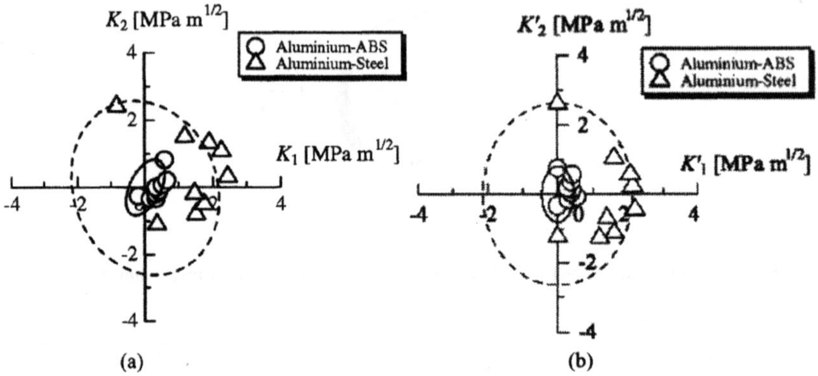

Figure 6: (a) The fracture boundary curves (L=2a). (b)The fracture boundary curves (L'=0.1d_0).

4. MULTI-STAGES PEEL TEST

The peel test is widely used to evaluate interfacial strength, since it is simple and easy to do the experiments. However, it is difficult to obtain the interfacial strength precisely due to the plastic dissipation W_p, residual stresses σ_{res} and mixed mode effect during the peel test. Recently, we developed a new device for peel test(in Figure 7), which make it possible to change the peel angles, i.e. mixed mode conditions at the peeling front, by introducing the tangential load P_h. The results of peel tests for Cu thin film(thickness h=10, 20μm) plated on PI/Si substrate are shown in Figure 8.

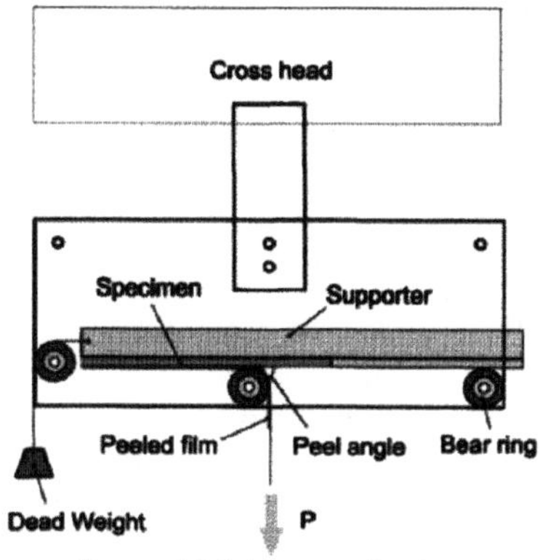

Figure 7: Multi-stages Peel Test.

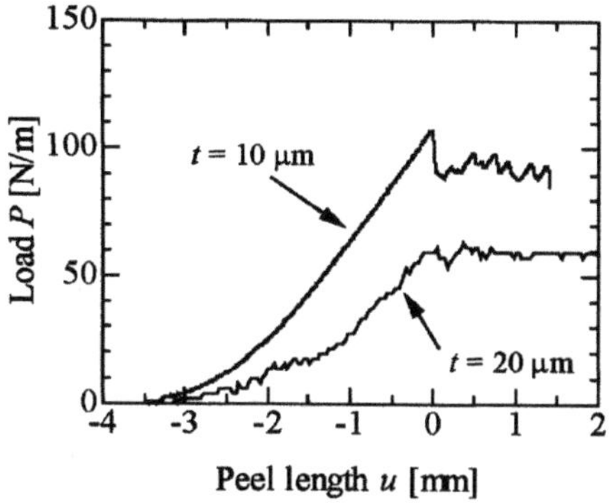

Figure 8: Peel load and peel length relationships.

When the delamination of Cu film begins, the peel load P becomes constant and the steady state peeling attained. During the steady state peeling, the energy balance may be expressed as[10],

$$P-P_h=\Gamma(\phi)+W_p+\frac{Eh^3}{24(1-\nu^2)}K_{coil}^2+\frac{1-\nu^2}{E}\left\{\frac{P^2}{h}\pm\int_0^h\sigma_{res}(t)^2dt\right\}. \quad (12)$$

where K_{coil} is the coiling curvature at the end of the film when it is released from the peeling grip. From this equation, the interfacial decohesion energy $\Gamma(\phi)$ can be evaluated. Using the present cohesive zone model, we carried out the simulations of peel tests, in which the parameters of the cohesive zone model were identified from Equation (7) by comparing with the experimental result of h=10μm. Figure 9 shows the numerical model of Multi-stages peel test. Cu and PI layers are assumed to be elastic-plastic material and Si layer is assumed to be rigid. Cohesive zone model is embedded along the interface. The deformation at the peel front is shown in Figure 10. As shown in Figure 11, the peel load becomes constant after the delamination commences. This tendency is similar to experimental results. When thermal stresses induced during fabrication process are taken into account, the peel load decreases and comes to agree well with the experiments.

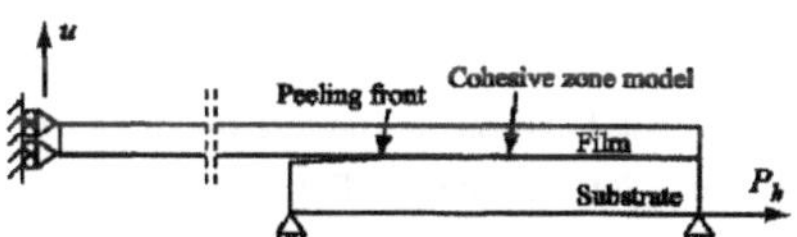

Figure 9: Numerical model.

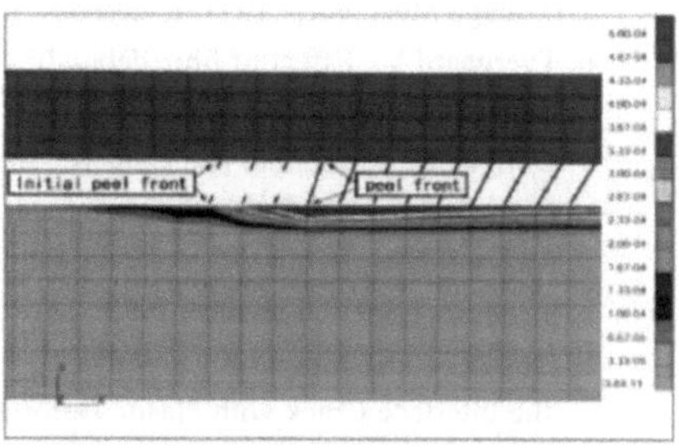

Figure 10: Deformation at the peeling front.

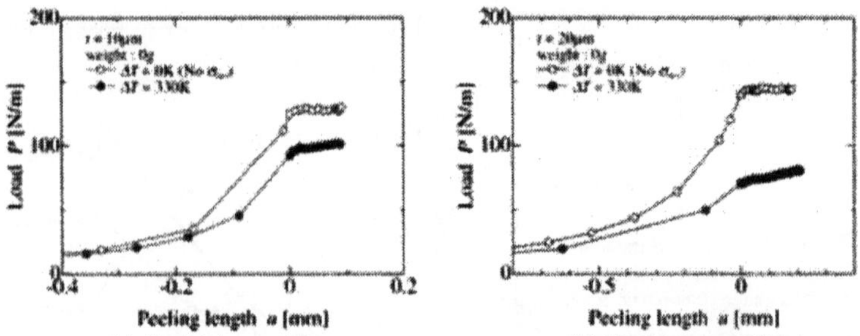

(a) Cu thickness:10μm (b) Cu thickness:20μm

Figure 11: Results of Multi-stages peel test by using the cohesive zone model.

5. CONCLUSION

A continuum interface constitutive model has been developed and applied to fracture problems of bi-materials and the peel test. It was shown that the present mode is useful to examine interface crack problems.

6. REFERENCES

1. Hutchinson J.W., Suo Z., Mixed mode cracking in layered materials, Adv. Appl. Mech., 1992: 29, 63-191.
2. Needleman A., A Continuum Model for Void Nucleation by Inclusion Debonding, J. Appl. Mech. 1987: 545,525-31.
3. Tvergaard V., Hutchinson J.W., The Relation between Crack Growth Resistance and Fracture Process Parameters in Elastic-Plastic Solids, J. Mech. Phys. Solids 1992: 40, 1377-97.
4. Ma F, Kishimoto K., A continuum interface debonding model and application to matrix cracking of composites, JSME Int. J. Series A, 1996: 39, 496-507.
5. Chaboche J.L., Girard, R., Levasseur, P., On the interface debonding models, Int. J. Damage Mechanics, 1997: 6, 220-57.
6. Tvergaard V., Effect of fibre debonding in a wisker-reinforced metal, Materials Science Engineering, 1990: A125, 203-13.
7. Omiya, M., Kishimoto K., Yang W., Interface debonding model and its application to the mixed mode interface fracture toughness, Int. J. Damage Mech., 2002: 11, 263-86.
8. Richard H.A., Benitz, K., A loading device for the creation of mixed mode in fracture mechanics, Int. J. Fracture, 1983: 22, R55-R58.
9. Omiya M., Kishimoto K., Shibuya T., Finite Element Analysis on Fracture Behavior of the Interface Crack with Plastic Deformation, Mat. Sci. Res. Int., STP-1, 2001: 238-43.
10. Kishimoto K., Omiya, M., Yang W., Muluti-stage peel tests and phase angle dependence of adhesion strength, to be submitted.

Molecular Dynamics Simulation of Influence of Grain Boundary on Near-Threshold Fatigue Crack Growth

S. Kubo* and M. Misaki
*Department of Mechanical Engineering and Systems, Graduate School of Engineering, Osaka University, 2-1, Yamadaoka, Suita, Osaka 565-0871 Japan
E-mail: kubo@mech.eng.osaka-u.ac.jp

Abstract In the near-threshold region of the fatigue crack growth, the amount of crack growth per cycle is in atomistic scale. This may suggest that the near-threshold fatigue crack growth is controlled by atomistic scale events. In previous papers the molecular dynamics was applied to the simulation of near-threshold fatigue crack growth. In this study, the influence of grain boundary on the near-threshold mode I fatigue crack growth behavior in BCC-iron was investigated using the molecular dynamics method. When the grain boundary existed perpendicular to the crack plane, the grain boundary served as an obstacle to crack growth. When dislocations got through the grain boundary, grain boundary had small influence.

Keywords: Near-threshold fatigue crack growth, BCC iron, Molecular dynamics, Fracture mechanics, Grain boundary

1. INTRODUCTION

Increasing number of plants require life evaluation for their safety assessment. For assessing their residual lives and strengths, fatigue is one of major issues to be considered. For structures in long term service the evaluation of very slow fatigue crack growth in the near-threshold regime and fatigue threshold is indispensable.

For explaining the mechanism of fatigue crack growth many models have been proposed. One of the well-known models for the fatigue crack growth is the Laird-Smith model[1-2]. In this model crack grows in every cycle of repeated loadings as a result of crack tip blunting and resharpening. This mechanism can explain the striation found on fracture surfaces when the amount of crack growth per

H. Kitagawa and Y. Shibutani (eds.), IUTAM Symposium on Mesoscopic Dynamics of Fracture Process and Materials Strength, 311–320.

cycle is in the order of around 10^{-6} (m).

In the near-threshold region the amount of crack growth per cycle is in the order of 10^{-9} to 10^{-10}(m) or smaller, which is comparable with the atomistic scale. This suggests that the macroscopic near-threshold fatigue crack growth in materials is determined by events in the atomistic scale. The existence of the threshold itself may be closely related to the discrete nature of atoms. Calculations taking account of the discrete nature of materials at the crack tip have advantages over the continuum mechanics approaches in naturally simulating the fracture events in the near-threshold region.

To simulate the deformation and fracture of materials in the atomistic scale under monotonic loadings, the molecular dynamics method and Monte Carlo method have been applied. Many investigations on the growth of cracks under monotonic loadings[3-14] have found that the molecular dynamics can be successfully applied to simulate the crack growth under monotonic loading.

The molecular dynamics method can therefore be a promising approach for understanding the fatigue crack growth in the near-threshold region. There are several reports on the application of the molecular dynamics to the fatigue damage[15-17], but no studies except ours[18-19] are available on its application to fatigue crack growth. The previous papers[18-19] showed the usefulness of the molecular dynamics approach in simulating the near-threshold fatigue crack growth in the BCC (body-centered cubic) iron

It has been reported that grain boundary has significant effect on strength under monotonic loading. In the present study the effect of grain boundary on the near-threshold fatigue crack growth in the BCC iron is studied by using the molecular dynamics method.

2. PROCEDURE OF ANALYSIS

2.1 Models Used in the Analyses

It has been reported that fatigue crack propagates predominantly in Mode I (opening mode), although it propagates in the mode II or mode III microscopically. A macroscopic crack under the mode I loading in the BCC iron at 600 (K) was considered. Due to the limitation of number of atoms treated in the simulation only a computational cell in the vicinity of the crack tip was modeled with atoms.

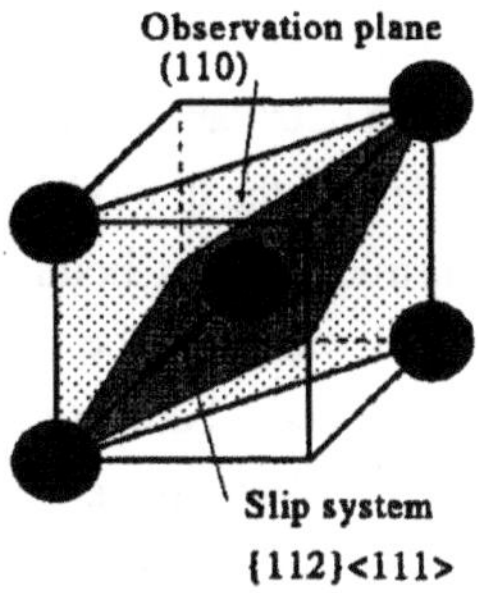

Figure 1: Slip system of BCC lattice and observation plane.

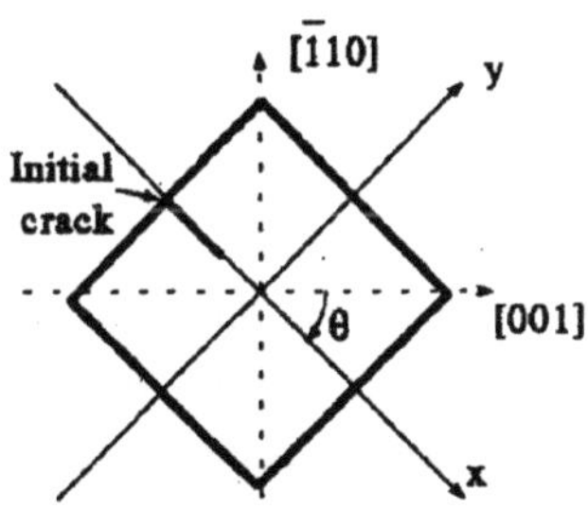

Figure 2: θ model in observation plane of (110).

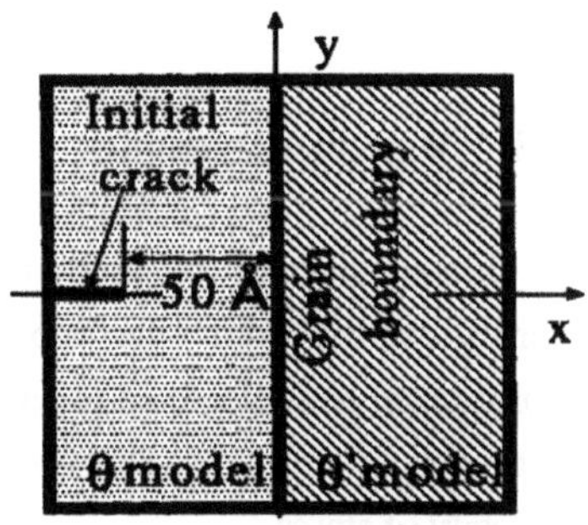

Figure 3: θ-θ' grain boundary model.

As is seen in Fig. 1, (110) plane is a close-packed plane for the BCC crystal, and the slip plane and the slip direction of the crystal are expressed by {112} plane and <111> direction. Considering that the slip system plays an important role in the fracture behavior, the close-packed plane (110), which contains the slip direction, is used as the observation plane. It has been reported that the fatigue crack preferred to propagate macroscopically in <100> direction with the crack plane of the close-packed plane {110} in experiments[20]. Then the crack expressed by the close-packed plane (110) is used as a reference model of the crack plane.

The coordinate with coordinates in [001] and [$\bar{1}$10] directions in (110) plane was taken as the reference coordinates. For investigating the effect of orientation of crystal, the θ model was defined by rotating the reference coordinates by angle θ as shown in Fig. 2. The model consists of about 6500 atoms within a square region of $-68.25\text{Å} < x < 68.25\text{Å}$, $-68.55\text{Å} < y < 68.55\text{Å}$.

To examine the effect of grain boundary, coincidence site lattice was employed. As is shown in Fig. 3, the θ model and the θ' model constituting the coincidence site lattice, were combined to construct a grain boundary model called θ-θ' model. The grain boundary in the model was set to be perpendicular to the crack plane. The effect of angle between the grain boundary and the crack was examined elsewhere[24]. In this study $0° - 50.48°$ model and $50.48° - 0°$ model, which correspond to Σ11 coincidence site lattice, were used as large angle grain boundary models. As small angle grain boundary models $0° - 168.46°$ model and $168.46° - 0°$ model corresponding to Σ99 coincidence site lattice were used.

2.2. Molecular Dynamics

In the molecular dynamics, force acting at an atom is calculated by summing up forces exerted from other atoms. In this study the equation of motion was constructed and integrated to calculate the location and velocity after a very small time step of 10^{-15} (s). The Verlet algorithm[21] was used to update the location and velocity of atoms.

2.3 Potential Describing Forces between Atoms

The atoms are allowed to interact with one another in accord with a force law between atoms. For modeling the force between atoms of the BCC iron, pair potentials are introduced. The force is determined by differentiating the potential expressed in terms of the distance between atoms with distance. Among various empirical pair potential functions describing the force between atoms, the Johnson potential[22] has been widely used and have demonstrated the usefulness and reliability in several studies[22,3,8,9]. In this study the Johnson potential was used.

The lattice constant was set at 2.873 Å, which was determined in such a way that the stress obtained from the molecular dynamics simulation at 600 (K) is close to 0. The cut-off distance of the Johnson potential was 3.44 (Å). To reduce the computation time, the book-keeping method[21] was used. The book-keeping distance was set at 6 (Å) in this study. Two layers of atoms were considered in the computational cell of the observations plane, and periodic boundary conditions were applied in the z direction, which was perpendicular to the observation plane. The dimension in the thickness direction is then about 4.06 (Å). Assuming the plane strain condition, the out-of-plane motion of atoms in the z direction was constrained.

2.4 Fatigue Loading

Cyclic load with stress ratio of $R = 0$ was applied to the model. The stress intensity range ΔK was set at 1.8 MPa·m$^{1/2}$. The atoms within the computational cell are surrounded by a boundary region, where displacements are given in accord with the linear elastic fracture mechanics. The Cartesian coordinates x, y, z are taken with the origin and the xy plane coinciding with the crack tip in the observation plane. The crack plane is parallel to the x axis. The displacements u and v in the x and y directions on the boundary of the computational cell under the plane strain condition are given by the near-tip fields expressed in terms of the stress intensity factor based on the fracture mechanics[23]. Initial crack was introduced by inactivating attractive forces for $-68.95\text{Å} < x < -48.95\text{Å}$, across the crack line. Irreversibility plays an important role in the fatigue crack growth. The forces between atoms departed from each other on the newly created fracture surface may be weakened by oxidation or deposition of foreign atoms on the surface. To introduce irreversibility due to oxidation, the attracting force was inactivated between atoms departed further than the cut-off distance 6.5 (Å)[19].

The current crack is defined by the surface, across which distance between atoms is larger than the cut-off distance 3.44 (Å). The displacement boundary conditions are updated by using the current crack tip[19]. The initial distribution of velocity of atoms were given in accord with the Maxwell distribution.

Quasi-static simulations with the temperature control were made. To conduct an isothermal computation, velocity of all atoms was scaled by a factor to fit the prescribed temperature of 600 (K). Initial relaxation of 1000 steps was introduced in the first place before loading. To realize the load increment, small amount of displacement change was given on the boundary of computational cell. Relaxation of 99 steps followed every step of the load increment. The set of loading of 1 step and relaxation of 99 steps is called 1 term. The increment of displacement in 1 term is equivalent to change of one-thousandth of the stress intensity range ΔK.

3. FATIGUE CRACK GROWTH BEHAVIOR

The crack growth behavior with cyclic loading was investigated using the molecular dynamics for the large angle grain boundary models of $\Sigma 11$ coincident site lattice, i.e. $0° - 50.48°$ model and $50.48° - 0°$ model. The locations of atoms at K_{max} were examined. The crack ceased to propagate in the 7th cycle for $0° - 50.48°$ model and in the 4th cycle for $50.48° - 0°$ model before the crack penetrate the grain boundary. The locations of atoms at K_{max} in the 3rd cycle for $0° - 50.48°$ model are shown in Fig. 4, and those in the 4th cycle for $50.48° - 0°$ model are shown in Fig. 5. In these figures the atoms, which are originally in the same plane perpendicular to $[\bar{1}10]$ direction, are interconnected by lines.

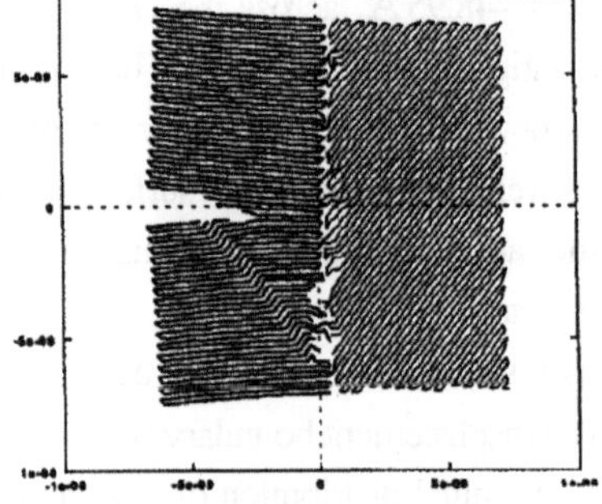

Figure 4: Locations of atoms at K_{max} in the 3rd cycle for $0° - 50.48°$ large angle grain boundary model, which has $\Sigma 11$ coincident site lattice.

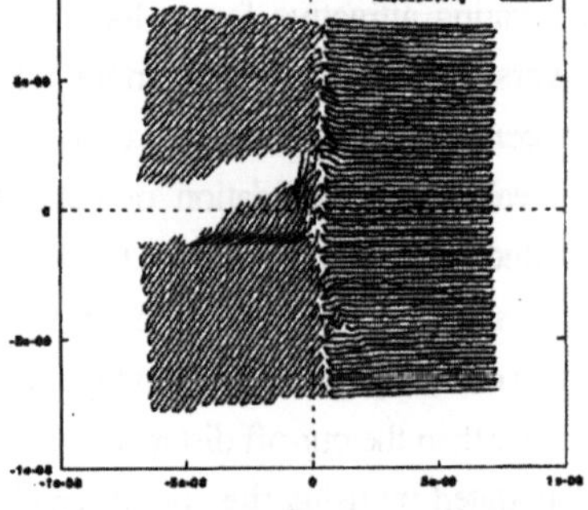

Figure 5: Locations of atoms at K_{max} in the 4th cycle for $50.48° - 0°$ large angle grain boundary model, which has $\Sigma 11$ coincident site lattice.

It is seen in Fig. 4 that the crack in the $0° - 50.48°$ model ceased to propagate before it reaches the grain boundary and dislocations are trapped on cavities formed on the grain boundary. Figure 5 shows that the crack in $50.48° - 0°$ model ceased to propagate when it reaches the grain boundary.

Molecular dynamics simulations were also made for the small angle grain boundary models of Σ99 coincident site lattice, i.e. $0° - 168.46°$ model and $168.46° - 0°$ model. The locations of atoms at K_{max} in the 6th cycle for $0° - 168.46°$ model are shown in Fig. 6, and those in the 3rd cycle for $168.46° - 0°$ model are shown in Fig. 7. It is seen in Fig. 6 that the crack tip grew beyond the grain boundary.

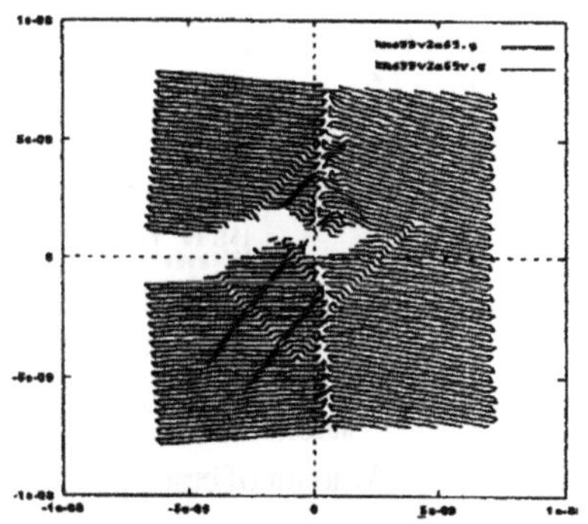

Figure 6: Locations of atoms at K_{max} in the 6th cycle for $0° - 168.46°$ small angle grain boundary model, which has Σ99 coincident site lattice.

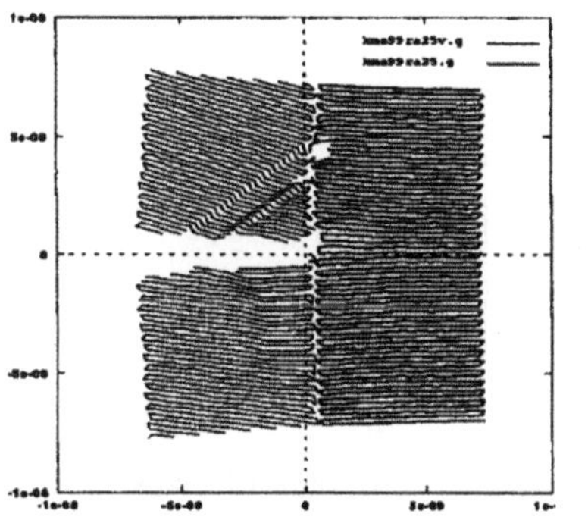

Figure 7: Locations of atoms at K_{max} in the 3rd cycle for $168.46° - 0°$ small angle grain boundary model, which has Σ99 coincident site lattice

The amount of crack growth obtained in the simulations are plotted against the number of cycles in Fig. 8 for the large angle grain boundary model (Σ11, $0° - 50.48°$ and $50.48° - 0°$ models). For comparison purpose the results for $0°$ model and $50.48°$ model without grain boundaries are shown in the figure. It is seen that the amount of crack growth decreases when the crack approaches the grain boundary and ceases to propagate.

Figure 9 shows the amount of crack growth obtained in the simulations against the number of cycles for the small angle grain boundary model (Σ99, $0° - 168.46°$

and 168.46°-0° models). For comparison purpose the results for 0° model and 168.46° model are shown. It is seen that the amount of crack growth per cycle decreases when the crack approaches and eventually the crack goes through the grain boundary. When dislocations or the crack got through the grain boundary, grain boundary had little influence on the crack growth behavior.

Since the present simulations were made only for limited conditions with small number of atoms, further researches are required.

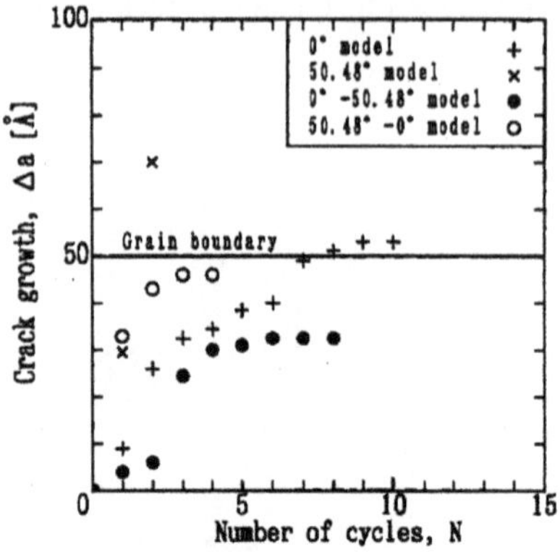

Figure 8: Amount of crack growth against number of cycles for 0°-50.48° and 50.48°-0° small angle grain boundary models, which has Σ11 coincident site lattice.

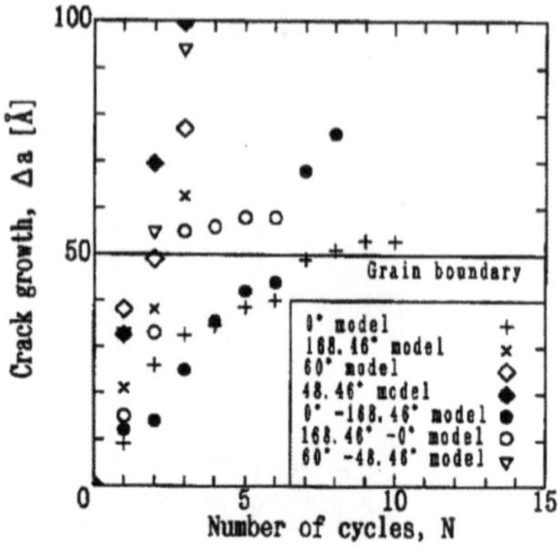

Figure 9: Amount of crack growth against number of cycles for 0°-168.46° and 168.46°-0° small angle grain boundary models, which has Σ99 coincident site lattice.

4. CONCLUSIONS

The effect of grain boundary on the near-threshold mode I fatigue crack growth behavior in BCC iron was investigated using the molecular dynamics method. The observation plane was set at (110), and the crack plane was assumed to be perpendicular to the observation plane. When the grain boundary existed perpendicular to the crack plane, the grain boundary served as an obstacle to crack growth. The effect of the grain boundary as a barrier was larger for the large angle grain boundary than for the small angle grain boundary. When dislocations got through the grain boundary, however, grain boundary had little influence on the crack growth behavior.

Acknowledgements

This study is partly supported by the Grant-in-Aid for Scientific Research, The Ministry of Education, Science, Sports and Culture.

REFERENCES

1. Laird C., Smith G.C. Crack propagation in high stress fatigue. Phil. Mag., 1962; 7: 847.
2. Pelloux R.M.N. Mechanisms of formation of ductile fatigue Striations. Trans. ASM 1969; 62: 281.
3. Chang R. An atomic study of fracture. Int. J. Fracture Mech. 1970; 6: 111.
4. Kanninen M.F., Gehlen P.C. Atomic simulation of crack extension in bcc iron. Int. J. Fracture Mech.. 1971; 7: 471.
5. Gehlen P.C., Hahn G.T., Kanninen M.F. Crack extension by bond rupture in a model of bcc iron. Scripta Metallurg. 1972; 6: 1087.
6. Mullins M., Dokainish M.A. Simulation of the (001) plane crack in α-Iron employing a new boundary scheme. Phil. Mag. A 1982; 46: 771.
7. Mullins M. Molecular dynamic simulation of propagating cracks. Scripta Metallurg. 1982; 16: 663.
8. deCelis B., Argon A.S., Yip S. Molecular dynamics simulation of crack tip processes in alpha-iron and copper. J. Appl. Phys. 1983; 54: 4864.
9. Sieradzki K., Dienes G.J., Paskin A., Massoumzadeh B. Atomistics of crack propagation. Acta Metallurg. 1988; 36: 651.
10. Kohlhoff S., Gumbsch P, Fishmeister H.F. Crack propagation in b.c.c. crystals, studied with a combined finite-element and atomistic model. Phil. Mag. A 1991; 64: 851.
11. Kitagawa H., Nakatani A., Shibutani Y. Molecular dynamic study of crack

processes associated with dislocation nucleated at the tip. Mater. Sci. Eng., A 1994; 176: 263.

12. Kitagawa H., Nakatani A. Computer modelling for materials with crystalline structure (molecular dynamic simulation of microscopic crack tip field under antiplane shear loading). JSME Int. J., Ser. A 1995; 38: 1.
13. Nakatani A., Kitagawa H., Shibutani Y., Nakatani K. Molecular dynamic study on ductile crack process (effect of temperature on dislocation nucleation). Mater. Sci. Research Int. 1995; 1: 11.
14. Kubo S., Nakata M., Ohji K., Ishii A. Analysis of deformation and fracture near interface crack tip using molecular dynamics method. Trans. Japan Soc. Mech. Engrs. (in Japanese) 1999; 65: 560.
15. Ikuno T., Sumida T., Tanaka K., Watanabe E. A molecular dynamic approach on fatigue damage. Preprint Japan Soc. Mech. Engrs. 1994; No. 940-39: 34.
16. Inoue H., Akahoshi Y., Harada S., Chobara H. Fracture simulation of meso-scale polycrystal of bcc iron by using molecular dynamic method. Preprint Japan Soc. Mech. Engrs. 1994; No. 940-39: 63.
17. Inoue H., Akahoshi Y., Harada S. An analysis of fatigue of nano-scale polycrystalline iron by using molecular dynamic method. Preprint Japan Soc. Mech. Engrs. 1995; No. 95-1: 154.
18. Kubo S., Yoshikawa M. Molecular dynamic simulation of near-threshold fatigue crack propagation. Fatigue, Fracture, and High Temperature Design Methods in Pressure Vessels and Piping, ASME 1998; PVP-365: 363.
19. Kubo S., Misaki M. Molecular dynamic study on the effect of crystallographic orientation on near-threshold fatigue crack propagation in iron. Material Science Research Int., Special Technical publications. 2001; 1: 220.
20. Jono M., Sugeta A. Direct SEM observations and mechanisms of fatigue crack growth. J. Mechanical Behavior of Materials 1991; 3: 157.
21. Verlet L. Computer 'experiments' on classical fluids. I. Thermodynamical properties of Lennard-Jones molecules. Phys. Rev. 1967; 159: 98.
22. Johnson R.A. Interstitials and vacancies in α Iron. Phys. Rev. 1964; 134: A1329.
23. Broek D. Elementary engineering fracture mechanics. Nordhoff ,1974
23. Kubo S., Misaki M. (to be presented).

PLASTICITY-INDUCED MARTENSITIC TRANSFORMATION AROUND SEMI-ELLIPTICAL SURFACE CRACKS IN FATIGUE OF AN AUSTENITIC STAINLESS STEEL

Y. Nakasone
Department of Mechanical Engineering, Faculty of Engineering, Tokyo University of Science 1-3 Kagurazaka, Shinjuku-ku, Tokyo 162-8601, Japan

S. Kasumi and Y. Iwasaki
Graduate Students, Graduate School of Engineering, Tokyo University of Science 1-3 Kagurazaka, Shinjuku-ku, Tokyo 162-8601, Japan

Abstract: The present study investigates plasticity-induced martensitic transformation around semi-elliptical cracks in an austenitic stainless steel SUS304 fatigued at room temperature. Distributions of the volume fraction of α' martensitic phase around semi-elliptical fatigue cracks were measured by ferrite scope. The results were compared with vertical magnetic flux density B_z distributions above and below the cracks in specimens magnetized by a strong electromagnet. It was revealed that the distance between the outermost peaks of the B_z distributions $2l$ showed good linear correlations with surface crack length $2a$. The maximum and the minimum values of B_z, B_{zmax} and B_{zmin}, also showed linear relations with maximum stress intensity factors at the surface tip K_{amax} and at the depth position K_{bmax}. These results imply that not only $2a$ but K_{amax} and K_{bmax} values can be estimated in an electromagnetic non-destructive way.

Key words: Martensitic Transformation, Semi-elliptical Crack, Stress Intensity Factor, Fatigue, Leakage Magnetic Flux Density, Non-Destructive Evaluation

1. INTRODUCTION

Type SUS304 stainless steel (equivalent of AISI304 stainless steel) is known as an austenitic stainless steel that has unstable γ austenitic phase

H. Kitagawa and Y. Shibutani (eds.), IUTAM Symposium on Mesoscopic Dynamics of Fracture Process and Materials Strength, 321–330.

around and below room temperature [1-7]. The stainless steel usually shows ductile and non-ferromagnetic nature, whereas the steel becomes brittle and ferromagnetic after it undergoes high stress or strain since γ austenitic phase in the steel is transformed into α' martensite.

In the present study, the volume fraction of α' martensitic phase induced in SUS304 plate specimens subjected to uniform tensile stress at room temperature was measured by three types of equipments; i.e., vibrating sample magnetometer (VSM), X-ray diffractometer and ferrite scope (FS) in order to obtain the applied strain level dependence of the volume fraction of the α' phase transformed. The distributions of α' volume fraction in the plastic wake regions produced around semi-elliptical cracks in fatigued SUS304 plates were then electromagnetically measured by flux gate (FG) sensor. The fatigued specimens were magnetized by a strong permanent magnet and the induced magnetic flux density B_z distributions above and below the surface cracks were measured by FG sensor in order to be correlated to the surface crack length and to the applied stress intensity factors at the depth position as well as at surface.

2. EXPERIMENTS

2.1 Fatigue crack propagation experiments

Table 1 shows the chemical composition of Type SUS304 stainless steel tested in the present study. Figure 1 shows the geometry of plate specimens used in this study. A part-through notch having a surface length of $2a$=2.5 mm and a depth of b=2 mm was made at the center of each specimen by electric discharge machining. Fatigue crack propagation tests were conducted on this type of specimens at maximum applied stress levels of

Table 1. Chemical composition of SUS304 stainless steel the material tested (mass %).

Material	C	Si	Mn	P	S	Ni	Cr
SUS304	0.06	0.50	0.87	0.01	0.037	8.10	18.21

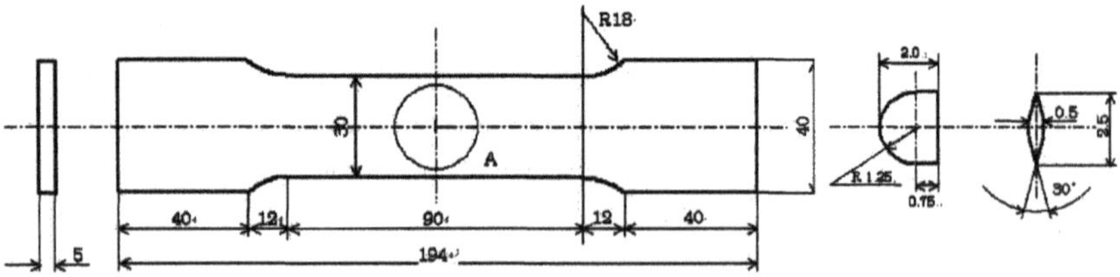

Figure 1. Specimen geometry (unit: mm).

σ_{max}=222-300MPa and at stress ratios of $R=\sigma_{min}/\sigma_{max}$=0.1-0.4 under load control condition. Fatigue propagation tests were intermitted at an surface crack length increment of Δa=0.5-1 mm and beach mark patterns were produced by changing the R-ratios with the maximum applied stress level kept constant in order to determine the profile of the part-through fatigue cracks.

2.2 Measurement of α' volume fraction by ferrite scope

α' volume fraction $\xi_{\alpha'}$ around the growing fatigue cracks was measured by ferrite scope (FS). The measurements of $\xi_{\alpha'}$ were made at grid points of approximately 1 mm by 1 mm in rectangular regions of 6 mm in the longitudinal direction by 21 mm in the width direction of the specimen with the crack placed at the center of the measurement regions on the front and the back surface sides of each specimen. The results of the measurements were calibrated by those obtained by a vibrating sample magnetometer (VSM) and by X-ray diffractometer [1]. γ austenitic phase was transformed into α' martensitic phase due to plasticity-induced martensitic transformation in the plastic wakes along the peripheries of surface cracks where stress was so high that severe plastic deformation took place.

2.3 Magnetic flux density distributions above and below semi-elliptical cracks in fatigue

Fatigued specimens were demagnetized intermittently during fatigue tests by a demagnetizer that applies ac magnetic field to the specimens by gradually reducing the amplitude of the magnetic field to zero. The demagnetized specimens were then magnetized by a high-field dc electromagnet of 0.6 T or higher in the directions of the x- and the z-axes, i.e., in the direction parallel to semi-elliptical cracks and that perpendicular to the specimen surface, respectively. A magnetic field of 0.5 T induced by electromagnet is considered sufficient to put fatigued specimens in a saturation magnetization state.

After the magnetization treatment, the magnetic flux leakage was measured by scanning a flux gate (FG) sensor supplied by Shimadzu Co., Ltd. above and below the part-through fatigue cracks with lift-offs of 3 mm and 4 mm for the x- and the z-direction magnetization, respectively. The magnetic flux leakage measurements were conducted in a magnetic shielding box covered with aluminum foil for the purpose of preventing environmental magnetic fields from disturbing the magnetic flux leakage measurements. The measurements were automatically made on the front and the back surface sides of the specimens by the use of a computer-controlled x-y table at grid points of 1 mm by 1 mm in rectangular regions of 23 mm in the longitudinal direction by 37 mm in the width direction of the specimens with semi-elliptical cracks placed at the center of the measurement regions.

3. RESULTS AND DISCUSSION

3.1 Martensitic phase transformation in Type SUS304 stainless steel

Figure 2 shows the nominal stress vs. nominal strain diagram of the present SUS304 steel at room temperature and the variation of the volume fraction of α' martensite $\xi_{\alpha'}$ with the applied strain level ε, indicating that higher α' phase content brings about higher work hardening ratio at ε=20 % and higher. The value of $\xi_{\alpha'}$ is determined by the ratio of the saturation magnetization of a deformed specimen to that of a fully transformed specimen of the present SUS304 steel whose chemical composition is shown in Table 1. The $\xi_{\alpha'}$ value of the fully transformed SUS304 specimen is theoretically calculated by the Slater-Pauling diagram [1, 7, 8]. γ austenite becomes so unstable that it is transformed into α' martensite above the yield stress level even at room temperature [1-4]. The volume fraction of α' martensite transformed is increased exponentially with the applied strain level ε in the plastic deformation region as shown by the solid line in Fig. 2. The $\xi_{\alpha'}$ vs. ε diagram indicated by the solid line can be expressed by the following equation:

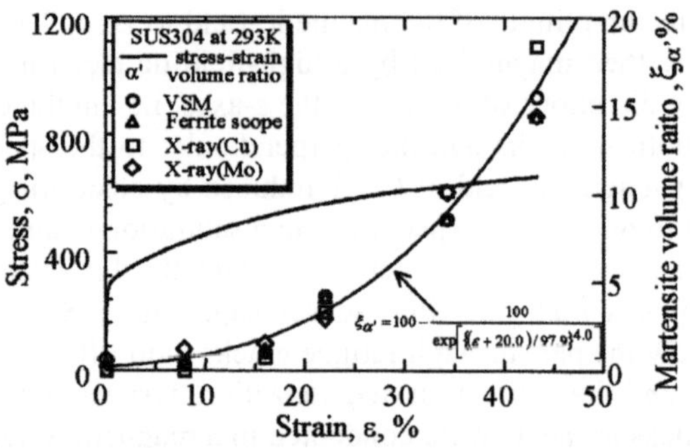

Figure 2. Comparison of the stress-strain diagram with the $\xi_{\alpha'}$ vs. strain diagrams for the present SUS304 showing that higher α' phase content brings about higher work hardening ratio.

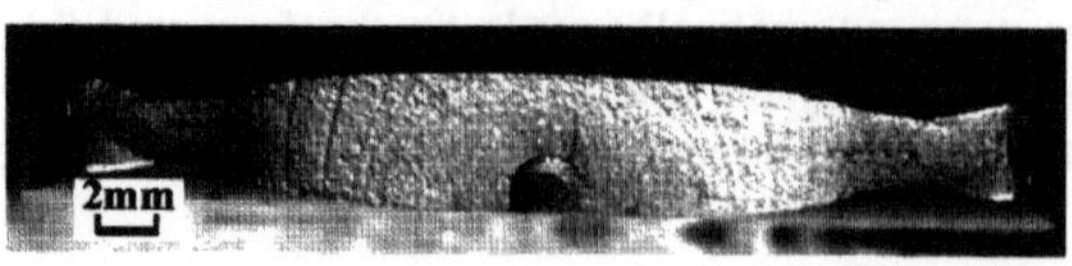

Figure 3. Crack morphology imprinted on a fracture surface.

$$\xi_{\alpha'} = 100 - \frac{100}{\exp\left[\left\{(\varepsilon + 20.0)/97.9\right\}^{4.0}\right]} \tag{1}$$

The value of $\xi_{\alpha'}$ remains small for strain levels lower than 20%, but it reaches as high as 15.5% or higher for ε=43.2 % at room temperature.

3.2 Part-through crack profiles

Figure 3 shows part-through crack profiles imprinted on the fracture surface by the beach mark method, showing the shape of the crack was semi-elliptical. The value of the aspect ratio b/a, the ratio of the crack depth b to the half surface length a varied from 1.61 to 0.87 during the experiments, i.e., the crack grew shallower and shallower.

3.3 α' martensitic phase distributions around a part-through fatigue crack

Figures 4(a) through (f) show the spatial distributions of the volume fraction of α' martensitic phase transformed around a growing semi-elliptical crack on the front and the back surfaces of a specimen fatigued at a constant stress range of $\Delta\sigma$=240MPa with R=0.1 and at different values of $K_{a\max}$ and $K_{b\max}$ where $K_{a\max}$ and $K_{b\max}$ are the values of the stress intensity factor (SIF) at the front surface tip and at the maximum point at depth, respectively. The semi-elliptical crack broke through the specimen back surface at the number of load cycles N=95,240, when $2a \fallingdotseq 11.0$ mm. The length of the crack on the back surface at the time of penetration was approximately 5.2 mm.

Figures 4 (a) through (c) and (d) to (f) illustrate contour maps of the volume fraction $\xi_{\alpha'}$ of α' phase transformed on the front and the back surfaces of the specimen respectively at different stages of the fatigue crack propagation process. Figure 4(a) is the $\xi_{\alpha'}$ distribution around the crack at N=51,000 when $2a$=7.8 mm and b=3.5 mm (70 % of the specimen thickness), showing that the transformation region was small and the value of $\xi_{\alpha'}$ was as low as 3 %. As the crack grew larger as shown in Figs. 4(b) and (c), the $\xi_{\alpha'}$ value around the surface crack tips became larger and the plastic wakes or the transformation region developed wider. Figure 4(b) illustrates the $\xi_{\alpha'}$ distribution around at N=56,000 when $2a$=10.5 mm and b=4.3 mm (86 % of the specimen thickness) where the value of $\xi_{\alpha'}$ reached a maximum value of about 8 % near the crack tips at surface and became lower away from the crack tips.

On the back surface, however, α' transformation had not become apparent until crack depth b reached 3.9 mm or 78 % of the specimen thickness. Figure 4(d) illustrates this situation. No indication of the transformation was

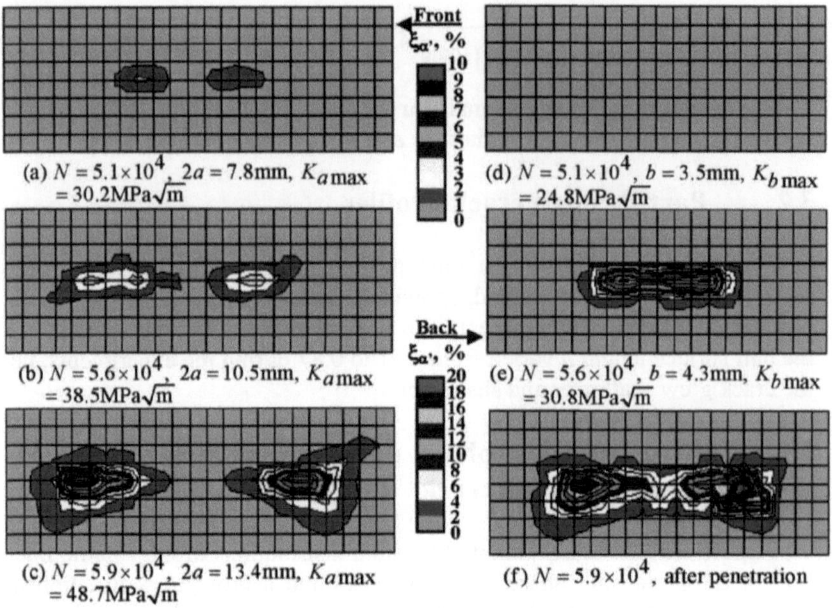

Figure 4. Evolution of contour maps of α' martensite volume fraction $\xi_{\alpha'}$ around a semi-elliptical crack on the front and back surfaces of a specimen: σ_{max}=300 MPa and R=0.2.

observed in the contour diagrams of $\xi_{\alpha'}$; i.e., the $\xi_{\alpha'}$ value remained about 0.3 %. A volume fraction of this amount is considered as that of retained δ-ferrite introduced during the production of this material. For a=10.5 mm and b=4.3 mm, a steep contour map of $\xi_{\alpha'}$ suddenly appeared on the back surface below the part-through crack (Fig. 4(e)). Due to severe plastic deformation in the uncracked ligament of the specimen, the $\xi_{\alpha'}$ value reached as high as 18 % and was much higher on the back surface than on the front, although the maximum value of the elastic SIF at depth K_{bmax} calculated by the Newman-Raju formula [9] was lower than that at surface K_{amax}.

3.4 Magnetic flux density distribution above and below a part-through fatigue crack

Figure 5(a) shows a representative 3D distribution of vertical magnetic flux density B_z, i.e., the component of leakage magnetic flux density $\vec{B}$ normal to specimen surface, above a semi-elliptical fatigue crack at K_{amax}=33.0 MPa$\sqrt{\text{m}}$ and K_{bmax}=24.9MPa$\sqrt{\text{m}}$ for R=0.1. The figure shows a B_z distribution measured on the front surface side for the case of the x-direction magnetization. The B_z distribution looks similar to those of the through-thickness crack cases [2-4]. The minimum negative peak was observed above the left crack tip near the north pole of a magnetizing electromagnet, whereas the maximum positive one above the right crack tip. A magnetized plastic wake is considered to play a role as a composite of permanent magnets having different magnetization properties which vary according to the volume fraction of α' phase transformed.

B_z distributions on the back surface sides are completely reversed but look similar in shape to those above through-thickness cracks for the x-direction magnetization in contrast to the contour maps of $\xi_{\alpha'}$ obtained by ferrite scope as shown in Figs. 4(d) to (f). Two small peaks in the middle are small compared to those observed above through-thickness cracks.

For the z-direction magnetization, B_z distributions look completely different from those for the x-direction magnetization as shown in Fig. 5(b). Two positive peaks of about the same height were detected on the front surface side, whereas two negative peaks on the back surface side when the north pole of the magnetizing electromagnet was placed facing against the front surface of a specimen and the south pole against the back surface. The reason for a pair of two peaks having almost the same height is that the SIF distribution along the crack periphery should be symmetric.

3.5 Surface crack length 2*a* vs. peak distance 2*l* relations

As indicated in Figs. 5(a) and (b), the vertical leakage magnetic flux density B_z reflects the distribution of the volume fraction of α' martensite transformed in the plastic wakes produced around two fatigue crack tips on the front surface side, thus the distance $2l$ between the outermost minimum and maximum peaks for the x-direction magnetization or the two maximum peaks for the z-direction magnetization can be correlated with real crack length $2a$. Figures 6(a) and (b) show relations between the real surface crack

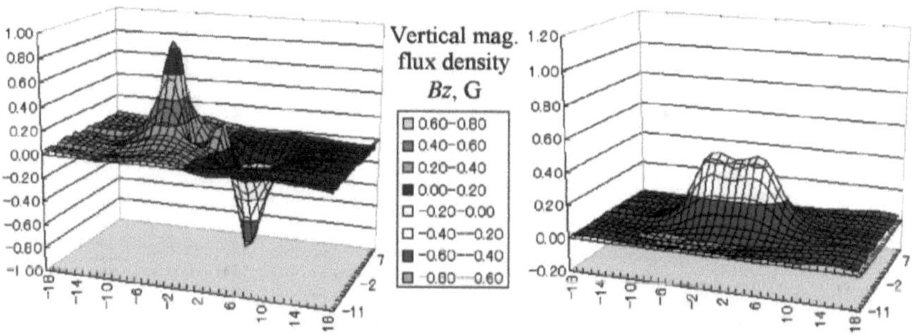

(a) x-direction magnetization (b) z-direction magnetization

Figure 5. 3D distributions of vertical magnetic flux density B_z above a semi-elliptical fatigue crack for $2a$=9.8 mm, b=4.4 mm, $K_{a\max}$=33.0 MPa$\sqrt{m}$, $K_{b\max}$=24.9 MPa$\sqrt{m}$, and R=0.1.

length $2a$ and the peak distance $2l$ obtained by FG sensor measurements for the x- and the z-direction magnetizations, respectively. Solid symbols in Figs. 6(a) and (b) indicate the $2a$ vs. $2l$ relations obtained for different values of the R-ratio from the front surface sides of fatigued specimens magnetized in the x- and z-directions, respectively. Good linear correlations can be found between $2a$ and $2l$ data obtained on the front surface side of the specimens for the two types of magnetization, and each relation can be represented by a single straight line irrespective of the R-ratio. The $2a$ vs. $2l$ relations can estimate real surface crack length $2a$ in service by measuring the peak distance $2l$. Even from the back sides of the specimens, the $2a$ vs. $2l$ relation can estimate real surface crack length $2a$, although the $2a$ vs. $2l$ plots measured on the back surface scatter as shown in Fig. 6(a).

For the z-direction magnetization, however, only a single negative peak was observed on the back surface side before crack penetration and the value of $2l$ cannot be obtained. Thus, it is concluded that surface crack estimation is difficult to make by the use of the $2a$ vs. $2l$ relations for the case of the z-direction magnetization from the back sides of the specimens.

3.6 Maximum or minimum peak value B_{zmax} or B_{zmin} vs. maximum stress intensity factor K_{amax} or K_{bmax}

The magnitude of the peaks varies according to the amount of plasticity-induced martensitic transformation, or fatigue damage specimens have suffered, in the plastic wakes around growing fatigue cracks. Figures 7(a) and (b) show the relations between the peak values of the vertical leakage magnetic flux density B_{zmax} and B_{zmin} and the values of K_{amax} and K_{bmax} for the case of the x-direction magnetization. Figures 7(a) and (b) were measured by FG sensor on the front and back surface sides, respectively. Open symbols in

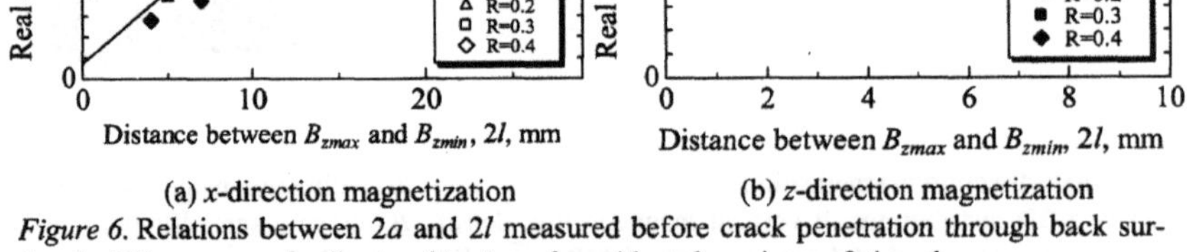

(a) x-direction magnetization (b) z-direction magnetization

Figure 6. Relations between $2a$ and $2l$ measured before crack penetration through back surface by FG sensor on the front and back surface sides of specimens fatigued.

the figures indicate the B_{zmax}- and B_{zmin}-K_{amax} diagrams and solid ones the B_{zmax}- and B_{zmin}-K_{bmax} diagrams. The values of K_{amax} and K_{bmax} can be linearly correlated with the peak values B_{zmax} and B_{zmin} irrespective of the R-ratio both on the front and the back surface sides. The data points obtained for different crack sizes but for the same values of K_{amax} or K_{bmax} fall on the same points on respective straight lines with positive and negative slopes.

Figures 8 (a) and (b) show the B_{zmax}-K_{amax} and -K_{bmax} relations for the z-direction magnetization obtained on the front and back surface sides, respectively. The results are the same as those obtained for the x-direction magnetization except that the B_z distributions show positive slopes only.

These results imply that the damage introduced in the plastic wake regions around fatigue cracks is controlled by maximum values of the SIFs and that the measurement of the leakage magnetic flux density can estimate

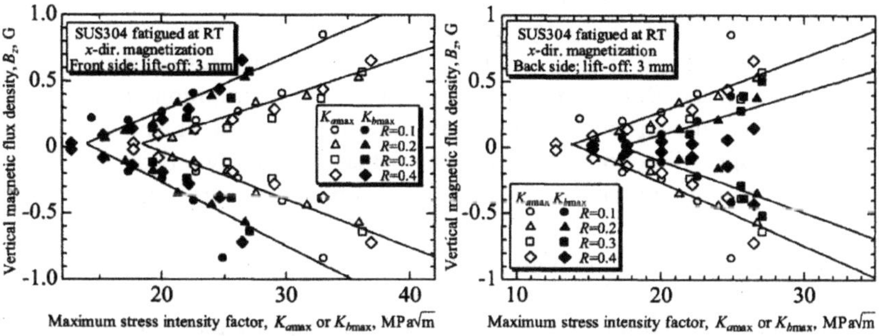

(a) Front surface side (b) Back surface side

Figure 7. Relations between B_{zmax} or B_{zmin} and K_{amax} or K_{bmax} obtained from FG sensor measurements on the front and the back surface sides of a fatigued specimen after the x-direction magnetization.

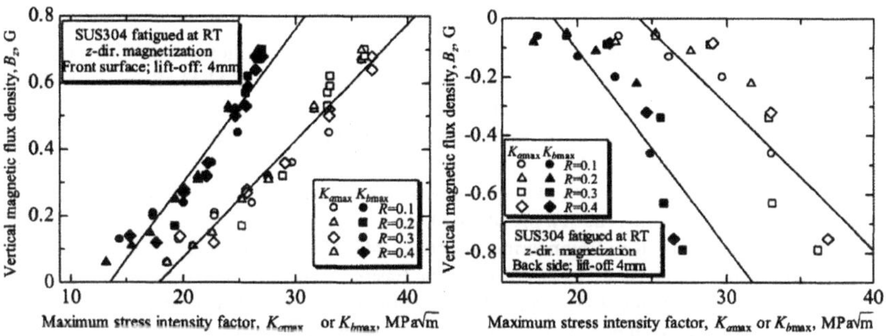

(a) Front surface side (b) Back surface side

Figure 8. Relations between B_{zmax} or B_{zmin} and K_{amax} or K_{bmax} obtained from FG sensor measurements on the front and the back surface sides of a fatigued specimen after the z-direction magnetization.

$K_{a\max}$ and $K_{b\max}$ from the back surface as well as from the front surface.

4. CONCLUSIONS

The present study investigates plasticity-induced martensitic transformation around semi-elliptical cracks in an austenitic stainless steel SUS304 fatigued at room temperature in air. Volume fraction $\xi_{\alpha'}$ of α' martensite transformed in uniformly stretched SUS304 plates was measured by VSM, X-ray diffractometer and ferrite scope, and expressed as a function of the applied strain level ε. Then, the spatial distributions of $\xi_{\alpha'}$ in the plastic wake regions around semi-elliptical cracks in fatigued SUS304 specimens were measured by ferrite scope. The results were compared with vertical magnetic flux density B_z distributions induced above and below surface cracks in the specimens magnetized by a strong electromagnet in the directions parallel to and perpendicular to the cracks, i.e., the x- and z-directions, respectively. It was revealed that the B_z distributions reflected the $\xi_{\alpha'}$ distributions in the wake regions; i.e., the distance $2l$ between outermost minimum and maximum peaks for the x-direction magnetization or that between two peaks of almost the same height and the same sign for the z-direction magnetization showed good linear relations with surface crack length $2a$. The magnitude of the maximum and/or the minimum peaks of B_z also showed good linear relations with the maximum values of the stress intensity factors at the surface tips $K_{a\max}$ and the depth position $K_{b\max}$. These results imply that not only $2a$ but $K_{a\max}$ and $K_{b\max}$ can be estimated in an electromagnetic non-destructive way.

REFERENCES

1. Y. Nakasone, et al., Report of Research Committee on the Application of Eletromagnetic Fracture Mechanics to the NDE of Damage and Degradation, JSAEM- R-9803, 9903 and 0005 (1999, 2000, 2001) (in Japanese).
2. Y. Nakasone, et al., Int. J. of Applied Electromagnetics and Mechanics, 15, Nos. 1-4 (2002), pp. 309-313.
3. Y. Nakasone, et al. In: Applied Electromagnetics and Mechanics, T. Takagi et al. (Eds.), IOS Press, Netherlands (2001), pp. 653-654.
4. Y. Nakasone, et al., Key Engineering Materials, **243-244** (2002), pp. 327-332.
5. Z. Nishiyama, Martensitic Transformations: the Basics, Academic Press, New York (1978).
6. Y. Tomita, et al., Trans. JSME, **60A**, No. 575 (1994), pp. 166-173 (in Japanese).
7. R. P. Reed and C. J. Gunter, TMA-AIME, **230** (1964), pp. 1713-1718.
8. J. Menard et al., Advances in Cryogenic Eng., **6** (1960), pp. 587-589.
9. K. S. Ravichandran, ASM Handbook, **19**, ASM International, Ohio (1997), pp. 159-160.
10. C. Kittel, Introduction to Solid State Physics, 2nd ed., John Wiley & Sons, Inc., New York (1953).

AN ATOMIC SIMULATION OF AFM-BASED NANO LITHOGRAPHY PROCESS FOR NANO PATTERNING

Y.S. Kim[1], S.O.Choi[2], S.R.Lee[1] and J. Kim[3]
[1] *Department of Mechanical Engineering, Kyungpook National University, Daegu 702-701, Korea; e-mail: caekim@knu.ac.kr*
[2] *Korea Institute of Industrial Technology, Chonan-Si 330-825, Korea*
[3] *Graduate School, Kyungpook National University, Daegu 702-701, Korea*

Abstract The atomic force microscopy (AFM)-based nano lithographic technique is currently used to directly machine material surfaces and fabricate nano patterning for MEMS devices. In this study, three-dimensional molecular dynamic (MD) simulations were conducted to evaluate the effect of crystallographic factors on the forming characteristics of the nano lithography process of monocrystalline copper. The simulation results revealed that the crystal orientation and ploughing direction had a significant influence on the ploughing force as well as the nature of the nano deformation and surface quality of the machined material.

Keywords: AFM, Molecular Dynamics Simulation, Nano Lithography, MEMS

1. INTRODUCTIONS

The rapid improvements in micro-electro-mechanical systems (MEMS) and micro-opto-electro-mechanical systems (MOEMS) in the last few decades have given rise to a wide variety applications, including mechanical elements, sensors, actuators, and electronics. In particular, MEMS devices, which are mostly manufactured through surface or bulk micromachining such as photolithography, electro-beam lithography or the LIGA process etc. on a silicon wafer, have many potential applications in the automotive industry, displays, printers, fluidics, optics, analytical instruments, communications and information, biomedical industry, and aerospace industry.[1-3]

As part of scanning probe microscope (SPM) technology, atomic force microscopy (AFM) has been used to evaluate and measure the mechanical and structural properties of various materials on a nanometer scale. When an AFM probe attached to the end of a flexible cantilever beam with a very low rigidity traces the surface of a material in a contact mode, deep scratching and several regimes, ranging from frictionless sliding to permanent wear, can be observed, depending on the applied load. In this way, AFM has been successfully used to characterize nano wear processes in materials of technological interest, such as the use of silicon for

H. Kitagawa and Y. Shibutani (eds.), IUTAM Symposium on Mesoscopic Dynamics of Fracture Process and Materials Strength, 331–341.

magnetic head sliders and polymers for electronic packaging and liquid crystals displays.

Recently, AFM-based nano lithography using the principle of ploughing has been proposed for machining material surfaces and fabricating nano structure components, such as nano patterning and nano wire.[4-6] AFM-based lithography is a technique that directly machines the surface of a material using a nano-sized hard probe (i.e. ploughing tool) attached to a rigid cantilever beam and can pattern nanometer structures, like holes and grooves. There, the material is removed from the substrate in a well-defined way, leaving behind deep trenches with the characteristic shape of the tool used. The advantages of the AFM-based lithographic technique are obviously the fabrication resolution,[7-9] the nondamaging process compared to conventional photolithography and electron-beam lithography, and the absence of additional processing steps, such as etching the substrate.

This process was recently adopted as a photomask repair tool to remove defects in mask pattern in semiconductor industry[9] where it is called atomic force microscopy guided nanomachining (AGN). To analyze the machining process and its characteristics on an atomistic scale, since it cannot be confirmed experimentally, numerical virtual simulation, such as molecular dynamic (MD) simulation,[10] is a very useful tool.

MD simulation has already been applied to a wide range of fields, including crystal growth, nano-indentation, tribology, fracture, and laser interactions etc, in order to simulate the atomic scale motion of the material.[11-14] Also the related studies by Ueda and Iwata,[15] Shimada *et al*,[16] Konig and Senrath,[17] Komanduri *et al*,[18] and Fang *et al*[19] are particularly noteworthy. Ueda and Iwata[15] investigated the mechanism of chip formation, the variation in the cutting forces, and the shear angle relative to the crystallographic orientation during the diamond cutting of β-brass, and reported on the formation of a discontinuous chip within a particular range of crystal orientations. Konig and Senrath[17] conducted cutting experiments using a monocrystalline copper OFHC substrate with {100}, {110}, and {111}-oriented crystals along different cutting directions, and observed a significant dynamic component in the cutting force along the <100> cutting direction, which was used to attribute the resulting poor surface quality of the machined surface. Recently, Fang *et al*[19] utilized a 3-dimensional MD simulation to study the effects of the scribing feed on the atomic-scale lithography process, and compared the MD simulation results with AFM experiments on a micro scale.

In this study, three-dimensional MD simulations for AFM-based nano lithography process of monocrystalline copper were conducted. The characteristics of the plastic deformation, dislocation generation with slip direction, chip formation, and force components (cutting force, thrust force, and width-direction force) as well as the surface quality of the machined material were then examined. The effects of crystallographic factors (crystal orientation and ploughing direction) for the AFM-based lithography process were also investigated.

2. SYNOPSIS OF MOLECULAR DYNAMICS

Based on the assumption that the behavior of every particle (atom or molecule) follows Newton's equation of motion, Molecular Dynamics is an analytical method that can identify the position vector $r_i(t)$, velocity vector $v_i(t)$, and other physical properties of every particle at each time step using the potential energy and force acting on each particle. The current study uses the Morse potential[20], which is known to effectively depict the interacting energy between two molecules combined by covalent bonding. The Morse potential has been used successfully in several s imilar studies[14,19] because it is simple, computationally inexpensive. The potential energy $\Phi(r_{ij})$ and force $F(r_j)$, calculated from the gradient(derivative) of the potential energy relative to a change in the atomic position r_{ij} $(=|r_i-r_j|)$ between molecular i and j, are represented by Eq. 1 and Eq. 2, respectively.

$$\Phi(r_{ij}) = D[e^{-2\alpha(r_{ij}-r_0)} - 2e^{-\alpha(r_{ij}-r_0)}] \tag{1}$$

$$F(r_i) = -\sum_{j=1}^{N} \nabla_i \Phi(r_{ij}) = m_i a_i \ , \qquad a_i = \frac{d^2 r_i(t)}{dt^2} \tag{2}$$

where F_i is the force on atom i resulting from the interaction of all other atoms, m_i is the mass of atom i, r_i is the position of atom i, and N is the total number of atoms.

Figure 1 shows the pair potential energies of the Morse potentials, while Table 1 shows the Morse potential parameters (D, α, r_o) used in the current study.[20]

The modified(velocity) Verlet scheme,[10] a numerical integration algorithm for Newton's equation of motion, Eq. 2, is used to trace the position of each atom throughout a specified period of time (typically in the order of picoseconds). There, position $r(t)$, velocity $v(t)$ and acceleration $a(t)$ at time $t+\Delta t$ are obtained from the same quantities at time t and $t+\Delta t/2$.

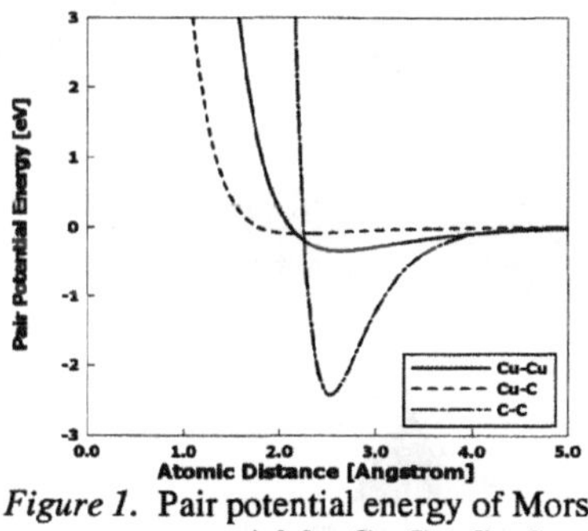

Figure 1. Pair potential energy of Morse potential for Cu-Cu, Cu-C, and C-C atoms

Table 1. Parameters used for Morse potential[20]

Pair	D [eV]	α [□]	r_o [□]	Lattice Constant [□]
Cu-Cu	0.343	1.359	2.626	3.6153
Cu-C	0. 100	1.700	2.200	-
C-C	2.423	2.555	2.522	3.5680

The initial velocities are assigned based on Maxwell-Boltzmann's distribution, where the direction of the initial velocity is assigned by uniformly random numbers and the magnitude of the initial velocity is chosen such that the average of the kinetic energy ($mv^2/2$) in each atom is equal to $3Nk_BT/2$, where k_B is Boltzmann's constant ($=1.38\times10^{-23} JK^{-1}$) and N is total number of atoms. To reduce the tremendous calculation time, the bookkeeping method is used. Furthermore, in the MD simulation of the AFM-based nano lithography process, since the temperature of the system tends to increase continuously, the velocity scale method, which rescales the velocities of the atoms during the simulation, is adopted to suppress the temperature at a specified temperature in the canonical system. The MD simulations are performed in the microcanonical ensemble or NVT ensembles: the number of molecules, the volume and the temperature are constant quantities.

3. MD SIMULATION PROCEDURE

MD simulations of the AFM-based nano lithography process were conducted on monocrystalline copper, an face-centered cubic (FCC) structure, to study the effect of crystallographic parameters on the nature of the material deformation during ploughing process. The tool as a AFM probe was modeled with a rigid cone-shape diamond with a hemispherical tip. The work material, as shown in Figure 2, was divided into three different zones, namely, the Newtonian atoms, thermostat atoms[10], and boundary atoms.

Figure 3 shows the various crystal orientations and corresponding cutting directions for the monocrystalline copper used in the current investigation. Table 2 summarizes the computational parameters used in MD simulation; tool and work material dimensions, simulation condition (crystal orientation and ploughing direction, ploughing speed, undeformed chip thickness (ploughing depth)). The size of the work material is a 201.6Å×201.6Å×39.6Å in width, length, and depth. The diamond tool has a 60° tool angle with a tip radius of 10Å. For example in Table 2, MD simulation for the the crystal setup A:(001)[100] indicates that the nano lithography was performed in ploughing direction [100] on ploughing plane (001) in crystallographic expression, where ploughing plane (001) becomes the top surface of the work material.

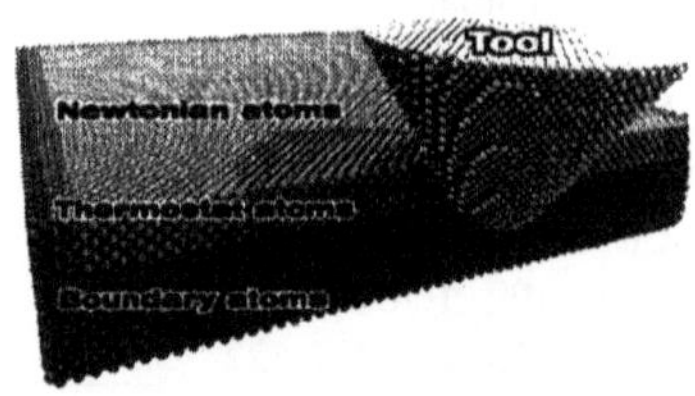

Figure 2. MD simulation model for AFM-based nano lithography

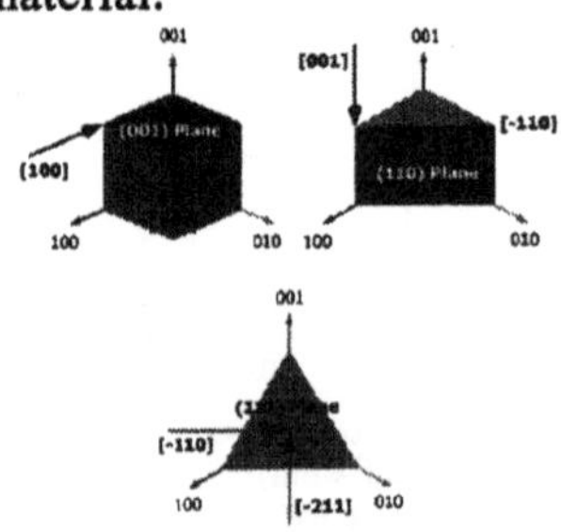

Figure 3. Crystal orientation and the ploughing directions

Table 2. Computational parameters used in MD simulation

Configuration	3-dimensional MD simulation conditions
Work-material dimensions	56a×56a×11a; a (= lattice constant) = 3.6153Å
Tool dimension	Tip radius: 10Å, Tool angle: 60°
No. of atoms in work material [Crystal setup/No. of atoms]	A:(001)[100]/129370, B:(001)[-110]/129332 C:(110)[001]/126420, D:(110)[-110]/126420 E:(111)[-211]/127638, F:(111)[-110]/127638
Ploughing(cutting) speed	200 m/sec
Bulk temperature	300 K
Undeformed chip thickness	10.8Å
Cutoff length	R_C = 10.8459Å
Time step	1fs (= 10^{-15}sec)

Thermal relaxation was performed on the model before the tool was advanced towards the work material. The model shown in Figure 2 is in thermal equilibrium when the initial temperature is set at 300K. The boundary atoms of the two layers on all sides except the top side are fixed, whereas the atoms in the three layers adjacent to the boundary atoms act as thermostat atoms, where the velocities of the atoms are adjusted for every specific time step of the computation in order to stabilize the average temperature of the thermostat layers at 300K. Moreover, the MD simulations were carried out under a ploughing speed of 200m/s. Even though this high cutting speed is unrealistic in comparison with AFM experiments, it was selected because previous studies[12,13,19] confirmed a little difference on deformation characteristics such as the surface quality between the effect of a 20m/s and 200m/s cutting speed. Also, Mehrez and Ciraci[21] supported this validity in that the MD simulation carried out with relatively higher speed could reveal the main features of the atomic rearrangements.

4. RESULTS AND DISCUSSIONS

The representative deformation behavior of the work material molecules during the AFM-based nano lithography process is shown in Figs. 4(a)-(c) for a crystal setup of A:(001)[100] with a tool travel (ploughing length) of 2nm, 6nm, and 10nm, respectively.

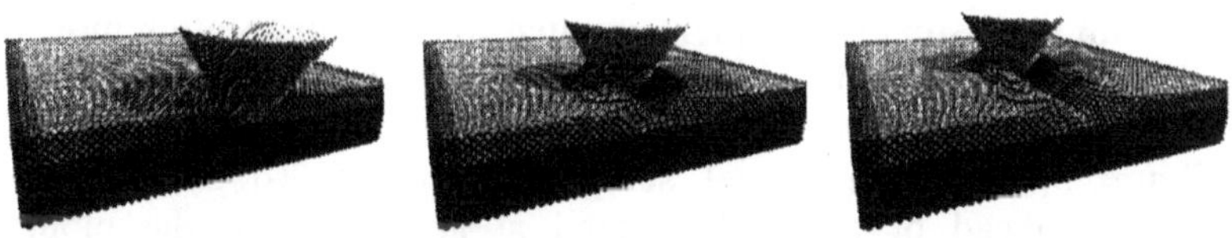

(a) tool travel: 2nm (b) tool travel: 6nm (c) tool travel: 10nm

Figure 4. Deformation behavior of molecules during AFM-based nano-lithography process at different tool travel distances for crystal setup A:(001)[100], where the ploughing operation conducted at crystal orientation of (001) and ploughing direction of [100]

All the simulations were conducted using a tip radius of 10Å, tool angle of 60°, and undeformed chip thickness of 10.8Å unless otherwise stated.

The figures were created using a special MD post program developed at the Nano/Micro Mechanics Laboratory, Kyungpook National University. It is clear that the range of plastic deformation was at most limited to the work material molecules surrounding the tool. Also, there was a clear accumulation and pileup of amorphous structural molecules ahead of the tool and a remarkable side-flow on the left and right sides of the tool as the tool moved forward. Since this inevitable side-flow of molecules in AFM-based nano lithography can deteriorate the surface quality, it should be removed by an additional operation such as surface cleaning if the pattern obtained is assumed to be used as MEMS device.

Figures 5(a)~(f) show the effects of crystallographic factors (crystal orientation and ploughing direction) on deformation characteristics of work material. MD simulation for various combinations of crystal orientation and ploughing direction at (001), (110), and (111) and the ploughing direction at [100], [-110], [001], and [-211] were conducted. Each figure shows cross-sectional views (upper figures) of the zx plane and plane views (lower figures) of the xy plane when the tool travelled 12nm along the ploughing direction. As shown in the upper figures, for each crystal setup, a clear accumulation and pile-up of atoms ahead of the tool were observed as the tool ploughed the work material. Moreover, the crystal orientation and ploughing direction were found to have a significant effect on the nano patterning deformation characteristic and machined surface quality. As such, the surface roughness, as a means of assessing the machined surface quality, plays a key role in determining whether nano components fabricated by AFM-based nano lithography have a fine surface quality suitable for MEMS applications.

The lower figures in each crystal setup show the slip direction of the work material molecules during the ploughing process. More rigorous speaking, the lower figures showed the atoms that have largely displaced and disordered from its original position. Of course, these atoms include both the amorphous structured atoms and the largely slipped atoms caused by the dislocation movement on the slip plane by the ploughing process. Even though the classification of each atom has not been identified in this figure, we can say that the atoms marked by the systematic pattern are those with large slip displacement resulted from the dislocation movement on the slip plane. In this sense, the patterned line corresponds to the slip direction of the work material.

Figure 5(a) with a crystal setup of A:(001)[100] shows that the molecules ahead of the tool sheared at about 45° to the ploughing direction, which is similar to conventional machining.[17] However, Figure 5(b) with a crystal setup of B:(001)[-110] shows that the dislocation propagated in parallel or perpendicular to the ploughing direction. For the crystal setup of E:(111)[-211] in Figure 5(e), the dislocation generated and propagated at about 60° to the ploughing direction, whereas the dislocation propagated along the ploughing direction for the crystal setup of D:(110)[-110] as shown in Figure 5(d).

The shear angle for F:(111)[-110] is for below the shear angle as observed with the case of E:(111)[-211]. In the case of crystal setup of D:(110)[-110], the dislocations are generated along the ploughing direction. Consequently, there is very little sub-surface (just beneath area of the top surface of the specimen) deformation. On the other hand, the secondary dislocations created perpendicular into the work material do result in some degree of subsurface deformation. The secondary dislocations are different from the primary dislocations created at the front and both side area of the tool.

In contrast, in the crystal setup of C:(110)[001], the dislocations are generated perpendicular to the cutting direction. This difference for the direction of the dislocation generation and propagation could also be explained in terms of the angle relation between the ploughing direction and the family of slip directions <110> for an FCC crystal of copper.

From the simulations, the dislocations in the case of A:(001)[100] seem to propagate at 45° to the ploughing direction as shown in Figure 5(a). In

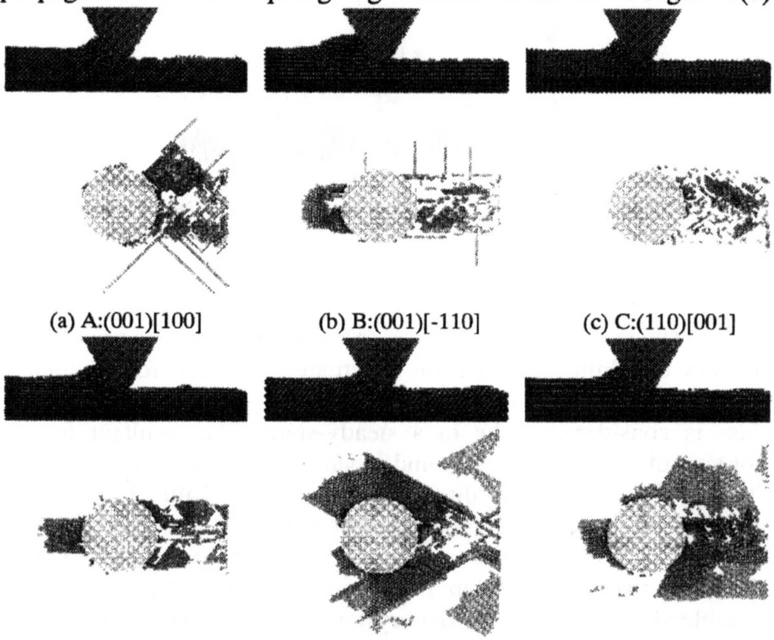

(a) A:(001)[100] (b) B:(001)[-110] (c) C:(110)[001]

(d) D:(110)[-110] (e) E:(111)[-211] (f) F:(111)[-110]

Figure 5. MD simulation results showing cross sectional views (upper figures) of zx plane and plane views (lower figures) of xy plane for various crystal setups at tool travel of 12nm along ploughing direction, where tip radius is 10Å, tool angle is 60°, and undeformed chip thickness is 10.8Å

the case of C:(110)[001], the dislocations seem to propagate perpendicular 90° to the ploughing direction as shown in Figure 5(c). In both these crystal setups, the ploughing direction corresponds to the <100> family.

In the case of ploughing direction of <110> family, the angle between the ploughing direction and the slip direction can be 0°, 60°, or 90°.[22]

Consequently, ploughing along this particular direction produces dislocations either parallel to (as shown in Figure 5(d)) or both parallel and perpendicular to (as shown in Figure 5(b)) the ploughing direction. Also, the angle between [-211] ploughing direction and the slip direction can be 30°, 54°, 73°, or 90°.[22] In the current study, the dislocations in the crystal setup of E:(111)[-211] were observed to propagate at an angle of 60° to the ploughing direction.

Figure 6 presents the force-ploughing distance curves during the AFM-based lithography process for the crystal setup of A:(001)[100]. After a rapid rise in the forces (cutting force and thrust force) until the tool travelled almost 2.0nm, large fluctuations in each force were then shown as the tool travelled further. These short-time scale fluctuations were due to the rapid approaching and separating motion of the work-material molecules surrounding the tool.

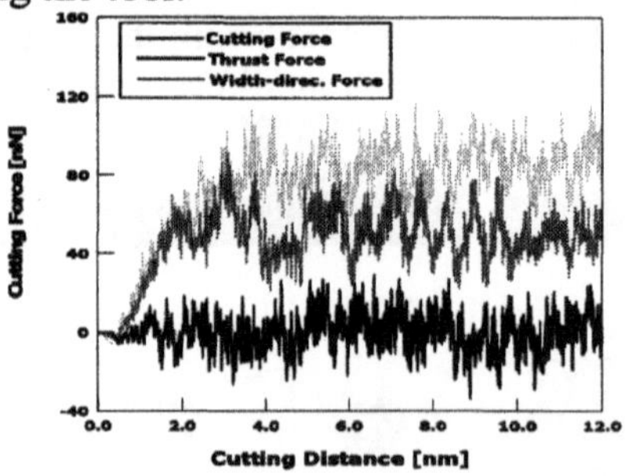

Figure 6. Force-ploughing distance curves in the case of A:(001)[100] simulation

Table 3 shows the averaged forces calculated for the tool travel distance from 4nm to 12nm, at where the fluctuation of each force component became saturated at an almost constant value and the nano lithography process is considered to be in a steady-state. The resultant force, i.e. vector sum of each force, was found to vary cyclically with the orientation of the crystal and ploughing direction. As shown in Table 3, the minimum value for the resultant force occurred with the crystal setup of C:(110)[001]. This is not in total agreement with the argument that the minimum cutting force for an FCC material should be along the most favourable slip system, i.e. in the crystal setup of F:(111)[-110]. Because the AFM tip used in this study has a negative rake angle and hemispherical tip-end shape, the slip system denoting the minimum cutting force does not match with that of the theoretical prediction based on crystal plasticity.[22] The differences in the results once again highlight the influence of the tool geometry on the resultant force system. However, the maximum value obtained with the crystal setup of E:(111)[-211].

This estimation of the forces exerted on the tool can be used when designing the strength and shape of the AFM cantilever beam.[23] Surface roughness of machined surface plays a key role in AFM-based lithography for MEMS devices with good finished surface.

To evaluate the surface quality of the machined surface, the height difference between the molecules at the lowest valley point and those at the highest side-flowing point of the machined surface was measured, as shown in Table 4. Table 4 includes cross-sectional views of the yz plane denoting the surface profile and elastic recovery of the molecules after the ploughing process for two only typical cases of the crystal setup of A:(001)[100] and D:(110)[-110]. The results for other cases are omitted due to space limitation. The figures of the first row show cross-sectional views of AA section after the ploughing distance of 10nm, where the tool is still on ploughing, whereas those of the second row show residual cross-sectional views of BB section at the ploughing distance of 5nm, where the tool has already passed.

Table 3. Calculated averaged forces for various crystal setups

Crystal setup	Cutting force [nN]	Thrust force [nN]	Resultant force [nN]
A: (001)[100]	85.20	50.10	99.79
B: (001)[-110]	80.37	58.98	100.66
C: (110)[001]	78.15	51.70	94.55
D: (110)[-110]	79.87	61.98	101.90
E: (111)[-211]	89.04	61.83	109.34
F: (111)[-110]	88.27	62.98	109.21

As shown in Table 4, the ploughed surface underwent a relaxation process, whereby the pile-up and side-flowed molecules on the machined surface became stabilized and elastically recovered. The results indicated that the ploughing process for the crystal setup of D:(110)[-110] had the minimum roughness and least pile-up of molecules after the passage of the tool.

Table 4. Surface profile and elastic recovery of ploughed surface (cross section view in yz plane

Section	A:(001)[100]	D:(110)[-110]
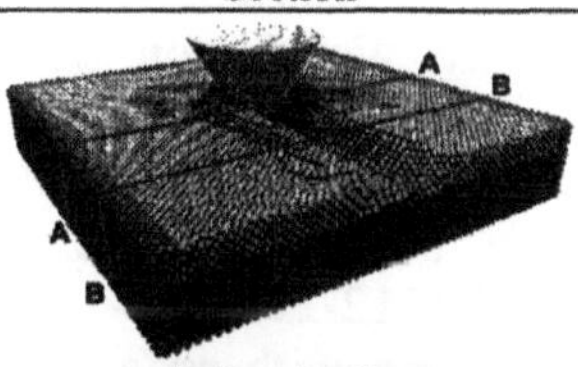	 Max. side-flow: 10.175Å Max. valley depth: 12.186Å	 Max. side-flow: 8.236Å Max. valley depth: 11.619Å
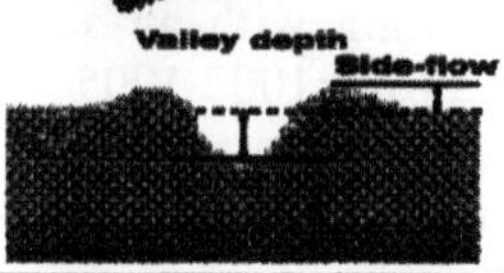	 Max. side-flow: 2.959Å Max. valley depth: 9.041Å	 Max. side-flow: 2.392Å Max. valley depth: 7.822Å

In this study, we have used the Morse potential function to represent the interacting force between the atoms of the simulation, but any other type of potential based on embedded atom method (EAM)[24] could be used as well. Using EAM potential would slightly affect the results.

5. CONCLUSIONS

The 3-dimensional molecular dynamic simulations for AFM-based nano lithography were performed to evaluate how crystallographic factors (crystal orientation and ploughing direction) influenced the nature of plastic deformation, dislocation generation relative to the slip direction, the resultant force, and surface quality. From this study, MD simulation for the AFM-based lithography of nanocrystalline copper revealed that different crystal orientations and ploughing directions have a significant impact on the ploughing force and nano deformation characteristics. For example, with the crystal setup of A:(001)[100], an extensive dislocation motion was found at about 45° to the ploughing direction, whereas with the crystal setup of E:(111)[-211], an extensive dislocation was propagated at about 60° to the ploughing direction. The resultant force varied with the orientation of the crystal and cutting direction. The maximum resultant force was with the crystal setup of E:(111)[-211], while the minimum was with the crystal setup of C:(110)[001].

ACKNOWLEDGEMENT

This project was sponsored by grants from the National R&D Project for Nano Science and Technology (M10214000237-02B1500-03810), Ministry of Science & Technology.

REFERENCES

[1] Werner M. et al, *Nanotechnology for applications in Microsystems*. Teltow: MST News, 2001.

[2] Yan Y., Yoshino M., Kuriagawa T., On the ductile machining of silicon for micro electro-mechanical systems (MEMS), opto-electronic and optical applications. Material Science and Engineering 2001; A297:230-234

[3] Rai-Choudhury P., *Handbook of Microlithography, Micromachining, and Microfabrication Volume 1:Microlithography*. Washington: SPIE Optical Engineering Press, 1997.

[4] van Loenen E.J., Dijkkamp D., Hoeven A.J., Lenssinck J.M., Dieleman J., Direct writing in Si with a scanning tunneling microscope. Applied Physics Letters 1989; 55-13:1312-1314

[5] Mamin H.J., Rugar D., Thermomechanical writing with an atomic froce microscope. Applied Physics Letters 1992; 61-8:1003-1005

[6] Samitsu Y., A study of atomic scale processing using an atomic force microscope. Journal of Japan Society for Precision Engineering 1995; 61-8:1121-1125

[7] Koin Kar V.N., Bhushan B., Scanning and transmission electron microscopies of single-crystal silicon microworn/machined using

atomic force microscopy. Journal of Material Research 1997; 12-12:3219-3224
[8] Notargiacomo A., Foglietti V., Cianci E., Capellini G., Adami M., Faraci P., Evangelisti F., Nicolini C., Atomic force microscopy lithography as a nanodevice development technique. Nanotechnology 2000; 10-4:458-463
[9] www.gennano.com
[10] Haile J.M., *Molecular Dynamics Simulation.* New York: John Wiley & Son, 1992
[11] Landman U., Luedtke W.D., Burnham N.A., Colton R.J., Atomistic mechanisms and dynamics of adhesion, nanoindentation, and fracture. Science 1990; 248:454-461
[12] Komanduri R., Chandrasekaran N., Raff L.M., Effect of tool geometry in nanometric cutting. Wear 1998; 219:84-97
[13] Kim Y.S., Lee Y.M., Choi D.Y., Kim C.I., Quasimolecular dynamic simulation for bending fracture of laminar composite materials. Journal of Material Science and Technology 2001; 17-5:547-552
[14] Kim Y.S., Choi D.Y., Park J.Y., Microscopic study for the behavior of grain boundary using molecular dynamics. Metals and Materials 2000; 6-2;81-87
[15] Ueda K., Iwata K., Chip formation mechanism in single crystal cutting of beta-brass. Annals of the CIRP 1980; 29-1:41-46
[16] Shimada S., Ikawa N., Tanaka H., Uchikoshi J., Structure of micro-machined surface simulated by molecular dynamics analysis. Annals of the CIRP 1994; 43-1:51-54
[17] Konig W.B., Senrath N., *Progress in Precision Engineering.* Heidelberg:Springer-Verlag, 1991.
[18] Komanduri R., Chandrasekaran N., Raff L.M., M.D. Simulation of nanometric cutting of single crystal aluminum-effect of crystal orientation and direction of cutting. Wear 2000; 242:60-88
[19] Fang T.H., Weng C.I., Chang J.G., Molecular dynamics simulation of nano-lithography process using atomic force microscopy. Surface Science 2002; 501:138-147
[20] Inamura T., Takezawa N., Atomic-scale cutting in a computer using crystal models of copper and diamond. Annals of the CIRP 1992; 41-1:121-124
[21] Mehrez,H., Ciraci,S., Yielding and fracture mechanisms of nanowire. Physical Review B 1997; 56-19; 12632-12642
[22] Wood E.A., *Crystal orientation manual.* New York:Columbia University Press, 1963
[23] Pedersen N.L., Design of cantilever probes for atomic force microscopy (AFM). Engineering Optics 2000; 32:373-392
[24] Foiles S.M., Baskes M.I., Daw M.S., Embedded atom method functions for the fcc metals Cu, Ag, Au, Ni, Pd, Pt, and their alloys. Physical Review B 1986; 33-12: 7983-7991

MOLECULAR DYNAMICS STUDY ON MORPHOLOGY AND STRENGTH OF COPPER ATOMIC-CLUSTER-ASSEMBLED STRUCTURE

Ken-ichi Saitoh, Syuichi Nagase, Noboru Shinke
Kansai University, 3-3-35 Yamate-cho, Suita-shi, Osaka, 564-8680 Japan
saitou@ipcku.kansai-u.ac.jp

Hiroshi Kitagawa
Osaka University, 2-1 Yamadaoka, Suita-shi, Osaka, 565-0871 Japan

Abstract Mechanical properties of unique-shaped nano materials called "atomic-cluster-assembled" structure are studied by using molecular dynamics (MD) simulation and adopting effective medium theory (EMT) potential function. First, copper clusters with stable globular shape are computationally arranged on 1-D, 2-D, or 3-D regular array in artificial manner and then their gathering behavior is observed. Piled-up clusters become one specimen, which is sometimes left with internal voids. It is found that morphology of the product is largely dependent on initial kinetic energy attached to clusters. Secondly, strength of the "atomic-cluster-assembled" structures is evaluated by tensile loading test. Analysis of tensile strength and strain energy shows that defects introduced at the generating process markedly alter the behavior of the structure in the breakage. Brittle fracture tends to occur when the specimen includes initial imperfection such as void structure between clusters.

Keywords: Molecular dynamics, Atomic cluster, Copper, Tensile strength, Strain energy, Void, Sintering, Nanotechnology

1. Introduction

An atomic cluster (sometimes called an ultra-fine particle) is a very tiny substance composed of less than a million copper atoms, holding a spherical shape with high symmetry. It is generated with relative ease by the gas condensation method or liquid phase condensation with present

H. Kitagawa and Y. Shibutani (eds.), IUTAM Symposium on Mesoscopic Dynamics of Fracture Process and Materials Strength, 343–353.

nanotechnology. From a nano-scale engineering point of view, atomic clusters can hopefully be recognized as nano-size building blocks for constructing nano-scale machines or materials. Nanocrystalline materials synthesized from these clusters are proposed and they are expected to possess favorable properties, but their mechanical properties are still in puzzle. Besides, numerous generating processes under various different conditions request consideration of strength from their morphological aspect. Sometimes clusters connect to each other jointlessly (that is, close-packing clusters with dense structure), and sometimes they are built up with hollows and voids, as often observed in zeolites. These materials obtained from atomic clusters can be generally called "atomic-cluster-assembled"(ACA) structures, even though their utility value has not been confirmed yet. Nowadays, computational mechanics can help us to study such materials with sub-nano-scale resolution. Here the molecular dynamics (MD) method is utilized to predict morphology, strength, or function of "ACA" structure.

MD simulations are conducted using an interatomic interaction based on effective medium theory (EMT), which is recognized as a precise many-body potential function. First, copper clusters (made up of 683 atoms) are arranged on 1-D, 2-D, or 3-D regular arrays in an artificial manner and then their gathering behavior and consequence thereof are investigated. The piled-up clusters become one specimen in which the ruggedness of cluster surface and internal voids often remain. Based on this morphological consideration, tensile strength and strain energy of the specimens are analyzed. It is found that inherent defects greatly influence the strength of the structure. Brittle fracture tends to occur when the specimen includes initial imperfection such as void structure or grain boundaries between clusters. Furthermore, broadening and thinning of grains are detected during deformation. This is largely due to grain rotation promoted by generation and motion of stacking faults. It is expected that grain rotation, grain migration, or GB migration plays dominant role when "ACA" structure deforms as similarly observed in the deformation mechanism of nanocrystalline structure (Schiøtz,J. et al., 1998) (Shimokawa, 2001) (Hasnaoui,A. et al., 2002) (Yamakov,V. et al., 2002) . It may be a novelty of this study that the strength of nano-size substance is checked up from its generating stage. It is discussed and concluded that the strength of atomic-cluster-assembled structures is tremendously dependent on their initial conformation (i.e. on the morphological aspect of the "ACA" structure).

2. Method

2.1 Preparation of atomic clusters

Copper atomic clusters are scrutinized in this study. Fig.1(a) shows a procedure to prepare stable atomic clusters. First, a spherical region with the required diameter is cut from a face-centered cubic (f.c.c.) bulk crystal which has been constructed with normal lattice constant at 0 K (=0.3466nm). Then, it experiences annealing process of MD up to 10 K to stabilize imbalances appearing mainly in surface region.

An interatomic potential derived from effective medium theory (EMT) (Jacobsen,K.W. et al., 1987) for copper crystal is adopted throughout this study. Fig.1(b) shows the relation between surface atom ratio and surface energy for several clusters (N is the number of atoms in one cluster). Surface atoms are determined by coordination number analysis, namely, by excluding twelve-coordinated atoms. They all have a spherical shape, except for very small cluster of less than a few hundreds of atoms. The 683 cluster, which is used as building blocks here, has 46% surface atoms and relatively low surface energy and its diameter is approximately 2.4 nm.

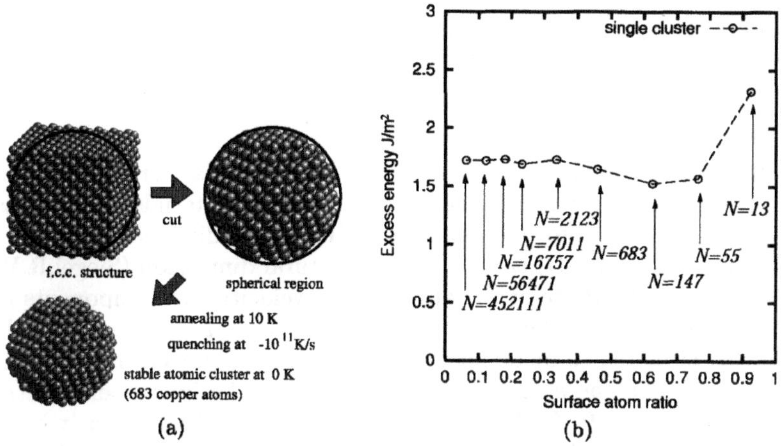

Figure 1. Procedure for making stable copper atomic cluster (a) and relation between surface atom ratio and surface energy for single atomic clusters (b)

2.2 Generation of atomic-cluster-assembled structure

Conceptually, "ACA" structures are able to be designed and be manufactured controllably as you like. Because of the inherent surface reactivity, clusters also show spontaneous gathering. We have only to prepare a

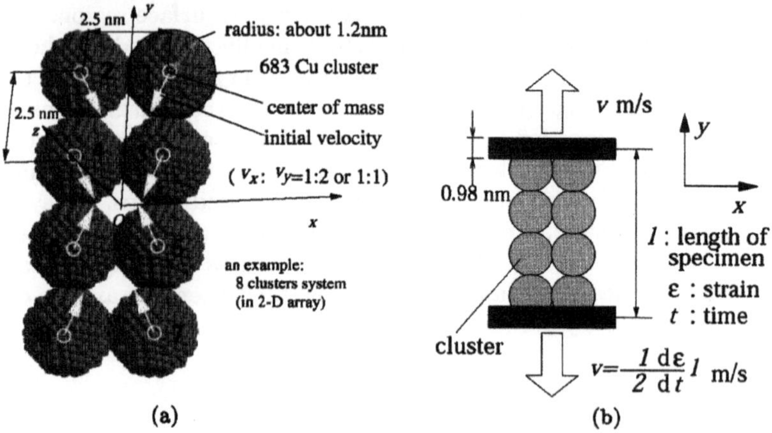

Figure 2. Computational generating process of atomic-cluster-assembled materials (an example) (a) and schematic of tensile test by MD (b)

multiple of equivalent clusters and give them a regular arrangement and initial velocity to get close to each other. "ACA" materials are supposed to be categorized into one-dimensional(1-D), two-dimensional(2-D), or three-dimensional(3-D) according to structural dimension (Siegel,R.W., 1993). The magnitude of the approaching velocity (its components are v_x, v_y, and v_z) must be below a few hundreds m/s, because allowable value is estimated to be about 700 m/s in order to prevent heavy lattice imperfection in the structure (Saitoh, 2002). The initial temperature of each cluster is 0 K, that is, there is no initial vibration in it before interacting. Limited arrangements to be examined are tabulated in Table 1 together with computation conditions. Probably cluster's size, the way of arrangement, or compatibility of crystalline direction will be also an important factor, but it is not unreasonable that the first discussion is done on these cluster systems of a single size with quite uniform arrangement.

2.3 Tensile testing procedure by molecular dynamics

Tensile testing of "ACA" materials is carried out as follows. Fig.2(b) shows the schematic of tensile loading by MD method. Rigid regions of each 0.98nm width (twice the cut-off distance of the potential function) are pulled to the opposite side with the relative velocity of 100 m/s. Temperature rise in MD loading is not restricted artificially, because it might be important in fracture process though it is small quantity.

Table 1. MD calculation conditions

Property	Value
Arrangement of clusters ($x \times y \times z$)	$1\times1\times1$ (0D), $1\times5\times1$ (1D), $1\times7\times1$ (1D), $2\times1\times1$ (1D), $3\times1\times1$ (1D), $2\times4\times1$ (2D), $2\times5\times1$ (2D), $3\times4\times1$ (2D), $4\times4\times1$ (2D), $2\times4\times2$ (3D), etc.
Integration scheme, Time increment	Gear method (4th order), 2.0×10^{-15} [s]
Periodic boundary	not applied in all directions
Tensile speed v (strain rate $d\varepsilon_y/dt$)	50.0 [m/s] ($\approx 1.2\times10^{10}$ [1/s])
Temperature control	not applied

2.4 Analysis method

Most defects appearing in "ACA" structure are either "under-coordinated atoms" (i.e. the coordination number $N_c < 12$) or stacking faults. The existence of these defects is detected by atom-by-atom topological analysis of bonding states (Honeycutt,J.D. and Andersen,H.C., 1987) , often referred to as topological medium-range order analysis or common neighbor analysis. Although this local atomic rearrangement is vital for fracture process, it is not a direct objective in the present treatise but will be treated in another one (Saitoh, 2002).

The tensile strength of "ACA" structure is evaluated from stress-strain curves obtained when elongation of the specimen occurs in the course of MD simulations. Each component of the stress tensor $\boldsymbol{\sigma}$ is calculated by averaging over the atomic stress tensors $\boldsymbol{\sigma}_i$ expressed as (Born,M. and Huang,K., 1954)

$$\boldsymbol{\sigma} \equiv \frac{1}{N_{\text{in}}}\sum_{i\in V_{\text{in}}}\boldsymbol{\sigma}_i = \frac{1}{N_{\text{in}}}\sum_{i\in V_{\text{in}}}\frac{1}{2\Omega_i}\sum_{j\neq i}\phi'(r_{ij})\frac{\boldsymbol{r}_{ij}\boldsymbol{r}_{ij}}{r_{ij}}, \quad (1)$$

where $\phi'(r)$ is the derivative of the effective potential for the EMT potential, $\boldsymbol{r}_{ij}$ is the difference vector directing from the i-th atom to the j-th atom, r_{ij} is the magnitude of $\boldsymbol{r}_{ij}$, and Ω_i is the atomic volume attributed to the i-th atom. Regional volume and the number of atoms to be averaged are defined by V_{in} and N_{in} respectively, which are selected for all region and atoms in this study. In the coordinate system of Fig.2, tensile stress is estimated from the y component, $\sigma_y = \sigma_{22}$.

3. Results & Discussions

3.1 Morphological Aspect

Fig.3 shows examples of united atomic lumps ("ACA" structures). Partly because of the low temperature, the clusters' arrangement in the original regular array is almost preserved. 1-D structures (Fig.3(a)) become like a wire. 3-D structures (Fig.3(b)) contain a lot of boundaries between clusters. On the other hand, 2-D structure is tremendously dependent on conditions of approaching velocity. In other words, the kinetic energy attributed to all clusters determines the morphology of "ACA" structures.

2-D structures arranged in $2 \times 4 \times 1$ (each integer corresponds to the number of clusters in x, y, or z directions, respectively) are divided into two major kinds of structure, as shown in Fig.3(c) and (d). One is a porous structure, which seems to be materialized with weak bondings in interface region. It includes voids inside of the structure and shows surface ruggedness originated from natural cluster's curvature. Another structure is a fully compact structure. We distinguish these two morphologically different structures by naming (A)void model (with internal void structure) and (B)compact model (in which void vanishes).

Generally speaking, as shown in Fig.4, surface atom ratio is closely related to initial kinetic energy as well as the structural dimension. Data obtained for all the structures from 1-D to 3-D are plotted in the figure. A single raw 683 cluster (so to speak, 0-D structure) has the surface atom ratio of 46% which is upper limit of the surface atom ratio. The border in surface atom ratio for distinguishing void model from compact one is roughly estimated at 30% in 2-D structure from this distribution map.

3.2 Aspect of Mechanical Properties (Strength)

Among possible numerous "ACA" structures, two types of 2-D specimen again show clear contrast in response to an external force. Fig.5 shows the transition of atomic configuration under tensile loading. Breakage occurs when total strain reaches 0.25 for model(A) and 0.45 for model(B). The specimen of model(B) presents large elongation and necking in the middle before breakage, whereas the model(A) shows inter-granular breakage not in the middle region. Ductile behavior of model(B) accom-

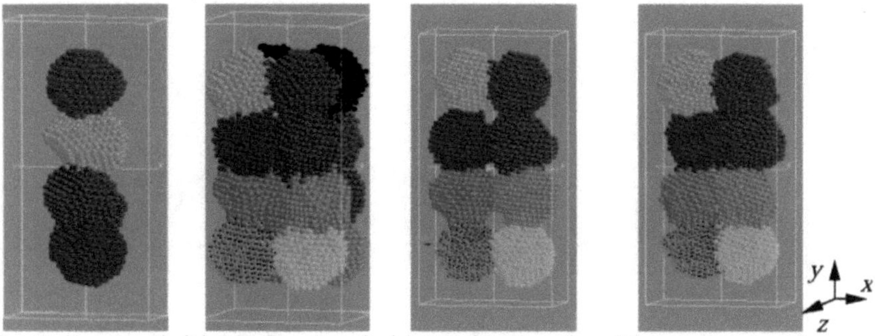

(a) 1-D (1 × 4 × 1) (v_y=10.0)
(b) 3-D (2 × 4 × 2) (v_x=v_z=10.0, v_y=20.0)
(c) 2-D (2 × 4 × 1) (v_x=5.0,v_y=10.0) →(A)void model
(d) 2-D (2 × 4 × 1) (v_x=v_y=10.0) →(B)compact model

Figure 3. **Atomic configuration of united clusters ("ACA" structures) obtained by MD simulation (different shading is used for distinguishing belongings to initial clusters; unit of initial velocity is [m/s].)**

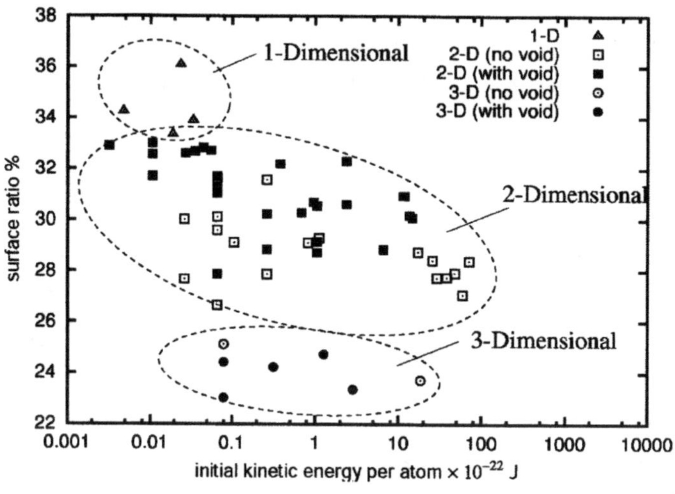

Figure 4. **Relation between surface atom ratio and initial kinetic energy**

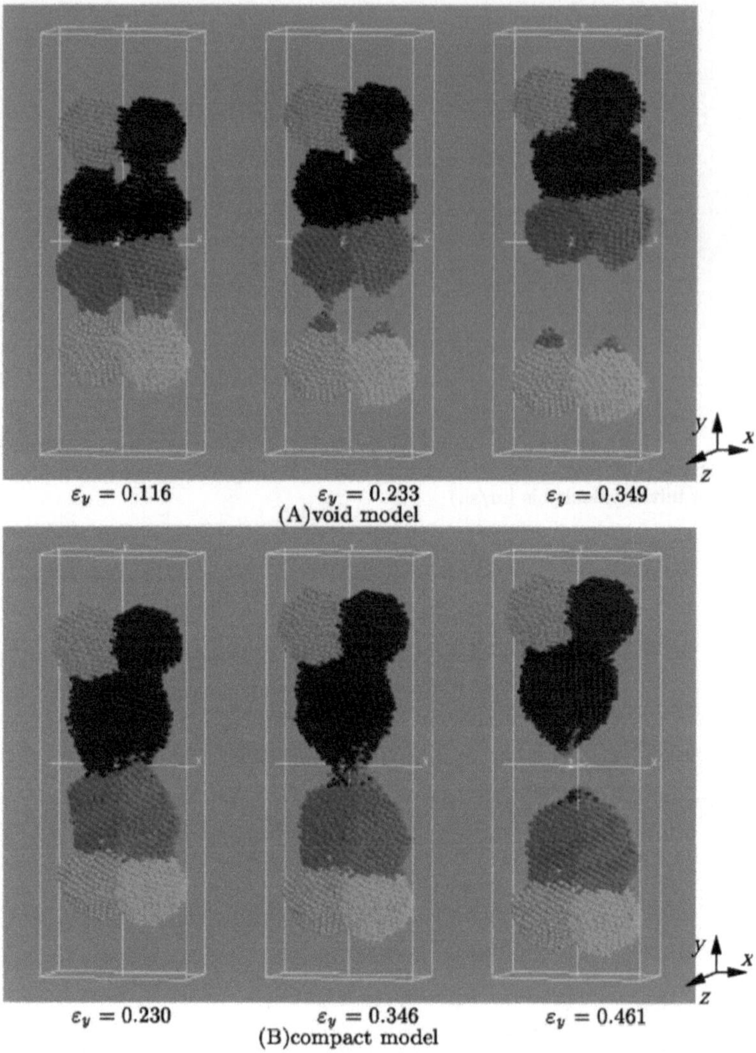

Figure 5. **Transition of atomic configuration of atomic-cluster-assembled structure under tensile loading**

panies unrecoverable thinning of the grain (originally, cluster), which is supposed to be caused by many intra-granular slips. On the other hand, model(A) presents a kind of brittle cleavage and its grains preserve initial spherical shape after the disassembly. It is also found that grain motion actively occurs in model(A).

Fig.6 shows the stress-strain curves for (A) and (B) specimens. Detachment

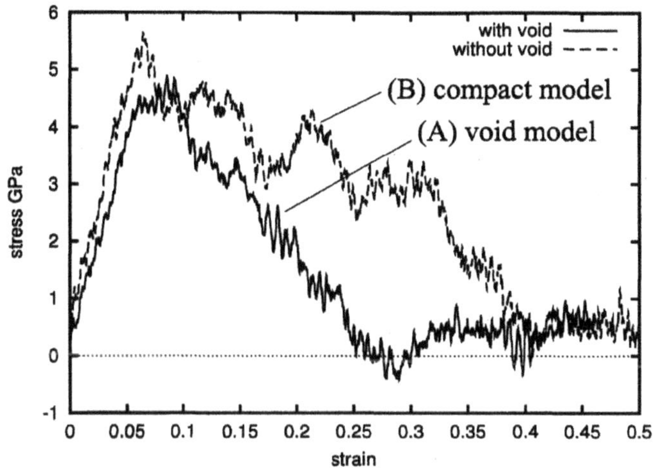

Figure 6. **Stress-strain curves obtained by MD tensile testing for two kinds of 2-D structures ((A)void model and (B)compact model)**

of the structure is found when full relaxation of stress is observed. Strain energy accumulated during loading are corresponding to the area surrounded by stress-strain curve and horizontal axis, so we can estimate strain energy from Fig.6. This quantity becomes an evaluation measure of ductility. It is obvious that strain energy of model(B) surpasses that of model(A).

We attempt to summarize general tendency about mechanical properties as follows. Fig.7 shows the relation between tensile strength (maximum stress appearing in stress-strain curves) and strain energy stored into the specimen, for all cases calculated. Tensile strength varies from 4 GPa to 7.5 GPa (except for 0-D structure or 1-D structure in latitudinal tension having an exceptional value such as 16 GPa). Dependence on morphology are described as following two points based on Fig.7. (1)3-D structure generally shows large ductility (plots of solid and open circles), (2)ductility of 2-D structure is sensitive to void structure (plots of solid(=void model) and open(=compact model) squares).

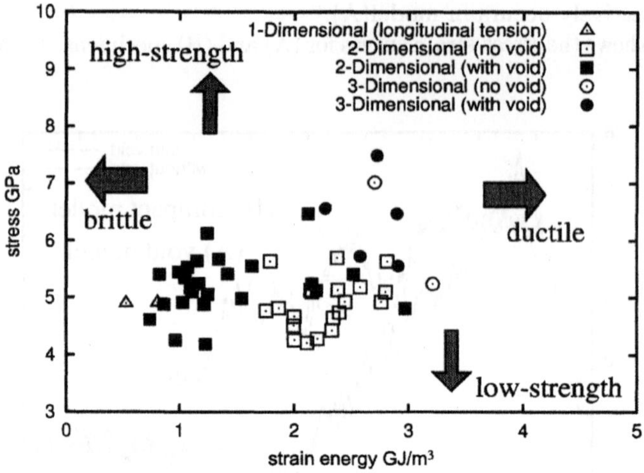

Figure 7. Relation between tensile strength and strain energy

4. Conclusion

We perform molecular dynamics simulation of copper atomic-cluster-assembled (ACA) materials. Points of view are on possible structural configuration and on their strength. The following observations are done.

- By adjusting approaching velocities between clusters, both dense and sparse structures are obtained. Surface atom ratio is one of deciding factors of the structure and has a certain relation with initial energy of clusters.

- A specimen without void structure possesses extremely higher toughness and exhibits ductile failure. On the other hand, a specimen with void transforms its shape retaining original cluster's geometrical feature and shows brittle breakage. Their difference is also understood by analysis of strain energy.

Acknowledgements

This work was supported in part by the Grant-in-Aid for Young Scientists (B) from the Japan Society for the Promotion of Science (No.14750063).

References

Born,M. and Huang,K., *Dynamical Theory of Crystal Lattices.* Oxford: 129, 1954.

Hasnaoui,A., Van Swygenhoven,H., and Derlet,P.M. On Non-equilibrium Grain Boundaries and Their Effect on Thermal and Mechanical Behaviour: A Molecular Dynamics Computer Simulation, *Acta Mater.*, 50 : 3927-3939, 2002

Honeycutt,J.D. and Andersen,H.C. Molecular Dynamics Study of Melting and Freezing of Small Lennard-Jones Clusters, *J.Phys.Chem.*, 91 : 4950-4963, 1987

Jacobsen,K.W., Nørskov,J.K., and Puska,M.J. Interatomic Interaction in the Effective-medium Theory, *Phys.Rev.,B*, 35 (14) : 7423-7442, 1987

Saitoh, K., "Molecular Dynamics Study of Interface Generated between Atomic Clusters." In *Engineering Plasticity from Macroscale to Nanoscale*, pp.591-596, L.C.Zhang and G.Lu, eds.: Trans Tech Publications, 2002.

Schiøtz,J., Di Tolla,F.D, and Jacobsen,K.W. Softening of Nanocrystalline Metals at Very Small Grain Sizes, *Nature*, 391 : 561-563, 1998

Shimokawa, T., Doctral thesis: Osaka University, Japan, 2001.

Siegel,R.W. Nanophase Materials Assembled from Atom Clusters, *Mat.Sci.Eng.,B*, B19 : 37-43, 1993

Yamakov,V., Wolf,D., Phillpot,S.R., and Gleiter,H. Deformation Twinning in Nanocrystalline Al by Molecular-dynamics Simulation, *Acta Mater.*, 50 : 5005-5020, 2002

COMPETING DEFORMATION MECHANISMS IN NANOCRYSTALLINE METALS

Jakob Schiøtz and Søren L. Frederiksen
Center for Atomic-scale Materials Physics (CAMP), Department of Physics, Technical University of Denmark, DK-2800 Kongens Lyngby, Denmark

Abstract: The deformation of nanocrystalline copper and molybdenum has been simulated using molecular dynamics. Competing deformation mechanisms are observed in the simulations, including grain boundary sliding, dislocation motion, twinning and intergranular fracture. The balance between these mechanisms depends on the grain size and the material, the most dramatic difference being that fracture is dominating in molybdenum while not observed in copper.
Keywords: Nanocrystalline metals, computer simulations, copper, molybdenum, deformation mechanism, dislocations, ductility, fracture.

1 INTRODUCTION

The properties of metals and alloys depend strongly on the grain size of the material, giving unusual properties to materials with unusual grain sizes. One example of this it the very high hardness of nanocrystalline metals, even metals such as copper which are normally rather soft become very hard when the grain size is reduced to the nanometer regime. This makes nanocrystalline metals interesting both from a practical and theoretical point of view - the materials may not only have (technologically) interesting properties, studying them as an extremal case of polycrystalline metals may reveal information about the behavior of metals which would otherwise be difficult to obtain.

The mechanical properties of nanocrystalline metals have been studied thoroughly both experimentally (see e.g. [1–6]) and with computer simulations [7–14]. In most experiments, the hardness and yield stress of nanocrystalline metals is seen to increase with decreasing grain size, roughly following the so-called Hall-Petch relation [15, 16] stating that the yield stress increases with the inverse of the square root of the grain size

$$\sigma_y = \sigma_{y,\infty} + \frac{k}{\sqrt{d}} \tag{1}$$

H. Kitagawa and Y. Shibutani (eds.), IUTAM Symposium on Mesoscopic Dynamics of Fracture Process and Materials Strength, 355–363.

Here σ_y is the yield stress, d is the grain size and k is called the Hall-Petch slope. The hardness and flow stress obey similar relations. In some cases, the Hall-Petch relation is observed to break down for the smallest grain sizes, leading to approximately constant or even decreasing hardness as the grain size is reduced sufficiently. Unfortunately, many of the measurements are probably dominated by defects such as nanovoids and impurities in the grain boundaries, especially for the smallest grain sizes [1, 4].

Most computer simulations give a different result: the flow stress (and thus the hardness) decrease with decreasing grain size [7–9]. This occurs because most computer simulations are limited to grain sizes slightly smaller than the range of grain sizes investigated experimentally. At these grain sizes (below approximately 10 nm), the large density of grain boundaries suppresses dislocation motion and favors grain boundary sliding as an alternative deformation mechanism. This shift in deformation mechanism results in a change in how the mechanical properties depend on grain size. Recently, this shift and the resulting maximum in flow stress for intermediate grain sizes was directly observed in computer simulations [12, 17].

Just as varying the grain size may cause a shift in deformation, so may varying the material. Metals with face-centered cubic (fcc) crystal structure typically show a ductile behavior, whereas metals with a body-centered cubic (bcc) structure often show a more brittle behavior, in particular at lower temperatures, and often show a brittle to ductile transition when the temperature is increased. This difference between fcc and bcc metals is generally attributed to different dislocation properties.

It is thus clear that many different factors may influence the deformation mechanism in polycrystalline metals, and that different deformation mechanisms may be dominating in different situations. Atomic-scale computer simulations are an excellent tool to study this, since the access to all atomic degrees of freedom in principle makes it possible to determine which mechanisms are active under which circumstances.

Molecular dynamics simulations of nanocrystalline copper and molybdenum were performed to investigate if different deformation mechanisms were active in different materials and for different grain sizes; and to investigate what effect such shifts in deformation mechanism may have on the mechanical properties.

2 METHODS

Polycrystalline "samples" are generated using a Voronoi construction [18], where a set of grain centers is chosen randomly, and the part of space nearer one center than any other center is filled with a randomly oriented crystal lattice. The resulting nanocrystalline sample is then briefly annealed by performing a molecular dynamics simulation at 300K for 50 ps, while allowing the system size to relax. The details of the generation procedure have been published elsewhere [8].

2.1 Molecular dynamics

The forces on the atoms are calculated using the Effective Medium Theory [19, 20] for copper, and a newly developed potential for molybdenum [21]. The Effective Medium Theory normally underestimates the stacking fault energy, but as the stacking fault energy is important for dislocation properties such as the splitting width of dislocations, we used a new set of potential parameters that reproduce the correct stacking fault energy [22]. The molybdenum potential is inspired by the Effective Medium Theory, but it is more complicated and includes effects of directional bonding. The parameters of the potential are optimized using the force matching procedure: the forces on all the atoms in a few representative systems are calculated using Density Functional Theory, and the potential parameters are adjusted to reproduce these forces as closely as possible. The details are described elsewhere [21].

The simulations are performed at constant temperature ($T = 300\,\mathrm{K}$) and zero transverse stress ($\sigma_{xx} = \sigma_{yy} = \sigma_{xy} = 0$) while applying a constant strain rate ($\dot{\varepsilon}_{zz}$) along the tensile axis. A combination of Parrinello-Rahman and Nosé-Hoover dynamics are used to obtain this [23, 24]; however $\dot{\varepsilon}_{zz}$ is excluded from the usual dynamics and remains a constant. Full periodic boundary conditions are applied both during the sample generation and during the simulations.

2.2 Parallel molecular dynamics

Due to the large system sizes, the molecular dynamics is performed on parallel computers. Molecular dynamics is usually parallelized by assigning a region of space to each processor. In the case of metallic systems, where the interactions are relatively short ranged, and where the density does not vary much, a simple decomposition of space into equally sized boxes can be very efficient. It is illustrated in Figure 1.

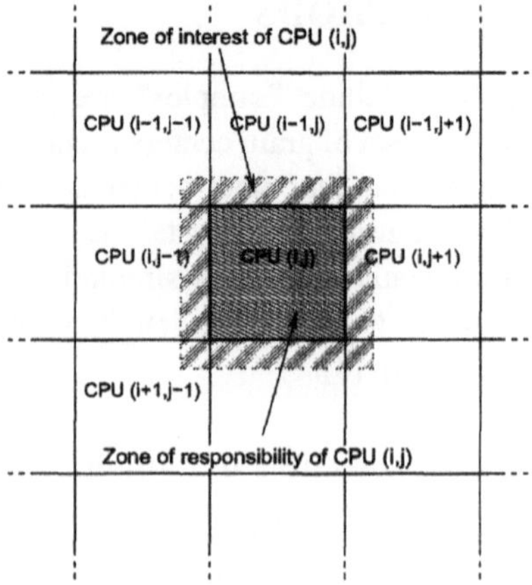

Figure 1: Spatial decomposition in two dimensions.

Each processor is responsible for all calculations pertaining to atoms in its region of space. To calculate the forces on these atoms, it needs information about all atoms in a slightly larger region of space (called the "zone of interest" in Figure 1). At the beginning of each time step, these positions must be received from the neighboring processors, requiring communication between the processors. Furthermore, with the potentials used here it is advantageous to communicate some intermediate quantities used in the force calculations. Occasionally, atoms move between regions of space assigned to different processors, and information pertaining to these atoms then needs to be moved between processors.

The large amounts of data generated by parallel simulations can be problematic. The positions and velocities of all atoms can only rarely be stored on disk. In order to be able to store configurations sufficiently often for the later analysis of the simulations, configurations are saved in a format where only information about "interesting" atoms are saved. Common Neighbor Analysis [25] can be used to decide which atoms are in a perfect crystal lattice, and which are near defects, and only atoms near defects are then saved.

The program used is available on the Internet under an Open Source license [26].

3 RESULTS

Simulations of nanocrystalline copper show a gradual shift in deformation mechanism as the grain size is varied. When the grain size is below 10 nanometers, grain boundary sliding is dominating; above 15 nm dislocation motion dominates. This is clearly seen in Figure 2, showing deformation maps from simulations with varying grain sizes. It is important to note that there is no sharp change in deformation mechanism as the grain size is increased; the shift from grain boundary mediated to dislocation-mediated plasticity is quite gradual.

The gradual shift in deformation mechanism as the grain size is varied re-

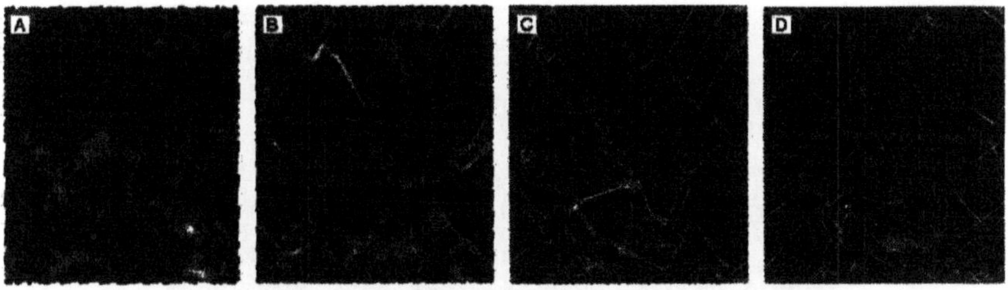

Figure 2: The deformation mode changes as the average grain diameter is increased from 7.5 nm (A) over 12 nm (B) and 24 nm (C) to 49 nm (D). The greyscale indicates the local strain as the global strain of the system is increased from 10% to 11%; lighter colors indicate higher strains. As the grain size is increased, the deformation moves from the grain boundaries into the grains. The simulations contain 0.38, 1.6, 12.7 and 102 million atoms, respectively.

sults in a maximum in the flow stress of the material when the grain size is around 10-15 nm [12]. When the grain size is reduced below 10 nm, increased grain boundary sliding softens the material, whereas above 15 nm the larger grains makes dislocation motion easier.

In the regime where the main deformation mechanism is dislocation motion, the small grain size still dominates the behavior of the dislocations. As there is barely space for dislocation tangles and Frank-Read sources inside the grains, the vast majority of the dislocations are nucleated at the grain boundaries. The dislocations thus nucleated eventually move through the grains and are absorbed by the grain boundary on the opposite side of the grain. On their way, many interactions with other dislocations occur. The elastic interactions between the dislocations are strong, and cause the formation of dislocation structures such as pile-ups (see Figure 3). Other types of interactions are also seen. Occasionally, two Shockley partial dislocations collide and form a stair-rod dislocation. Although the stair-rod dislocations are sessile, they do not significantly hinder dislocation motion, as they are short-lived and quickly split into two mobile Shockley partials.

In molybdenum, the deformation mechanism is completely different [21]. Although both grain boundary sliding and dislocation motion are observed, the imposed strain is mainly accommodated through intergranular fracture (See Figure 4). Dislocation motion is known to be more difficult in bcc metals than in fcc metals due to different structures of the dislocation cores. Grain boundary sliding also appears to be less important in nanocrystalline molybdenum than in nanocrystalline copper. In spite of the much larger stresses, less grain boundary sliding is observed in the molybdenum simulations, this may be due

Figure 3: Dislocation structure formed on a glide plane in a large grain. The figure shows a thin slice of the grain, containing the slip planes where a dislocation pile-up has formed. The dislocations were nucleated in the grain boundary in the upper left corner, and are moving towards the lower right corner. The elastic repulsion between the dislocations keep them apart and prevent them from moving into the grain boundary, the dislocation thus form a pile-up. A few other dislocation on nearby slip planes are also seen, some of them appear to end where they move out of the slice included in the figure.

to the directional bonding in molybdenum. In copper the grain boundaries slide through a large number of small shuffling events where a few atoms move with respect to each other. Directional bonding is likely to make this less favorable.

Although dislocation motion and twinning are seen both in copper and in molybdenum, there are significant differences due to the different crystallography. In fcc metals dislocations split into two Shockley partials separated by a stacking fault [27], and simulations often show the emission of only one partial. The core structure is extended on the glide plane, making it difficult for screw dislocations to move onto other glide planes (cross-slip). This is seen in the simulations, where even for the largest grain sizes single Shockley partials often travel through the grain before a second one is nucleated. In the molybdenum simulations screw dislocations are not bound to a specific glide plane, as can be seen by the trace left by the dislocation in Figure 4.

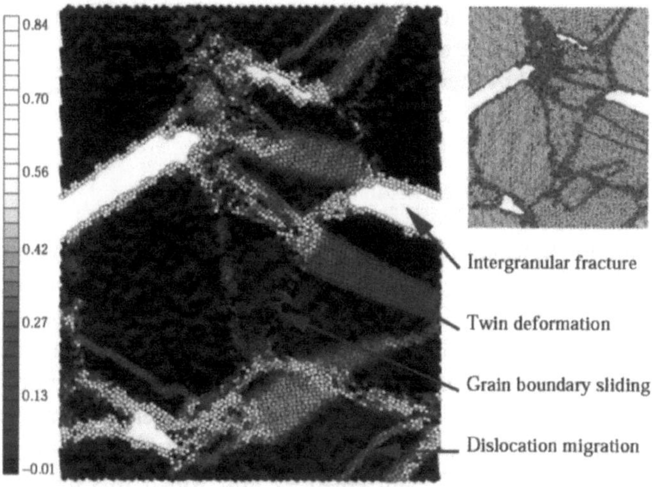

Figure 4: Competing deformation mechanisms in nanocrystalline molybdenum. The main deformation mechanism is intergranular fracture, but twinning, dislocation motion and grain boundary sliding are also observed. The large plot shows the local strain during the simulation, the inset shows the crystal structure, white atoms are in bcc structure, dark atoms are in other structures. Adapted from Ref. [21].

Twinning is seen in both copper and molybdenum, but the mechanism is quite different. In the simulations of copper (and in other simulations of fcc metals) twinning typically starts by a single Shockley partial traveling through a grain, leaving a stacking fault behind. Instead of the matching trailing partial, a second leading partial is later nucleated at an adjacent slip plane, presumably because the resolved shear stress is larger on the leading than on the trailing partial. This process can then be repeated, creating a narrow twin. The twins created by this process are only a few layers wide. Their role as barriers for dislocation motion is reduced by the fact that the twin planes are parallel to the primary slip plane of the grains. Several other twinning mechanisms have been observed in nanocrystalline aluminum with a columnar grain structures [13, 14], although it should be noted that the quasi-two-dimensional nature of the simulations might have increased the tendency to form twins [28].

In the simulations of molybdenum, twinning is also seen, but here wider twins are created as arrays of dislocations are nucleated on adjacent glide planes. The Burgers vector of these dislocations do not correspond to a lattice vector of the bcc lattice, they therefore transform the lattice [27]. As the motion of a single such dislocation does not result in a low-energy planar fault, they must be nucleated in a sufficiently large number to create a twin. Their nucleation is favored by their much small burgers vector, reducing their energy to approximately 11% of the energy of a normal lattice dislocation.

4 CONCLUSIONS

The deformation mechanisms in nanocrystalline copper and molybdenum were studied using molecular dynamics on parallel computers. The simulations show several competing deformation mechanism. In both metals, we observed grain boundary sliding, dislocation motion and twinning. Furthermore, intergranular fracture was observed in molybdenum. In copper, grain boundary sliding was the dominant deformation mechanism when the grain size is below approximately 10 nm, and dislocation motion was dominating for grain sizes from approximately 15 nm up to 49 nm, which was the largest grain size studied. In molybdenum, intergranular fracture was the most important deformation mechanism, but the effect of varying the grain size has not yet been studied. Nanocrystalline molybdenum is thus brittle, whereas we find that nanocrystalline copper is intrinsically ductile, i.e. large plastic strains can be obtained locally. This does not mean that macroscopic samples of nanocrystalline copper will be ductile, as the lack of work hardening will cause failure by strain localization, unless special techniques are used to prevent it [29].

ACKNOWLEDGMENTS

The authors wish to thank Prof. Karsten W. Jacobsen for many useful discussions. The Center for Atomic-scale Materials Physics is sponsored by the Danish National Research Foundation. The authors gratefully acknowledge support from the Materials Research Program of the Danish Research Agency through grant no. 5020-00-0012 and from the Danish Center for Scientific Computing.

REFERENCES

[1] P. G. Sanders, J. A. Eastman, and J. R. Weertman, *Acta mater.* **45**, 4019 (1997).
[2] C. A. Schuh, T. G. Nieh, and H. Iwasaki, *Acta Mater.* **51**, 431 (2003).
[3] H. Conrad and J. Narayan, *Appl. Phys. Lett.* **81**, 2241 (2002).
[4] S. R. Agnew, B. R. Elliott, C. J. Youngdahl, K. J. Hemker, and J. R. Weertman, *Mater. Sci. Eng. A* **285**, 391 (2000).
[5] M. Ke, S. A. Hackney, W. W. Milligan, and E. C. Aifantis, *NanoStructured Mater.* **5**, 689 (1995).

[6] R. C. Hugo, H. Kung, J. R. Weertman, R. Mitra, J. A. Knapp, and D. M. Follstaedt, *Acta Mater.* **51**, 1937 (2003).

[7] J. Schiøtz, F. D. Di Tolla, and K. W. Jacobsen, *Nature* **391**, 561 (1998).

[8] J. Schiøtz, T. Vegge, F. D. Di Tolla, and K. W. Jacobsen, *Phys. Rev. B* **60**, 11971 (1999).

[9] H. Van Swygenhoven, M. Spaczer, A. Caro, and D. Farkas, *Phys. Rev. B* **60**, 22 (1999).

[10] H. Van Swygenhoven, A. Caro, and D. Farkas, *Mater. Sci. Eng. A* **309-310**, 440 (2001).

[11] T. Shimokawa, *Mechanical Properties, Deformation and Fracture Mechanism of Nanocrystalline Metals by Molecular Dynamics Simulation*, Ph.D. thesis, Osaka University (2001).

[12] J. Schiøtz and K. W. Jacobsen, *Science* (2003), in press.

[13] V. Yamakov, D. Wolf, S. R. Phillpot, and H. Gleiter, *Acta Mater.* **50**, 5005 (2002).

[14] Y. Yamakov, D. Wolf, S. R. Phillpot, A. K. Mukherjee, and H. Gleiter, *Nature Materials* **1**, 45 (2002).

[15] E. O. Hall, *Proc. Phys. Soc. London* **B64**, 747 (1951).

[16] N. J. Petch, *J. Iron Steel Inst.* **174**, 25 (1953).

[17] A. Nakatani, T. Shimokawa, R. Matsumoto, and H. Kitagawa, *An atomistic study of ideal strength of polycrystalline metals*, presented at the IUTAM Symposium on Mesoscopic Dynamics in Fracture Processes and Materials Stength, Osaka, Japan (2003).

[18] A. Okabe, B. Boots, and K. Sugihara, *Spatial Tesselations: Concepts and Applications of Voronoi Diagrams* (Wiley, Chichester, 1992).

[19] K. W. Jacobsen, J. K. Nørskov, and M. J. Puska, *Phys. Rev. B* **35**, 7423 (1987).

[20] K. W. Jacobsen, P. Stoltze, and J. K. Nørskov, *Surf. Sci.* **366**, 394 (1996).

[21] S. L. Frederiksen, K. W. Jacobsen, and J. Schiøtz, *Simulations of intergranular fracture in nanocrystalline molybdenum*, submitted for publication.

[22] T. Rasmussen, private communication.

[23] S. Melchionna, G. Ciccotti, and B. L. Holian, *Mol. Phys.* **78**, 533 (1993).

[24] S. Melchionna, *Phys. Rev. E* **61**, 6165 (2000).

[25] D. Faken and H. Jónsson, *Comput. Mater. Sci.* **2**, 279 (1994).

[26] *The CAMP Open Source project.*
URL `http://www.fysik.dtu.dk/CAMPOS/`

[27] D. Hull and D. J. Bacon, *Introduction to Dislocations* (Butterworth Heinemann, Oxford, 1984), 3rd edn.

[28] P. M. Derlet and H. Van Swygenhoven, *Scripta Mater.* **47**, 719 (2002).

[29] Y. Wang, M. Chen, F. Zhou, and E. Ma, *Nature* **419**, 912 (2002).

ATOMISTIC STUDY ON IDEAL STRENGTH OF NANOCRYSTAL AND DEFORMATION INDUCED NANOSTRUCTURES

A. Nakatani[1], T. Shimokawa[2], R. Matsumoto[3], H. Kitagawa[1]

[1]*Department of Adaptive Machine Systems, Osaka University*
2-1 Yamadaoka, Suita, Osaka 565-0871, Japan

[2]*Department of Mechanical Systems Engineering, Kanazawa University*
2-40-20 Kodatsuno Kanazawa, Ishikawa 920-8667, Japan

[3]*Department of Mechanical Systems Engineering, Kyushu Institute of Technology*
680-4 Kawazu, Iizuka-City, Fukuoka 820-8502, Japan

Abstract Molecular dynamics simulations are carried out to obtain fundamental knowledge of the mechanical properties of nanocrystalline metal and mechanisms of grain refinement. First, tensile tests of aluminum nanocrystals with different grain sizes are conducted. As the result, an ideally high flow strength is shown in the case of 40 nm grain size. The deformation mechanisms are discussed based on the detail investigation of atomic configuration and the ideal grain size corresponds to the threshold, at which the deformation mode is switched from the dislocation controlled deformation to the grain boundary controlled deformation and vice versa. Second, single- and poly-crystalline iron nanowires in torsion are examined by using twisted periodic boundary condition (TPBC) which is newly proposed. Various mechanical properties of nanowires, e.g. anisotropy, grain size dependence, and specimen size effect, are studied through torque versus twist relationship, and the process of polycrystallization from single crystal and internal structural change of nanocrystals are clarified. Third, an tensile test of double-notched test-piece of amorphous iron is examined for the deformation induced crystallization mechanism.

Keywords: Atomistic simulation, Nanocrystal, Amorphous, Tensile deformation, Torsion, Twisted periodic boundary condition, Grain boundary, Dislocation, Stacking fault

1. INTRODUCTION

Grain refinement of polycrystalline metals is a promising process for strengthening, since it has the advantage of recycling. Above all, severe plastic deformation (SPD, or intense straining) processes are expected for obtaining the ultrafine grained structure. However, the mechanisms of grain subdivision of

H. Kitagawa and Y. Shibutani (eds.), IUTAM Symposium on Mesoscopic Dynamics of Fracture Process and Materials Strength, 365–380.

single crystal and grain refinement in such process have not necessarily been clarified. In addition, there seems to be no decisive answer to the question what structure shows ideally high strength and high ductility. In the present study, to obtain the essential knowledge for structure and deformation of polycrystalline metals, molecular dynamics simulations are carried out. We deal with three topics in the present paper.

First, the mechanical property of aluminum nanocrystals with different grain size under tensile loading is examined. There are lots of studies dealing with the deformation of nanocrystalline materials from atomistic point of view (Schiøtz et al., 1998; Van Swygenhoven et al., 1999; Yamakov et al., 2001; Shimokawa et al., 2002). Here, we will focus on the grain size dependency of strength and mechanical properties by using an idealized model as simple as possible.

Second, the mechanical properties and structural changes of iron nanowire in torsion are investigated. Various practical-use-oriented SPD processes have been invented so far, such as the equal-channel angular pressing (ECAP), high pressure torsion (HPT), accumulative roll-bonding (ARB) (Saito et al., 1999) and so on. A characteristic common to above all is that the process succeeds in giving large deformation to materials, keeping the macroscopic shape during the process. In such meaning, torsion is promising and it is also convenient to exclude the shape change of the test-piece in atomistic simulation carried out under a limited size. We consider the deformation in both single crystals and polycrystals and discuss the polycrystalization and nanostructural changes occurring during torsion.

Third, deformation induced crystallization in amorphous iron under tensile loading is considered. There are some reports regarding experimental observation for nano-sized particles at the severe plastic zone in amorphous metal (Nieh et al., 2001; Tarumi et al., 2000). Both temperature field caused by heating by plastic deformation and mechanical field affect the process, and there is some possibility that controlling them is applicable to new production methodology. Here, we discuss the structural changes in severely deformed double-edge-notched specimen and nanocrystallization.

2. MODELS AND METHODOLOGIES

2.1 Aluminum Nanocrystal in Tension

The embedded atom method potential proposed by Mishin et al. (1999) is adopted as the interatomic potential for the aluminum nanocrystals. A unit structure composed of 8 regular hexagonal column-shape grain is considered. Size of the unit structure is determined uniquely by grain size d nm, and whole analysis model consists of M and N arrangements of this unit structure in x and y directions, respectively. The periodic boundary condition is applied in the direction of all three Cartesian axes. Each grain has perfect fcc structure of which the lattice is rotated in xy-plane around $\langle 110\rangle$ direction oriented to z axis.

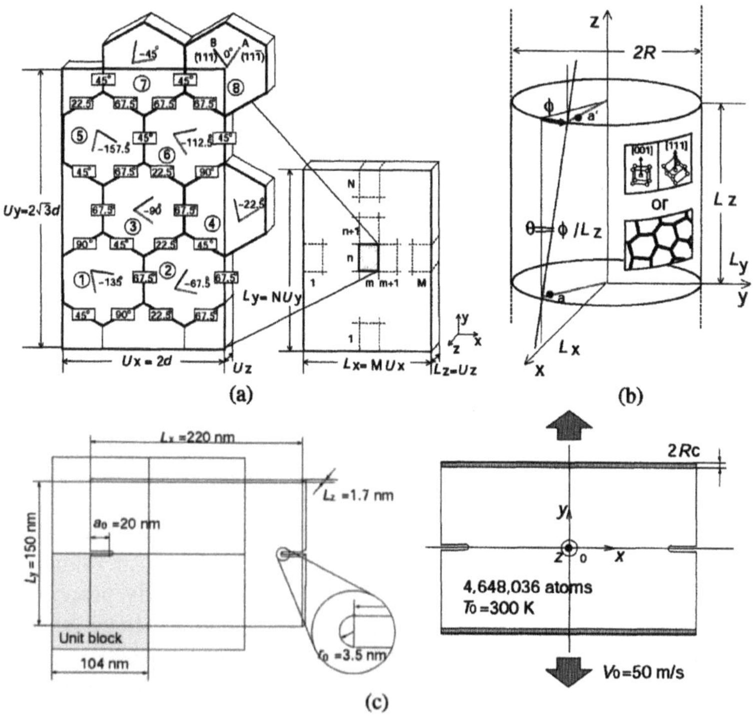

Figure 1. (a) Model for the uniaxial tensile simulation of aluminum nanocrystal. (b) Cylindrical iron nano-wire model for torsion simulation using twisted periodic boundary condition. (c) Double-edge-notch amorphous iron model for tensile simulation.

In Figure 1 (a), figures described in each grain show the rotation angle and the figures on the grain boundary show the misorientation angle between grains composing the grain boundary. Since the unit cell length in z direction $L_z = 1.1$ nm is very small, the periodic boundary condition brings an artifact, i.e. active slip systems are limited to only two 60° dislocations on A $(11\bar{1})$ and B (111) planes. Moreover, we should take into consideration the fact that the stable width of extended dislocation core under external force depends on crystallographic orientation of each grain using this model, since Schmid factors of 2 Shockley partial dislocations decomposed from each 60° dislocations are different from each other.

Specimens with different grain size ($d = 5, 10, 20, 40, 60, 80$ nm) are investigated as shown in Table 1, in which the numbers of unit structures and atoms are shown. After a relaxation calculation which is carried out during 20 ps, tensile deformation is realized with a nominal strain increment 0.0004 which is given to unit cell length in y direction L_y every 0.5 ps, then the atomic system is relaxed under the fixed L_y condition. Consequently, the nominal strain rate is 8×10^8 s^{-1} and unit cell length in x direction L_x is changed under a pressure free condition during the relaxation. The temperature of whole atomic system is controlled to a room temperature 300 K

Table 1. Summary of information with respect to specimen geometries

Grain size d, nm	5	10	20	40	60	80
Number of basic unit (M, N)	(12, 12)	(6, 6)	(4, 4)	(2, 2)	(1, 1)	(1, 1)
Number of atoms	1,683,120	1,699,632	1,711,236	3,049,680	1,717,768	3,055,716

2.2 Iron Nanowire in Torsion

The material models for the torsion problem of nanowire are single- and poly-crystalline iron in which interatomic interaction is introduced by Finnis-Sinclair type potential (Finnis and Sinclair, 1984; Finnis and Sinclair, 1986). Solid cylindrical wire with radius R is modeled as shown in Figure 1 (b). Twisted periodic boundary condition (TPBC) which is newly proposed here is employed. The lateral dimensions L_x and L_y of the unit cell are taken to be sufficiently bigger than the diameter $2R$ of the wire, so that periodic boundary condition is actually assumed only in z direction with dimension L_z of the unit cell.

Summary of TPBC is shown as following. A coordinates transformation between natural coordinates component $(Q^{(\alpha)})^I$ and twisted coordinates component $(q^{(\alpha)})^i$ is considered for an atom α. The canonically conjugate momenta $(P^{(\alpha)})_I$ and $(p^{(\alpha)})_i$ with coordinates $(Q^{(\alpha)})^I$ and $(q^{(\alpha)})^i$ are introduced as follows.

$$(q^{(\alpha)})^i = h^i{}_J (Q^{(\alpha)})^J, \qquad (p^{(\alpha)})_i = [(\boldsymbol{h}^{-1})^T]_i^J (P^{(\alpha)})_J, \tag{1}$$

When we assume a Cartesian coordinates and $\boldsymbol{Q}^{(\alpha)}$ and a specific twist angle θ along the z axis are taken into account, the transformation matrix $\boldsymbol{h}$ is expressed as follows,

$$\boldsymbol{h} = \begin{bmatrix} \cos(\theta L_z (Q^{(\alpha)})^z) & -\sin(\theta L_z (Q^{(\alpha)})^z) & 0 \\ \sin(\theta L_z (Q^{(\alpha)})^z) & \cos(\theta L_z (Q^{(\alpha)})^z) & 0 \\ 0 & 0 & 1 \end{bmatrix} \begin{bmatrix} L_x & 0 & 0 \\ 0 & L_y & 0 \\ 0 & 0 & L_z \end{bmatrix}. \tag{2}$$

Note the metric $g^I{}_J = (\boldsymbol{h}^T)^I{}_k h^k{}_J$ specifies the macroscopic deformation.

Implementation to the numerical code is as follows. The natural coordinate $\boldsymbol{Q}^{(\alpha)}$ is transformed to real coordinate $\boldsymbol{q}^{(\alpha)}$ as the contravariant manner. The interaction force $\boldsymbol{F}^{(\alpha)}$ at atomic site α is calculated, at which the contribution of atoms in image cell is correctly taken into account. The acceleration $\ddot{\boldsymbol{q}}^{(\alpha)} = \boldsymbol{F}^{(\alpha)}/m^{(\alpha)}$ is transformed to acceleration $\ddot{\boldsymbol{Q}}^{(\alpha)}$ as the covariant manner. $\boldsymbol{Q}^{(\alpha)}$ is updated using the velocity version of Verlet scheme.

In the present study, the specimen having the dimensions R = 10 nm and L_z = 20 nm is mainly considered. Two single crystal models are examined, in which $\langle 001 \rangle$ and $\langle 111 \rangle$ are set as a twist axis (z direction), respectively. Polycrystal models, which consist of average grain size is d = 5 nm and 10 nm, are also examined. In addition, polycrystal models with grain size $d = 5$ nm with geometry $R = 20$ nm, $L_z = 10$ nm and with grain size $d = 10$ nm with geometry $R = 20$ nm, $L_z = 40$ nm are considered in order to study size effect.

The wire is twisted monotonically by the increment of specific twist angle $\Delta\theta = 1.25 \times 10^6$ [deg/m] which is added to the specific twist angle θ every 0.5 ps. The nominal surface shear strain $\gamma_0 = R\theta$ is used to express the strain level. The system temperature is set as 300 K during the calculation.

2.3 Amorphous Iron in Tension

The simulation of amorphous iron in the tensile loading is carried out using the Finnis-Sinclair potential which is the same one used in nanowire torsion mentioned above. Beforehand, an amorphous block, which is composed of $1,555,200$ atoms with dimensions of 104.6 nm $\times$104.6 nm $\times$1.74 nm, is produced by simulation of heating-rapid quenching process (Nakatani et al., 2000). The heating-rapid quenching simulation is carried out under full periodic boundary conditions and the obtained amorphous block, which is called as unit block, can cover the whole space without any gaps.

A specimen is cut out as shown in Figure 1 (c). The periodic boundary condition is applied in z direction and the unit cell length L_z of the specimen is as same as that of the original unit block. As shown in Figure 1 (c), the shape of notch-bottom is an arc with radius $r_0 = 3.5$ nm and the initial notch-depth is $a_0 =$ 20 nm. The number of atoms is $4,648,036$, but the displacement component of atoms within boundary regions are specified during the simulation. As shown in Figure 1 (c), the boundary region, in which the displacement component of atoms is specified during the simulation, corresponds to the shaded areas within $2R_c$ (where R_c is the cut-off length: 0.357 nm) from the head or bottom.

The calculation for relaxation is performed during 12 ps under 300 K. Then, the tensile deformation is given by specifying the y-component of displacement corresponding to the opposite constant velocities ± 50 [m/s] to the boundary regions, while x and z components are specified as zero. The nominal average strain rate is $\dot{\bar{\varepsilon}} = 6.7 \times 10^8$ s^{-1}. The loading test is performed until $t = 1200$ ps and the final nominal strain (average strain) $\bar{\varepsilon}$ is 0.800 at the final stage. The initial temperature is set as $T_0 = 300$ K, but the temperature is not controlled during the loading test. As the influence of the periodic boundary condition which is applied in the thickness-direction (z-direction), the deformation is restricted under plane-strain condition in a broad sense ($\varepsilon_z = 0$).

3. RESULTS AND DISCUSSIONS

3.1 Aluminum Nanocrystal in Tension

3.1.1 Grain Size dependency of Strength.

Stress versus strain curves of each aluminum specimen under the uniaxial tensile deformation are shown in Figure 2 (a). The stress is estimated as the average normal component of atomic stress $(\sigma^{(\alpha)})_{yy}$. The curve for all specimens is composed of the linear elastic region, the overshoot region including maximum stress point, and the constant flow stress region. We characterize the curves by three strength parameters, i.e. σ_{max}: the strength the maximum stress, $\sigma_{0.02}$: the stress at strain $\varepsilon = 0.02$, and σ_{flow}: the flow stress that is evaluated as the average value in an flat region ($\varepsilon > 0.09$).

Figure 2 (b) shows the strength parameters as functions of inverse square root of grain size $d^{-1/2}$. Proportion of grain boundary parts which are distinguished from crystal parts by using the CNA is also shown with open triangles in Figure 2 (b). The broken curve in the figure indicates the grain boundary proportion estimated under an ideal assumption that the polycrystal consists of perfectly regular hexagonal grains and that all grain boundaries have a constant thickness $b_{\mathrm{GB}} = 0.7$ nm. It is clearly found that the plots are lying on the curve and grain size dependence of them is closely related with the inverse Hall-Petch relation.

According to Figure 2 (a), tangent modulus in elastic regime below maximum stress point decreases with decrease of the grain size. This can be straightforwardly understood from the fact that the proportion of grain boundary region increases with decrease of the grain size. That is to say, the apparent modulus, which can be estimated approximately based on the law of mixture, decreases with increase of the volume fraction of grain boundary in which the stiffness is less than that in perfect crystal region. It is also easier in the model with smaller grain size that irreversible deformation mechanisms, such as grain boundary diffusion, grain boundary sliding and grain boundary migration, are thermally activated.

The maximum stress increases with increase of grain size. Since there is no lattice defect within the grains at initial state, grain boundaries behave as a dislocation source in the present models, so the result about the grain size dependence of maximum stress means that the much energy is required to yield plastic deformation.

Next, relation between the flow stress and grain size is discussed. When grain size decreases from 80 nm to 40 nm, the flow stress level increases. On

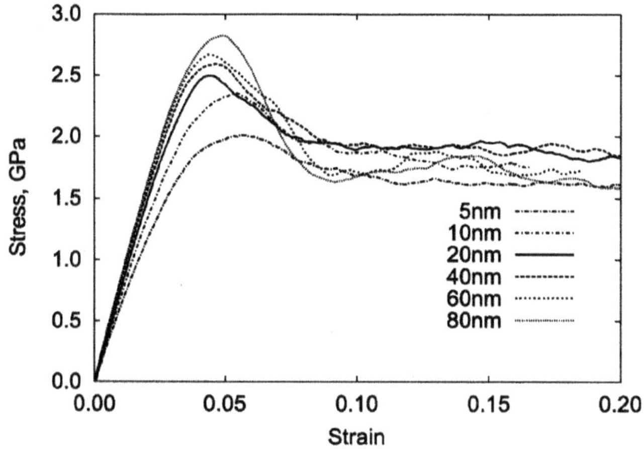

(a)

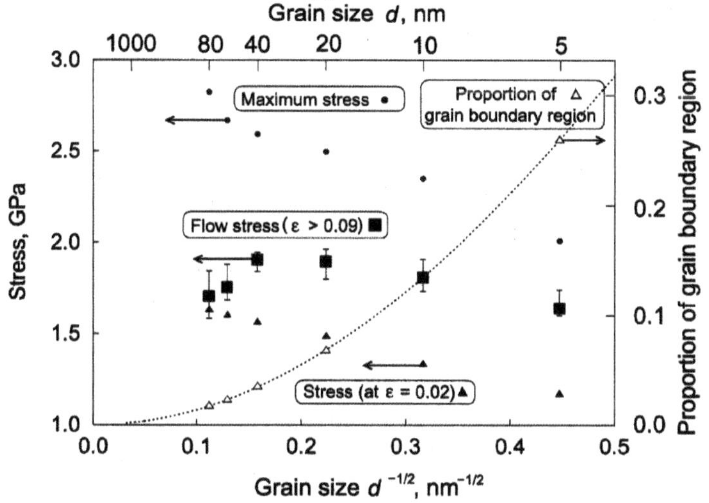

(b)

Figure 2. (a) Stress-strain curves under tensile deformation for different grain sizes. (b) Stress and proportion of grain boundary region versus grain size $d^{-\frac{1}{2}}$. Closed triangles, discs and squares are the stress value at $\varepsilon = 0.02$, maximum stress and flow stress evaluated as average over $\varepsilon \geq 0.09$, respectively. Open triangles which mean the proportion of GB region are lying on a theoretical curve, estimated with the assumption of constant GB thickness.

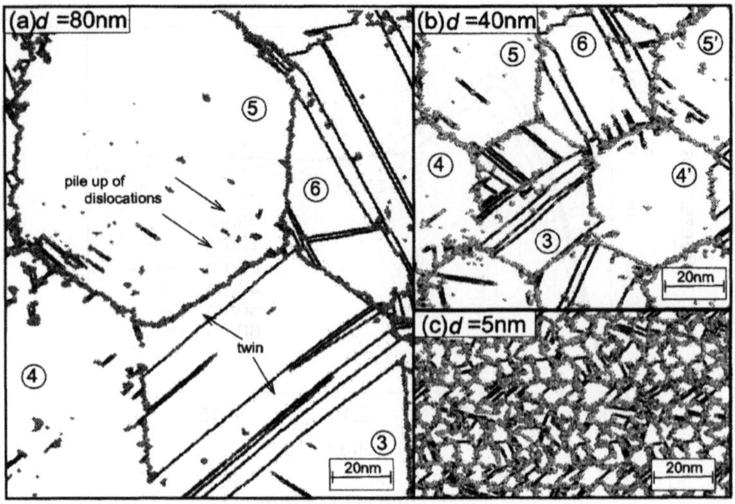

Figure 3. Internal structure of nanocrystalline metals with different grain sizes. Light gray atoms and dark gray atoms show atoms with defect and hcp local structure, respectively. (a) $d=80$ nm. (b) $d=40$ nm. (c) $d=5$ nm.

the contrary, when grain size further decreases from 20 nm to 5 nm, the corresponding flow stress decreases. That is to say, the result implies that there is an optimal grain size between 20 nm and 40 nm, at which the nanocrystal shows the maximum flow stress.

3.1.2 Deformation Mechanisms.

Figures 3 (a), (b) and (c) show the atomic configurations in the cases of $d=$ 80 nm, 40 nm and 5 nm at nominal strain $\varepsilon=0.12$, respectively. Development of stacking faults and deformation twins are remarkably observed in grains 3 and 6 in both cases of $d=80$ nm and $d=40$ nm, although it is well known that the stacking fault energy of aluminum is relatively large and deformation twins are hardly observed in coarse grain aluminum. It is fact that the deformation mode depends on the orientation of grain and it is due to an artifact at the present model i.e. the resolved shear stress contributed on the leading partial dislocation is different from that on the trailing dislocation. However, recent experimental observation has insisted the possibility of deformation twin in an aluminum nanocrystal (Chen et al., 2003). As another reason why the stacking faults are found, it is deduced that the partial dislocations prefer to exist on the grain boundary in a nanocrystal, in a meaning of energetics.

In the case of grain size $d=80$ nm, the plural dislocations can exist simultaneously in a grain (Grain 5 in Figure 3 (a)). However, in the case of small grain size $d = 5$ nm, the probability to find some lattice defects within a grain (Figure 3 (c)) is relatively low. A grain boundary behaves as a barrier against the dislocation motion and the mean free path of dislocation in $d = 80$ nm is larger than that in $d = 40$ nm grain. When a dislocation crosses a grain boundary, the dislocation disappears at a point on the grain boundary and a new dislocation has to be nucleated at the point. This is just one of the reasons of strengthening by grain refinement.

Figure 3 (a), it is found that some dislocations are piled up and interact each other near a boundary within grain 5. In the case of $d = 5$ nm, local deformation in grain boundary region strongly affects the global deformation. As we see in Figure 2 (b), the proportion of grain boundary region remarkably increases as the grain size decreases, hence, the distance between junctions at which grain boundaries meet become short, so intergranular deformation is much easier than intragranular deformation.

Consequently, in the case of small grain size, grain boundary migration and grain boundary sliding become main mechanism. According to Figure 3 (c), it can be found that the regular hexagonal profile of grain at the initial set-up model is lost after the deformation.

3.2 Iron Nanowire in Torsion

3.2.1 Grain Refinement of Single Crystal by Torsion.

To investigate the deformed internal structure in detail, unloading process is simulated, in which the starting point of the unloading is $\gamma_0 = 1$ and the process is realized by the reverse twist up to torque free state. Figures 4 and 5 show cross-section views after the unloading. The common neighbor analysis (CNA) is employed and atomic sites are regarded as belonging to bcc when more than 10 bcc-characteristic atoms are found among the 8 first neighbors and the 6 second neighbors.

In Figure 4, undetermined orientation atoms and lattice defect atoms except for the bcc atoms are indicated in dark color and coloring for bcc atoms means the crystallographic orientation of the twist axis, which is determined by using the second neighbor atoms, in the unit triangle of the reverse stereographic pole figure. In Figure 5, lattice defect atoms are shown in dark color and coloring for bcc atoms shows the distribution of cylindrical shear component $\tau_{\theta z}$ of residual stress.

In Figure 4, the difference of colors at arbitrary two different sites implies the misorientation between them. It is observed that a polycrystal structure is developed near the surface in the cases of twisted single crystal (Figures 4 (a) and (b)). Defect atoms are highly concentrated around the center of twist axis in the case of $\langle 001 \rangle$ torsion, but the elastic core which drags in some dislocations remains around the twist axis in the case of $\langle 111 \rangle$ torsion. Twin

Figure 4. Cross-section views after unloading. Defect atoms are colored by black, and colors correspond to the crystallographic orientation of twist axis in the unit triangle of the reverse stereographic pole figure. (a) single crystal in ⟨001⟩ torsion. (b) single crystal in ⟨111⟩ torsion. (c) nanocrystal with $d = 5$ nm ($R = 10$ nm). (d) nanocrystal with $d = 10$ nm ($R = 10$ nm).

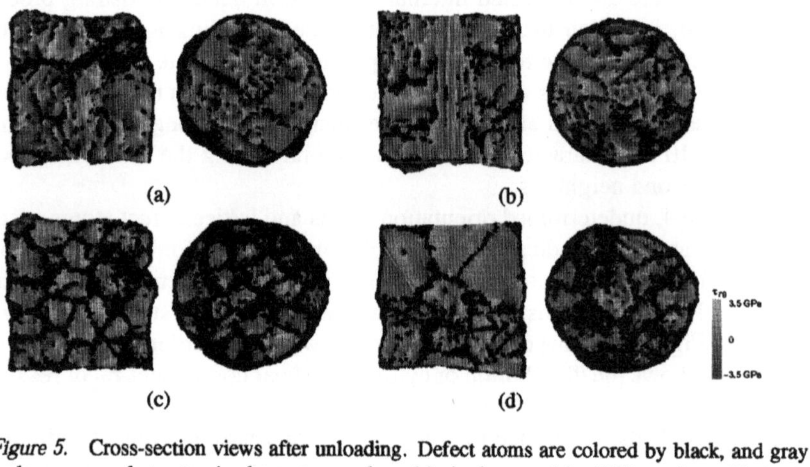

Figure 5. Cross-section views after unloading. Defect atoms are colored by black, and gray scale corresponds to atomic shear stress value. (a) single crystal in ⟨001⟩ torsion. (b) single crystal in ⟨111⟩ torsion. (c) nanocrystal with $d = 5$ nm ($R = 10$ nm). (d) nanocrystal with $d = 10$ nm ($R = 10$ nm).

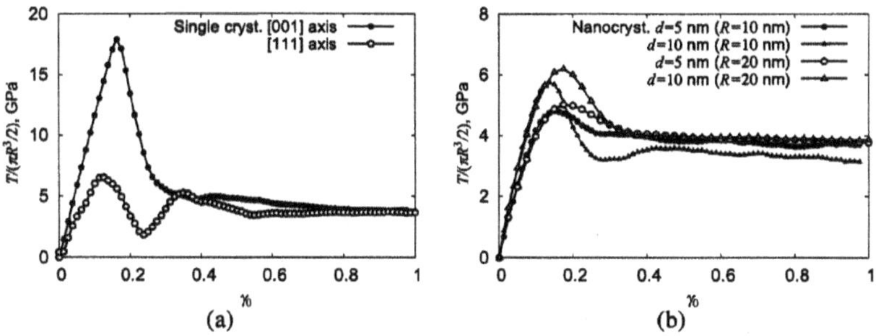

Figure 6. Torque versus twist curves for (a) single crystals and (b) nanocrystals. Torque value is converted to a representative value having the dimension of stress with multiplication by factor of $2/\pi R^3$ and twist is expressed by nominal surface strain $\gamma_0 \equiv R\theta$.

boundaries and low angle grain boundaries which consist of grain boundary dislocation array are also observed. On the other hand, in the case of polycrystalline case (Figures 4 (c) and (d)), remarkable difference between surface and interior is not recognized and there is no significant enlargement or refinement of grains.

According to Figures 5 (a) and (b), large gradient of stress is observed around the dislocations. So many dislocations are not observed within the grains in the polycrystal cases (Figures 5 (c) and (d)), but high stress gradient regions are spreading near grain boundary and this means the grain boundaries have some dislocation character.

3.2.2 Torque versus Twist Curves.

Figures 6 (a) and (b) show the torque versus twist curves for single crystals and for nanocrystals, respectively. The torque is calculated as the resultant moment of the transverse shear component of atomic stress $(\tau^{(\alpha)})_{\theta z}$ around the torsion axis. Anisotropy in the maximum torque is significant in the cases of single crystal and the maximum torque in $\langle 001 \rangle$ torsion is twice and half as large as in $\langle 111 \rangle$ torsion. The torque in $\langle 001 \rangle$ torsion monotonically decrease after the $\gamma_0 = 0.2$, but an oscillation is observed in $\langle 111 \rangle$ torsion. In both cases, the torque is almost constant within the range $\gamma_0 > 0.5$ and the value of torque standardized by a factor of $\pi R^3/2$ is about 4 GPa.

In the polycrystal cases, it is clear that the elastic modulus of nanocrystal with $d = 5$ nm is smaller than $d = 10$ nm, but the elastic modulus does not hardly depend on the radius of specimen R. According to comparison between radius $R = 10$ nm models, the peak value of torque in $d = 5$ nm model is about 80 percent of that in $d = 10$ nm model. Comparing the peak stress between different radius R models for same grain size models, peak stress in $R = 20$ nm model is slightly bigger than that in $R = 10$ nm model. This result is discrepant with well known size effect that is observed by experiment and is explained by the strain gradient theory (Fleck and Hutchinson, 2001).

One reason for the discrepancy is concerning the strain rate effect caused by the analysis condition in the present simulation, i.e. surface strain rate in $R = 20$ nm model is twice as big as $R = 10$ nm. Another reason is that the deformation mechanism of nanocrystal may be different from the coarse grain polycrystals. As we mentioned in the previous section for aluminum case, the global deformation behavior of nanocrystal with small grain size is mainly subjected to the deformation on grain boundary rather than dislocation behavior which is important in the coarse grain polycrystals (Weertman, 2002).

3.2.3 Shear Stress Distribution.

Average shear stress $\tau_{\theta z}$ in an annular ring is shown as a function of twist in Figures 7 (a), (b) and (c). In elastic deformation range, shear stress monotonically increases as radial coordinates r increases, except for the value in a few layers from outer surface, in which the stress value almost does not depend on r. Since inelastic deformation occurs near the surface and it evolves toward interior, the peak stress of interior annular ring is smaller than outer ring. The average stress distribution in the annular rings with different radial coordinates r becomes uniform and the value is almost constant as 3 GPa during the deformation after $\gamma_0 = 0.4$ in every model.

Figure 7 (d) shows the distribution of cylindrical shear component $\tau_{\theta z}$ of residual stress which is averaged in the annular rings after unloading. The stress component is positive in inside rings and is negative in outside rings and this trend agrees qualitatively with the conventional continuum prediction.

3.3 Nanocrystallization in Amorphous Iron

3.3.1 Macroscopic Deformation.

The distribution of the atomic site, at which the atomic equivalent strain ε^{a}_{eq} is greater than 3.38 at strain level $\bar{\varepsilon} = 0.2$ (300 ps), is shown in Figure 8. Here, ε^{a}_{eq} is evaluated using relative displacement of neighboring atoms (Nakatani et al., 1998) with assuming a normal distribution function as the weight function. The distribution of the macroscopic band-like deformation zones which grow from the notch-bottoms and the four corners are almost symmetry to x-axis and y-axis. Micro-shear-bands are formed to the two conjugate directions of maximum shear stress in deformation zones, and lattice like shear-band-structures are formed. Similar micro-shear-bands are recognized in the result of Falk (1999) .

3.3.2 Nanocrystallization Process.

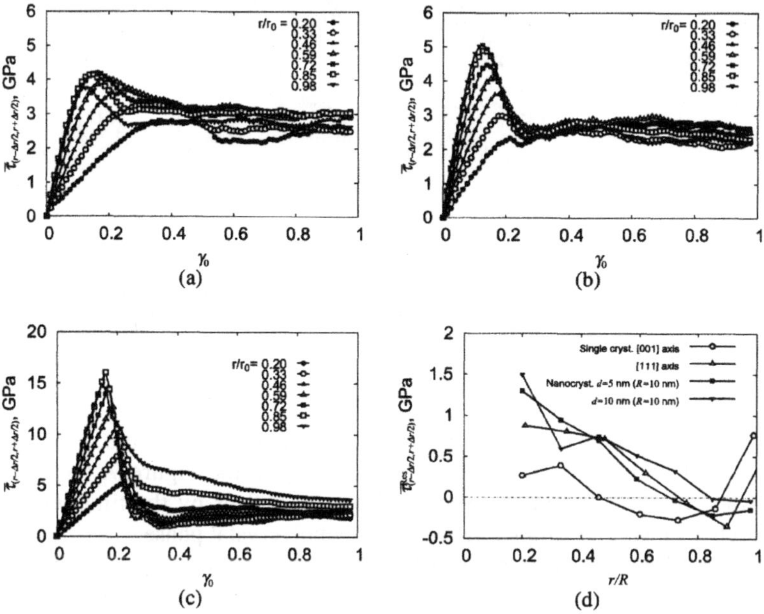

Figure 7. Shear stress versus twist curves for (a) nanocrystal d = 5 nm (R = 10 nm), (b) nanocrystal d = 10 nm (R = 10 nm) and (c) single crystal in ⟨001⟩ torsion. (d) Residual stress distribution

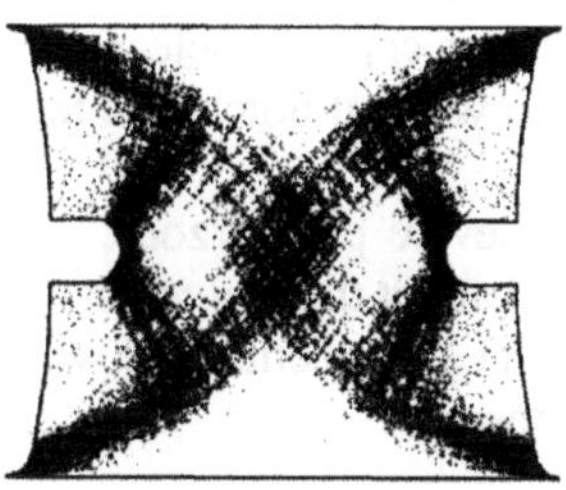

Figure 8. Distribution of the severe deformed atomic sites where the local equivalent strain $\varepsilon_{eq}^{a} \geq 3.38$ at nominal strain $\bar{\varepsilon} = 0.2$ (time $t = 300$ ps).

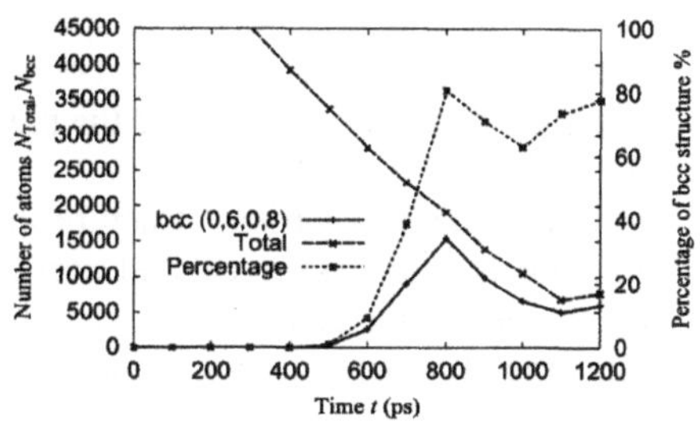

Figure 9. The number of total atoms and bcc-structure atoms having (0, 6, 0, 8) Voronoi's polyhedron, and percentage of bcc-structure atoms during the total atoms. These numbers are investigated in a center region of the specimen (-1.25 nm $\leq y \leq 1.25$ nm)

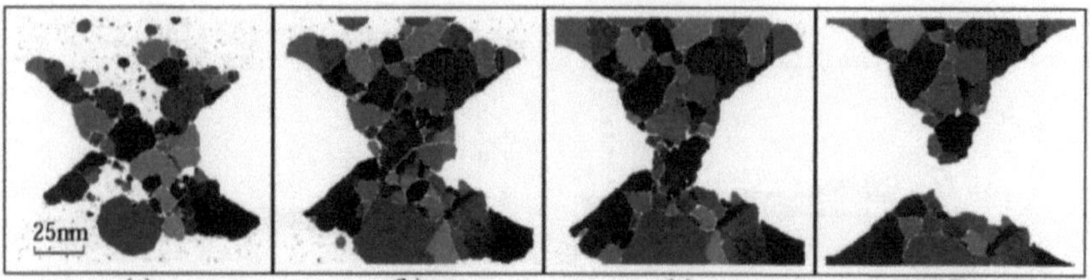

Figure 10. Distribution of the bcc-structural atoms having (0, 6, 0, 8) Voronoi's polyhedron in a center region of the specimen. Atomic colors correspond to the crystallographic orientation of tensile axis in the unit triangle of the reverse stereographic pole figure. (a) $\bar{\varepsilon} = 0.53$ ($t = 800$ ps). (b) $\bar{\varepsilon} = 0.60$ ($t = 900$ ps). (c) $\bar{\varepsilon} = 0.67$ ($t = 1000$ ps). (d) $\bar{\varepsilon} = 0.80$ ($t = 1200$ ps).

Figure 9 show the time evolution of proportion of bcc-structure in the center region (-1.25 nm $\leq y \leq 1.25$ nm). According to the Figure 9, the crystallization process proceeds between 600 ps and 800 ps by the time when the amorphous phase is completely replaced by polycrystal phase.

Figure 10 shows snapshots regarding distributions of bcc-structure. Voronoi's polyhedra which are calculated from snapshots of coordinate are used to distinguish bcc-structure from others. In order to distinguish each grain, all displayed atoms are colored by the position of tensile axis in the inverse stereographic pole figure. A lot of crystalline clusters has been nucleated in front of the notches and notch-bottom regions are occupied by polycrystal phase by the time 800 ps.

After the nanocrystallization, some hollows appear around grain boundaries and the fracture process of specimen starts with nucleation and growth of voids and cracks in the grain boundaries. Similar crystal nucleation and crystal growth is observed using the plate specimen without notches in tensile (Matsumoto et al., 2003).

Both of the geometrical structure of localized severe plastic zone, which corresponds to a heat source, and stress distribution affect this polycrystallization process. If the pattern of localized deformation can be controlled, we can design the ultrafine grain texture answering a purpose.

4. CONCLUSIONS

In this paper, aluminum nanocrystals with different grain sizes in tension, single- and poly-crystalline iron nanowire in torsion, and amorphous iron in tension have been studied by employing molecular dynamics method, for the aim of understanding the basic knowledge concerning what the optimal microstructure is and how we obtain it.

As the result of aluminum nanopolycrystal in tension, the existence of optimal grain size for the highest flow stress has been shown. Around the optimal grain size, the dominant deformation mechanisms interchange, i.e. the optimal grain size seems to correspond to transition size between dislocation dominant and grain boundary dominant behaviors. It should be specially interesting that the limited idealized model, which has only two active slip-systems, shows such differences in mechanisms.

Twisted periodic boundary condition is proposed and it was applied to torsion problem of microscopic iron specimen and initial internal structure dependence of the grain refinement has been found by identifying the atomistic structure.

Deformation induced nanocrystallization of amorphous iron has been observed and it has been found that the nanocrystalization process depends on the localized deformation pattern and stress state. There is possibility that the texture of nanocrystals up to our aims can be obtained by controlling the deformation in amorphous structure. Further studying the internal structural changes obtained by atomistic simulations can clarify the mechanism of grain refinement and can make quite a few proposals for material design.

Acknowledgments

AN thanks Nobuhiro Tsuji for fruitful discussions and appreciates support by Grant-in-Aid for Scientific Research from Ministry of Education, Culture, Sports, Science and Technology, Japan. TS is supported by Grant-in-Aid from the Japan Society for the Promotion of Science. AN and HK gratefully appreciate support from New Energy and Industrial Technology Development Organization (NEDO) of Japan and support from the Strategic Research Base, Handai Frontier Research Center supported by the Japanese Government's Special Coordination Fund for Promoting Science and Technology.

References

Chen, M., Kevin, E. M., Hemker, J., Sheng, H., Wang, Y., and Cheng, X. (2003). Deformation twinning in nanocrystalline aluminum. *Science*, 300:1275–1277.

Falk, M. L. (1999). Molecular-dynamics study of ductile and brittle fracture in model noncrystalline solids. *Phys. Rev., B*, 60(10):7062–7070.

Finnis, M. W. and Sinclair, J. E. (1984). A simple empirical *n*-body potential for transition metals. *Philos. Mag., A*, 50(653):45–55.

Finnis, M. W. and Sinclair, J. E. (1986). Erratum. *Philos. Mag., A*, 53(1):161.

Fleck, N. A. and Hutchinson, J. W. (2001). A reformulation of strain gradient plasticity. *J. Mech. Phys. Solids*, 49:2245–2271.

Matsumoto, R., Kitagawa, H., and Nakatani, A. (2003). Molecular dynamics simulation of deformation-induced nanocrystallization in an amorphous metal. *J. Soc. Mat. Sci., Japan*, 52(3):235–240 (in Japanese).

Mishin, Y., Farkas, D., Mehl, M. J., and Papaconstantopoulos, D. A. (1999). Interatomic potentials for monoatomic metals from experimental data and *ab initio* calculations. *Phys. Rev. B*, 59(5):3393–3407.

Nakatani, K., Nakatani, A., and Kitagawa, H. (1998). Molecular dynamics study on fracture mechanism of fe-amorphous metal (j integral near mode i crack tip in amorphous metal). *Adv. Mat. Res.*, 1:88.

Nakatani, K., Nakatani, A., Sugiyama, Y., and Kitagawa, H. (2000). Molecular dynamics study on mechanical properties and fracture in amorphous metal. *AIAA Journal*, 38(4):695–701.

Nieh, T. G., Wadsworth, J., Liu, C. T., Ohkubo, T., and Hirotsu, Y. (2001). Plasticity and structural instability in a bulk metallic glass deformed in the supercooled liquid region. *Acta Mater.*, 49:2887.

Saito, Y., Utsunomiya, H., Tsuji, N., and Sakai, T. (1999). Novel ultra-high straining process for bulk materials — development of the accumulative roll-bonding (arb) process —. *Acta Mater.*, 47:579–583.

Schiøtz, J., Di Tolla, F. D., and Jacobsen, K. W. (1998). Softening of nanocrystalline metals at very small grain size. *Nature*, 391:561–563.

Shimokawa, T., Nakatani, A., and Kitagawa, H. (2002). Mechanical properties depending on grain sizes of fcc nanocrystalline metals by using molecular dynamics simulation (investigation of stacking fault energy's influence). *Trans. JSME, Ser. A*, 68(676):1708–1715 (in Japanese).

Tarumi, A., Ogura, A., Shimojo, M., Takashima, K., and Higo, Y. (2000). Molecular dynamics simulation of crystallization in an amorphous metal during shear deformation. *Jpn. J. Appl. Phys.*, 39(6):L611–L613.

Van Swygenhoven, H., Spaczer, M., and Caro, A. (1999). Microscopic description of plasticity in computer generated metallic nanophase samples: A comparison between cu and ni. *Acta Mater.*, 47(10):3117–3126.

Weertman, J. (2002). Anomalous work hardening, non-redundant screw dislocations in a circular bar deformed in torsion, and non-redundant edge dislocations ina bent foil. *Acta Mater.*, 50:673–689.

Yamakov, V., Wolf, D., Salazar, M., Phillpot, S. R., and Gleiter, H. (2001). Length-scale effects in the nucleation of extended dislocations in nanocrystalline Al by molecular-dynamics simulation. *Acta Mater.*, 49:2713–2722.

CONTINUUM DESCRIPTION OF INHOMOGENEOUSLY DEFORMING POLYCRYSTALLINE AGGREGATE BASED ON FIELD THEORY

Tadashi Hasebe
Doshisha University

Abstract This paper presents a field theoretical approach toward a continuum description of polycrystalline media. Collective behavior of the composing grains is shown to produce large stress fluctuation which can be an essential feature of polycrystalline aggregates. Stress function tensor is proposed as a new mechanical parameter describing such fluctuated fields. Implementation to the constitutive model based on crystal plasticity is extensively discussed.

Keywords: Polycrystal, Crystal Plasticity, Field Theory, Strain Gradient Plasticity

1. INTRODUCTION

One of the ultimate goals of the multiscale modeling in plasticity would be to predict ductility (i.e., uniform and total elongations) as well as hardening properties of a material concerned (e.g., yield stress, hardening coefficient, tensile strength, etc.) under given loading conditions from their fundamental or microscopic properties transcending the hierarchy of scales. This, however, has not been satisfactorily accomplished to date partially because of the absence of rational theoretical framework describing evolutions of multi-level inhomogeneous fields relating to substructure evolution, intra-grain deformation, etc.

On the other hand, FEM-based simulations have been widely conducted on polycrystalline plasticity even with Voronoi tessellation models having realistic grain morphology. But they are rather limited to discussing macroscopic (averaged) responses with limited number of grains. Fujita, et

H. Kitagawa and Y. Shibutani (eds.), IUTAM Symposium on Mesoscopic Dynamics of Fracture Process and Materials Strength, 381–390.

al.[1] reported a series of implicative experimental findings for polycrystalline metals demonstrating a number effect of grains on macroscopic response as well as the existence of an affected zone (AZ) beyond the grain size. This evidently means deficiencies of the frequently used "coarse-grain" models in simulating the real behavior of polycrystals.

This paper presents a field theoretical approach toward a continuum description of polycrystalline aggregates based on the field theory of plasticity advocated recently by the author[2,3]. In order to examine the collective effect of a large number of composing grains on meso- and macroscopic deformation behavior of polycrystals, finite element simulations are conducted on systematically designed polycrystalline models. A new parameter for describing the inhomogeneous field as a result of the collective behavior is proposed based on the simulation results. The theoretical background and a constitutive modeling incorporating the above parameter are extensively discussed.

2. FIELD THEORY OF PLASTICITY

The field theory of plasticity, advocated by the author[2,3], is roughly composed of two aspects, i.e., (A)continuum description of inhomogeneous fields including dislocations and defects based on the differential geometrical field theory, often referred to as "non-Riemannian plasticity," and (B)field theoretical descriptions of collective behavior of interacting dislocations, defects and crystal grains. The (A) covers all the aspects of the strain gradient plasticity accommodating "scale effects" of the fields, while the (B) describes "evolution" of the fields. In the (B), the method of quantum field theory (QFT), which is equivalent to classical statistical mechanics, has been successfully applied to construct a model for dislocation cell structure formation. In there, rigorous derivation of the effective theory yielding cellular patterning from a microscopic description of dislocation interaction through systematic "coarse-graining" and renormalization of scales is achieved. The theory-based simulation results are exemplified in Figure 1, where *gem* measures the magnitude of internal stress in the cell interior region. The findings obtained there are, e.g., (1)the collective effect of the key process (pair annihilation) results in a singularity of the system to be emerged leading to a modulation of dislocation distribution rather than uniform distribution, and (2)existence of long range

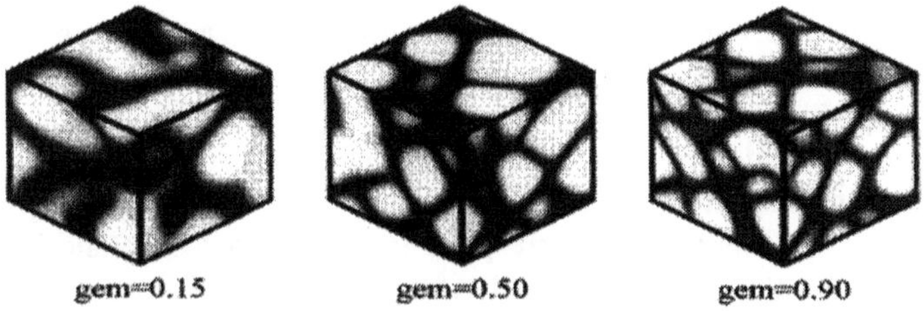

Figure 1 Examples of simulated results for 3D cell formation based on field theory-based effective model.

internal stress field in the cell interior region is responsible for both the cell size and its morphology to be evolved, without which "far-from-cell" structures will possibly result. Note that the long range internal stress field has been confirmed to exist both by experiment and dislocation dynamics simulation.

3. SSS HYPERTHESIS

A large number of crystal grains comprising a polycrystalline medium is expected to yield a certain collective behavior ultimately controlling its mechanical behavior. As a candidate of the collective effect, the evolution of

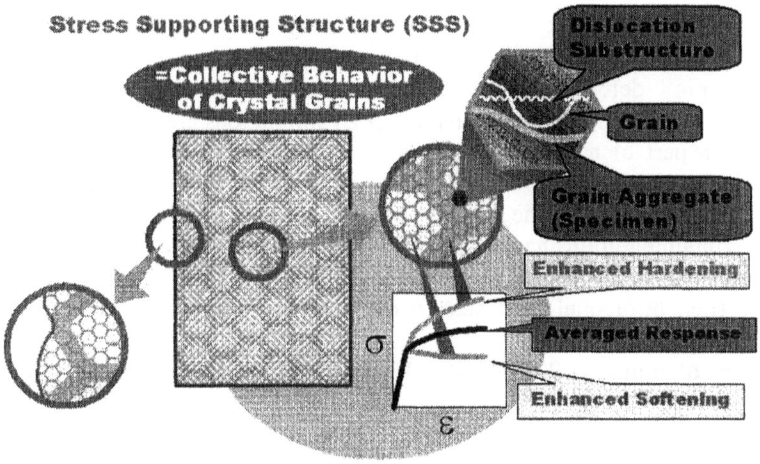

Figure 2 Schematics of mesoscopic structure as a result of collective behavior of crystal grains, together with "role sharing" of grains.

a mesoscopic structure through which the applied force and the deformation are transmitted is hypothesized as schematically shown in Figure 2. This meso-structure will be organized so as to support the externally applied stress, implying a “role sharing” of composing grains between load transmission and deformation transmission relating to enhanced hardening and softening, respectively. The former is called stress supporting structure (SSS) whose wavelength will correspond to the affected zone and it will be responsible for the “remote effect.” One possible realization of the SSS would be the evolution of “surface roughness” during plastic deformation, where its wavelength is reported to be basically unchanged during deformation implying the SSS is constructed relatively in the early stage of deformation as far as the deformation mode is unaltered. The morphology as well as the characteristics of the SSS cannot be determined simply by the number of composing grains but will be greatly affected by the absolute grain size and the materials properties such as stacking fault energy (SFE). This is why the “multiscale” viewpoint is indispensable

Note, in this case, the direct application of the method of QFT is inappropriate because the “grains” are distinguishable and deformable as well different from dislocations.

4. ANALYTICAL RESULTS AND DISCUSSION

To identify the key factor of modeling polycrystalline aggregate, especially the collective effect of a large number of composing grains on the mesoscopic deformation behavior including the SSS formation, systematically designed finite element (FE) simulations are made (Figure 3). This is a part of a recent collaboration in Research Group on Multiscale Crystal Plasticity, Kansai[1].

The FE polycrystalline models (Fig.3(a)) have a common representative area, containing grains having a same orientation distribution, surrounded by different numbers of grains. All the grains here are assumed to have the same shape (i.e., hexagonal) and size to eliminate additional factors (e.g., size distribution and morphology) than the "number" effect at this stage. The total numbers of grains are 23, 77, 168, 613 and 2328, where each grain is divided

---

[1] Co-organized with Prof. Y. Tomita’s group in Kobe University and Dr. Y. Higa in Osaka University, Japan since 2002.

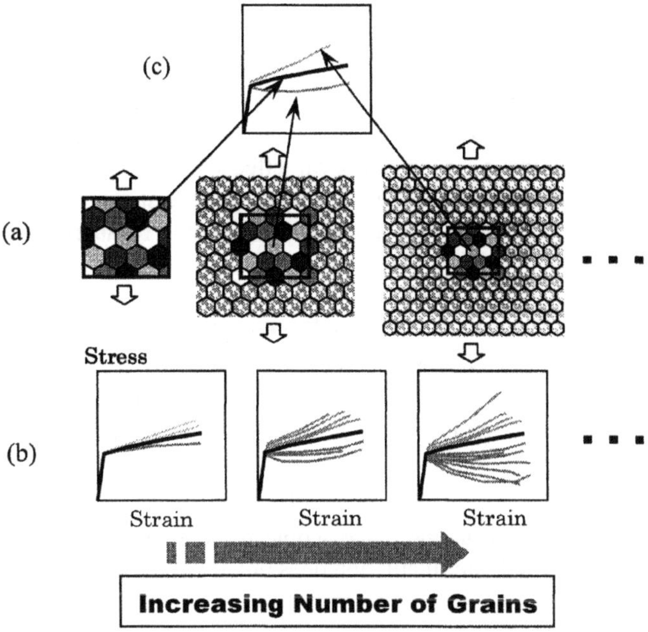

Figure 3 Schematic polycrystal behavior obtained in recent crystal plasticity FE analysis.

into 576, 256, 256 and 64 triangular elements, respectively. Plane stress or plane strain condition is assumed to take into account the in-plane heterogeneity only.

Figure 3(b) schematically illustrates the obtained results showing stress-strain curves for the individual grains together with the averaged response. There observed a large scatter of the stress response for the individual grains, which increases greatly with increasing number of grains while the macroscopic stress-strain curve stays unchanged. Also demonstrated (Figure 3(c)) is that even with the same combination of neighboring grains in the common area, stress response of a grain in it tends to show different trend depending on the number of surrounding grains and its randomness, meaning that the stress response of a grain is not determined only by its immediate neighbors but is greatly affected by the grains in the distances. This implies that there exists a "remote effect" activated through the SSS. Therefore, the representative region or volume on which every mesoscale simulation puts its basis cannot be determined so easily as expected and is substantially controlled by the macroscopic information.

The above results also show that even for the constant-size isotropic grain model eliminating morphological effects, still it naturally have the number effect leading to the large fluctuation of the stress field. Therefore, the existence of the stress fluctuation is considered to be an essential feature of the polycrystal plasticity. The effect of the morphological features on the fluctuation needs to be examined in the next step.

The macro stress is thus expressed formally as,

$$\boldsymbol{\sigma}_{macro}(\boldsymbol{x}) = \langle \boldsymbol{\sigma}_{micro}(\boldsymbol{x}) \rangle + \delta\boldsymbol{\sigma}_{micro}(\boldsymbol{x})$$

where < > represents spatial average and $\delta\boldsymbol{\sigma}$ expresses the stress fluctuation. Since the fluctuation tends to vanish under spatial average, $\langle \delta\boldsymbol{\sigma}_{micro}(\boldsymbol{x}) \rangle \approx 0$, we have,

$$\langle \boldsymbol{\sigma}_{macro}(\boldsymbol{x}) \rangle = \langle \boldsymbol{\sigma}_{micro}(\boldsymbol{x}) \rangle$$

This means that, as far as the macroscopic response like overall hardening behavior is concerned, the fluctuation does not always have to be considered.

5. DESCRIPTION OF FLUCTUATED FIELDS

For a continuum description of the fluctuated fields, use is made of the graph theory. Figure 4 illustrates schematics of a particle and a void graphs based on the graph theory, responsible for force and deformation transmissions, respectively. A particle in the particle graph should be regarded, in the present context, as an aggregate of hardened grains rather than an individual grain, whereas a void as that of softened grains. For more details about the graph theory, refer to [4]. Void force $\boldsymbol{F}_v$ and void displacement $\boldsymbol{U}_v$ are, respectively, related with particle force $\boldsymbol{F}_p$ and particle displacement $\boldsymbol{U}_p$ through the fundamental matrices;

$$\boldsymbol{F}_P = -(\tilde{D}_{PC}\tilde{L}_{CV})\boldsymbol{F}_V = 0, \quad \boldsymbol{U}_V = -(\tilde{L}_{VC}\tilde{D}_{CP})\boldsymbol{U}_P = 0$$

which express balance of force and compatibility of displacement, respectively. Here $\tilde{D}_{PC}, \tilde{L}_{CV}$ respectively indicate incidence and loop matrices[4]. There is a mathematical correspondence between the fundamental matrices in the graph theory and the differential operators, motivating us to introduce incompatibility tensor and stress function tensor for the continuum description of the inhomogeneous fields. The corresponding relationships to the above equations are,

$$\boldsymbol{\sigma} = \nabla \times \nabla \times \boldsymbol{\chi}, \quad \boldsymbol{\eta} = -\nabla \times \nabla \times \boldsymbol{\varepsilon}^{p}$$

The stress function tensor, corresponding to $\boldsymbol{F}_v$, characterizes the deviation of the stress field from its equilibrium. The fluctuated stress, to be vanished macroscopically, can be regarded as this sort. Therefore, this can be a suitable mechanical parameter describing why and how the stress fluctuation

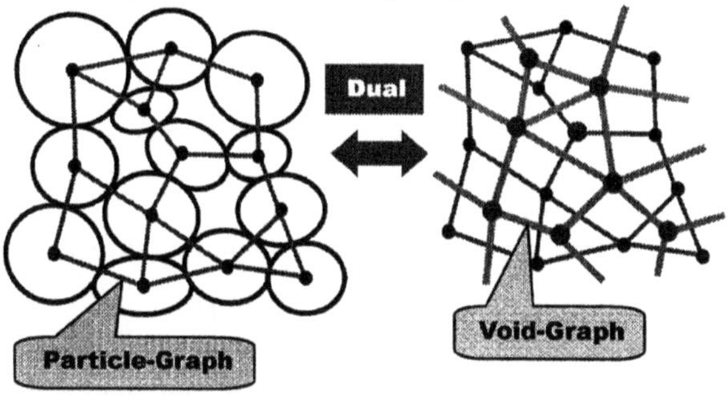

Figure 4 Graph theoretical representation of crystal grain aggregates (particle graph) and its dual (void graph).

evolves. The incompatibility tensor, on the other hand, is given as a second derivative (more precisely, double curl) of strain tensor, defining, as it stands, degree of breaking the strain compatibility.

The quality as well as the quantity of the stress fluctuation can be discussed in terms of correlation function of $\delta\sigma$;

$$G(\delta\sigma(\boldsymbol{x})) = \frac{1}{V}\int_{\Delta V'} \langle \delta\sigma(\boldsymbol{x})\delta\sigma(\boldsymbol{x}+\boldsymbol{x}') \rangle d\boldsymbol{x}'$$

The correlation length ξ will be used in explicitly evaluating χ to be discussed in the following section. Here ξ is expected to be coincide with the "affected zone" and is determined as a result of micro-macro interactions through the collective behavior.

Further decomposition of $\delta\sigma$ into the deviatoric and the hydrostatic components can clarify the roles for "local plasticity" and the SSS evolution, respectively.

$$\delta\sigma = \delta\left(\sigma' + \frac{1}{d}\sigma_m \boldsymbol{I}\right) = \delta\sigma' + \frac{\boldsymbol{I}}{d}\delta\sigma_m$$

6. NEW PARAMETER FOR INHOMOGENIOUS FIELDS

As discussed above, macroscopically vanishing stress fluctuation can be the essential feature of polycrystal plasticity e.g., dominating transition from macroscopically uniform deformation field to the non-uniform one leading to instability and, ultimately, fracture. The stress tensor is given by double curl of the stress function, as in the conventional definition;

$$\begin{aligned}\sigma &= \nabla\times\nabla\times\chi \\ &= \nabla(\nabla\cdot\chi)-\Delta\chi\end{aligned}$$

In the present case, the stress tensor should be replaced by the fluctuation $\delta\sigma$. With the additional condition $\nabla\cdot\chi=0$, for the stress components to be uniquely determined, the above equations is reduced to the Poisson's equation, i.e., $\delta\sigma(x)=-\Delta\chi(x)$. The special solution of this is given by,

$$\chi(x)=\frac{1}{4\pi}\int_{\Delta V'}\frac{\delta\sigma(x')}{|x-x'|}dV'$$

where $\Delta V'$ should be specified by ξ based on simulated results. Thus the stress function tensor can be explicitly evaluated from a distribution of the stress fluctuation under given stress boundary conditions.

The above discussion makes the rational continuum description of such multi-level fluctuated fields based on the field theory possible. The evolution of the stress function field is expected to be responsible for the onset of instability of plastic deformation and, ultimately, transition from deformation to fracture. This motivates us to investigate, as a next step, the evolution of a singularity from the fluctuated fields

The evolution of the singularity of the underlying field can be detected by evaluating the energy-momentum (E-M) tensor. Three singularities corresponding to the elastic, dislocation and defect fields in terms of the E-M tensor can be defined based on the field theory of plasticity. Note, in this case, the transition from the singularity due to dislocated or defected fields to that due to the existence of a crack must be separately discussed.

7. CRYSTAL PLASTICITY AS A TENTATIVE FRAMWORK

Three scales relating to inhomogeneous fields are considered to be essential in polycrystal plasticity, i.e., orders of (1)dislocation substructures,

(2)individual grain and (3)aggregate of grains (or apecimen). The orders of (1) and (3) are related to "collective behavior" of individual elements composing the system, i.e., dislocations and grains.

The author[5,6] has proposed a constitutive model in crystal plasticity applicable to wide ranges of strain rate and strain history, which also accommodates scale effects through effective cell size reflecting dislocation substructure evolutions depending on loading histories. General form of the microscopic constitutive equation can be written as,

$$\dot{\gamma}^{(\alpha)} = f\left(\frac{\tau^{(\alpha)} - \Omega^{(\alpha)}}{K^{(\alpha)}}\right)$$

where $K^{(\alpha)}$ and $\Omega^{(\alpha)}$ are drag stress and back stress, respectively. The drag stress is further equated as

$$\dot{K}^{(\alpha)} = Q_{\alpha\beta} H(\gamma) \left|\dot{\gamma}^{(\beta)}\right|,$$

where $H(\gamma)$ is a referential hardening modulus. Here $Q_{\alpha\beta} \equiv \delta_{\alpha\beta} + f_{\alpha\kappa} S_{\kappa\beta}$ represents hardening ratio, responsible for the hardening characteristics, expressing the substructure-order inhomogeneity. Anisotropic evolutions of the history is expressed by taking account of the anisotropy of the dislocation-dislocation interactions expressed by the interaction matrix $f_{\alpha\beta}$ specifying kinds and strengths of interactions between arbitrary two slip systems, further given as a function of SFE. Their contributions to the hardening characteristics are determined by the history of the underlying system expressed by history matrix $S_{\alpha\beta}$ given as a function of plastic work for FCC metals and slip strain for BCC metals. Since all the information concerning the evolved dislocation substructures during deformation is incorporated in $Q_{\alpha\beta}$, the effective cell size is assumed to be given as a function of a quadratic invariance of $Q_{\alpha\beta}$, e.g., as,

$$d_{cell} = d_{Gr} \left(Q_{\alpha\beta} Q_{\alpha\beta}\right)^{-1/2}$$

d_{cell} can be regarded as mean free path of dislocations. Other two inhomogeneous fields can be taking into account by simply adding their contributions to $Q_{\alpha\beta}$, i.e.,

$$Q_{\alpha\beta} = \delta_{\alpha\beta} + f_{\alpha\kappa} S_{\kappa\beta} + \sum_{k}^{Gr,Spe} F_k(\alpha^{(\alpha)}, \eta^{(\alpha)})$$

where $F_k(\alpha^{(\alpha)}, \eta^{(\alpha)})$ gives functions describing the inhomogeneous effects of grain-size order (k=Gr) and specimen order (k=Sp). The explicit form of

$F_k(\alpha^{(\alpha)},\eta^{(\alpha)})$ can be obtained by simple considerations based on the differential geometrical field theory in terms of length as,

$$\alpha_k^{(\alpha)}=\frac{1}{b}\cdot\frac{\Delta\gamma_k^{(\alpha)}}{d_k},\quad \eta_k^{(\alpha)}=\frac{1}{d_k^{(2)}}\cdot\left(\frac{1}{b}\cdot\frac{\Delta\gamma_k^{(\alpha)}}{d_k^{(1)}}\right)\cdot l_k$$

where $\alpha_k^{(\alpha)}$, $\eta_k^{(\alpha)}$ are dislocation density and incompatibility tensors of the underlying orders k. Therefore, for the two levels of inhomogeneous fields,

$$F_k(\alpha_k^{(\alpha)},\eta_k^{(\alpha)})=d_{Gr}\delta_{\alpha\beta}\left(\frac{b\left|\alpha_k^{(\beta)}\right|}{\Delta\gamma_k^{(\beta)}}+\text{sgn}(\eta_k^{(\beta)})\sqrt{\frac{b\left|\eta_k^{(\beta)}\right|}{l_k\cdot\Delta\gamma_k^{(\beta)}}}\right)\quad (no\ sum\ on\ \beta)$$

Here $\text{sgn}(\eta_k^{(\alpha)})$ indicates the sign of $\eta_k^{(\alpha)}$ accommodating not only additional hardening but also softening. All the effects of the inhomogeneous fields are embodied as the effective cell size d_{cell}, without using ill-defined dislocation density as in the conventional models. Note, in this case, $\alpha_k^{(\alpha)}$, $\eta_k^{(\alpha)}$ should be the resolved components, i.e., $\alpha^{(\alpha)}=P_{ij}^{(\alpha)}\alpha_{ij}$, $\eta^{(\alpha)}=P_{ij}^{(\alpha)}\eta_{ij}$, where $P_{ij}^{(\alpha)}$ is the Schmid tensor.

Thus evaluated d_{cell} is also used to express the back stress evolution, i.e.,

$$\dot{\Omega}^{(\alpha)}=-\left\{\left\langle d_{cell}-\bar{x}_N^{(\alpha)}\right\rangle+\alpha\right\}^{-2}\left\langle 1-\frac{K^{(\alpha)}}{K_{sat}}\right\rangle$$

where $\bar{x}_N^{(\alpha)}$ is mean moving distance of dislocations evaluated from the slip strain $\bar{x}_N^{(\alpha)}=l^*\cdot\gamma^{(\alpha)}$ with l^* being a free flying distance of dislocations.

REFERENCES

1. Miyazaki, S., Shibata, K. and Fujita, H., Effect of Specimen Thickness on Mechanical Properties of polycrystalline Aggregates with Various Grain Sizes, *Acta Metall.*, (1979), Vol.27, pp.855-862.
2. Hasebe, T. and Imaida, Y., Construction of Quantum Field Theory of Dislocations Based on the Non-Riemannian Plasticity, *Acta. Metall. Sinica*, (1998), Voll.11, No.6, pp.405-411.
3. Hasebe, T. and Imaida, Y., Gauge Field Theory of Dislocations and the Non-Riemannian Plasticity, *Mechanical Behavior of Materials*, (1999), Vol.III, pp.1077-1082.
4. Oda, M., Iwashita, K. (eds), Mechanics of Granular Materials, (1999) A.A. Balkema.
5. Hasebe, T. and Imaida, Dislocation Pileup Model for Back Stress Evolution in Crystalline Plasticity, *Trans. Struct. Maters. Reactor Technol.* (1999),Vol. 15, pp.801-808.
6. Hasebe, T., Kumai, S. and Imaida, Y., Impact Compression Behavior of FCC Metals with Pre-Torsion Behavior, *J. Maters. Process. Technol.*, (1998), Vol. 85, pp.184-187.

STUDY ON STRENGTH OF MICROSCOPIC MATERIAL BY SIMULATIONS WITH ATOM AND ELECTRON MODELS

Yoshitaka Umeno and Takayuki Kitamura
Graduate School of Engineering, Kyoto University, Japan

Abstract The "ideal strength" or the "theoretical strength" is originally defined as the stress at which a perfect crystal becomes mechanically unstable, and is a fundamental mechanical parameter of material. The definition can be extended to the mechanical instability of inhomogeneous structures. This paper reviews the simulations on the "ideal strength" conducted by our groups based on quantum mechanics, and three cases are presented. (1)The ideal shear strength of silicon is precisely evaluated by *ab initio* calculations. (2)The ideal deformation of carbon nanotubes and its relationship with the electric conductivity are examined by semi-empirical band calculation. (3)The structure and strength of interface between aluminum and silicon are investigated by *ab initio* molecular dynamics simulations.

Keywords: Ideal strength, Inhomogeneous structure, Deformation, Ab initio, Quantum mechanics, Atomic simulation

1 INTRODUCTION

The "ideal strength" or "theoretical strength" of materials is a fundamental mechanical parameter of material behavior[1,2]. It is necessary to evaluate the strength of the materials based on the atom and electron model in order to understand the property in detail from the atomic scale. Therefore, atomic simulations on the mechanical behavior of the materials have been carried out[3–5]. In this paper, simulations conducted by our groups on the "ideal strength" based on quantum mechanics are presented.

2 SIMULATION METHODS

2.1 *Ab Initio* (First Principles) Calculation

Among *ab initio* calculations, which non-empirically simulate the electron structure of the materials, without experimental data, the norm-conserving pseudopotential method[6] based on the density functional theory (DFT)[7] is appropriate for the evaluation of the mechanical property because the interatomic force can be precisely evaluated. According the theory, the total energy of a system, E, is expressed as a functional of the charge density, ρ,

$$E[\rho] = T[\rho] + \int V_{\rm ext}(\boldsymbol{r})\rho(\boldsymbol{r})d\boldsymbol{r} + \frac{1}{2}\int\int \frac{\rho(\boldsymbol{r}')\rho(\boldsymbol{r})}{|\boldsymbol{r}' - \boldsymbol{r}|}d\boldsymbol{r}'d\boldsymbol{r} + E_{\rm xc}[\rho]. \quad (1)$$

Here, $\boldsymbol{r}$ indicates the coordinate vector in the real space and the terms in the right hand are the kinetic energy of electrons, the potential energy of electrons induced by nuclei, the coulomb interaction between electrons, and the exchange-correlation energy of electrons, respectively. Under the local density

H. Kitagawa and Y. Shibutani (eds.), IUTAM Symposium on Mesoscopic Dynamics of Fracture Process and Materials Strength, 391–399.

approximation, the solution of Eq.(1) is obtained by solving the one-electron Schrödinger equation (the Kohn-Sham equation)[8],

$$H\psi_i = E\psi_i, \tag{2}$$

where i indicates the electron state, ψ is the wave function, and $H = -\frac{1}{2}\nabla^2 + V$. Since V includes the potential acting on the electron, which is a functional of the charge density, ρ,

$$\rho(\boldsymbol{r}) = \sum_i |\psi_i(\boldsymbol{r})|^2, \tag{3}$$

where the sum runs over all the occupied states, the above equations should be solved self-consistently. It is able to obtain not only mechanical but also electronic properties with high accuracy by means of the *ab initio* method.

2.2 Semi-Empirical Band Calculation

Although the *ab initio* method has high accuracy and transferability, it requires enormous computational time and resources. Therefore, semi-empirical band calculation methods, where the Schrödinger equation is approximately solved under some assumptions, have been developed. The tight-binding (TB) method is one of the semi-empirical methods and is recently used for molecular dynamics simulations. In the method, the hamiltonian is constructed with the effective potential, V,

$$H = -\frac{1}{2}\nabla^2 + V, \tag{4}$$

where the first term in the right hand means the kinetic energy, and V is obtained by simple functions with parameters (TB potential). The wave functions are expressed as linear combination of atomic orbitals (LCAO)[9,10]. The hamiltonian is solved once and no self-consistent iteration is performed. Although the semi-empirical band calculations have less transferability than the *ab initio* ones, they are still reliable because they take into account electron on the basis of the Schrödinger equation. Moreover, electron structures can be simulated so that the electron properties of the materials is evaluated.

2.3 Atomic Simulaion with Potential Function

In conventional procedures for atomic simulations, atomic interactions are calculated with the functions of the coordinate vectors of the atoms (potential functions)[11]. Thus, the total energy of the atomic system, E, is expressed as

$$E = E(\boldsymbol{R}), \qquad \boldsymbol{R} = (\boldsymbol{r}_1, \boldsymbol{r}_2, \cdots), \tag{5}$$

where $\boldsymbol{r}_I$ denotes the coordinate vector of the I-th atom. In general, the expression is in the form,

$$E(\boldsymbol{R}) = \sum_{I,J} \phi^{(2)}_{IJ} + \sum_{I,J,K} \phi^{(3)}_{IJK} + \cdots, \tag{6}$$

where the first term in the right hand is for two-body interactions and the second one is for three-body, and so on. Two-body (pair) potentials have been widely used due to their simplicity, but they have low transferability. Many-body potentials such as EAM (Embedded Atom Method)[12] and EMT (Effective Medium Theory)[13], which take into account effect of electron density, have been suggested for better transferability. However, their transferability still need to be examined because they are not versatile[14]. Moreover, in case that there is no interatomic potential available for the simulation object, development of an interatomic potential is needed[15,16].

3 STRENGTH OF MATERIALS

The ideal strength of a crystal is defined as the stress at which the crystal lattice becomes mechanically unstable under uniform deformation, and is a fundamental mechanical property[1]. Atomic simulations using interatomic potential functions for evaluating the ideal strength of crystals have been carried out for a long time[1,2,17–19]. On the other hand, since it is important to evaluate the ideal strength with high accuracy, precise evaluations on the ideal strength with *ab initio* calculations have started in recent years[14,20–26]. Moreover, not only the mechanical property but also its relationships with other properties, e.g. electric conductivity, are of interest and they have been investigated[22,27].

Materials in general include inhomogeneous structures, which strongly affects the strength of the whole system. For example, surfaces and grain boundaries are the typical inhomogeneous structures and they have significant influence in the strength of microscopic materials. This means that it is essential to elucidate the strength at which the structure becomes mechanically unstable. It can be termed "ideal strength" or "theoretical strength" of a material with a surface (grain boundary). Atomic models for simulations on such inhomogeneous structures usually contain many atoms so that it is difficult to employ the *ab initio* method. The mechanical instability criterion of atomic structures is derived taking into account the freedom of the system[28]. Simulation procedures using interatomic potential functions are adopted in such cases. Attention should be paid on the validity of the potential functions for the object system as mentioned above.

While the surface and the grain boundary are the inhomogeneous structure consisting of identical crystals, interfaces between dissimilar materials have more complicated properties. For example, extremely high stress and strain can emerge in the region near the interfaces due to deformation mismatch. In order to simulate atomic behavior at the interfaces, *ab initio* non-empirical calculation methods is inevitable because there are few valid potential functions for interactions between dissimilar materials. Although the calculations are extremely time consuming because of complexity of the systems, they are becoming feasible with the use of the parallel computation and the development of the calculation algorithm[29].

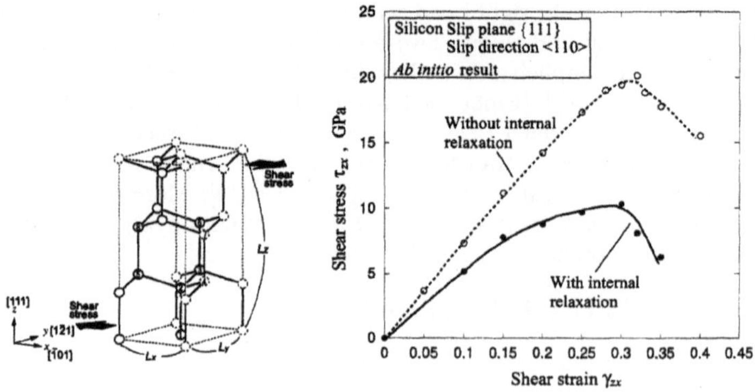

Figure 1 Stress-strain curves in ideal shear deformation of silicon single crystal.

4 SIMULATION RESULTS

4.1 Ideal Strength of Single Crystal

Because the silicon single crystal has the diamond structure, which is highly anisotropic and causes large internal displacement during deformation, it is necessary to carefully evaluate the ideal strength of the crystal. We have conducted *ab initio* calculations on the ideal shear strength of silicon[22]. The shear direction is $[\bar{1}01]$. The shear stress-strain curves are shown in Fig.1. In the shear deformation, the normal stresses are controlled to be zero by adjusting the cell size. In the figure, simulation results with and without the internal relaxation are plotted. It is obvious that the effect of the internal displacement is large and it should not be ignored. The ideal shear strength is evaluated about 10GPa at the shear strain of 0.3. The result corresponds well with an extrapolation from the critical shear stresses of dislocation nucleations at high temperatures evaluated with experiments by Ohta *et al*[30]. Change in the bond structure during the shear is depicted in Fig.2. The bond breaking at $\gamma > 0.3$ corresponds to the decrease of the shear stress.

4.2 Deformation of Carbon Nanotube

Carbon nanotubes (CNTs) have been attracting attention because of their characteristic properties. We have carried out tight-binding simulations on the axial tensile deformation of CNTs. Figure 3 shows relationships between the tensile load and the strain in (8,8) armchair and (14,0) zigzag CNTs. Ideal strengths of the CNTs are almost the same but the strain at the fracture differs. This indicates that the atomic structure in the CNT affects its mechanical property. Changes in the electronic conductivity of various CNT structures are examined[27]. Figure 4 shows the changes obtained by evaluating the band gap

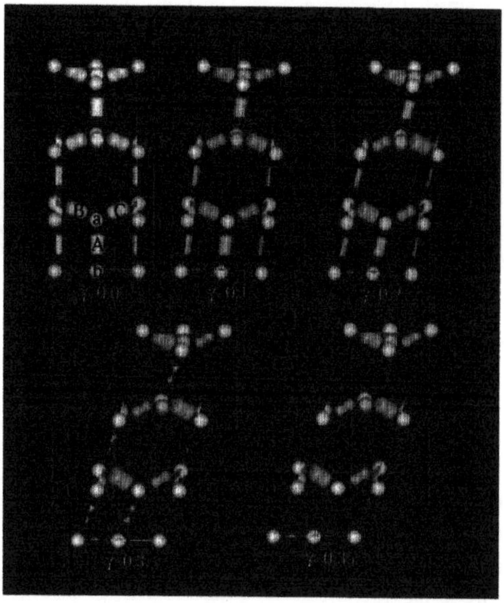

Figure 2 Change in bond of silicon single crystal during shear deformation.

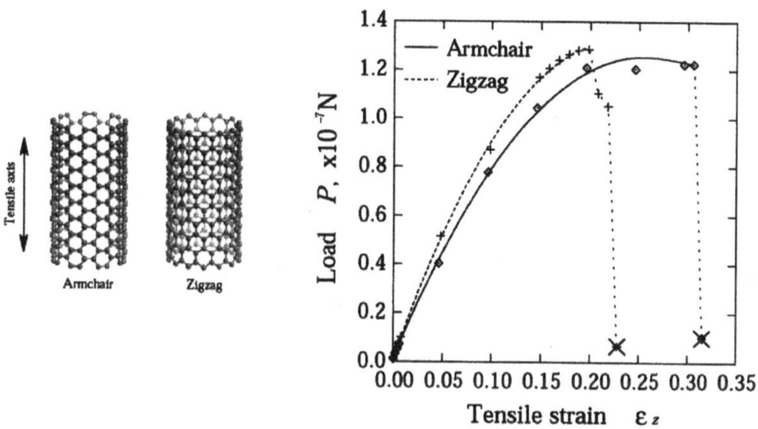

Figure 3 Relationships between tensile load and strain in (8,8) armchair and (14,0) zigzag CNTs.

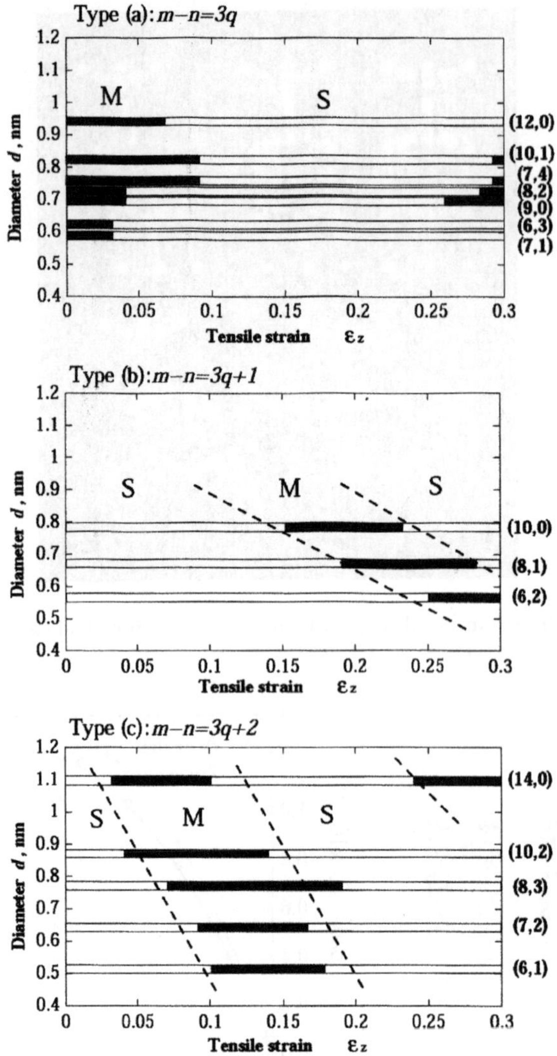

Figure 4 Transition of electronic conductivity of CNTs. (m, n) indicates the chiral vector and q is an integer. S and M indicate "semiconducting" and "metallic", respectively.

energy of the CNTs during the axial tension. CNTs show various patterns of transition between metallic and semiconducting properties depending on the structure.

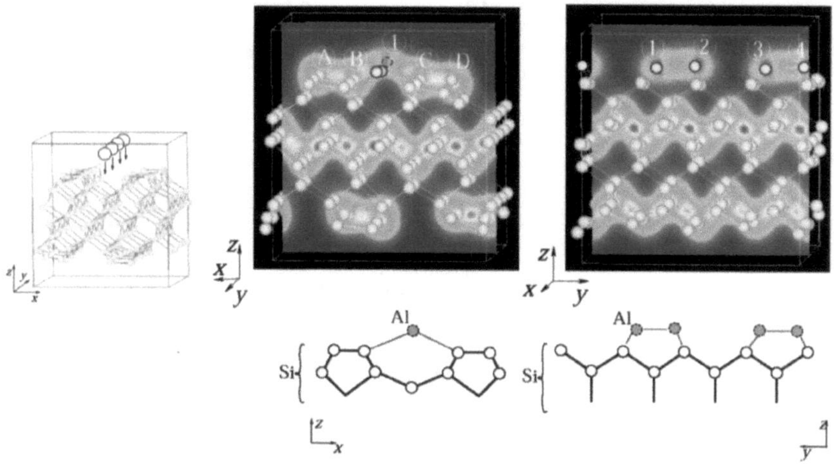

Figure 5 Atom configuration and bond structure after precipitation of Al atoms onto (001) Si surface.

4.3 Structure and Strength of Interface between Dissimilar Materials

The atom and electron structures at an interface between dissimilar materials which have different bond structures in bulk have to be clarified by a non-empirical method because interactions at the interface are complicated. We have conducted *ab initio* molecular dynamics simulations on the structure and the adhesive strength of aluminum on a (001) surface of silicon[31]. After reproducing the (001) silicon surface by relaxation with the *ab initio* MD, aluminum atoms are precipitated from above the surface as shown in Fig.5. Figure 6 shows the atom configuration and the bond structure after the precipitation, where the Al atoms are arranged in a line between Si dimers. The Al atoms are in dimer pairs with strong bonds and construct weak bonds with the Si dimers. The adhesive strength is evaluated by pulling an Al atom from the surface. Only the atom pulled is detached from the surface and the strength obtained is about 1.3nN. Further simulations should be performed to elucidate the structure and the strength in detail.

5 SUMMARY

With the aim to understand the mechanical property of materials in detail from the atomic scale, we have conducted simulations based on quantum mechanics to investigate the "ideal strength" as a fundamental mechanical parameter, which can be defined not only for perfect crystals but also for materials with inhomogeneous structures.

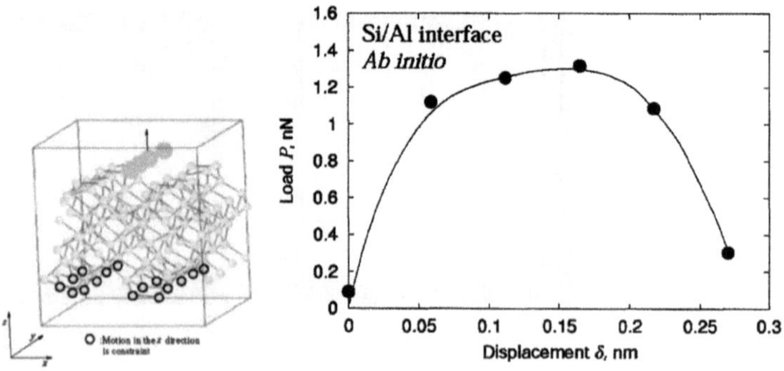

Figure 6 Load-displacement curve of Al atom detached from Si surface.

1) The ideal strength of a crystal, which is a fundamental mechanical property, is precisely evaluated by *ab initio* calculations.

2) Relationships between the mechanical property and other ones, e.g. electronic one, can be investigated by simulations with the atom and electron model such as *ab initio* and semi-empirical band calculations.

3) Due to the complexity of the structures, the ideal strength of typical inhomogeneous structures with surface or grain boundary is evaluated by conventional simulations with interatomic potential functions, the validity of which is examined based on *ab initio* calculations beforehand.

4) For an arbitrary inhomogeneous structure, the criterion of the mechanical instability is derived taking into account the freedom of the system.

5) The structure and strength of interfaces between dissimilar materials have to be examined by *ab initio* calculations because the atomic interactions are complicated. This becomes feasible with parallel computation and efficient algorithm.

ACKNOWLEDGMENT

This study is partly supported by the Asahi Glass Foundation, Industrial Technology Research Grant Program from New Energy and Industrial Technology Development Organization (NEDO), and Grant-in-Aid for Scientific Research of (B)(No.14350055) of Japan Society of the Promotion of Science.

REFERENCES

[1] Born,M. and Huang,K., *Dynamical Theory of Crystal Lattices*, Oxford UP, (1954).

[2] Milstein,F., *Phys. Rev.*, B, **3**(1971), 1130–1141.

[3] Nielsen,O.H. and Martin,R.M., *Phys. Rev.*, B, **32**(1985), 3792–3805.

[4] Karki,B.B., Ackland,G.J. and Crain,J., *J. Phys. Condens. Matter*, **9**(1997), 8579–8589.

[5] Kohyama,M. and Hoekstra,J., In Kitagawa,H, Aihara,T. and Kawazoe,Y., editors, *Mesoscopic dynamics of fracture*, 166–175. Springer-Verlag, (1998).

[6] Bachelet,G.B., Hamann,D.R. and Schlüter,M., *Phys. Rev.*, B, **26**(1982), 4199–4228.

[7] Hohenberg,P. and Kohn,W., *Phys. Rev.*, **136**(1964), B864–871.

[8] Kohn,W. and Sham,L.J., *Phys. Rev.*, **140**(1965), A1133–1138.

[9] Slater,J.C. and Koster,G.F., *Phys. Rev.*, **94**(1954), 1498–1524.

[10] Sutton,A.P., Finnis,M.W., Pettifor,D.G. and Ohta,Y., *J. Phys.*, C, **21**(1988), 35–66.

[11] Rapaport,D.C., *The Art of Molecular Dynamics Simulation*, Cambridge University Press, (1995).

[12] Daw,M.S. and Baskes,M.I., *Phys. Rev.*, B, **29**(1984), 6443–6452.

[13] Jacobsen,K.W., Nørskov,J.K. and Puska,M.J., *Phys. Rev.*, B, **35**(1987), 7423–7442.

[14] Kitamura,T. and Umeno,Y., *Modelling Simul. Mater. Sci. Eng.*, **11**(2003), 127–136.

[15] Umeno,Y., Kitamura,T., Date,K., Hayashi,M. and Iwasaki,T., *Comp. Mater. Sci.*, **25**(2002), 447–456.

[16] Izumi,S., Sato,Y., Hara,S. and Sakai,S., *J. Soc. Mat. Sci., Japan (in Japanese)*, **52**(2003), 225–230.

[17] Wang,J., Yip,S., Phillpot,S. and Wolf,D., *Phys. Rev.*, B, **71**(1993), 4182–4185.

[18] Wang,J., Yip,S., Phillpot,S. and Wolf,D., *Phys. Rev.*, B, **52**(1995), 12627–12635.

[19] Kitamura,T., Yashiro,K. and Ohtani,R., *JSME Int. J.*, A, **40**(1997), 430–435.

[20] Šob,M., Wang,L.G. and Vitek,V., *Comp. Mater. Sci.*, **8**(1997), 100–106.

[21] Šob,M., Wang,L.G. and Vitek,V., *Mater. Sci. Eng.*, A, **234-236**(1997), 1075–1078.

[22] Umeno,Y. and Kitamura,T., *Mater. Sci. Eng.*, B, **88**(2001), 79–84.

[23] Ogata,S., Hirosaki,N., Kocer,C. and Kitagawa,H., *Phys. Rev.*, B, **64**(2001), 172102.

[24] Ogata,S., Li,J. and Yip,S., *Science*, **298**(2002), 807–811.

[25] Hirosaki,N., Kocer,C. and Ogata,S., *Phys. Rev.*, B, **67**(2003), 035210.

[26] Yashiro,K., Oho,M., Yamagami,K. and Tomita,Y., *J. Soc. Mat. Sci., Japan (in Japanese)*, **52**(2003), 241–246.

[27] Umeno,Y., Kitamura,T. and Kushima,A., *Modelling Simul. Mater. Sci. Eng.*, submitted.

[28] Kitamura,T., Umeno,Y. and Tsuji,N., *Comp. Mater. Sci.*, submitted.

[29] Kresse,G. and Furthmüller,J., *Phys. Rev.*, B, **54**(1996), 11169–11186.

[30] Ohta,H., Miura,H. and Kitano,M., *J. Soc. Mat. Sci., Japan*, **45**(1996), 1322–1327.

[31] Umeno,T. and Kitamura,T., *Proc. of NANOTECH2002: ACRS Joint Meeting/MSM&ICCN, San Juan, Puerto Rico*, 405–408.

AB INITIO STUDY OF IDEAL SHEAR STRENGTH

Shigenobu Ogata[1,2,3], Ju Li[4], Yoji Shibutani[2,3] and Sidney Yip[1]

[1] *Department of Nuclear Engineering and Department of Materials Science and Engineering, Massachusetts Institute of Technology, Cambridge, MA 02139, USA*
ogata@mech.eng.osaka-u.ac.jp, sogata@mit.edu

[2] *Department of Mechanical Engineering and Systems, Osaka University, Osaka 565-0871, Japan*

[3] *Handai Frontier Research Center, Osaka University, Osaka 565-0871, Japan*

[4] *Department of Materials Science and Engineering, Ohio State University, Columbus, OH 43210*

Abstract Ideal strength, which can be defined as the stress necessary to induce permanent deformation in a material without prior imperfections, is one of the important materials characterizations. In this study we calculate the ideal pure shear and simple shear strengths of fcc (Al, Cu, Ni, Ag) and bcc (Fe, Mo, W) metals in their common slip systems using density functional theory. We find the critical shear strains (γ_m) of bcc metals are narrowly distributed ($\sim$0.18), and are higher than fcc metals except Al. In contrast, the γ_m of fcc metals spread over a wide range (0.13$\sim$0.2), with Al having an extremely high γ_m (0.2). As a result, although Al has smaller moduli than Cu in $\{111\}\langle 11\bar{2}\rangle$ shear, its ideal pure shear and simple shear strengths are higher than Cu. By comparative analyses of the generalized stacking fault energy and valence charge distribution in Al, Ag and Cu, we conclude that the large intrinsic stacking fault energy, γ_m and ideal shear strengths of Al are all related to directional bonding. Cu and Ag do not have strong directional bonding. Bcc metals have stronger directional bonding than the fcc metals except Al. By turning off spin polarization in the calculations, we find magnetism is a main source of bond directionality in Ni and Fe.

Keywords: Ideal shear strength, density functional theory, directional bonding, fcc metals, bcc metals, magnetism

H. Kitagawa and Y. Shibutani (eds.), IUTAM Symposium on Mesoscopic Dynamics of Fracture Process and Materials Strength, 401–410.

1. INTRODUCTION

The minimum shear stress to destabilize a crystal lattice without imperfections is fundamental to our concept of materials strength and its theoretical limits under large strains [Wang et al. (1995); Morris and Krenn (2000)]. With the possible exception of recent nanoindentation measurements [Gouldstone et al. (2000); Li et al. (2002); Van Vliet et al. (2003)], it has not been feasible to directly measure the ideal shear strength of materials. The demonstration that this property can be reliably determined by first-principles calculations[Ogata et al. (2002)] therefore would have significant implications for the understanding of behavior of solids at the limit of structural stability or mechanical failure.

Here we report and substantiate our findings by probing in detail the energetics of shear deformation and valence charge redistribution during deformation for several fcc and bcc metals.

2. METHOD

We perform density functional theory (DFT) calculations on the following systems: fcc Ag, Cu, ferromagnetic (FM) Ni, Al and bcc W, Mo, Fe (FM), using the Vienna Ab-initio Simulation Package (VASP) [Kresse and Hafner (1993); Kresse and Furthmuller (1996)]. The exchange-correlation density functional adopted is the Perdew-Wang generalized gradient approximation (GGA) [Perdew and Wang (1992)]; except for Ag, where the Ceperley-Alder local density approximation (LDA) [Ceperley and Alder (1980); Perdew and Zunger (1981)] is used. Basically, the ultrasoft pseudopotentials [Vanderbilt (1990)] are used, but we switch to the projector augmented-wave (PAW) method [Kresse and Joubert (1999)] for the difficult system of Fe. Brillouin zone k-point sampling is performed using the Monkhorst - Pack algorithm [Monkhorst and Pack (1976)], and integration follows the Methfessel-Paxton scheme [Methfessel and Paxton (1989)] with the smearing width chosen so the entropic free energy ("$-TS$" term) is less than 0.5 meV/atom. Incremental affine shear strains are imposed on each crystal along the experimentally determined common slip systems to obtain the corresponding unrelaxed and relaxed energies and stresses. The unrelax and relax conditions are defined by ϵ_{ij}=0 excluding $\gamma \equiv x/d_0$ (d_0 is the equilibrium separation between two adjacent atomic planes and x is the shear displacement along the Burgers vector) and σ_{ij}=0 excluding the resolved shear stress, respectively.

3. RESULTS AND DISCUSSIONS

fcc metals

At equilibrium, Ag has a stiffness comparable to Al in $\{111\}\langle 11\bar{2}\rangle$ shear, and Cu is considerably stiffer, with simple and pure shear moduli greater by

Table 1. Ideal $\{111\}\langle 11\bar{2}\rangle$ shear strains and stresses of fcc metals.

	relaxed			unrelaxed		
material	γ_m^r	σ_m^r GPa	σ_m^r/G_r	γ_m^u	σ_m^u GPa	σ_m^u/G_u
Al	0.200	2.84	0.110	0.210	3.73	0.147
Ni	0.140	5.05	0.084	0.160	6.29	0.079
Cu	0.137	2.16	0.070	0.157	3.45	0.084
Ag	0.145	1.65	0.066	0.156	2.57	0.079

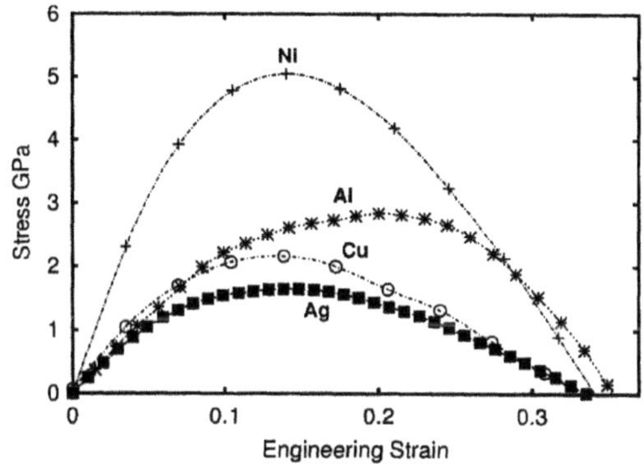

Figure 1. Resolved $\{111\}\langle 11\bar{2}\rangle$ shear stress vs. strain curves of fcc metals (relaxed).

65% and 25%, respectively, than Al. However, Al ends up with 70% and 32% larger ideal pure shear strength σ_m^r than Ag and Cu, respectively, because it has a longer range of strain before softening (see Fig. 1): γ_m=0.200 in Al, γ_m=0.145 in Ag and γ_m=0.137 in Cu. The σ_m^r/G_m^r ratio shows a similar trend the two are in fact almost linearly correlated (see Table 1).

Fig. 2 shows the iso-surfaces of valence charge density ($h \equiv V_{\text{cell}}\rho_{\text{v}}$, V_{cell} and ρ_v are the supercell volume and valence charge density, respectively). We select two h-contour values for each metal, and for Ni (FM) the difference between spin-up and -down densities ($h_{\text{diff}} \equiv V_{\text{cell}}(\rho_{\text{v}}\uparrow - \rho_{\text{v}}\downarrow)$) is also shown. At the octahedral interstice in Al (Fig.2(a)), the pocket of charge density has cubic symmetry and is angular in shape, with a volume comparable to the pocket centered on every ion. In contrast, Figs. 2(c), 2(d), 2(e), 2(f) show that in Cu and Ag there is no such interstitial charge pocket, the charge density being nearly spherical about each ion. Al has an inhomogeneous charge distribution in the interstitial region because of bond covalency [Feibelman (1990)] and directional bonding [Grossman et al. (1999)], while Cu and Ag have relatively homogeneous charge distributions and little bond directionality. For Ni, the total charge density (spin-up plus -down) shows a spherical distribution (Fig.2(g),2(h)). However, the difference between spin-up and -down (2(j)), which results in magnetization, shows a cube-shaped distribution centered on the ion, similar to that in Al (Fig.2(b)), even though the volume is smaller. This suggests that magnetization promotes directional bonding and causes the γ_m, σ_m^r/G_m^r values of Ni to deviate from those of Cu and Ag.

The generalized stacking fault (GSF) energy, the energy increase when two adjacent atomic planes in the crystal are sheared relative to each other, is known to play an important role in the structure and energetics of dislocations. While it is known experimentally that the intrinsic stacking fault energy is much larger in Al than in Ag and Cu, this fact has not been related to their ideal shear strengths. For this purpose, we introduce a general function (Fig. 3),

$$\gamma_n(x) \equiv \frac{E_n(x)}{nS_0}, \quad n = 1, 2, ... \tag{1}$$

where x is the relative displacement in the slip direction between two adjacent atomic planes (we focus on $\{111\}\langle 11\bar{2}\rangle$ slip here), $E_n(x)$ is the increase in total energy relative to its value at $x = 0$, with $n + 1$ being the number of planes involved in the shearing and S_0 being the cross-sectional area at $x = 0$. The series of functions $\gamma_1(x), \gamma_2(x), ..., \gamma_\infty(x)$, may be called the multi-plane generalized stacking fault energy, with $\gamma_1(x)$ being the conventional GSF [Zimmerman et al. (2000)], and $\gamma_\infty(x)$ being the affine strain energy. The intrinsic stacking fault energy $\gamma_{\rm sf}$ is $\gamma_1(b_p)$, where $\vec{b}_p$=$[11\bar{2}]a_0/6$ is the partial Burgers vector. The unstable stacking energy $\gamma_{\rm us}$, an important material parameter in the analysis of dislocation nucleation, [Rice and Beltz (1994)], is $\gamma_1(x_0)$, where $d\gamma_1/dx(x_0 < b_p)$=0. It is instructive to compare different $\gamma_n(x)$ for the same slip system as n varies. The difference should be relatively small from a local "glue" (shaded region in Fig. 3) point of view where we take the valence electron cloud to be the glue. We also have the asymptotic behavior at large n,

$$\gamma_n(x) = \gamma_\infty(x) + \frac{2\gamma^{\rm twin}(x)}{n} + \mathcal{O}\left(n^{-2}\right), \tag{2}$$

where $\gamma^{\rm twin}(b_p)$ is the unrelaxed twin boundary energy. The rate of convergence to (2) reflects the localization range of metallic bonding in a highly deformed bulk environment.

Unrelaxed $d\gamma_1(x)/dx$ and $d\gamma_\infty(x)/dx$ are compared in Fig. 4. First we note that for Ag and Cu, $d\gamma_1(x)/dx$ and $d\gamma_\infty(x)/dx$ are not very different across the entire range of shear. The fact that the sliding of a layer is effectively decoupled from that of adjacent layers indicates that bonding in Ag and

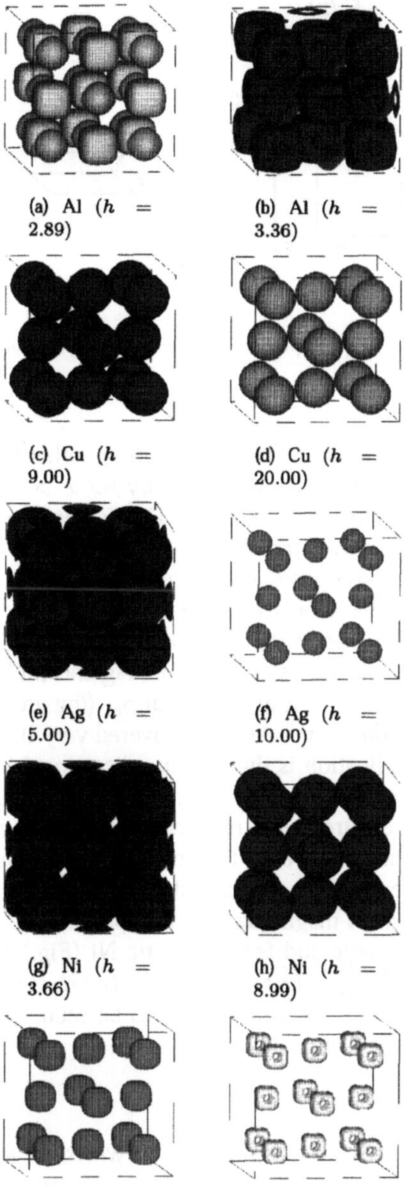

(a) Al (h = 2.89)
(b) Al (h = 3.36)
(c) Cu (h = 9.00)
(d) Cu (h = 20.00)
(e) Ag (h = 5.00)
(f) Ag (h = 10.00)
(g) Ni (h = 3.66)
(h) Ni (h = 8.99)
(i) Ni ($h_{\rm diff}$ = 4.0)
(j) Ni ($h_{\rm diff}$ = 6.0)

Figure 2. Valence charge density iso-surfaces in fcc metals.

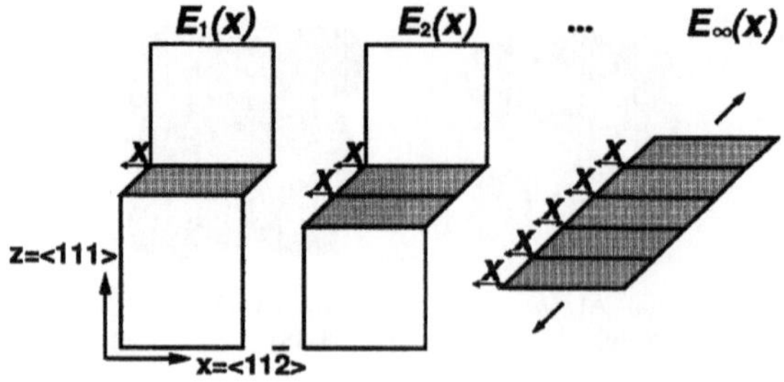

Figure 3. Multi-plane generalized stacking fault energy

Cu has nearly no bond-angle dependence. On the other hand, the same functions behave very differently in Al, especially for $x > x_m$, where the stress (generalized force) reaches its maximum. This is because of the coupling between two or more interlayer regions through directional bonding. Note also that $d\gamma_1(x)/dx$ for Al stays positive for an extended range, whereas it becomes for of Ag and Cu negative sooner and goes to lower value. Thus, while Al, Ag and Cu all have approximately the same γ_{us}, when x reaches b_p and the configuration becomes an intrinsic stacking fault, Ag and Cu have recovered most of their losses in the sense of a low value of γ_{sf} (the area under the curve of $d\gamma_1(x)/dx$). On the other hand, Al has recovered very little as its γ_{sf} remains close to γ_{us}. The implication is that when a directional bond is broken, it is more difficult for the electrons to re-adapt. In contrast, in non-directionally bonded systems, the electrons can redistribute well and the system does not incur a large energy penalty even if the bond angles are wrong, as long as the volumes fit as in the case of the intrinsic stacking fault.

To isolate the effects of magnetization, we compare the stress-displacement functions for paramagnetic and ferromagnetic Ni (Fig. 4(d) and Fig. 4(c)). In paramagnetic Ni, $d\gamma_1(x)/dx$ and $d\gamma_\infty(x)/dx$ are similar, whereas in ferromagnetic Ni relatively large differences can be seen. In other words, spin-polarization promotes directional bonding in Ni.

bcc metals

Bcc metals have three common slip systems which are almost equally likely (pencil glide). We performed the same shear deformation calculations in the three slip systems as for fcc metals. The ideal shear strains are rather narrowly distributed ($\sim$ 0.18)[Paxton et al. (1991)] and in good agreement with the previous Mo result [Luo et al. (2002)]. Moreover, the values of σ^r_m/G^r_m for the

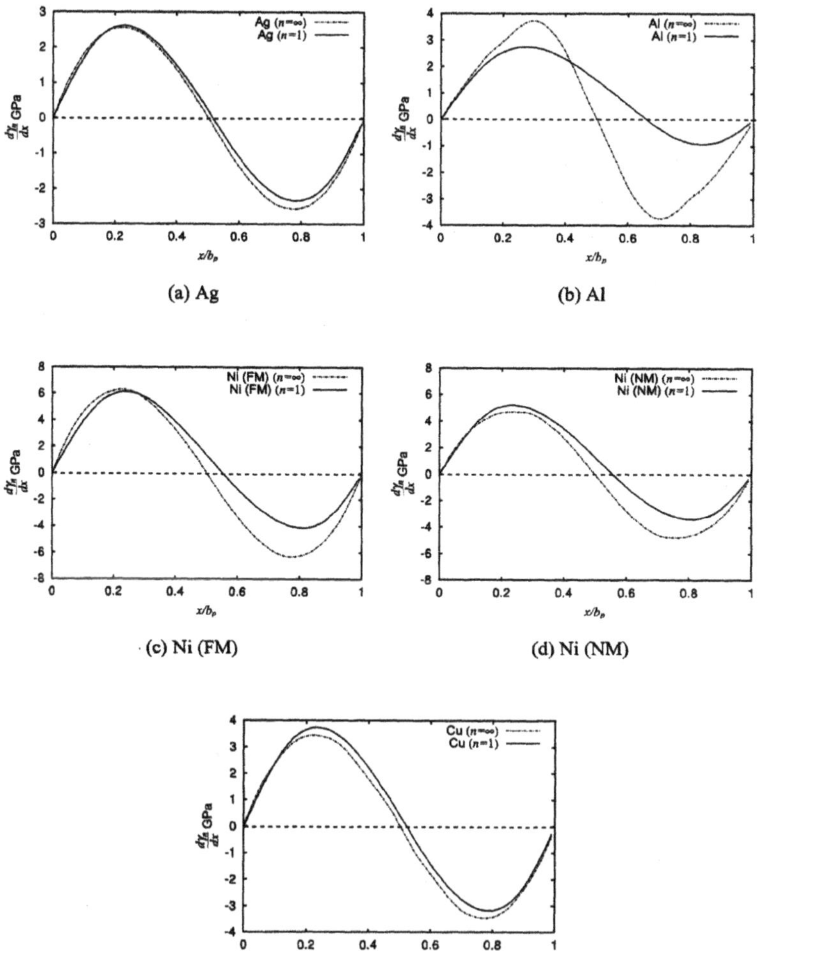

(a) Ag (b) Al

(c) Ni (FM) (d) Ni (NM)

(e) Cu

Figure 4. $d\gamma_1(x)/dx$ vs. $d\gamma_\infty(x)/dx$ (unrelaxed) in $\{111\}\langle 11\bar{2}\rangle$ shear of fcc metals.

three metals are almost equal (~ 0.11) and are also close to that of Al (Fig. 5, Table. 2). This confirms that bcc metals have more bond directionality than fcc metals except Al. Fig. 6 shows the valence charge density iso-surfaces. In W and Mo, we see cuboidal distortions of the ion-centered charge density which

Table 2. Ideal shear strains and stresses of bcc metals.

material	relaxed			unrelaxed		
	γ_m^r	σ_m^r GPa	σ_m^r/G_r	γ_m^u	σ_m^u GPa	σ_m^u/G_u
W $\{110\}\langle\bar{1}\bar{1}1\rangle$	0.179	17.52	0.114	0.196	17.63	0.113
W $\{211\}\langle\bar{1}\bar{1}1\rangle$	0.176	17.37	0.113	0.175	17.28	0.111
W $\{321\}\langle\bar{1}11\rangle$	0.176	17.33	0.113	0.175	17.27	0.111
Mo $\{110\}\langle\bar{1}11\rangle$	0.190	15.18	0.120	0.192	16.52	0.123
Mo $\{211\}\langle\bar{1}\bar{1}1\rangle$	0.175	14.84	0.117	0.177	15.99	0.119
Mo $\{321\}\langle\bar{1}\bar{1}1\rangle$	0.176	14.87	0.117	0.175	15.93	0.119
Fe $\{110\}\langle\bar{1}11\rangle$	0.178	8.14	0.106	0.234	11.43	0.142
Fe $\{211\}\langle\bar{1}\bar{1}1\rangle$	0.184	7.51	0.099	0.236	9.95	0.124
Fe $\{321\}\langle\bar{1}11\rangle$	0.181	7.57	0.100	0.197	9.43	0.118

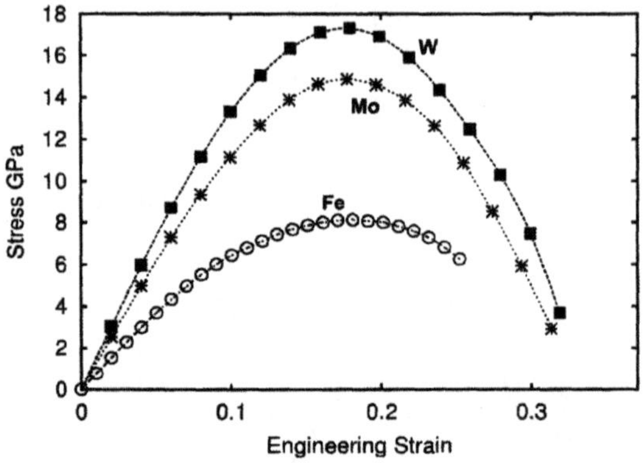

Figure 5. Resolved shear stress vs. strain curves of bcc metals in $\{211\}\langle\bar{1}\bar{1}1\rangle$ shear for Mo and Fe and $\{321\}\langle\bar{1}11\rangle$ shear for W, respectively (relaxed).

can be used to explain the bond directionality. In Fe, the total charge density (spin-up plus spin-down) is almost spherical. However, the difference between spin-up and spin-down (Fig. 6(h)) clearly shows angular distortion as well. So in Fe, like in Ni, magnetism promotes directional bonding. This agrees with the general observation that magnetism is important for phase stability and elasticity [Cohen et al. (2002)]; it is responsible for stabilizing the bcc phase of Fe at ambient conditions which would otherwise take a close-packed structure.

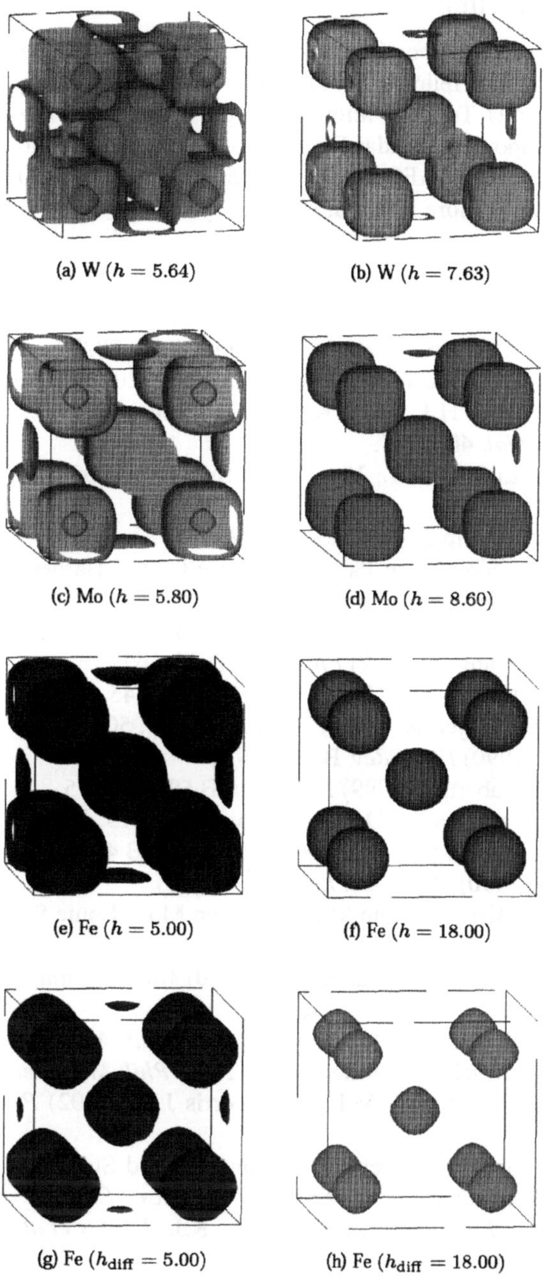

Figure 6. Valence charge density iso-surfaces in bcc metals.

Acknowledgments

SO acknowledges support by Hattori-Houkoukai fellowship and Handai Frontier Research Center fellowship. JL acknowledges support by Honda R&D, Inc., and the OSU Transportation Research Endowment Program. YS acknowledges support by Handai Frontier Research Center fellowship. SY acknowledges support by AFOSR, Honda R&D, Inc., DARPA/ONR, NSF-ITR, and Lawrence Livermore National Laboratory.

References

Wang J., Li J., Yip S., Phillpot S., and Wolf D. (1995) *Phys. Rev.* B **52**, 12627.
J.W. Morris and C.R. Krenn (2000) *Phil. Mag.* A **80**, 2827.
Li J., Van Vliet K.J., Zhu T., Yip S., and Suresh S. (2002) *Nature* **418**, 307.
Gouldstone A., Koh H.J., Zeng K.Y., Giannakopoulos A.E., and Suresh S. (2000) *Acta Mat.* **48**, 2277.
Van Vliet K.J., Li J., Zhu T., Yip S., and Suresh S. (2003) *Phys. Rev.* B **67**, 104105.
Ogata S., Li J., and Yip S. (2002) *Science* **298**, 807.
Vienna Ab-initio Simulation Package: Kresse G. and Hafner J. (1993) *Phys. Rev.* B **47**, RC558.
Kresse G. and Furthmuller J. (1996) *Phys. Rev.* B **54**, 11169.
Perdew J.P. and Wang Y. (1992) *Phys. Rev.* B **46**, 6671.
Ceperley, D. and Alder, B.J. (1980) *Phys. Rev. Lett.* **45**, 566.
Perdew, J.P. and Zunger, A. (1981) *Phys. Rev.* B **23**, 5048.
Vanderbilt, D. (1990) *Phys. Rev.* B **41**, 7892.
Kresse, G. and Joubert, J. (1999) *Phys. Rev.* B **59**, 1758-75.
Monkhorst, H.J. and Pack, J.D. (1976) *Phys. Rev.* B **13**, 5188.
Methfessel, M. and Paxton, A.T. (1989) *Phys. Rev.* B **40**, 3616-21.
Feibelman P.J. (1990) *Phys. Rev. Lett.* **65**, 729.
Grossman J.C., Mizel A., Cote M., and Cohen M.L., Louie S.G. (1999) *Phys. Rev.* B **60**, 6343.
Zimmerman J.A., Gao H., Abraham F.F. (2000) *Model. Simul. Mater. Sci. Eng.* **8**, 103.
Rice J.R. and Beltz G.E. (1994) *J. Mech. Phys. Solids* **42**, 333.
Paxton A.T., Gumbsch P., Methfessel M. (1991) *Phil. Mag. Lett.* **63**, 267.
Luo W., Roundy D., Cohen M.L., and Morris J.W. (2002) *Phys. Rev.* B **66**, 094110.
Cohen R.E., Gramsch S., Steinle-Neumann G., and Stixrude L. (2002) *Proceedings of the International School of Physics "Enrico Fermi"*, Volume CXLVII, *High-Pressure Phenomena*, eds. R.J. Hemley *et al.*. `cond-mat/0110025`.

INVESTIGATION OF GLIDE PROPERTIES IN HEXAGONAL TITANIUM AND ZIRCONIUM: AN AB INITIO ATOMIC SCALE STUDY

Christophe Domain[1] and Alexandre Legris[2]

[1] *EDF R&D, Département MMC, Les Renardières, F-77250 Moret sur Loing, France*

[2] *Université de Lille 1, Laboratoire de Métallurgie Physique et Génie des Matériaux - UMR8517, Bât C6, F-59655 Villeneuve d'Ascq Cedex, France*

Abstract: In this work we present *ab initio* atomic-scale simulations based on the density functional theory of stacking faults and of the structure of the ⟨a⟩ screw dislocation core in hexagonal Zr and Ti. The basal, prismatic, pyramidal π_1 and π_2 gamma-surfaces were investigated and the energy profiles along ⟨a⟩ and ⟨c+a⟩ Burgers vectors were determined. The results clearly indicate preferential prismatic spreading of screw dislocation cores suggesting a primary prismatic glide. The *ab initio* simulations are in an overall good agreement with previous tight binding ones[1] although differences concerning the atomic relaxation around prismatic faults have been observed. Some environment effects on dislocation glide properties have been investigated through the study of the hydrogen effect on the stacking fault excess energies. Hydrogen in solid solution induces significant reductions of the stacking fault energy and should promote enhanced planar prismatic glide.

Key words: zirconium, titanium, hydrogen embrittlement, ab initio, screw dislocation core, stacking faults

H. Kitagawa and Y. Shibutani (eds.), IUTAM Symposium on Mesoscopic Dynamics of Fracture Process and Materials Strength, 411–420.

1. INTRODUCTION

The hexagonal transition metals of the IVB column (Ti, Zr, Hf) have glide properties at low temperature that remind those of bcc metals: below room temperature, the lattice friction is significant, and the deformation is controlled by the movement of screw dislocations[2]. These metals have a prismatic principal slip system and the basal, pyramidal π_1 and π_2 systems have been observed as secondary ones. The pyramidal $\langle a \rangle$–π_1 is the main secondary slip system in Zr, the basal slip being the less active one for all kinds of loadings. The observed $\langle c+a \rangle$–π_1 dislocations, which accommodate deformation components along the $\langle c \rangle$ direction, have most often a mixed character[3].

There are still open questions concerning elementary glide mechanisms in these materials, and the determination of the equilibrium structure of the screw dislocation core may provide a first step towards a better understanding of them. A direct observation of the screw dislocation cores using atomic-scale resolution techniques such as High Resolution Transmission Electron Microscopy is not possible in this case since the main atomic displacements are parallel to the electron beam. The only available experimental observation concerns the$\langle a \rangle$ edge dislocation in Ti[4] for which HRTEM indicates a prismatic spreading of 12 Å. Accurate atomic–scale simulation should therefore be very useful to elucidate the core structure and elementary glide mechanisms. Most of the available empirical potentials for Zr and Ti do not reproduce correctly the stacking fault excess energies and therefore should give poor results regarding the screw dislocation core structure, that was analyzed by Legrand[5] using a tight binding scheme. Here we present results obtained using *ab initio* calculations based on the density functional theory. We evaluated the gamma-surface profiles along $\langle a \rangle$ and $\langle c+a \rangle$ directions corresponding to the Burgers vector in principal (prismatic) and secondary (basal, π_1 and π_2 pyramidal) slip planes in α Zr and α Ti. In addition, a simulation of the screw dislocation core structure was performed using up to 127 atoms. These last results should be considered as qualitative owing the reduced number of atoms and the boundary conditions applied.

The DFT-based methods being able to treat on a equal footing different atomic species, we investigated the effect of hydrogen in solid solution on the stacking fault excess energy in order to asses its influence on the dislocations mobility. In transition metals, hydrogen in solid solution entails an increase of the dislocation mobility[6] either by interacting with the core or by a modification of the long-range elastic interactions between dislocations (shielding effect). In addition, Ferreira *et al.*[7] have shown that hydrogen hinders cross slip in Al.

2. METHODS

The calculations were performed using the Vienna *ab initio* Simulation Package (VASP)[8] based on the density functional theory. The ionic Zr (Ti) cores are modelled by ultra soft pseudopotentials of the Vanderbilt type[9, 10] that explicitly consider the valence electrons $5s^2$ ($4s^2$) and $4d^2$ ($3d^2$). Since it has been shown[11] that the semi-core $4p^6$ ($3p^6$) electrons have to be taken into account to reproduce the correct phase stability for the pure metals and their oxides, the $4p^6$ ($3p^6$) electrons were included in the calculations except when mentioned (Zr_{np} and Ti_{np} results). The cut-off energy for the plane waves is 225 eV and we used the exchange–correlation functional developed by Perdew and Wang (PW91)[12] in the framework of the Generalised Gradient Approximation (GGA). Further details can be found in [13].

To calculate the excess stacking fault energy for given fault plane and vector, a rigid translation of two crystal blocks surrounding the fault plane is applied. The atoms are relaxed only in the direction perpendicular to the fault plane until the forces are smaller than 0.02 eV/Å. Free surfaces parallel to the fault plane and periodic boundary conditions in the remaining directions were used in order to investigate all possible fault planes and vectors.

To simulate a screw ⟨a⟩ dislocation, the line is placed at the center of a supercell containing a stacking of two $(2\bar{1}\bar{1}0)$ atomic planes. An initial atomic displacement field derived from linear isotropic elasticity is then applied. The atoms are then relaxed except those located far from the line that are kept fixed, surrounded by a vacuum ribbon (see Figure 1). A periodic boundary condition is applied along the dislocation line. The relaxed dislocation core is analyzed using the arrow method developed by Vitek[14].

The influence of hydrogen on the stacking fault excess energy at 0 K was evaluated by subtracting to the energy of a supercell containing H in the fault plane the energy of a supercell containing the same number and type of atoms and no fault. The amount of H in the fault is measured by the coverage ratio $\Theta = n_{oc}/n_{av}$, where n_{oc} and n_{av} are the occupied and available number of tetrahedral sites in between the atomic planes surrounding the fault plane before the fault vector is applied.

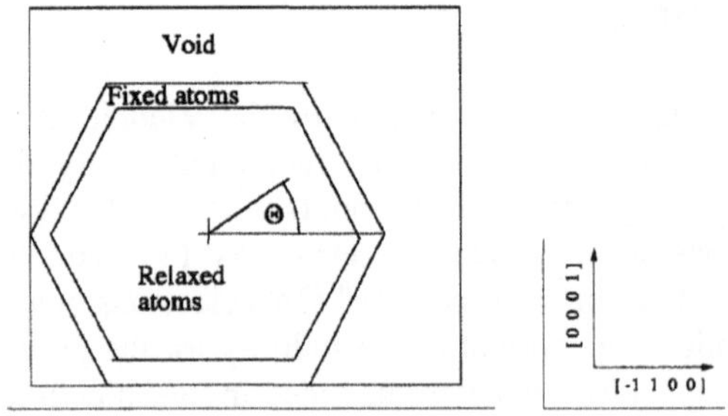

Figure -1. Schematics of the supercell used to simulate screw dislocation cores.

3. RESULTS

In this section we present and discuss the results starting with the pure elements. The influence of H on the stacking fault energy is then exposed and discussed.

3.1 Stacking faults excess energy for pure Zr and Ti

The pseudopotentials used satisfactory reproduce the structural equilibrium parameters of hcp Zr and Ti, results closer to the experimental values are obtained when the semi-core electrons are taken into account. A similar remark can be made concerning the calculated shear constants (see Tables 1 and 2).

Table -1. Lattice parameters of hexagonal Zr and Ti

	a (Å)	c (Å)	c/a
Zr	3.239	5.163	1.594
Zr_{np}	3.204	5.141	1.605
Zr exp	3.232	5.149	1.593
Ti	2.940	4.672	1.589
Ti_{np}	2.926	4.641	1.586
Ti exp	2.95	4.68	1.586

The calculated values of the basal and prismatic stacking fault excess energies (Table 2) are in qualitative agreement with the tight binding calculations, the basal stacking fault energy (Zr: 200 mJ/m^2, Ti: 291 mJ/m^2) being larger than the prismatic one (Zr: 145 mJ/m^2, Ti: 174 mJ/m^2). Both types (with or without p electrons) of pseudopotential give results within 10% and 20% for basal and prismatic faults respectively, which means that despite their deficiency the "ligther" pseudopotentials are accurate enough to model the dislocation cores. Atomic relaxations are essential to obtain the γ_{pris} value while their effect is less pronounced to obtain γ_{bas} (they account for only 15% of the value).

Our results are in agreement with the only available experimental value 150 mJ/m^2 obtained for Ti[4]. Assuming a dislocation core dissociation in two partials, the previous results entail a preferential prismatic spreading.

Table -2. Zr and Ti excess stacking fault energies, shear elastic constants and Legrand ratio. (in parenthesis values of the stacking fault energy without atomic relaxations).

		γ_{bas} (mJ/m^2)	γ_{prism} (mJ/m^2)	C_{44} (GPa)	C_{66} (GPa)	R
Zr	GGA	200 (237)	145 (455)	29	39	1.9
Zr_n	GGA	216	111	21.3	43.3	
p						
	TB [5]	340	150	42	42	2.3
	exp. [15]	—	—	36.3	44	prismatic
Ti	GGA	291	174	43	45	1.8
Ti_n	GGA	322	142	36.6	50.8	
p						
	TB [5]	290	110	45	43	2.5
	exp. [15]	—	150 [4]	46.7	35	prismatic

Beside the stable stacking fault energy, we calculated the gamma-surface excess energy path along the Burgers vectors ⟨a⟩ (for the prismatic, basal and pyramidal π_1 planes) and ⟨c+a⟩ (for the pyramidal π_1 and π_2 planes). For both Ti and Zr, the profile corresponding to the prismatic plane has a local minimum for a translation vector of ⟨a⟩/2, which corresponds to the stable prismatic fault configuration. This minimum is obtained neither by existing empirical potentials[16] nor by the TB model used by Legrand[5] or that of Girshick *et al.*[16].

The analysis of the evolution of the reticular distance between the two planes surrounding the fault plane as a function of the fault distance along the ⟨a⟩ vector (Figure 2 d) shows an inward relaxation close to the prismatic fault configuration. This behavior may be related to the existence of a local energy minimum. Previous calculations always lead to outward relaxation as well as to a local maximum. The sign of the relaxation cannot be easily predicted using simple arguments based on local coordination. For example, in hcp Zr, each Zr atom has 12 nearest neighbor (nn) atoms located at ~3.2 Å, and 6 second nn atoms at ~4.5 Å. For the configuration corresponding to the prismatic fault, a Zr atom close to the fault plane as one nn at ~2.95 Å, 8 second nn at ~3.2 Å and 4 third nn at ~3.45 Å.

For translations along ⟨a⟩, the basal and π_1 profiles have a similar shape, and the values obtained for the excess energy are of the same order of magnitude.

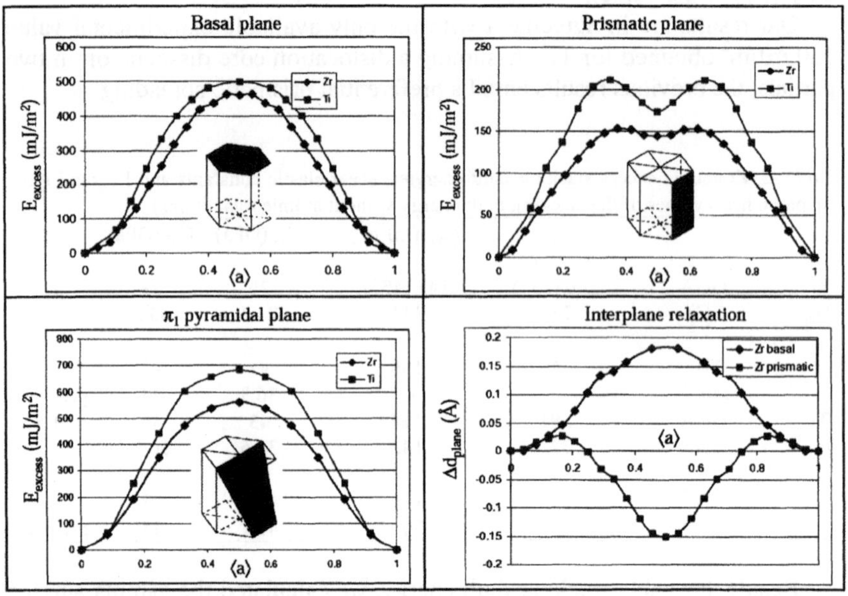

Figure -2. Gamma-surface excess energy (in mJ/m²) profile along ⟨a⟩ in the basal, prismatic and pyramidal π_1 slip planes for Zr and Ti. The relaxation of the distance between the atomic planes surrounding the fault plane (in Å) is plotted in Zr for the basal and prismatic planes.

The excess energy profile along the ⟨c+a⟩ vector for the π_1 and π_2 pyramidal planes look similar (Figure 3). Both profiles have a local minimum close to ⟨c+a⟩/2, the excess energy being nevertheless high enough to prevent from a significant dislocation core dissociation.

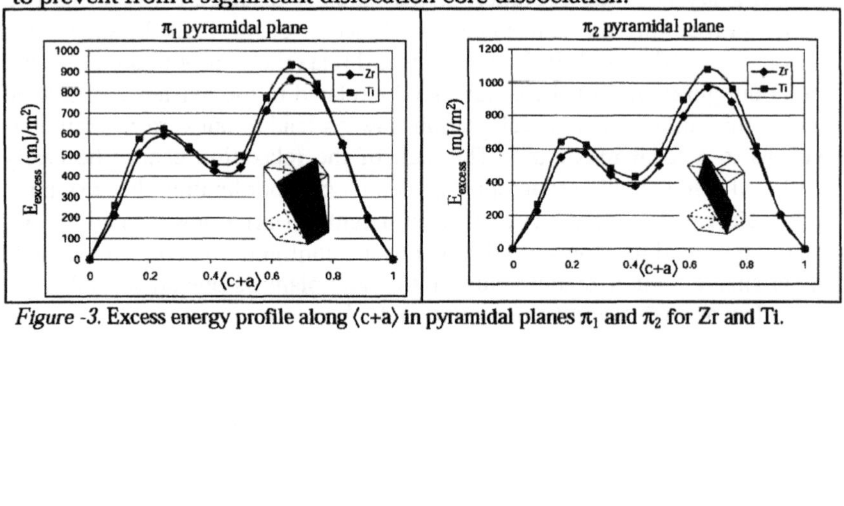

Figure -3. Excess energy profile along ⟨c+a⟩ in pyramidal planes π_1 and π_2 for Zr and Ti.

According to Legrand[5], the stacking fault excess energy values are determined by the filling of the d band and are well described by a simple tight binding model. In spite of a qualitative agreement between the TB results and those presented here, there are at least two discrepancies:

i- the TB results predict a decreasing value of γ_{pris} along the series (from Ti to Hf) and the contrary is observed here ($\gamma^{Zr}_{pris} > \gamma^{Ti}_{pris}$)

ii- inner (see Figure 2) relaxations (decrease of the distance between the atomic planes surrounding the fault) are necessary to obtain the γ_{pris} values in our case while the contrary is obtained by Legrand.

These differences may be due either to the level of description of the electronic interactions (DFT vs. TB) and/or to s electrons effects not considered by Legrand.

3.2 Screw dislocation core structure in pure Zr and Ti

The structure of the screw dislocation cores was determined for Zr and Ti using the same simulation conditions. Whatever the cell simulation size and pseudopotential used (Figure 4 to 6), both Zr and Ti present similar trends: the core has a marked prismatic spreading with screw character while a secondary spreading with and edge component is observed along the basal planes.

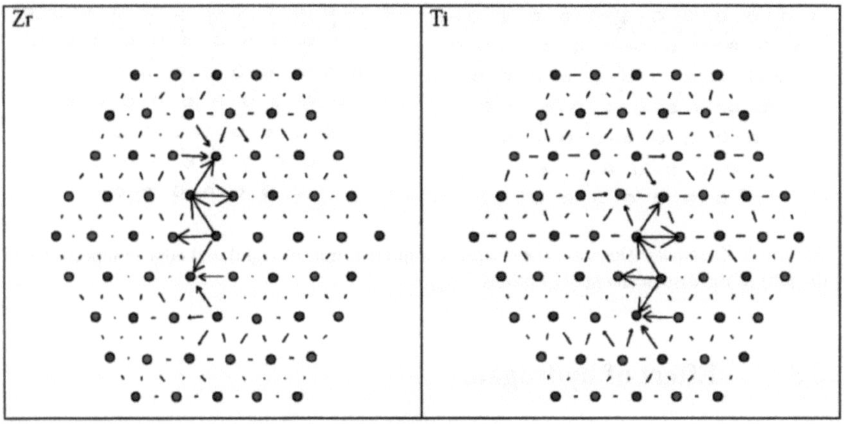

Figure -4. Screw dislocation cores for Zr and Ti (61 atoms)

Starting from the geometrical configuration, the fully relaxed structure depicted in Figures 4 to 6 is obtained only if the atoms are allowed to move along and perpendicularly to the dislocation line. The sole relaxation parallel to the line does not entail significant modifications with respect to the geometric configuration. This fact illustrates the importance of the edge character atomic relaxation to obtain a correct core structure. As illustrated by Figure 5 (b), the strongest edge relaxation component is parallel to the c axis. Both types of pseudopotentials give very similar core structures as can be seen comparing Figure 5 (a) (Zr_{np}) and 6 (a) (Zr).

These trends are almost independent on the cell size for cells containing between 61 to 127 atoms. The prismatic spreading, of the order of 3c, is quite well converged even using a 61-atom supercell. The secondary basal spreading, smaller than the primary prismatic one, seems to be long-range as shown by Figure 5 (b) obtained using the larger cells. It is important to point out that the previously mentioned discrepancies between atomic scale potentials concerning the atomic relaxations around the prismatic fault may induce significant differences in the secondary spreading, modifying not only the core structure but also the dislocation mobility.

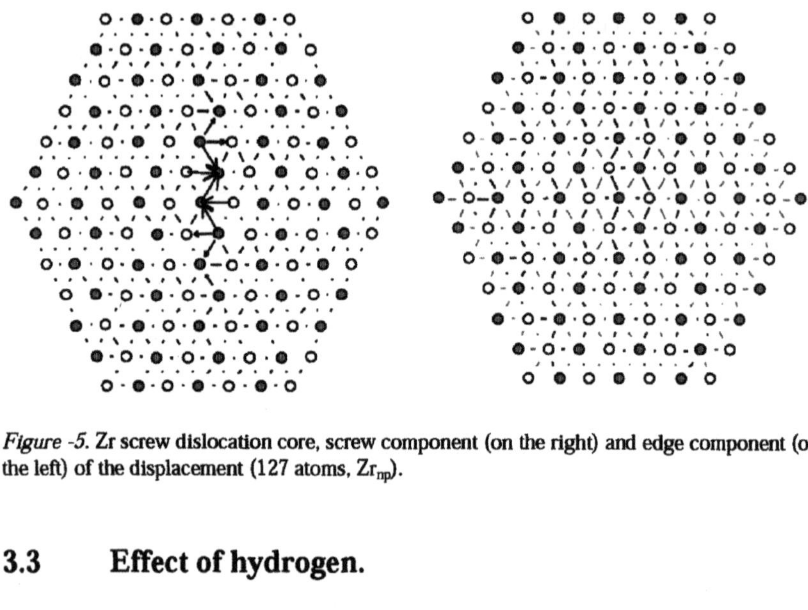

Figure -5. Zr screw dislocation core, screw component (on the right) and edge component (on the left) of the displacement (127 atoms, Zr_{np}).

3.3 Effect of hydrogen.

The hydrogen influence on dislocation mobility has been investigated on an indirect way, placing H atoms at the stacking fault plane using different coverage ratios[17]. For both Zr and Ti, in basal and prismatic fault planes, the presence of H induces a significant fault energy decrease (see Table 3) enhanced by the H coverage. In some cases, the stacking fault energy may become negative, a result that has been associated to the existence of hydrides with high formation enthalpy[17]. Since the screw dislocation core may be seen as an "incipient" prismatic fault, it is likely that H atoms should segregate to the dislocation core (in this case the driving force is purely chemical since there is no first order elastic interaction between screw dislocations and H atoms). Due to the H-induced stacking fault energy reduction the prismatic spreading will increase leading to an enhanced planar glide.

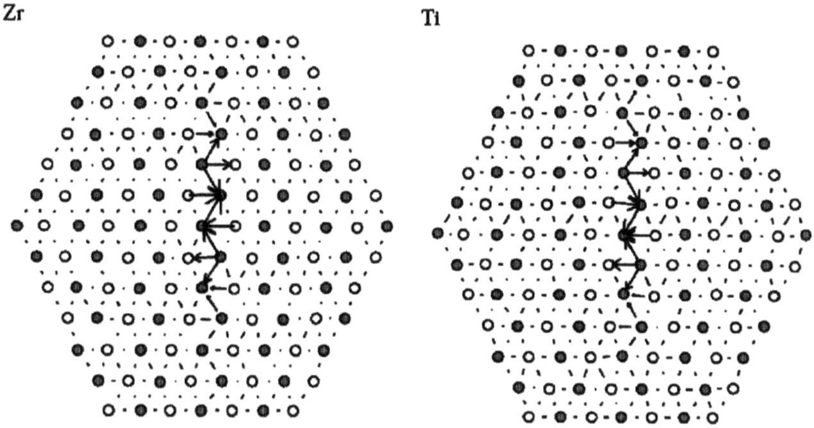

Figure -6. Screw dislocation core of Zr and Ti (screw component of the displacement, 127 atoms, Zr and Ti pseudopotentials)

Table -3. Effect of hydrogen on basal and prismatic stacking fault energy in Zr and Ti (in mJ/m^2)

Θ	Basal plane		Prismatic plane	
	Zr	Ti	Zr	Ti
0	200	291	145	174
0.25	80	102	73	72
0.5	-60	-105	67	66
1	-14	-54	-90	-190

4. CONCLUSIONS

From an experimental point of view Zr and Ti have a similar behavior regarding plasticity. Our *ab initio* determination of the stacking fault energies, elastic constants and structure of the screw dislocation core support these observations. The prismatic stacking fault energy is lower than the basal one and the prismatic fault configuration is located at a local minimum of the gamma-surface excess energy.

The excess stacking fault energy is always larger for Ti than for Zr, the differences are more significant for profiles along the ⟨a⟩ vector than along the ⟨c+a⟩ one.

Despite the limited supercell size and approximate boundary conditions used, screw dislocation cores in Ti and Zr show a clear spreading in the prismatic plane. The secondary spreading of edge character along the basal planes should control the lattice friction at low temperature.

Hydrogen reduces the stacking fault energy in Ti and Zr, and should induce a larger prismatic spreading of the screw dislocation cores favoring planar glide and hindering cross slip .

ACKNOWLEDGEMENTS

The research was supported by the C.I.N.E.S. and I.D.R.I.S. French national computational centres as well as by the C.R.I. of the U.S.T.L. supported by the Fonds Européens de Développement Régional. Part of this work was done on the supercomputers at CEA Grenoble in the framework of an EDF-CEA contract.

REFERENCES

1. Legrand B., Phil. Mag. B 1984; 49: 171
2. Farenc S., Caillard D., and Couret A, Acta metall. mater 1993; 41: 2701-2709; ibid. 1995; 43: 3669-3678
3. Geyer P., Comportement élasto-viscoplastique de tubes de zircaloy-4, approche expérimentale et modélisation micromécanique, Thesis, Ecole des Mines, 1999
4. De Crecy A, Bourret A., Naka S., and Lasalmonie A., Phil Mag A 1983; 47: 245-254
5. Legrand B., Influence de la structure électronique sur la facilité relative des glissements dans les métaux de structures hc, Thesis, Orsay University, 1884
6. Roberston I.M., Engineering Fracture Mechanics 1999; 64: 649-673
7. Ferreira P.J., Robertson I.M., and Birnbaum H.K., Acta mater. 1999; 10: 2991-2998
8. Kresse, G. and Hafner, J., *Phys. Rev.* B 1993; **47**: 558; ibid. 1994; **49**: 14251
9. Vanderbilt, D., 1990, *Phys. Rev.* B 1990; **41**: 7892
10. Kresse, G. and Hafner, J., 1996, *J. Phys. C. Condens. Mater.* 1996; **6**: 8245
11. Jomard G., Etude ab initio de processus d'oxydation du Zircaloy-4, Thesis, INPG, 2000
12. Perdew, J. P. and Wang, Y., *Phys. Rev.* B 1991; **45**: 13244
13. Domain, C., Legris, A., and Besson, R., *Acta mater.* 2002; **50**: 3513
14. Vitek V., Perrin R.C., and Bowen D.K., Phil Mag 1970; 21: 1049
15. Allard S., Int. table of selected constants, Oxford Pergamon Press, 1969; 19
16. A. Girshick, D. G. Pettifor and V. Vitek, Phil. Mag. A, 1998; 77: 999-1012.
17. Domain, C., Legris, A., and Besson, R., submitted to *Acta mater.*

COMPUTATIONAL STUDY OF THE MECHANICAL PROPERTIES OF ALUMINA-COPPER INTERFACES: AB INITIO CALCULATIONS AND COMBINATION WITH MESOSCOPIC SIMULATIONS

Shingo Tanaka, Rui Yang and Masanori Kohyama †

Special Division for Green Life Technology, National Institute of Advanced Industrial Science and Technology, 1-8-31, Midorigaoka, Ikeda, Osaka 563-8577, Japan

Abstract Adhesive and mechanical properties of the $Al_2O_3(0001)/Cu$ interface have been examined by using the *ab initio* calculations based on the density-functional theory, and the development of interfacial interatomic potentials has been discussed. Stable configurations for the Al-terminated and O-terminated interfaces have been examined. The O-terminated interface has quite larger adhesive energy. The rigid-type *ab initio* tensile test has been applied to the O-terminated interface. The Cu-O interface is twice as strong as the Cu-Cu interlayer. The effective interatomic potentials have been constructed by converting the interlayer potential curves in the tensile test into the contribution of each atomic pair.

Keywords: *ab initio* calculation, multi-scale modeling, ceramic/metal interface, interatomic potential, adhesive energy, tensile strength

†Corresponding Author: m-kohyama@aist.go.jp

H. Kitagawa and Y. Shibutani (eds.), IUTAM Symposium on Mesoscopic Dynamics of Fracture Process and Materials Strength, 421–430.

1. INTRODUCTION

Ceramic/metal interfaces are important in various applications such as thermal-barrier coating, corrosion or wear-resistance coating, composites, electronic packaging, electrode and catalysts. It is of great scientific and technological importance to understand the basic properties of ceramic/metal interfaces, such as bonding nature, origins of adhesion, and structural, mechanical, thermal, electronic and chemical properties. Recently, *ab initio* calculations based on the density-functional theory have been applied to ceramic/metal interfaces[1-7]. It has been shown that the behavior of electrons dominates the bonding between dissimilar materials. Atomic bonding between ceramics and metals has features like chemical bonds as well as peculiar features of image interactions [1].

In order to understand the mechanical behavior of cetamic/metal interfaces, it is desirable to deal with large-scale regions including the behavior of interface cracks or interface dislocations. *Ab initio* calculations can deal with only narrow and simple configurations because of huge computational efforts. Therefore, multi-scale modeling should be applied to ceramic/metal systems. At the micro/nano scales, atomic bonding between ceramics and metals should be dealt with accurately, ideally through the behavior of electrons by using the *ab initio* scheme. At the nano/meso scales, atomistic simulations such as classical molecular dynamics are desirable so as to deal with the dynamics of cracks or dislocations. And at the meso/macro scales, continuum models such as the finite element method can deal with practical macroscopic distributions of stresses and strains. Of course, the technique to link different length scales is essential. For the link between the molecular-dynamics method and the finite element method, the quasicontinuum method can be used, for example. However, the link between the *ab initio* scheme and the classical molecular dynamics is not so easy.

In this paper, we consider the link between the latter two schemes. We adopt an indirect link. Namely, we consider the possibility to develop reliable interatomic potentials between ceramics and metals for classical molecular dynamics using the data from *ab initio* calculations. There have been several attempts to develop the interatomic potentials for ceramic/metal interfaces based on the data from *ab initio* calculations [8-10]. However, there seems to exist only few successful examples. This seems to be caused by rather complex interfacial interactions and no general functional forms of potentials between ceramics and metals theoretically [1].

In this paper, we deal with the $Al_2O_3(0001)/Cu(111)$ interface. This is a typcal ceramic/metal system frequently observed in structural and electronic applications. We perform *ab initio* calculations of this interface and provide *ab initio* data for the development of interfacial interatomic potentials. First, we obtain the stable configurations of the coherent interface by *ab initio* calculations. We examine the interfaces with different stoichiometry, namely O-terminated (O-rich) and Al-terminated (stoichiometric) interfaces. Second, we perform the *ab initio* tensile test of *rigid type*, where total-energy calculations are iterated by stretching a selected interlayer distance in a small increment. This can clarify the effective interlayer potentials. Interfacial interatomic potentials should be constructed so as to reproduce such *ab initio* results of stable configurations and interlayer potentials.

2. THEORETICAL METHOD

2.1. Ab Initio Method

Ab initio calculations are perfomrd by using the plane-wave psedopotential method [11] based on the density-functional theory with the local density approximation [12,13]. We use the TM-type optimized pseudopotentials [14] with the plane-wave cut-off energy of 80 Ry. The RMM-DIIS (residual minimization/direct inversion in the iterative subspace) scheme [15] is used for the electronic optimization. By this scheme, we can perform parallel computations with respect to each band very efficiently [16], because the optimization of each wave function can be performed almost independently in each processor with minimum execution of mutual orthogonalization. We use the Kerker or Pulay charge mixing scheme [17,18].

2.2. Atomic Models

We deal with coherent $Al_2O_3(0001)/Cu(111)$ interfaces, where two-dimensional periodicity is attained by expanding Cu layers along the interface. About the position of the Cu(111) layer on the $Al_2O_3(0001)$ surface, we deal with three kinds of models, Al-site, O-site, and hollow-site models [6,7], where the Cu atoms are located above the Al sites, O sites and hollow sites of the hexagonal unit cell of the Al_2O_3 surface, respectively. These configurations should correspond to the energy extrema. We deal with both Al-terminated (stoichiometric) and O-terminated (O-rich) interfaces [6,7]. In the former case, the stoichiometric Al-terminated (0001) surface of alumina forms the interface, and in the latter case, the O-terminated surface without the surface Al atoms forms the interface.

We have constructed the supercell configurations of the models, which contain the Al_2O_3 (0001) slab consisting of 4 O layers and 8(6) Al layers for the stoichiometric (O-rich) case with 5 or 7 Cu (111) layers on both surfaces. There exist no vacuum regions in the supercells. Each supercell has the C_{3i} symmetry, and the two interfaces in the cell are symmetrically equivalent to each other.

3. RESULTS AND DISCUSSION

3.1. Stable Configurations

Fig. 1 shows the most stable configurations for the O-terminated (O-rich) and Al-terminated (stoichiometric) interfaces. The hollow-site model is the most stable for the O-terminated interface, and the O-site model is the most stable for the Al-terminated one. These points are consistent with other *ab initio* results [7]. The adhesive energy defined from the relaxed surfaces is $6.8\,\mathrm{Jm}^{-2}$ for the O-terminated one, and is $0.9\,\mathrm{Jm}^{-2}$ for the Al-terminated one. The O-terminated interface has very large adhesive energy, and the Cu-O distance is rather small (2.06Å). There exists substantial charge transfer between the interfacial Cu and O layers, and seems to exist strong Cu-O orbital hybridization.

This kind of covalent and ionic interactions should be the origin of the strong adhesion [19]. On the other hand, for the Al-terminated interface, there is no significant charge transfer between alumina and Cu layers. However, the charge redistribution by the interface formation from the surfaces shows the electron accumulation at the interstitial sites of the 1st Cu layer near the surface Al atoms, in contrast to the decrease of electrons at the Cu-atom sites near the surface O atoms. This kind of charge redistribution is similar to that observed in MgO/metal interfaces [2] and is consistent with the image model [20]. In addition, we have found the interactions between the Al surface orbitals and Cu orbitals. Thus the origin of adhesion for the Al-terminated case seems to be the electrostatic effect and the Al-Cu orbital hybridization [21].

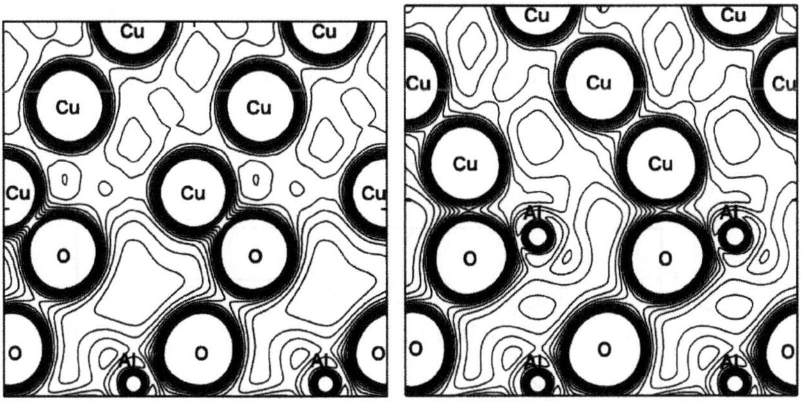

Figure 1. Stable configurations of the Al_2O_3(0001)/Cu(111) interface. O-terminated (left) and Al-terminated (right) interfaces.

The relative stability between the Al-terminated and O-terminated interfaces can be theoretically analysed by calculating the free energy containing the atomic chemical potentials. Recently, such an analysis has been performed for the same system [7], which indicates the relative stability of the O-terminated interface in usual atmosphere in air. And recent electron microscopy observation [22] as well as the comparison with our *ab initio* results [19] indicates the real presence of the

O-terminated interface. Therefore, we deal with the mechanical properties of the O-terminated interface in this paper.

3.2. Ab Initio Tensile Tests

We have performed the *ab initio* tensile test of the O-terminated interface. This is an ideal cleavage simulation, where *ab initio* total-energy calculations are iterated for the increase of a selected interlayer or interface distance in a small increment. We call this a *rigid-type* tensile test in contrast to a *relaxed-type* tensile test with full atomic relaxation [23]. Of course, the realistic mechanical behavior is obtained by the *relaxed-type* test, although such a test only reveals the weakest point originating failure. The *rigid-type* test can clarify the local strength of each selected interlayer or interface, which provides rich information for the development of interatomic potentials. In the *rigid-type* test, the size of the supercell is stretched in accordance with the cleavage of selected two symmetrical interlayers in the cell.

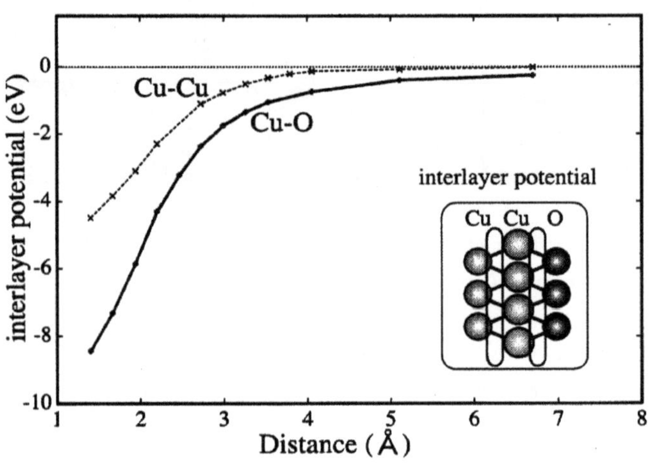

Figure 2. Energy changes in the rigid-type tensile tests for the Cu-O interface and for the Cu-Cu interlayer.

We have performed the tensile tests for the Cu-O interface and for the interlayer between the 1st and 2nd Cu layers at the interface. Fig. 2 shows the energy changes in the tensile tests, which correspond to effective interlayer potential curves. In Fig. 2, the energy gain is plotted against the energy of cleaved (unrelaxed) two surfaces. The bottom value of each curve corresponds to the ideal fracture energy of each interface or interlayer, although the relaxation of the two surfaces should decrease the practical fracure-energy value. The energy value is 8.4 eV for the Cu-O interface and is 4.5 eV for the Cu-Cu interlayer. And the maximum gradient of each curve corresponds to the ideal strength of each interface or interlayer. The strength is about 50 GPa for the Cu-O interface and is about 26 GPa for the Cu-Cu interlayer. It can be said that the Cu-O interface is twice as strong as the Cu-Cu interlayer. It can be said that the failure occurs in the Cu side for the O-terminated interface.

Fig. 3 shows the changes of the valence-charge density distribution during the tensile test of the Cu-O interface. It is clear that the bond charge between the Cu and O atoms disappears after the distance of the maximum gradient point of the curve in Fig. 2.

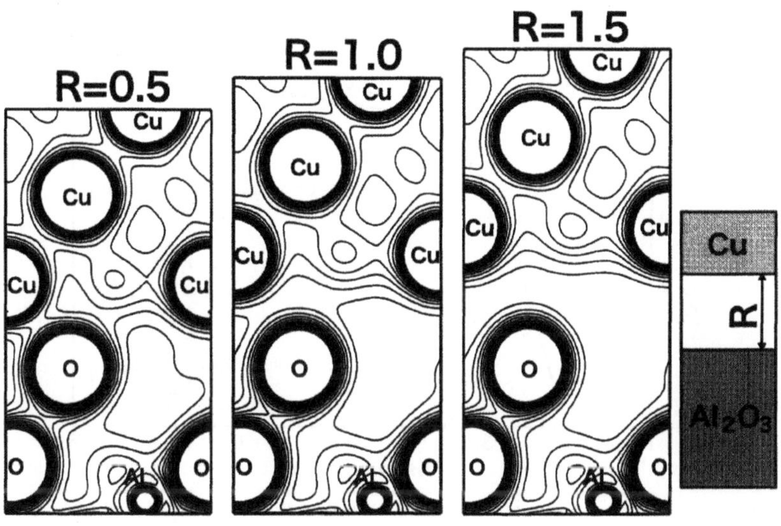

Figure 3. Charge distribution in the tensile test of the Cu-O interface. R means the cleavage distance in unit of Å.

3.3. Construction of Effective Interatomic Potentials

About the development of the interfacial interatomic potentials for the O-terminated interface, the necessary conditions are, first, to reproduce the configuration in Fig. 1 as the most stable one, and second, to reproduce the interlayer potential curves in Fig. 2. Of course, there eixst a lot of selections for the potential forms such as pair-wise or many-body ones, ranges of interations, functional forms and so on. And it is desirable to prepare much more *ab initio* data. Here, we would like to show a simple example of the potentials. From the curves in Fig. 2, we can construct effective interfacial Cu-O and Cu-Cu interatomic potentials as shown in Fig. 4. These potentials can be constructed by converting the interlayer potential curves in Fig. 2 into the contributions of the interfacial first-neighbor Cu-O pairs and of the first-neighbor Cu-Cu pairs at the 1st and 2nd Cu layers. This can be performed because of the simple symmetric configuration of the hollow-site model. Of course, if we include the second-neighbor interactions, the potential forms are changed. And it is necessary to complement repulsive potential forms for shorter distances.

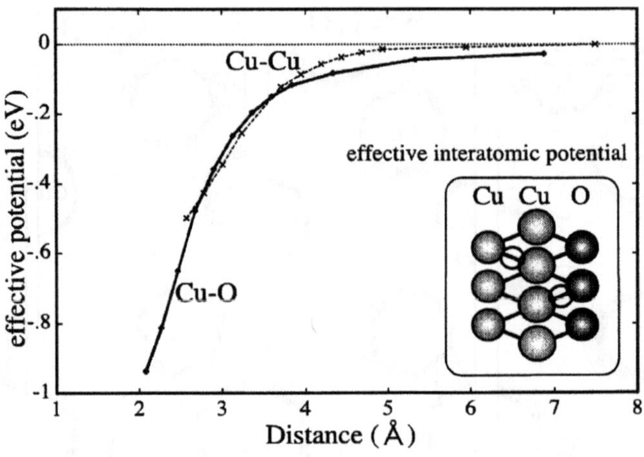

Figure 4. Effective Cu-O and Cu-Cu interatomic potentials constructed from the effective interlayer potentials in Fig. 2.

The trial potentials in Fig. 4 can be used as the guideline so as to develop better potentials satisfying all the necessary conditions in the near future. It is interesting to construct this kind of effective interatomic potentials from results of *ab initio* tensile tests of other interfacial configurations such as the O-site or Al-site models of the O-terminated interface. And of course, it is quite interesting to perform similar analysis for the Al-terminated interface in the near future.

4. Summary

Adhesive and mechanical properties of the $Al_2O_3(0001)/Cu(111)$ interface have been examined by using the *ab initio* calculations based on the density-functional theory, and the development of effective interatomic potentials at the interface has been discussed. First, stable configurations for the Al-terminated (stoichiometric) and O-terminated (O-rich) interfaces have been obtained. The O-terminated interface has quite larger adhesive energy than that of the Al-terminated one, and the origins of adhesion are quite different for the two kinds of interfaces. Second, the mechanical properties of the O-terminated interface have been examined through the *rigid-type* tensile test, which can clarify the local strength of each interface or interlayer. It has been shown that the Cu-O interface is twice as strong as the Cu-Cu interlayer. Third, the effective interatomic potentials at the interface have been constructed by converting the interlayer potential curves in the tensile test into the contribution of each atomic pair, which can be used as the guideline for the future development of reliable potentials.

Acknowledgement

The present study was supported by NEDO as the Nano-Coating Project. The authors whould like to thank Dr. S. Dmitriev, Dr. Y. Hangai, Prof. N. Yoshikawa, Dr. Y. Liu, and Prof. Y. Kagawa for fruitful discussions on the development of interatomic potentials and the multi-scale techniques. The authors also thank Mr. T. Sasaki, Dr. K. Matsunaga and Prof. Y. Ikuhara for helpful discussions on the electron microscopy observations of the alumina/copper interfaces.

References

1. M.W. Finnis, J. Phys. Condens. Matter **8**, 5811 (1996).
2. J.R. Smith, T. Hong and D.J. Srolovitz, Phys. Rev. Lett. **72**, 4021 (1994); T. Hong, J.R. Smith and D.J. Srolovitz, J. Adhesion Sci. Tech. **8**, 837 (1994).
3. R. Benedek, M. Minkoff and L.H. Yang, Phys. Rev. B **54**, 7697 (1995).
4. J. Hoekstra and M. Kohyama, Phys. Rev. B **57**, 2334 (1998); M. Kohyama and J. Hoekstra, *ibid.* **61**, 2672 (2000); S. Tanaka and M. Kohyama, *ibid.* **64**, 235308 (2001).
5. I.G. Batyrev, A. Alavi, M.W. Finnis and T. Deutsch, Phys. Rev. Lett. **82**, 1510 (1999); I.G. Batyrev and L. Kleinman, Phys. Rev. B **64**, 033410 (2000).
6. W. Zhang and J.R. Smith, Phys. Rev. B **61**, 16883 (2000); Phys. Rev. Lett. **85**, 3225 (2000); J.R. Smith and W. Zhang, Acta Mater. **48**, 4395 (2000).
7. W. Zhang, J.R. Smith and A.G. Evans, Acta Mater. **50**, 3803 (2002).
8. R. Benedek, D.N. Seidman, M. Minkoff, L.H. Yang and A. Alavi, Phys. Rev. B **60**, 16094 (1999).
9. K. Albe, R. Benedek, D.N. Seidman and R.S. Averback, Mat. Res. Soc. Symp. Proc. **654**, AA4.3.1 (2001).
10. C.X. Guo, D.E. Ellis, V.P. Dravid and L. Brewer, Mat. Res. Soc. Symp. Proc. **654**, AA4.5.1 (2001).
11. J.R. Chelikowsky and M.L. Cohen, *Handbook on Semiconductors Vol.1*, edited by P.T. Landsberg (Elsevier, 1992), p.59.
12. P. Hohenberg and W. Kohn, Phys. Rev. **136**, B864 (1964); W. Kohn and J.L. Sham, *ibid.* **140**, A1133 (1965).
13. J.P. Perdew and A. Zunger, Phys. Rev. B **23**, 5048 (1981).
14. N. Troullier and J.L. Martins, Phys. Rev. B **43**, 1993 (1991).
15. P. Pulay, Chem. Phys. Lett. **73**, 393 (1980); G. Kresse and J. Furthmüller, Phys. Rev. B **54**, 11169 (1996).
16. M. Kohyama, S. Tanaka and K. Okazaki, J. Soc. Mater. Sci. Jpn. **52**, 260 (2003).
17. G.P. Kerker, Phys. Rev. B **23**, 3082 (1981).
18. P. Pulay, J. Comp. Chem. **3**, 556 (1982).
19. S. Tanaka, R. Yang and M. Kohyama, to be submitted.
20. P.W. Tasker and A.M. Stoneham, J. Chimie Phys. **84**, 149 (1987).
21. R. Yang, S. Tanaka and M. Kohyama, to be submitted.
22. T. Sasaki, K. Matsunaga, H. Ohta, H. Hosono, T. Yamamoto and Y. Ikuhara, J. Soc. Mater. Sci. Jpn. **52**, 555 (2003).
23. M. Kohyama, Phil. Mag. Lett. **79**, 659 (1999); Phys. Rev. B **65**, 184107 (2002).

DEVELOPMENT OF HYBRID ELECTRONIC-DENSITY-FUNCTIONAL/MOLECULAR-DYNAMICS SIMULATION SCHEMES FOR CERAMICS AND SEMICONDUCTORS

Shuji Ogata

Graduate School of Engineering, Nagoya Institute of Technology

Gokiso, Showa-ku, Nagoya 466-8555, Japan

Abstract Recent development in hybrid elecronic-density-functional/molecular-dynamics simulation schemes is reviewed. In the hybrid scheme, a total system is partitioned into the quantum-mechanical (QM) region treated by the electronic-density-functional theory and the molecular dynamics (MD) region in which atoms are interacting through the empirical inter-atomic potential. In the former hybrid scheme [Ogata *et al.*, Comp. Phys. Comm. 149 (2002) 30], appropriate selection of QM atoms for seamless coupling between the QM and MD regions is limited. Novel hybrid scheme that is free from the limitation is presented.

Keywords: hybrid simulation, density-functional theory, molecular dynamics, multiscale simulation

1. INTRODUCTION

Various engineering processes in materials, such as oxidation and fracture, involve chemical reactions between constituent atoms [1,2]. Empirical inter-atomic potentials used in molecular dynamics (MD) simulations often fail to describe such processes. Inter-atomic interaction in the reactive region needs to be calculated by a quantum mechanical (QM) method that can describe breaking and formation of bonds. Furthermore heat produced in the QM region by reactions should be correctly transferred to surrounding regions. Large-scale atomistic simulations are therefore

H. Kitagawa and Y. Shibutani (eds.), IUTAM Symposium on Mesoscopic Dynamics of Fracture Process and Materials Strength, 431–440.

required, in which reacting atoms are described by a reliable QM method. There have been growing interests in developing hybrid QM/MD simulation schemes, in which a reactive region treated by a QM method is embedded in a classical system of atoms interacting via an empirical inter-atomic potential [3].

During the past decade, computationally efficient QM approaches based on the density-functional theory (DFT) [4,5] have been advanced. Recently a hybrid QM/MD simulation scheme [6] employing the DFT for the QM calculation has been developed for simulations of biological molecules in complex solvent. In this scheme, plane waves are used as the basis to solve the Kohn-Sham equations for a QM cluster.

Ogata *et al.* [7] have proposed an improved hybrid QM/MD scheme (hereafter referred to O02 scheme) for materials simulation by employing real-space DFT [8] in which wave functions and pseudopotentials are represented on uniform real-space mesh points in Cartesian coordinates. Its parallel implementation based on spatial decomposition is straightforward, and a few thousand atoms can be simulated quantum mechanically using 100~1000 processors [7,9,10]. Ogata *et al.* [11] successfully applied the hybrid QM/MD scheme to a Si system with environmental water molecules to study moisture effects on fracture initiation in Si. A (110) crack under tension (mode-I opening) was simulated with multiple H_2O molecules around the crack front. Electronic structure near the crack front was calculated with the DFT. The DFT description was embedded in a large classical MD simulation. The hybrid simulation results showed that the reaction of H_2O molecules at a silicon crack tip is sensitive to the stress intensity factor.

In Sec. 2, we explain O02 scheme and its limitation in appropriate selection of the QM region. A novel hybrid scheme that overcomes the limitation and is applicable to a wide range of semiconductors and ceramics is introduced in Sec. 3. Concluding remarks are given in Sec. 4.

2. O02 SCHEME: FORMULATION AND LIMITATION

To partition a total system into the cluster and its environmental regions, Ogata *et al.* [7,9] followed the modular approach by Svensson *et al.* [12] and Eichler *et al.* [13] that is based on a linear combination of QM and MD potential energies. Dynamics of all atoms are determined by the following hamiltonian:

$$H = H_{\rm MD}^{\rm system}(\vec{R}_{\rm all}, d\vec{R}_{\rm all}/dt) + \sum_{\rm cluster}(E_{\rm QM}^{\rm cluster} - E_{\rm MD}^{\rm cluster}). \quad (1)$$

Here $\vec{R}_{\rm all}$ represents a set of all atoms. The $H_{\rm MD}^{\rm system}$ in Eq. (1) is the MD hamiltonian of the total system. The last two terms on the right hand side of Eq. (1) represent the QM correction by the DFT to the MD potential energy for the cluster of atoms in the QM region.

All the atoms in the total system are grouped into MD and QM atoms. Positions of the MD and QM atoms are denoted as $\{\vec{r}_{\rm MD}\}$ and $\{\vec{r}_{\rm QM}\}$, respectively. The MD atoms contain handshake (HS) atoms, which are the nearest neighbor MD atoms of the surface atoms of the QM cluster. Their positions are denoted as $\{\vec{r}_{\rm QM/MD}^{\rm HS}\}$. In O02, the last two terms in the right hand side of Eq. (1) are functions of both $\{\vec{r}_{\rm QM}\}$ and $\{\vec{r}_{\rm QM/MD}^{\rm HS}\}$:

$$E_{\rm QM}^{\rm cluster} = E_{\rm QM}^{\rm cluster}(\{\vec{r}_{\rm QM}\};\{\vec{r}_{\rm QM/MD}^{\rm HS}\}), \quad (2)$$

$$E_{\rm MD}^{\rm cluster} = E_{\rm MD}^{\rm cluster}(\{\vec{r}_{\rm QM}\};\{\vec{r}_{\rm QM/MD}^{\rm HS}\}). \quad (3)$$

Termination atoms are introduced for dangling bonds of the cluster in both QM and MD calculations: H atoms, for the QM calculation; constituting atoms, for the MD calculation. Positions of the termination-H atoms are determined from those of the HS atoms and their bonding QM-atoms as follows. Let $\vec{r}_{\rm QM}(j;i)$ be the positions of the QM atoms bonding to a HS atom at $\vec{r}_{\rm QM/MD}^{\rm HS}(i)$; depending on selection of QM atoms multiple numbers of different j may exist for each i. A termination H is placed at $\vec{x}^{\rm H} = \beta\vec{r}_{\rm QM/MD}^{\rm HS}(i) + (1-\beta)\vec{r}_{\rm QM}(j;i)$ for each (i,j) pair with a scaling parameter $\beta = 0.6$. In the case of MD cluster, termination-Si atoms are set at the positions of the HS atoms, *i.e.*, $\vec{x}^{\rm Si} = \vec{r}_{\rm QM/MD}^{\rm HS}(i)$. Hence the MD and QM regions interact each other through $\{\vec{r}_{\rm QM/MD}^{\rm HS}\}$. The gradient of H with respect to the position of the i-th atom $\vec{r}(i)$ gives the corresponding force $\vec{F}_i$ as a summation of three partial forces corresponding to the three terms of H in Eq. (1): $\vec{F}_i = -\partial H / \partial \vec{r}(i) = \vec{F}_{\mathrm{MD},i}^{\mathrm{system}} + \vec{F}_{\mathrm{QM},i}^{\mathrm{cluster}} - \vec{F}_{\mathrm{MD},i}^{\mathrm{cluster}}$.

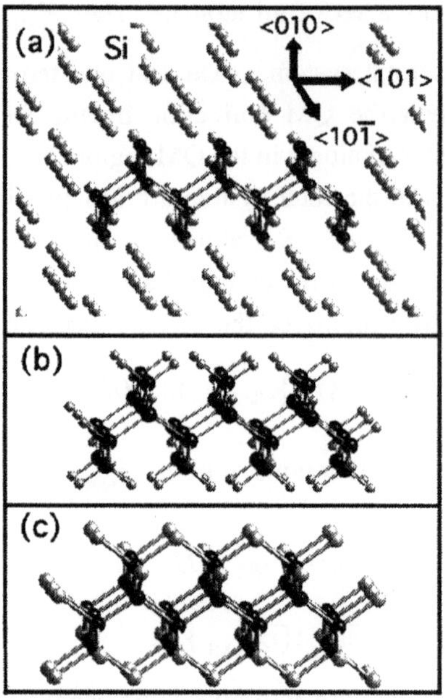

Figure 1. An example of inappropriate selection of QM atoms in bulk Si in O02 scheme. (a) The QM atoms are depicted as black spheres, the MD atoms, as white spheres. (b) The QM cluster used in the DFT calculation for the QM region in (a). (c) The MD atoms in the MD calculation for the QM region in (a).

We find that appropriate selection of QM atoms for seamless coupling of the QM and MD regions is limited in O02 scheme. When one selects an inappropriate set of QM atoms in the total system, QM configuration at the ground state is highly distorted from the initial one. An example of such an inappropriate selection is depicted in Fig. 1 for bulk Si. Figure 1(a) shows a set of QM Si-atoms (black spheres) surrounded by MD Si-atoms (white spheres) in bulk Si. The cluster configurations used to calculate $E_{\mathrm{QM}}^{\mathrm{cluster}}$ and $E_{\mathrm{MD}}^{\mathrm{cluster}}$ in Eq. (1) are shown in Fig. 1(b) and Fig. 1(c), respectively. After relaxing positions of all the atoms to minimize the total energy H, the QM cluster distorts significantly. Since the total system is bulk-Si, such distorting is undesirable in the hybrid scheme. We have tried with various selections of the QM cluster and find that such undesirable distortion may occur if a HS atom relates to multiple numbers of termination-H atoms in the DFT calculation, that is, those H atoms reside at close proximately as seen in Fig. 1(b). For instance, if we minimize the potential energy of an atomic cluster in Fig. 1(b) through the DFT calculation, we obtain highly distorted atomic cluster as depicted in Fig. 2. Some of the Si-Si distances at the peripheral in Fig. 2 are larger than 3.5Å; clearly, no bonding between the two Si atoms exists.

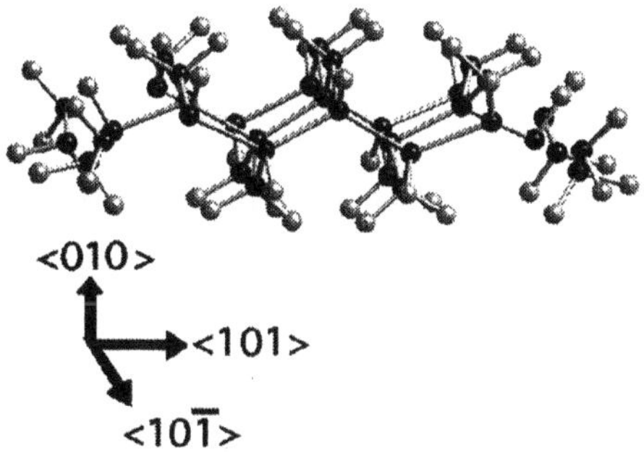

Figure 2. Atomic configuration of the QM cluster obtained through the DFT calculation after minimizing its potential energy. Initial configuration is same with that in Fig. 1(b).

3. A NOVEL HYBRID SCHEME

We here propose a novel hybrid scheme, which requires no HS atoms and therefore free from such contacting termination-H atoms in the DFT calculation. The present scheme is applicable to virtually any selection of the QM cluster in bulk Si including the case in Fig. 1. We find that the present scheme is not limited to Si in which empirical inter-atomic potential is short-ranged. It is applicable also to alumina in which semi-empirical inter-atomic potential is long-ranged.

In the present scheme, the cluster terms in the right hand side of Eq. (1) are functions of $\{\vec{r}_{QM}\}$:

$$E_{QM}^{cluster} = E_{QM}^{cluster}(\{\vec{r}_{QM}\}), \tag{4}$$

$$E_{\mathrm{MD}}^{\mathrm{cluster}} = E_{\mathrm{MD}}^{\mathrm{cluster}}(\{\vec{r}_{\mathrm{QM}}\}). \tag{5}$$

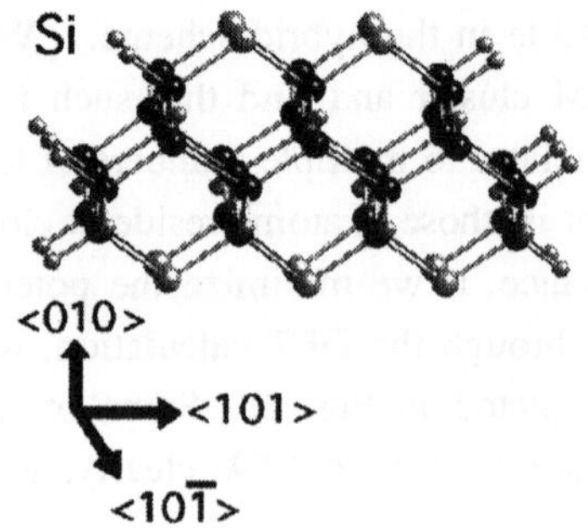

Figure 3. The QM cluster used in the DFT calculation to obtain $E_{\mathrm{QM}}^{\mathrm{cluster}}$ in the present scheme.

Positions of the termination-Si atoms in the MD cluster calculation are determined dynamically to minimize the potential energy of the cluster in the present scheme, while in O02 scheme positions of the termination-Si atoms are same with those of the handshake atoms $\vec{r}_{\mathrm{QM/MD}}^{\mathrm{HS}}$. In the DFT calculation, the QM cluster is terminated by either H or Si atoms as shown in Fig. 3 for the case of the QM cluster in Fig. 1(a). If the coordination number of a termination-Si in the corresponding MD cluster is one, such a termination-Si relates to a termination-H with the scaling factor $\beta = 0.6$ in the DFT calculation; if the coordination number of a termination-Si is equal or larger than two, its position is used as that of a termination-Si in the DFT calculation. Therefore the QM atoms interact with the MD atoms through the MD potentials in the present scheme.

Degrees of coupling between the QM and the MD regions are analyzed through spatial variation of recoil forces felt by an atom when the atom is shifted from its equilibrium position by a same amount. We select QM atoms in bulk Si as shown in Fig. 4(a). Atoms-1, 2, and 3 in Fig. 4(a) are equivalent in crystal periodicity. Atom-1 is located at the center of the QM region; atom-2, at the QM/MD boundary; atom-3, in the MD region. Figure 4(b) shows the recoil force $\vec{F}_i$ of atom-*i* as a function of displacement

$d\vec{r}_i$ from its equilibrium position. As we observe in Fig. 4(b), the recoil forces are similar between the atoms-1, 2, and 3 when $dr_i \leq 0.1$ a.u. When $dr_i > 0.2$ a.u., the recoil forces differ by about 10%. Such differences are reasonable since the Stillinger-Weber intet-atomic potential [14] that we use in the present analyses is constructed to reproduce the elastic behaviors of Si and therefore limited to small atomic displacements.

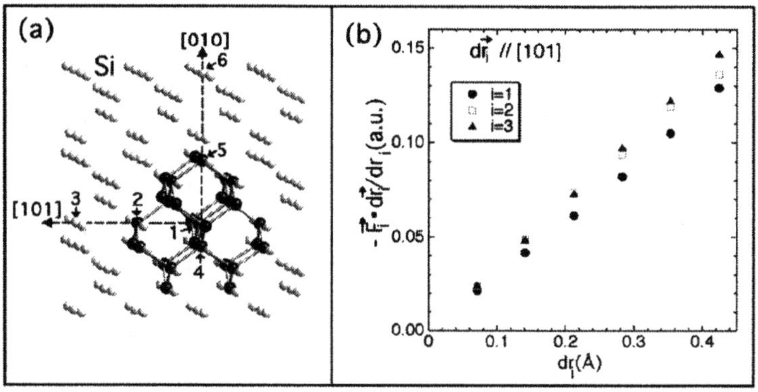

Figure 4. (a) The QM atoms (black spheres) selected in bulk Si. The atoms-1, 2, and 3 are equivalent in crystalline periodicity. (b) Recoil forces $\bar{F}_i$ felt by atom-*i* as a function of its displacement dr_i from its equilibrium position.

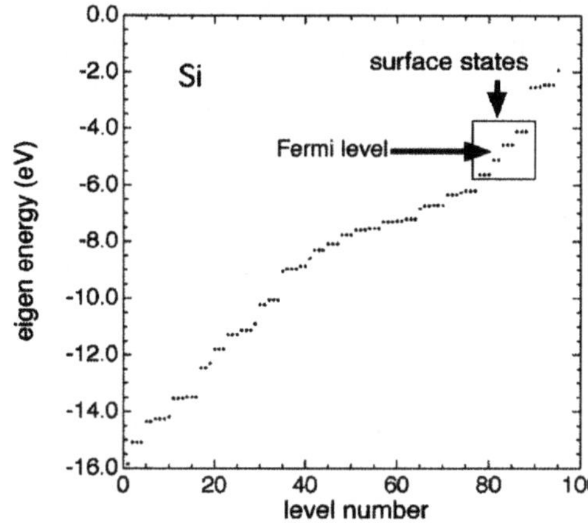

Figure 5. Energy levels obtained in DFT calculation to obtain $E_{\mathrm{QM}}^{\mathrm{cluster}}$ for the QM cluster shown in Fig. 4(a).

In the present hybrid scheme, the DFT calculation in Fig. 4 is performed for the atomic cluster that contain dangling bonds of termination-Si atoms. Therefore eigenstates relating to such dangling bonds appear near the Fermi level as shown in Fig. 5. Detailed analyses show that those states near the Fermi level are localized at the peripheral region of the cluster, *i.e.*, surface states. We expect that those surface states do not cause any significant effects on the eigenstates localized around the central of the QM region.

To confirm conservation of the hamiltonian H during a simulation run, we shift atom-2 by 1.0 a.u. toward [010]-direction at the initial and observe subsequent evolution of H. Figure 6 depicts time evolutions of $E_{\mathrm{MD}}^{\mathrm{system}}$ and H. While $H_{\mathrm{MD}}^{\mathrm{system}}/N$ oscillate by about 0.005eV, variation of H/N is as small as 0.0001eV.

The present scheme is applicable also to α-Al_2O_3 system. Variable-charge potential [15] is used for MD calculations, in which atomic charges vary dynamically following the generalized electro-negativity equalization principles. We find that seamless coupling between the QM and MD regions in α-Al_2O_3 requires that (i) the outermost QM atoms *i.e.*, connecting to MD atoms, should be Al, and (ii) the QM atoms should be surrounded by the first neighboring O atoms and the second neighboring Al or "virtual" atom with its core charge $Z_{\mathrm{core}} = 0.5\,|e|$ as shown in Fig. 7.

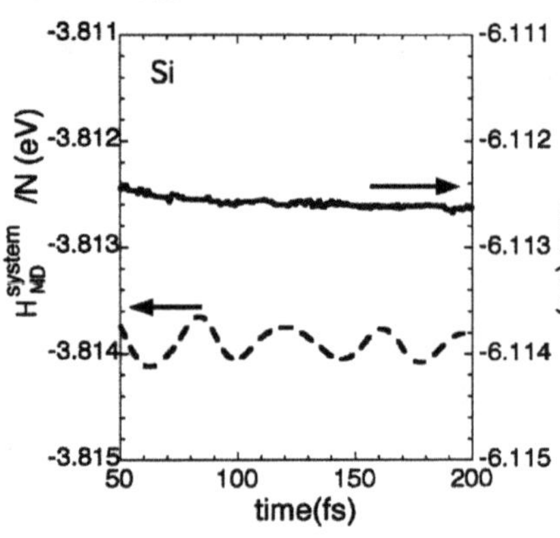

Figure 6. Time evolutions of $H_{\mathrm{MD}}^{\mathrm{system}}/N$ and H/N after shifting atom-2 by 1.0 a.u. to [010] direction at time = 0.

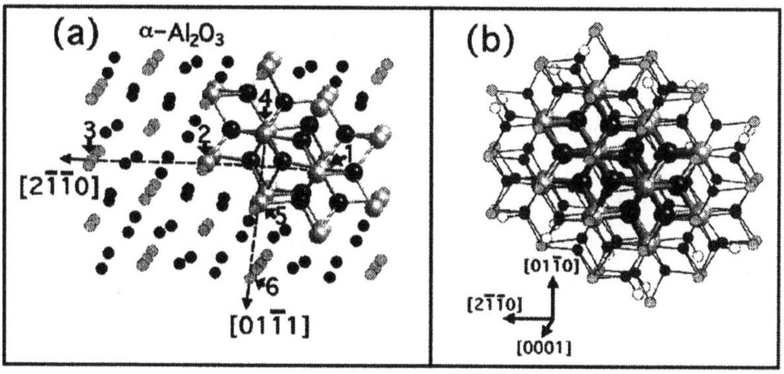

Figure7. (a) A selection of QM atoms in bulk α-Al_2O_3. The QM-O atoms are depicted as black spheres, QM-Al atoms as white spheres, MD-O atoms as black dots, MD-Al atoms as gray dots. (b) The QM cluster used in DFT calculation. The QM-O atoms are depicted as black spheres, QM-Al atoms as white spheres, surrounding O atoms as black dots, surrounding Al atoms as gray dots

4. CONCLUDING REMARKS

Reactive region moves as simulation proceeds in practical hybrid simulations, and hence re-selection of QM atoms to trace the reactive region is required. There exists virtually no limitation in selection of QM atoms in the present hybrid scheme. Therefore the present scheme is well suited to automatic, adaptive re-selection of QM atoms. Further development is in progress to install such adaptive features in the hybrid scheme.

It is well known experimentally that α-Al_2O_3 shows the stress corrosion cracking when it is exposed to moisture environment. There exists no microscopic theory to explain its mechanisms. We are now performing hybrid QM/MD simulations to study environmental effects of water molecules on fracture initiation of α-Al_2O_3.

ACKNOWLEDGMENT

This work is supported by ACT-JST.

REFERENCES

1. *Nanoparticles and Nanostructured Films,* edited by J.H. Fendler (Wiley-VCH, NY, 1998).
2. Y.-M. Chiang, D. Birnie III, and W.D. Kingery, *Physical Ceramics* (Wiley & Sons, New York, 1997).
3. A. Nakano, M.E. Bachlechner, P. Branicio, T.J. Campbell, I. Ebbsjö, R.K. Kalia, A. Madhukar, S. Ogata, A. Omeltchenko, J.P. Rino, F. Shimojo, P. Walsh, and P. Vashishta, IEEE Trans. Electron Devices 47 (2000) 1804.
4. P. Hoenberg and W. Kohn, Phys. Rev. 136 (1964) B864; W. Kohn and L. J. Sham, Phys. Rev. 140 (1965) A1133.
5. See, *e.g.*, M.C. Payne, M.P. Teter, D.C. Allan, T.A. Arias, and J.D. Joannopoulos, Rev. Mod. Phys. 64 (1992) 1045.
6. M. Eichinger, P. Tavan, J. Hutter, and M. Parinello, J. Chem. Phys. 110 (1999) 10452.
7. S. Ogata, F. Shimojo, R.K. Kalia, A. Nakano, and P. Vashishta, Comp. Phys. Comm. 149 (2002) 30-38.
8. J.R. Chelikowsky, N. Troullier, and Y. Saad, Phys. Rev. Lett. 72 (1994) 1240.
9. S. Ogata, E. Lidorikis, F. Shimojo, A.Nakano, P. Vashishta, and R.K. Kalia, Comp. Phys. Comm. 138 (2001) 143.
10. F. Shimojo, T.J. Campbell, R.K. Kalia, A. Nakano, S. Ogata, P. Vashishta, and K. Tsuruta, Future Generation Comp. Sys. 17 (2000), 279.
11. S. Ogata, F. Shimojo, A. Nakano, P. Vashishta, and R.K. Kalia, submitted to J. Appl. Phys.
12. M. Svensson, S. Hymbel, R.D.F. Froese, T. Matsubara, S. Sieber, and K. Morokuma, J. Comp. Chem. 100 (1996) 19357.
13. U. Eichler, C. M. Kölmel, and J. Sauer, J. Comp. Chem. 18 (1996) 463.
14. F.H. Stillinger and T.A. Weber, Phys. Rev. B 31 (1985) 5262.
15. F.H. Streitz and J.W. Mintmire, Phys. Rev. B 50, (1994) 11996

AUTHOR INDEX

Aoyagi, Y.	87
Bacon, D.J.	173, 193
Bartelt, M.	1
Beauchamp, P.	129
Bitzek, E.	45
Brechet, Y.	163
Brochard, S.	129
Bulatov, V.V.	1, 183
Cai, W.	1
Chaboche, J.-L.	235
Chang, J.	223
Choi, S.O.	331
Choi, Y.	203
Demenet, J	139
Deshpande, V.S.	79
Devincre, B.	35, 235
Dmitriev, S.V.	279
Domain, C.	411
Dubuisson, P.	163
Feyel, F.	235
Frederiksen, S.L.	355
Van der Giessen, E.	79
Godet, J.	129
Groh, S.	235
Gumbsch, P.	45
Hashimoto, S.	13
Hara, S.	117
Hasebe, T.	381
Higa, Y.	255
Higashida, K.	153
Hiratani, M.	1
Izumi, S.	117
Kaneko, Y.	13
Kim, J.	331
Kim, Y.S.	331
Kishimoto, K	301
Kitagawa, H.	107, 255, 365
Kitamura, T.	391
Kohyama, M.	421
Kubin, L.P.	35, 69, 235
Kubo, S.	311
Kumagai, T.	117
Kysar, J.W.	289
Lee, S.R.	331
Legris, A.	411
Li, J.	203, 223, 401
Lin, X.	223
Madec, R.	35, 69
Maksimov, I.L.	269
Massoud, J.P.	163
Matsumoto, R.	365
Misaki, M.	311
Nagase, S.	343
Nakasone, Y.	321
Nakatani, A.	365
Nakatani, K.	107
Needleman, A.	79
Ogata, Shigenobu	401
Ogata, Shuji	431
Ohashi, T.	97

Omiya, M. 301
Osetsky, Y.N. 173, 193
Pierce, T.G. 1
Pirouz, P. 139
Pizzagalli, L. 129
Pokor, C. 163
Qian, X.-F, 223
Rhee, M. 1
Rodney, D. 163
Rong, Z. 173

Roos, A 235
de la Rubia, T.D. 183
Saitoh, K 343
Sakai, S. 117
Schiotz, J. 355
Shehadeh, M.A. 183
Shibutani, Y. 213, 401
Shimokawa, T. 365
Shinke, N. 343
Shizawa, K. 87

Sugiyama, Y. 107
Suresh, S. 203
Tabata, Y. 59
Tanaka, M 153
Tanaka, S. 421
Tang, M. 1
Tapasa, K. 173
Tomita, Y. 59, 245, 255
Tsuru, T. 213
Uchida, M. 245

Umeno, Y. 391
Vasiliev, A.A. 279
Van Vliet, K.J. 203
Wang, S. 139
Weygand, D. 23, 45
Yang, R. 421
Yashiro, K 59
Yip, S. 203, 223, 401
Yoshikawa, N. 279
Zbib, H.M. 183

Zhang, M. 139
Zhu, T. 203, 223

Mechanics

SOLID MECHANICS AND ITS APPLICATIONS

Series Editor: G.M.L. Gladwell

Aims and Scope of the Series

The fundamental questions arising in mechanics are: *Why?*, *How?*, and *How much?* The aim of this series is to provide lucid accounts written by authoritative researchers giving vision and insight in answering these questions on the subject of mechanics as it relates to solids. The scope of the series covers the entire spectrum of solid mechanics. Thus it includes the foundation of mechanics; variational formulations; computational mechanics; statics, kinematics and dynamics of rigid and elastic bodies; vibrations of solids and structures; dynamical systems and chaos; the theories of elasticity, plasticity and viscoelasticity; composite materials; rods, beams, shells and membranes; structural control and stability; soils, rocks and geomechanics; fracture; tribology; experimental mechanics; biomechanics and machine design.

1. R.T. Haftka, Z. Gürdal and M.P. Kamat: *Elements of Structural Optimization.* 2nd rev.ed., 1990 ISBN 0-7923-0608-2
2. J.J. Kalker: *Three-Dimensional Elastic Bodies in Rolling Contact.* 1990 ISBN 0-7923-0712-7
3. P. Karasudhi: *Foundations of Solid Mechanics.* 1991 ISBN 0-7923-0772-0
4. *Not published*
5. *Not published.*
6. J.F. Doyle: *Static and Dynamic Analysis of Structures.* With an Emphasis on Mechanics and Computer Matrix Methods. 1991 ISBN 0-7923-1124-8; Pb 0-7923-1208-2
7. O.O. Ochoa and J.N. Reddy: *Finite Element Analysis of Composite Laminates.* ISBN 0-7923-1125-6
8. M.H. Aliabadi and D.P. Rooke: *Numerical Fracture Mechanics.* ISBN 0-7923-1175-2
9. J. Angeles and C.S. López-Cajún: *Optimization of Cam Mechanisms.* 1991 ISBN 0-7923-1355-0
10. D.E. Grierson, A. Franchi and P. Riva (eds.): *Progress in Structural Engineering.* 1991 ISBN 0-7923-1396-8
11. R.T. Haftka and Z. Gürdal: *Elements of Structural Optimization.* 3rd rev. and exp. ed. 1992 ISBN 0-7923-1504-9; Pb 0-7923-1505-7
12. J.R. Barber: *Elasticity.* 1992 ISBN 0-7923-1609-6; Pb 0-7923-1610-X
13. H.S. Tzou and G.L. Anderson (eds.): *Intelligent Structural Systems.* 1992 ISBN 0-7923-1920-6
14. E.E. Gdoutos: *Fracture Mechanics.* An Introduction. 1993 ISBN 0-7923-1932-X
15. J.P. Ward: *Solid Mechanics.* An Introduction. 1992 ISBN 0-7923-1949-4
16. M. Farshad: *Design and Analysis of Shell Structures.* 1992 ISBN 0-7923-1950-8
17. H.S. Tzou and T. Fukuda (eds.): *Precision Sensors, Actuators and Systems.* 1992 ISBN 0-7923-2015-8
18. J.R. Vinson: *The Behavior of Shells Composed of Isotropic and Composite Materials.* 1993 ISBN 0-7923-2113-8
19. H.S. Tzou: *Piezoelectric Shells.* Distributed Sensing and Control of Continua. 1993 ISBN 0-7923-2186-3
20. W. Schiehlen (ed.): *Advanced Multibody System Dynamics.* Simulation and Software Tools. 1993 ISBN 0-7923-2192-8
21. C.-W. Lee: *Vibration Analysis of Rotors.* 1993 ISBN 0-7923-2300-9
22. D.R. Smith: *An Introduction to Continuum Mechanics.* 1993 ISBN 0-7923-2454-4
23. G.M.L. Gladwell: *Inverse Problems in Scattering.* An Introduction. 1993 ISBN 0-7923-2478-1

Mechanics

SOLID MECHANICS AND ITS APPLICATIONS

Series Editor: G.M.L. Gladwell

24. G. Prathap: *The Finite Element Method in Structural Mechanics*. 1993 ISBN 0-7923-2492-7
25. J. Herskovits (ed.): *Advances in Structural Optimization*. 1995 ISBN 0-7923-2510-9
26. M.A. González-Palacios and J. Angeles: *Cam Synthesis*. 1993 ISBN 0-7923-2536-2
27. W.S. Hall: *The Boundary Element Method*. 1993 ISBN 0-7923-2580-X
28. J. Angeles, G. Hommel and P. Kovács (eds.): *Computational Kinematics*. 1993 ISBN 0-7923-2585-0
29. A. Curnier: *Computational Methods in Solid Mechanics*. 1994 ISBN 0-7923-2761-6
30. D.A. Hills and D. Nowell: *Mechanics of Fretting Fatigue*. 1994 ISBN 0-7923-2866-3
31. B. Tabarrok and F.P.J. Rimrott: *Variational Methods and Complementary Formulations in Dynamics*. 1994 ISBN 0-7923-2923-6
32. E.H. Dowell (ed.), E.F. Crawley, H.C. Curtiss Jr., D.A. Peters, R. H. Scanlan and F. Sisto: *A Modern Course in Aeroelasticity*. Third Revised and Enlarged Edition. 1995 ISBN 0-7923-2788-8; Pb: 0-7923-2789-6
33. A. Preumont: *Random Vibration and Spectral Analysis*. 1994 ISBN 0-7923-3036-6
34. J.N. Reddy (ed.): *Mechanics of Composite Materials*. Selected works of Nicholas J. Pagano. 1994 ISBN 0-7923-3041-2
35. A.P.S. Selvadurai (ed.): *Mechanics of Poroelastic Media*. 1996 ISBN 0-7923-3329-2
36. Z. Mróz, D. Weichert, S. Dorosz (eds.): *Inelastic Behaviour of Structures under Variable Loads*. 1995 ISBN 0-7923-3397-7
37. R. Pyrz (ed.): *IUTAM Symposium on Microstructure-Property Interactions in Composite Materials*. Proceedings of the IUTAM Symposium held in Aalborg, Denmark. 1995 ISBN 0-7923-3427-2
38. M.I. Friswell and J.E. Mottershead: *Finite Element Model Updating in Structural Dynamics*. 1995 ISBN 0-7923-3431-0
39. D.F. Parker and A.H. England (eds.): *IUTAM Symposium on Anisotropy, Inhomogeneity and Nonlinearity in Solid Mechanics*. Proceedings of the IUTAM Symposium held in Nottingham, U.K. 1995 ISBN 0-7923-3594-5
40. J.-P. Merlet and B. Ravani (eds.): *Computational Kinematics '95*. 1995 ISBN 0-7923-3673-9
41. L.P. Lebedev, I.I. Vorovich and G.M.L. Gladwell: *Functional Analysis*. Applications in Mechanics and Inverse Problems. 1996 ISBN 0-7923-3849-9
42. J. Menčik: *Mechanics of Components with Treated or Coated Surfaces*. 1996 ISBN 0-7923-3700-X
43. D. Bestle and W. Schiehlen (eds.): *IUTAM Symposium on Optimization of Mechanical Systems*. Proceedings of the IUTAM Symposium held in Stuttgart, Germany. 1996 ISBN 0-7923-3830-8
44. D.A. Hills, P.A. Kelly, D.N. Dai and A.M. Korsunsky: *Solution of Crack Problems*. The Distributed Dislocation Technique. 1996 ISBN 0-7923-3848-0
45. V.A. Squire, R.J. Hosking, A.D. Kerr and P.J. Langhorne: *Moving Loads on Ice Plates*. 1996 ISBN 0-7923-3953-3
46. A. Pineau and A. Zaoui (eds.): *IUTAM Symposium on Micromechanics of Plasticity and Damage of Multiphase Materials*. Proceedings of the IUTAM Symposium held in Sèvres, Paris, France. 1996 ISBN 0-7923-4188-0
47. A. Naess and S. Krenk (eds.): *IUTAM Symposium on Advances in Nonlinear Stochastic Mechanics*. Proceedings of the IUTAM Symposium held in Trondheim, Norway. 1996 ISBN 0-7923-4193-7
48. D. Ieşan and A. Scalia: *Thermoelastic Deformations*. 1996 ISBN 0-7923-4230-5

Mechanics

SOLID MECHANICS AND ITS APPLICATIONS

Series Editor: G.M.L. Gladwell

49. J.R. Willis (ed.): *IUTAM Symposium on Nonlinear Analysis of Fracture*. Proceedings of the IUTAM Symposium held in Cambridge, U.K. 1997 ISBN 0-7923-4378-6
50. A. Preumont: *Vibration Control of Active Structures*. An Introduction. 1997 ISBN 0-7923-4392-1
51. G.P. Cherepanov: *Methods of Fracture Mechanics: Solid Matter Physics*. 1997 ISBN 0-7923-4408-1
52. D.H. van Campen (ed.): *IUTAM Symposium on Interaction between Dynamics and Control in Advanced Mechanical Systems*. Proceedings of the IUTAM Symposium held in Eindhoven, The Netherlands. 1997 ISBN 0-7923-4429-4
53. N.A. Fleck and A.C.F. Cocks (eds.): *IUTAM Symposium on Mechanics of Granular and Porous Materials*. Proceedings of the IUTAM Symposium held in Cambridge, U.K. 1997 ISBN 0-7923-4553-3
54. J. Roorda and N.K. Srivastava (eds.): *Trends in Structural Mechanics*. Theory, Practice, Education. 1997 ISBN 0-7923-4603-3
55. Yu.A. Mitropolskii and N. Van Dao: *Applied Asymptotic Methods in Nonlinear Oscillations*. 1997 ISBN 0-7923-4605-X
56. C. Guedes Soares (ed.): *Probabilistic Methods for Structural Design*. 1997 ISBN 0-7923-4670-X
57. D. François, A. Pineau and A. Zaoui: *Mechanical Behaviour of Materials*. Volume I: Elasticity and Plasticity. 1998 ISBN 0-7923-4894-X
58. D. François, A. Pineau and A. Zaoui: *Mechanical Behaviour of Materials*. Volume II: Viscoplasticity, Damage, Fracture and Contact Mechanics. 1998 ISBN 0-7923-4895-8
59. L.T. Tenek and J. Argyris: *Finite Element Analysis for Composite Structures*. 1998 ISBN 0-7923-4899-0
60. Y.A. Bahei-El-Din and G.J. Dvorak (eds.): *IUTAM Symposium on Transformation Problems in Composite and Active Materials*. Proceedings of the IUTAM Symposium held in Cairo, Egypt. 1998 ISBN 0-7923-5122-3
61. I.G. Goryacheva: *Contact Mechanics in Tribology*. 1998 ISBN 0-7923-5257-2
62. O.T. Bruhns and E. Stein (eds.): *IUTAM Symposium on Micro- and Macrostructural Aspects of Thermoplasticity*. Proceedings of the IUTAM Symposium held in Bochum, Germany. 1999 ISBN 0-7923-5265-3
63. F.C. Moon: *IUTAM Symposium on New Applications of Nonlinear and Chaotic Dynamics in Mechanics*. Proceedings of the IUTAM Symposium held in Ithaca, NY, USA. 1998 ISBN 0-7923-5276-9
64. R. Wang: *IUTAM Symposium on Rheology of Bodies with Defects*. Proceedings of the IUTAM Symposium held in Beijing, China. 1999 ISBN 0-7923-5297-1
65. Yu.I. Dimitrienko: *Thermomechanics of Composites under High Temperatures*. 1999 ISBN 0-7923-4899-0
66. P. Argoul, M. Frémond and Q.S. Nguyen (eds.): *IUTAM Symposium on Variations of Domains and Free-Boundary Problems in Solid Mechanics*. Proceedings of the IUTAM Symposium held in Paris, France. 1999 ISBN 0-7923-5450-8
67. F.J. Fahy and W.G. Price (eds.): *IUTAM Symposium on Statistical Energy Analysis*. Proceedings of the IUTAM Symposium held in Southampton, U.K. 1999 ISBN 0-7923-5457-5
68. H.A. Mang and F.G. Rammerstorfer (eds.): *IUTAM Symposium on Discretization Methods in Structural Mechanics*. Proceedings of the IUTAM Symposium held in Vienna, Austria. 1999 ISBN 0-7923-5591-1

Mechanics

SOLID MECHANICS AND ITS APPLICATIONS

Series Editor: G.M.L. Gladwell

69. P. Pedersen and M.P. Bendsøe (eds.): *IUTAM Symposium on Synthesis in Bio Solid Mechanics*. Proceedings of the IUTAM Symposium held in Copenhagen, Denmark. 1999 ISBN 0-7923-5615-2
70. S.K. Agrawal and B.C. Fabien: *Optimization of Dynamic Systems*. 1999 ISBN 0-7923-5681-0
71. A. Carpinteri: *Nonlinear Crack Models for Nonmetallic Materials*. 1999 ISBN 0-7923-5750-7
72. F. Pfeifer (ed.): *IUTAM Symposium on Unilateral Multibody Contacts*. Proceedings of the IUTAM Symposium held in Munich, Germany. 1999 ISBN 0-7923-6030-3
73. E. Lavendelis and M. Zakrzhevsky (eds.): *IUTAM/IFToMM Symposium on Synthesis of Nonlinear Dynamical Systems*. Proceedings of the IUTAM/IFToMM Symposium held in Riga, Latvia. 2000 ISBN 0-7923-6106-7
74. J.-P. Merlet: *Parallel Robots*. 2000 ISBN 0-7923-6308-6
75. J.T. Pindera: *Techniques of Tomographic Isodyne Stress Analysis*. 2000 ISBN 0-7923-6388-4
76. G.A. Maugin, R. Drouot and F. Sidoroff (eds.): *Continuum Thermomechanics*. The Art and Science of Modelling Material Behaviour. 2000 ISBN 0-7923-6407-4
77. N. Van Dao and E.J. Kreuzer (eds.): *IUTAM Symposium on Recent Developments in Non-linear Oscillations of Mechanical Systems*. 2000 ISBN 0-7923-6470-8
78. S.D. Akbarov and A.N. Guz: *Mechanics of Curved Composites*. 2000 ISBN 0-7923-6477-5
79. M.B. Rubin: *Cosserat Theories: Shells, Rods and Points*. 2000 ISBN 0-7923-6489-9
80. S. Pellegrino and S.D. Guest (eds.): *IUTAM-IASS Symposium on Deployable Structures: Theory and Applications*. Proceedings of the IUTAM-IASS Symposium held in Cambridge, U.K., 6–9 September 1998. 2000 ISBN 0-7923-6516-X
81. A.D. Rosato and D.L. Blackmore (eds.): *IUTAM Symposium on Segregation in Granular Flows*. Proceedings of the IUTAM Symposium held in Cape May, NJ, U.S.A., June 5–10, 1999. 2000 ISBN 0-7923-6547-X
82. A. Lagarde (ed.): *IUTAM Symposium on Advanced Optical Methods and Applications in Solid Mechanics*. Proceedings of the IUTAM Symposium held in Futuroscope, Poitiers, France, August 31–September 4, 1998. 2000 ISBN 0-7923-6604-2
83. D. Weichert and G. Maier (eds.): *Inelastic Analysis of Structures under Variable Loads*. Theory and Engineering Applications. 2000 ISBN 0-7923-6645-X
84. T.-J. Chuang and J.W. Rudnicki (eds.): *Multiscale Deformation and Fracture in Materials and Structures*. The James R. Rice 60th Anniversary Volume. 2001 ISBN 0-7923-6718-9
85. S. Narayanan and R.N. Iyengar (eds.): *IUTAM Symposium on Nonlinearity and Stochastic Structural Dynamics*. Proceedings of the IUTAM Symposium held in Madras, Chennai, India, 4–8 January 1999 ISBN 0-7923-6733-2
86. S. Murakami and N. Ohno (eds.): *IUTAM Symposium on Creep in Structures*. Proceedings of the IUTAM Symposium held in Nagoya, Japan, 3-7 April 2000. 2001 ISBN 0-7923-6737-5
87. W. Ehlers (ed.): *IUTAM Symposium on Theoretical and Numerical Methods in Continuum Mechanics of Porous Materials*. Proceedings of the IUTAM Symposium held at the University of Stuttgart, Germany, September 5-10, 1999. 2001 ISBN 0-7923-6766-9
88. D. Durban, D. Givoli and J.G. Simmonds (eds.): *Advances in the Mechanis of Plates and Shells The Avinoam Libai Anniversary Volume*. 2001 ISBN 0-7923-6785-5
89. U. Gabbert and H.-S. Tzou (eds.): *IUTAM Symposium on Smart Structures and Structonic Systems*. Proceedings of the IUTAM Symposium held in Magdeburg, Germany, 26–29 September 2000. 2001 ISBN 0-7923-6968-8

Mechanics

SOLID MECHANICS AND ITS APPLICATIONS

Series Editor: G.M.L. Gladwell

90. Y. Ivanov, V. Cheshkov and M. Natova: *Polymer Composite Materials – Interface Phenomena & Processes*. 2001 ISBN 0-7923-7008-2
91. R.C. McPhedran, L.C. Botten and N.A. Nicorovici (eds.): *IUTAM Symposium on Mechanical and Electromagnetic Waves in Structured Media*. Proceedings of the IUTAM Symposium held in Sydney, NSW, Australia, 18-22 Januari 1999. 2001 ISBN 0-7923-7038-4
92. D.A. Sotiropoulos (ed.): *IUTAM Symposium on Mechanical Waves for Composite Structures Characterization*. Proceedings of the IUTAM Symposium held in Chania, Crete, Greece, June 14-17, 2000. 2001 ISBN 0-7923-7164-X
93. V.M. Alexandrov and D.A. Pozharskii: *Three-Dimensional Contact Problems*. 2001 ISBN 0-7923-7165-8
94. J.P. Dempsey and H.H. Shen (eds.): *IUTAM Symposium on Scaling Laws in Ice Mechanics and Ice Dynamics*. Proceedings of the IUTAM Symposium held in Fairbanks, Alaska, U.S.A., 13-16 June 2000. 2001 ISBN 1-4020-0171-1
95. U. Kirsch: *Design-Oriented Analysis of Structures*. A Unified Approach. 2002 ISBN 1-4020-0443-5
96. A. Preumont: *Vibration Control of Active Structures*. An Introduction (2nd Edition). 2002 ISBN 1-4020-0496-6
97. B.L. Karihaloo (ed.): *IUTAM Symposium on Analytical and Computational Fracture Mechanics of Non-Homogeneous Materials*. Proceedings of the IUTAM Symposium held in Cardiff, U.K., 18-22 June 2001. 2002 ISBN 1-4020-0510-5
98. S.M. Han and H. Benaroya: *Nonlinear and Stochastic Dynamics of Compliant Offshore Structures*. 2002 ISBN 1-4020-0573-3
99. A.M. Linkov: *Boundary Integral Equations in Elasticity Theory*. 2002 ISBN 1-4020-0574-1
100. L.P. Lebedev, I.I. Vorovich and G.M.L. Gladwell: *Functional Analysis*. Applications in Mechanics and Inverse Problems (2nd Edition). 2002 ISBN 1-4020-0667-5; Pb: 1-4020-0756-6
101. Q.P. Sun (ed.): *IUTAM Symposium on Mechanics of Martensitic Phase Transformation in Solids*. Proceedings of the IUTAM Symposium held in Hong Kong, China, 11-15 June 2001. 2002 ISBN 1-4020-0741-8
102. M.L. Munjal (ed.): *IUTAM Symposium on Designing for Quietness*. Proceedings of the IUTAM Symposium held in Bangkok, India, 12-14 December 2000. 2002 ISBN 1-4020-0765-5
103. J.A.C. Martins and M.D.P. Monteiro Marques (eds.): *Contact Mechanics*. Proceedings of the 3rd Contact Mechanics International Symposium, Praia da Consolação, Peniche, Portugal, 17-21 June 2001. 2002 ISBN 1-4020-0811-2
104. H.R. Drew and S. Pellegrino (eds.): *New Approaches to Structural Mechanics, Shells and Biological Structures*. 2002 ISBN 1-4020-0862-7
105. J.R. Vinson and R.L. Sierakowski: *The Behavior of Structures Composed of Composite Materials*. Second Edition. 2002 ISBN 1-4020-0904-6
106. Not yet published.
107. J.R. Barber: *Elasticity*. Second Edition. 2002 ISBN Hb 1-4020-0964-X; Pb 1-4020-0966-6
108. C. Miehe (ed.): *IUTAM Symposium on Computational Mechanics of Solid Materials at Large Strains*. Proceedings of the IUTAM Symposium held in Stuttgart, Germany, 20-24 August 2001. 2003 ISBN 1-4020-1170-9

Mechanics

SOLID MECHANICS AND ITS APPLICATIONS

Series Editor: G.M.L. Gladwell

109. P. Ståhle and K.G. Sundin (eds.): *IUTAM Symposium on Field Analyses for Determination of Material Parameters – Experimental and Numerical Aspects.* Proceedings of the IUTAM Symposium held in Abisko National Park, Kiruna, Sweden, July 31 – August 4, 2000. 2003 ISBN 1-4020-1283-7
110. N. Sri Namachchivaya and Y.K. Lin (eds.): *IUTAM Symposium on Nonlnear Stochastic Dynamics.* Proceedings of the IUTAM Symposium held in Monticello, IL, USA, 26 – 30 August, 2000. 2003 ISBN 1-4020-1471-6
111. H. Sobieckzky (ed.): *IUTAM Symposium Transsonicum IV.* Proceedings of the IUTAM Symposium held in Göttingen, Germany, 2–6 September 2002, 2003 ISBN 1-4020-1608-5
112. J.-C. Samin and P. Fisette: *Symbolic Modeling of Multibody Systems.* 2003 ISBN 1-4020-1629-8
113. A.B. Movchan (ed.): *IUTAM Symposium on Asymptotics, Singularities and Homogenisation in Problems of Mechanics.* Proceedings of the IUTAM Symposium held in Liverpool, United Kingdom, 8-11 July 2002. 2003 ISBN 1-4020-1780-4
114. S. Ahzi, M. Cherkaoui, M.A. Khaleel, H.M. Zbib, M.A. Zikry and B. LaMatina (eds.): *IUTAM Symposium on Multiscale Modeling and Characterization of Elastic-Inelastic Behavior of Engineering Materials.* Proceedings of the IUTAM Symposium held in Marrakech, Morocco, 20-25 October 2002. 2004 ISBN 1-4020-1861-4
115. H. Kitagawa and Y. Shibutani (eds.): *IUTAM Symposium on Mesoscopic Dynamics of Fracture Process and Materials Strength.* Proceedings of the IUTAM Symposium held in Osaka, Japan, 6-11 July 2003. Volume in celebration of Professor Kitagawa's retirement. 2004 ISBN 1-4020-2037-6
116. E.H. Dowell, R.L. Clark, D. Cox, H.C. Curtiss, Jr., K.C. Hall, D.A. Peters, R.H. Scanlan, E. Simiu, F. Sisto and D. Tang: *A Modern Course in Aeroelasticity.* 4th Edition, 2004 ISBN 1-4020-2039-2
117. T. Burczyński and A. Osyczka (eds.): *IUTAM Symposium on Evolutionary Methods in Mechanics.* Proceedings of the IUTAM Symposium held in Cracow, Poland, 24-27 September 2002. 2004 ISBN 1-4020-2266-2
118. D. Ieşan: *Thermoelastic Models of Continua.* 2004 ISBN 1-4020-2309-X

Kluwer Academic Publishers – Dordrecht / Boston / London

MIX
Papier aus verantwortungsvollen Quellen
Paper from responsible sources
FSC® C105338

If you have any concerns about our products,
you can contact us on
ProductSafety@springernature.com

In case Publisher is established outside the EU,
the EU authorized representative is:
Springer Nature Customer Service Center GmbH
Europaplatz 3, 69115 Heidelberg, Germany

Printed by Libri Plureos GmbH
in Hamburg, Germany